a LANGE medical book

Fluid & Electrolytes

Physiology & Pathophysiology

a LANGE medical book

Fluid & Electrolytes

Physiology & Pathophysiology

First Edition

Martin G. Cogan, MD
Professor of Medicine
University of California, San Francisco
Chief, Nephrology Section
Department of Veterans Affairs Medical Center
San Francisco

APPLETON & LANGE
Norwalk, Connecticut/San Mateo, California

0-8385-2546-6

Notice: Our knowledge in clinical sciences is constantly changing. As new information becomes available, changes in treatment and in the use of drugs become necessary. The author and the publisher of this volume have taken care to make certain that the doses of drugs and schedules of treatment are correct and compatible with the standards generally accepted at the time of publication. The reader is advised to consult carefully the instruction and information material included in the package insert of each drug or therapeutic agent before administration. This advice is especially important when using new or infrequently used drugs.

91 92 93 94 95 / 10 9 8 7 6 5 4 3 2 1

Prentice Hall International (UK) Limited, *London*
Prentice Hall of Australia Pty. Limited, *Sydney*
Prentice Hall Canada, Inc., *Toronto*
Prentice Hall Hispanoamericana, S.A., *Mexico*
Prentice Hall of India Private Limited, *New Delhi*
Prentice Hall of Japan, Inc., *Tokyo*
Simon & Schuster Asia Pte. Ltd., *Singapore*
Editora Prentice Hall do Brasil Ltda., *Rio de Janeiro*
Prentice Hall, *Englewood Cliffs, New Jersey*

ISBN: 0-8385-2546-6
ISSN:1053-7910

Production Editor: Charles F. Evans

PRINTED IN THE UNITED STATES OF AMERICA

This book is dedicated to
John R. Collins, PhD, and Floyd C. Rector, MD

Table of Contents

SECTION II: WATER HOMEOSTASIS & DISORDERS: HYPONATREMIA & HYPERNATREMIA

SECTION III: POTASSIUM HOMEOSTASIS & DISORDERS: HYPOKALEMIA & HYPERKALEMIA

SECTION IV: ACID-BASE HOMEOSTASIS & DISORDERS: ACIDOSIS & ALKALOSIS

SECTION VI: PHOSPHORUS HOMEOSTASIS & DISORDERS: HYPOPHOSPHATEMIA & HYPERPHOSPHATEMIA

SECTION VII: MAGNESIUM HOMEOSTASIS & DISORDERS: HYPOMAGNESEMIA & HYPERMAGNESEMIA

Preface

Few topics are more fundamental to understanding human biology than fluid and electrolyte metabolism. My goal in writing this book has been to provide a modern, scientifically based, clinically relevant, short, and lucid overview of normal and abnormal fluid and electrolyte regulation. To the extent that I have succeeded, this book should be useful to the beginning student of human physiology, the medical student starting clinical practice, and even the practicing physician wishing to review any aspect of this complex subject.

In each section, the normal renal and nonrenal handling of the common electrolytes or water is first summarized, setting the stage for critical analysis of the pathophysiology and clinical expression of related disorders. An important fundamental tenet of this field is that knowledge of normal homeostasis is essential for understanding the mechanisms that underlie the derangements that can occur and thus for providing a rational framework for diagnosis and therapy.

Throughout the book, I have stressed the specific roles and integrative function of the individual segments of the nephron that mediate the filtration and reabsorption of solutes and water. Whenever possible, the cellular and molecular mechanisms that form the basis of the transport processes involved have been emphasized. Because of the summary nature of this text, extensive literature references have not been included, and complex or controversial topics have in some cases been discussed only briefly.

I wish to thank various individuals who have contributed substantially to creation of this work—and especially Jack D. Lange, MD, for his inspiration and encouragement; James Ransom, PhD, for thoughtful and meticulous textual editing; Don Ramie, for the excellent illustrations; and David Pearce, MD, and Mervin Goldman, MD, for helpful criticisms of specific chapters.

This book will be revised every three years. Comments, suggestions, and corrections of factual matters will be gratefully received by the author at the address following.

Martin G. Cogan, MD
Nephrology Section (111J)
Department of Veterans Affairs Medical Center
4150 Clement St.
San Francisco, CA 94121

January, 1991

Section I: Sodium Homeostasis & Disorders: Hypovolemia & Hypervolemia

Normal Sodium & Extracellular Volume Homeostasis

1

INTRODUCTION

TOTAL BODY SODIUM

Sodium is the most abundant cation in the body, averaging approximately 60 meq/kg body weight, or about 4200 meq in a 70-kg person. As a percentage of body weight, total body sodium does not vary significantly as a function of sex or age in the adult.

Exchangeable Sodium

About 30% of total body sodium is in bound form, primarily within bone. This bound sodium is inaccessible for diffusional exchange with other stores and is physiologically "silent." The remaining 70% of total body sodium—about 40 meq/kg body weight—is termed "exchangeable sodium." Exchangeable sodium is dissolved chiefly in the extracellular volume or in fluid compartments that functionally communicate with the extracellular volume. The exchangeable sodium—rather than total body sodium—is physiologically regulated. For simplicity, the term "sodium" will be used henceforth to mean exchangeable sodium.

Sodium & Chloride

Although the major focus in this chapter is sodium, a cation of course never exists in isolation. Sodium's major anionic partner is chloride. In general—exceptions will be mentioned as they arise—intake, absorption, distribution, and renal handling of chloride parallel that of sodium. Another (quantitatively minor) anionic partner of sodium in the body fluids is bicarbonate, the physiology of which is discussed in Chapter 10.

ANATOMY OF THE BODY VOLUMES

In broad terms, total body water is divided into 2 unequal parts, as shown in Fig 1-1. About 55% of total body water is inside cells—the **intracellular volume;** the remaining 45% is outside cells—the **extracellular volume.**

Sodium as the Major Extracellular Cation

The electrolyte compositions of the extracellular and intracellular fluids differ substantially. For practical purposes, potassium is the main cation of the intracellular fluid and sodium of the extracellular fluid, as illustrated in Fig 1-1.

This electrolyte concentration polarity—the potassium concentration is high and the sodium concentration low within cells, whereas sodium concentration is high and potassium concentration low in the extracellular fluid—is maintained by the powerful action of the **Na^+/K^+-ATPase.** Resident on all cell membranes, this ATP-requiring enzyme extrudes sodium from the cell while pumping potassium into the cell with a coupling ratio of 3:2.

Sodium Content as a Determinant of Extracellular Volume

Since water is in osmotic equilibrium between the extracellular and intracellular spaces, the volume distribution of water between these two compartments is determined by the relative quantities of sodium and of potassium.

For example, consider the consequence of adding sodium to the body, as shown in Fig 1-2. Before the addition of sodium (top panel), the concentrations of extracellular sodium and of intracellular potassium are equivalent, so there is no net water flow across the cell membrane. When

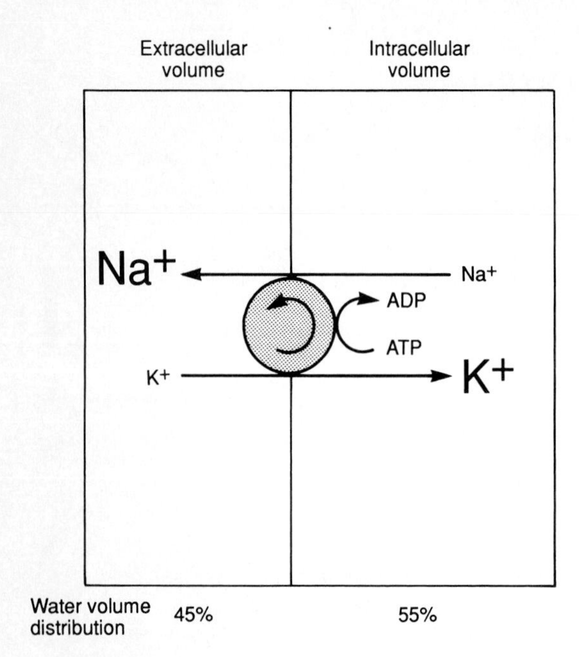

Figure 1-1. Water volume distribution.

sodium is added, it is excluded from cells (any that leaks into cells is immediately pumped out by the Na^+/K^+-ATPase) and confined to the extracellular space. The increase in extracellular sodium concentration osmotically draws water from the intracellular space into the extracellular space (Fig 1-2, center panel). This redistribution of water occurs until osmotic equilibrium is reestablished; extracellular sodium concentration is then again equal to intracellular potassium concentration (Fig 1-2, lower panel). Compared to the original situation, the extracellular volume has been expanded at the expense of shrinkage of the intracellular volume. This example demonstrates how the ratio of sodium content to potassium content establishes the distribution of body water between the extracellular and intracellular spaces.

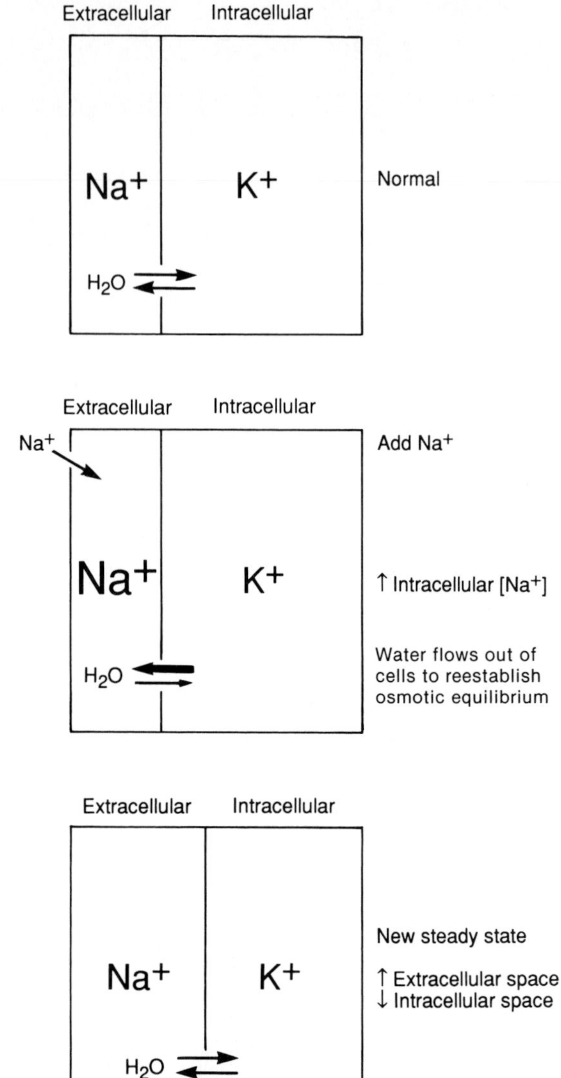

Figure 1-2. Effect of sodium addition on water volume distribution between extracellular and intracellular spaces.

DISTRIBUTION OF EXTRACELLULAR SODIUM

Exchangeable sodium is mainly in three interdependent extracellular volume subunits, as shown in Fig 1-3. The central subcompartment (41% of the total exchangeable sodium) is the **interstitial volume,** the fluid that bathes cells, plus lymphatic fluid. There are also a group of fluid spaces, known as **transcellular volumes** (40% of total exchangeable sodium), that are in functional contiguity with the interstitial space, though diffusional equilibrium with plasma usually requires a longer period of time. The transcellular sodium includes a fraction of sodium in bone, dense connective tissue, and cartilage that is exchangeable, plus sodium in specialized and partially sequestered spaces, including saliva, gastrointestinal fluids, bile, cerebrospinal fluid, aqueous humor of the eye, and sweat. The sum of the interstitial plus transcellular volumes represents about 81% of total sodium. Finally, in diffusional equilibrium with the interstitial fluid is the **plasma volume** (16% of total exchangeable sodium), which is the critical subcompartment for signaling hemodynamic and renal processes that adjust total

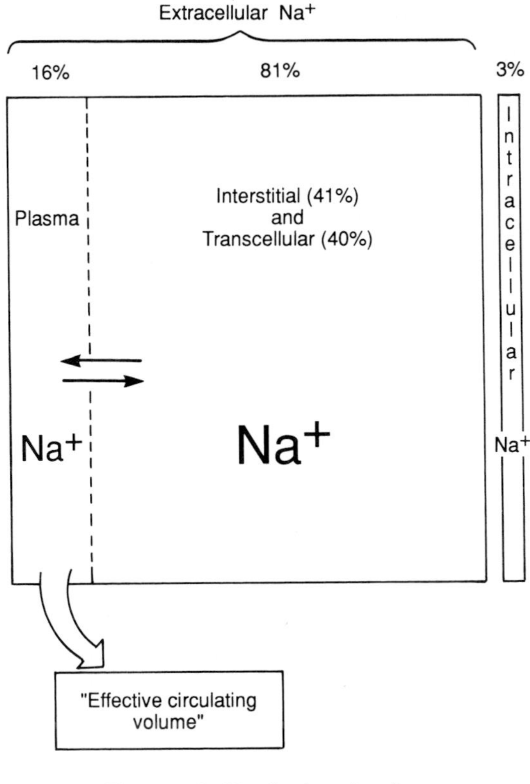

Figure 1-3. Distribution of sodium.

body sodium. The total extracellular sodium (interstitial, transcellular, and plasma) accounts for 97% of total exchangeable sodium; only about 3% of total body sodium is intracellular.

Note that in Fig 1-3, body volumes are seen from the perspective of *sodium* distribution, whereas in Fig 1-1, volumes were shown as a function of *water* distribution. The important difference is that the large intracellular volume is defined by potassium but not sodium. As a corollary, body volumes seen from a *potassium* perspective would emphasize the intracellular volume (discussed in greater detail in Chapter 7 and illustrated in Fig 7-1).

Plasma Volume as a Critical Determinant of Effective Circulating Volume & Cardiovascular Homeostasis

From a physiologic viewpoint, the extracellular fluid volume, though representing less than half of total body water, is extraordinarily important for normal cardiovascular function. As shown in Fig 1-3, extracellular fluid volume is divided into two parts in a ratio of about 1:5—a small plasma volume and a larger compartment representing mainly the interstitial and transcellular spaces. It is the smaller plasma volume component that is crucial for normal cardiovascular function.

A complex and incompletely understood interplay between the arterial plasma volume and arterial pressure defines the "**effective circulating volume.**" This effective circulating volume is sensitively regulated by an array of neurohumoral and renal mechanisms.

Thus, plasma volume, critically important for optimal cardiovascular health, is actually but a small proportion—less than 10%—of total body water and only about 4.5% of body weight. In fact, the cardiovascular system does not even "know" how large are the total extracellular or intracellular volumes. Thus, regulation of total body sodium—by defining the extracellular volume and its important component, the plasma volume and hence the effective circulating volume—is of paramount physiologic importance.

This chapter describes the complex neurogenic, endocrinologic, and renal mechanisms that maintain the body's sodium content at a normal level, allowing the plasma volume and effective circulating volume to remain at an optimal set-point for cardiovascular function.

Plasma Sodium Concentration

As will be discussed more fully in Chapter 4, the osmolality of body fluids is governed by a ratio: the total exchangeable solute in the body (mainly sodium plus potassium) divided by the total body water. In practice, the maintenance of normal plasma osmolality and hence the concentration of extracellular sodium is due to the regulation of the denominator of this ratio, total body water. Within limits, when total body sodium increases or decreases, very sensitive mechanisms adjust total body water proportionately to maintain constant extracellular sodium concentration. With large changes in total body sodium, the extracellular sodium concentration may sometimes be low or high depending on the concurrent water balance. The plasma sodium concentration itself gives no insight into the *quantity* of sodium or of water—only their ratio. Assessment of total body sodium is based on clinical criteria and cannot be estimated from the plasma sodium concentration alone.

SODIUM INTAKE, ABSORPTION, & DISTRIBUTION

SODIUM INTAKE

Diet

Large population surveys of food intake in industrialized Western countries have reported dietary sodium intake to be approximately 150 meq/d (equivalent to about 3.5 g/d of sodium) in adult males. Such surveys underestimate true sodium consumption, however, since they do not include added table salt. The range of dietary sodium intake is amazingly broad: the limits that describe sodium intake by 95% of people extend from 50 to 350 meq/d. Sodium intake in women averages about two-thirds that of men, owing to their generally lower food and calorie consumption. In both sexes, sodium consumption declines modestly as a function of age—again in parallel with total food intake. The sodium ingested in natural foods is usually in the form of the chloride salt; hence, sodium and chloride intakes are generally equivalent.

There is no "normal" sodium intake. Sodium ingestion is greatly influenced by cultural factors. Unlike the thirst mechanism for water, there is very little evidence that a "sodium appetite" exists to regulate sodium consumption. Dietary sodium intakes of less than 10 meq/d and greater than 1000 meq/d have been recorded by individuals within a similar culture or in different cultures.

It has been argued—based on relatively weak epidemiologic evidence related to predisposition to hypertension and cardiovascular disease—that the average sodium intake in industrialized Western societies may exceed the kidney's normal capacity to excrete sodium, thereby causing a rise in the effective circulating volume and blood pressure. However, there is little direct experimental evidence that shows any significant relationship of blood volume or blood pressure with dietary sodium intake spanning the usual values (50–350 meq/d) in either normal individuals or those with preexisting hypertension.

SODIUM ABSORPTION & DISTRIBUTION

Sodium chloride is rapidly and completely absorbed from the intestine. Following intestinal absorption, sodium diffuses rapidly into the plasma volume and equilibrates with the interstitial volume. The time course of distribution within the extracellular space—within hours—is far less than the time necessary for the kidneys to excrete the sodium load, as shown in Fig 1–4.

The increase in extracellular sodium content and concentration initially draws fluid from the intracellular space to increase the entire extracellular volume (Fig 1–2). With time, increased extracellular sodium concentration stimulates thirst, augments renal water reabsorption to restore body osmolality and intracellular volume, and thereby further increases extracellular volume, as discussed in Chapter 4.

Plasma-Interstitial Fluid Exchange: Capillary Starling Forces

The distribution of sodium and water between the plasma and interstitial spaces—normally a ratio of about 1:3—is of great importance in defining how much of a given sodium load is retained within the vasculature. The interstitial volume becomes a more significant reservoir for the ingested sodium than the plasma volume.

The interface of the plasma and interstitial spaces occurs at the level of the capillary. The distribution of sodium and water between these two compartments is determined by the Starling forces.

As illustrated in Fig 1–5 (top panel), four components of the Starling forces govern net transcapillary sodium and water distribution: the **hydraulic** pressures within the capillary (P_c) and the interstitium (P_i) and the **oncotic** pressures in the capillary (Π_c) and the interstitium (Π_i). The capil-

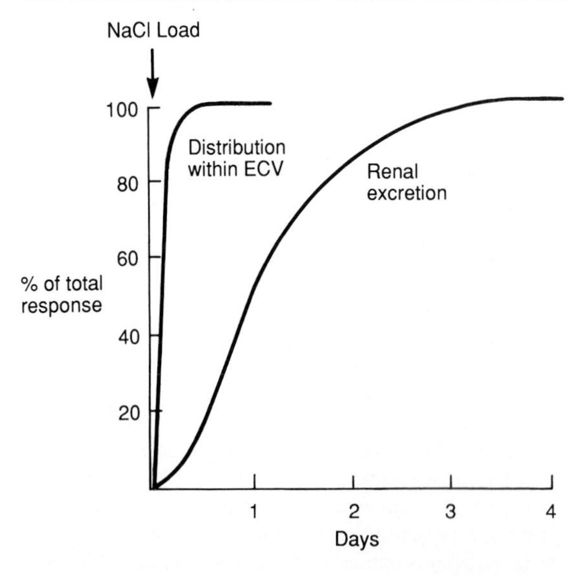

Figure 1–4. Time course of distribution and renal excretion following a sodium load.

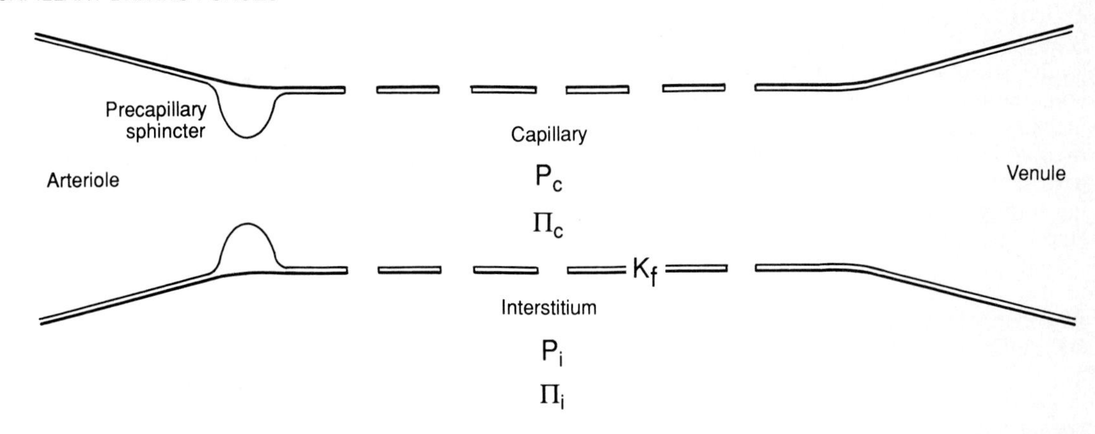

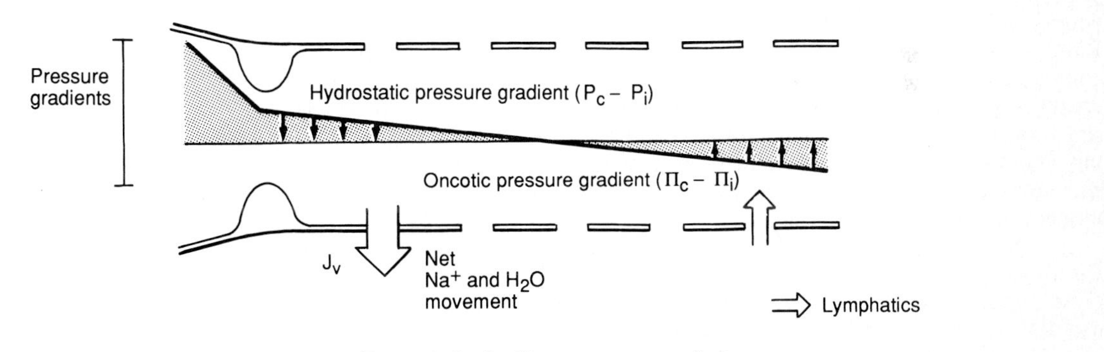

Figure 1-5. Capillary pressure gradients.

lary wall is sodium- and water-permeable, and the leakiness or porosity of the vessel is defined by the hydraulic water permeability coefficient (K_f).

The capillary hydraulic pressure is less than its parent arteriole owing to the resistance afforded by the precapillary sphincter. The interstitial hydraulic pressure is normally very low in subcutaneous tissues—actually a negative value (subatmospheric; about −2 mm Hg). Oncotic pressures are attributable primarily to the albumin concentrations in the two compartments. Owing to the semipermeable nature of the capillary endothelium that restricts macromolecular passage, the albumin concentration and hence the oncotic pressure in the interstitium are normally much lower than the respective plasma values (about one-fifth).

As shown in Fig 1-5 (bottom panel), the hydraulic pressure gradient ($P_c - P_i$) tends to move sodium and water out of the capillary, while the oncotic pressure gradient ($\Pi_c - \Pi_i$) is in the absorptive direction. The coordinated movement of sodium and water is determined by the sum of these two gradients—the net driving force (shaded area in Fig 1-5, bottom panel—as well as by the capillary hydraulic permeability coefficient (K_f). Thus, net sodium and water movement (J_v) is expressed by the following equation:

$$J_v = K_f \times [(P_c - P_i) - (\Pi_c - \Pi_i)]$$

The net driving force at the beginning of the capillary is generally in the direction of forcing water and sodium out of the capillary, since the hydraulic pressure gradient (about 35 mm Hg) exceeds the oncotic pressure gradient (about 25 mm Hg). Toward the end of the capillary, there is inversion of the net driving force, so that net sodium and water uptake occurs. This axial change in net driving forces arises because P_c declines by about half—so the hydraulic pressure gradient falls to about 17 mm Hg—owing to the hydraulic resistance of the capillary, while Π_c rises slightly owing to the ultrafiltration of water without protein.

Interstitial fluid is returned to the circulation by lymphatics. About 25-50% of the total circulating plasma protein enters the lymphatics daily and is returned to the systemic circulation. Lymphatic drainage increases if fluid entry to the interstitium is augmented.

The steady-state plasma volume-to-interstitial volume ratio is maintained by the balance of the Starling forces. A perturbation of one or more of the four component Starling forces or of capillary permeability can upset the normal balance, allowing sodium and water to enter or leave the capillary.

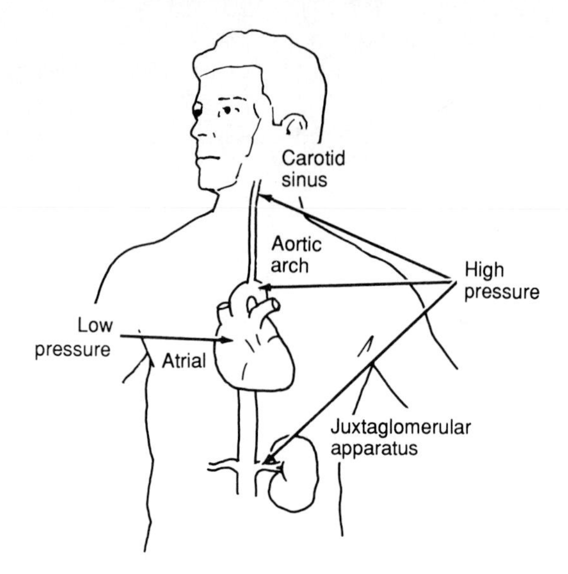

Figure 1-6. Low and high pressure baroreceptors.

AFFERENT LIMB OF SODIUM HOMEOSTASIS

The discussion thus far has described the distribution between the plasma and interstitial compartments of ingested sodium. The "buffering" effect of the interstitium ameliorates the plasma volume expansion that otherwise would occur. Ridding the body of this excess sodium requires complex interacting systems that "sense" the increase in plasma sodium and water content and signal the kidney to excrete the sodium. The sensing systems are called the **afferent limb** and the excretory mechanisms the **efferent limb** of sodium homeostasis.

Baroreceptors

As mentioned above, a concept that has been useful in defining the perception of the "fullness" of the circulation has been that of "effective circulating volume." In broad terms, the effective circulating volume represents a complex and incompletely understood product of arterial plasma volume and pressure.

Monitoring of the effective circulating volume is performed by a network of baroreceptors. As shown in Fig 1–6, there are two general groups of baroreceptors, categorized according to the part of the circulation they monitor: the **low-pressure** and **high-pressure** baroreceptor systems. Whether it is stretch, pressure, or tension that activates these baroreceptors is not completely understood.

Low-Pressure Baroreceptor Control of the Sympathetic Nervous System, Antidiuretic Hormone (ADH), & Atrial Natriuretic Factor (ANF)

Low-pressure baroreceptors are located in the **cardiac atria,** principally the left atrium (Fig 1–6). Stretch of these structures by increased effective circulating volume cause signaling via unmyelinated fibers that travel up the vagus nerve to cardiovascular centers in the brain stem and medulla. Stimulation by these baroreceptors tends to inhibit tonic discharge of the sympathetic nervous system involved in vasoconstriction. Low-pressure baroreceptor activation also excites the cardioinhibitory center to augment vagal and parasympathetic neural discharge. A decrease in effective circulating volume has the opposite effects, as summarized in Fig 1–7.

The low-pressure baroreceptor system also controls release of antidiuretic hormone (ADH) from the hypothalamic-posterior pituitary axis. ADH can have powerful hemodynamic effects via type 1 (or V_1) receptors, which cause vasoconstriction and diminish GFR. ADH also increases renal water reabsorption via type 2 (or V_2) receptors, discussed more fully in Chapter 4. The routine control of ADH is usually not by baroreceptors but rather by the central osmoreceptor that senses extracellular osmolality. However, this osmoregulation of ADH can be overridden when a large ($>$ 10%) decrement in effective circulating volume activates the baroreceptors. Baroreceptor-induced ADH release can then occur despite normal or even low osmolality.

The atria—the right more so than left—are also the site of synthesis and regulated release of the 28-amino-acid vasodilatory hormone atrial natriuretic factor (ANF), also known as atrial natriuretic peptide (ANP). Stretch of the atria by an increase in atrial volume or pressure releases ANF. To a lesser extent, an increase in the rate of atrial beating also causes ANF release.

ANF has many actions that are mediated following binding to a specific receptor on target cells. The receptor is also a guanylate cyclase, and the intracellular second messenger for ANF is cGMP. A different receptor effects tje demyelination and clearance of ANF but is without biologic

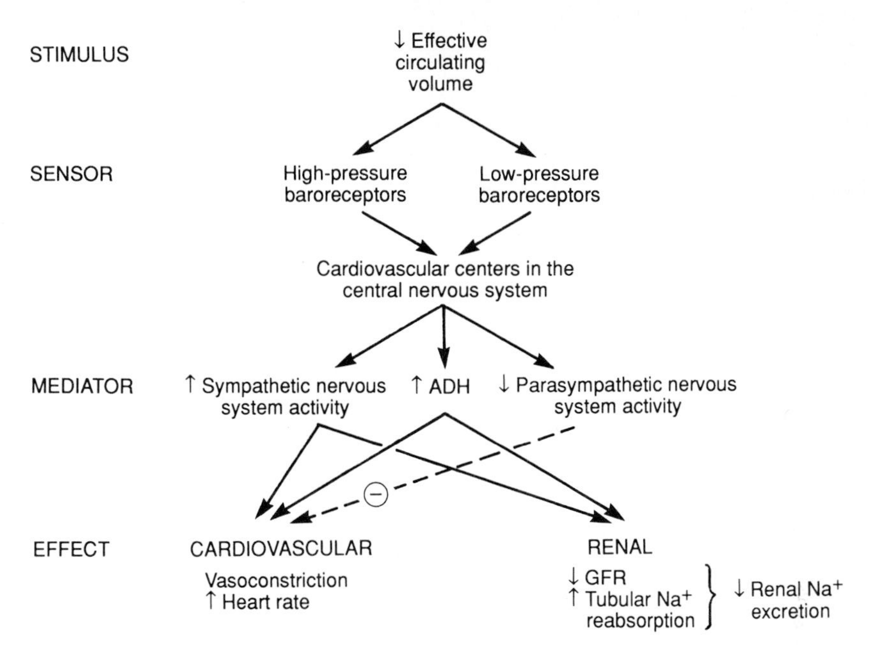

Figure 1-7. Physiologic responses to decreased effective circulating volume.

effect. As summarized in Fig 1-8, ANF causes dilation of vascular smooth muscle even when the muscle has been preconstricted by potent vasoconstrictors. ANF antagonizes the major antinatriuretic and antidiuretic hormonal systems by inhibiting the release of renin by granular cells of the renal juxtaglomerular apparatus, of aldosterone by zona glomerulosa cells of the adrenal, and of ADH by cells of the hypothalamus-posterior pituitary. Finally, ANF causes natriuresis directly by inducing glomerular hyperfiltration and reduction in soium transport in the medullary collecting tubule.

High-Pressure Baroreceptor Control of the Sympathetic Nervous System & the Renin-Angiotensin II System

The high-pressure baroreceptors are found in several locations (Fig 1-6). There are nerve endings in the **aortic arch** and in the **carotid sinus** with afferent fibers that travel via the glossopharyngeal nerve to the brain stem and medulla. Like the low-pressure baroreceptors, they are activated by increase in effective circulating volume to cause efferent sympathetic nervous system depression and parasympathetic nerve activation while depressing ADH release. The opposite changes occur in response to a decrease in effective circulating volume (Fig 1-7).

Another high-pressure baroreceptor resides in the **afferent arteriole of the renal juxtaglomerular apparatus** (Fig 1-6). As summarized in Fig 1-9, this baroreceptor responds by a direct myogenic response to increase renin secretion when renal arterial perfusion or pressure is decreased. Prostacyclin is involved in mediating this primary myogenic response. Renin secretion is also inversely related to the delivery of sodium to cells in the macula densa region of the thick ascending limb of Henle (a response related to tubuloglomerular feedback; see discussion below). Finally, renin secretion is neurogenically regulated by a $beta_2$-adrenergic receptor-mediated mechanism.

Renin, a glycoprotein hormone with a molecular weight of 37,326 in humans, is an aspartyl protease which is the rate-limiting factor in the conversion of angiotensinogen to the decapeptide angiotensin I. Subsequent cleavage of physiologically inactive angiotensin I to the active octapeptide angiotensin II is then effected by converting enzyme resident on endothelial cells, especially in the lung.

Angiotensin II has many cardiovascular, endocrine, and renal effects. It powerfully constricts vascular smooth muscle; increases adrenal release and biosynthesis of aldosterone, which controls sodium reabsorption in the cortical collecting tubule; stimulates ADH release and thirst (discussed in Chapter 4); and potently regulates GFR and sodium transport directly in the early proximal tubule.

It is now recognized that local angiotensin II formation in the kidney and other organs can occur. While the endocrine system of regulating the

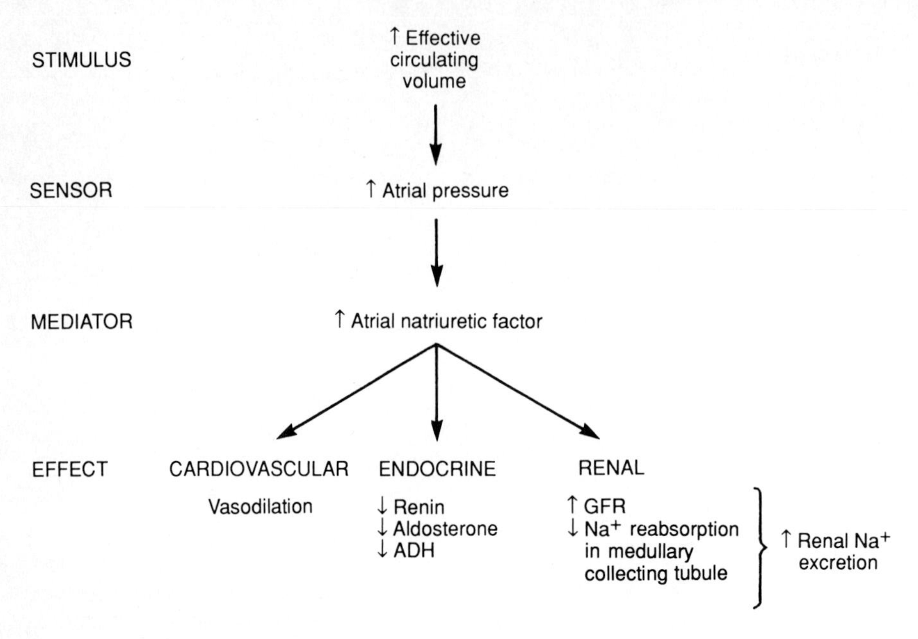

Figure 1–8. Regulation and actions of atrial natriuretic factor.

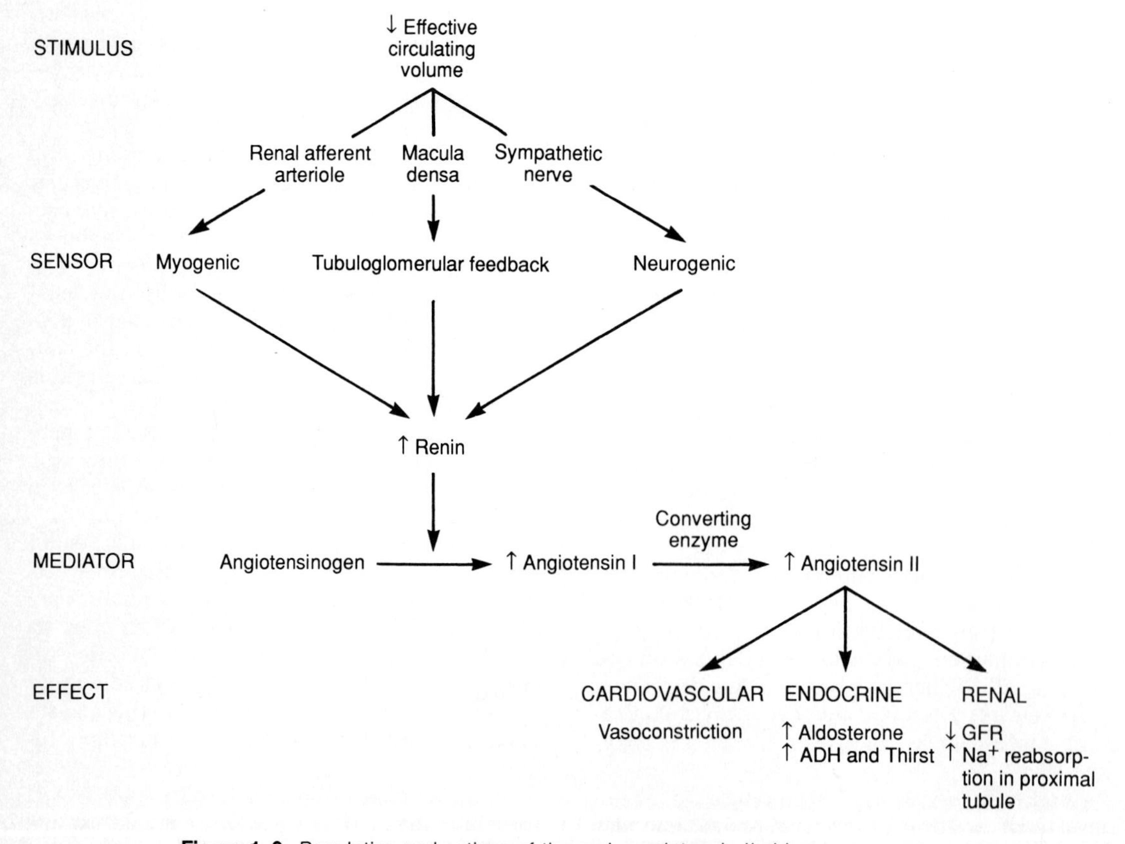

Figure 1–9. Regulation and actions of the renin-angiotensin II-aldosterone system.

circulating level of angiotensin II is certainly an important controller of systemic hemodynamics and adrenal function, a paracrine system also exists by which angiotensin II is produced in tissues and acts locally. For instance, renal tissue angiotensin II levels may be substantially higher—and more important from a functional and physiologic perspective—than circulating levels.

Integrated Role of the Baroreceptors

The low- and high-pressure baroreceptors sensitively monitor the effective circulating volume and coordinately regulate sympathetic and parasympathetic neural as well as endocrine (renin-angiotensin II-aldosterone and ANF) activities. Controversy exists about whether the high- or the low-pressure baroreceptors are more important in the feedback control of the circulation. Clinical situations exist in which contradictory signals are sent, activating one system while inhibiting the other. Both systems appear to be important. Resetting of the baroreceptors also occurs. For example, adaptation occurs in the state of chronic hypertension, when an elevated level of blood pressure becomes physiologically defended.

Other baroreceptors exist with less well documented significance. For example, afferent sympathetic signals that ascend to the central nervous system to modulate sympathetic efferents originate in the liver and biliary system, the pulmonary circulation, the cardiac ventricles, and the renal medulla (activated by increased renal pelvic pressure). The biliary baroreceptors may play an important role in the circulatory changes and antinatriuresis that attend portal hypertension and cirrhosis.

RENAL SODIUM EXCRETION

EFFERENT LIMB OF SODIUM HOMEOSTASIS

With plasma volume expansion caused by dietary sodium intake, there is direct cardiovascular feedback control by the baroreceptors. Diminished sympathetic nervous system activity, decreased angiotensin II concentration, and increased ANF cause peripheral vasorelaxation and cardiodeceleration, blunting the increase in blood pressure that would otherwise ensue. But these cardiovascular "buffering" effects are temporizing. To reestablish normal extracellular volume and rid the body of the sodium surfeit, it is ultimately necessary to excrete excess sodium, as summarized in Fig 1–10.

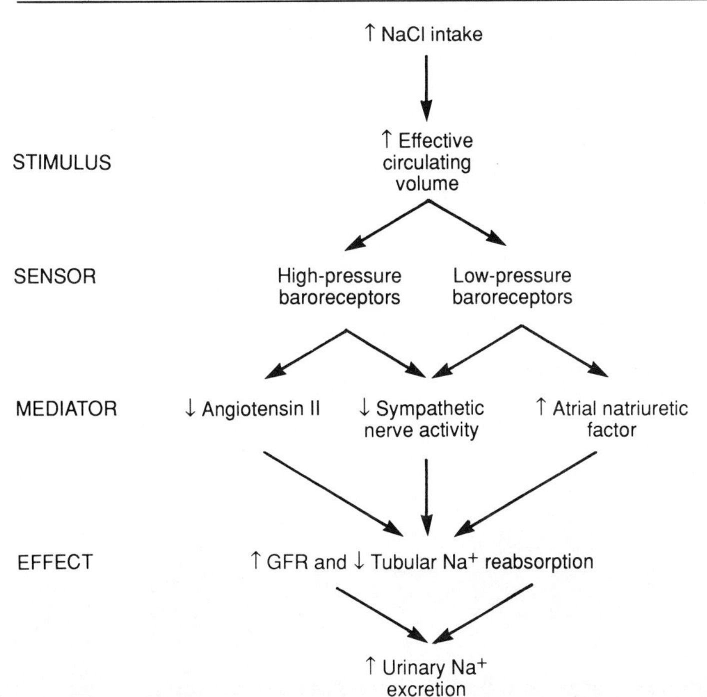

Figure 1–10. Response to increased NaCl intake.

Nonrenal Sodium Excretion

Before we examine in detail the complex mechanisms that control renal sodium excretion, it should be mentioned that a small amount of sodium—usually less than 10% of the usual dietary sodium (150 meq/d)—is not excreted by the kidneys. About 0.1 L of water is lost daily via the **stool** and a similar volume in **sweat.** Since the sodium concentration in both stool and sweat water is 50 meq/L, about 5 meq/d of sodium is lost by each route, as shown in Fig 1-11. As part of the response to hypovolemia, aldosterone can act on both the colon and the sweat glands to decrease the sodium concentration and minimize sodium loss.

Renal Sodium Excretion

The kidneys excrete sodium by means of differential filtration and reabsorption; sodium is not secreted. The purpose of the following sections is to review the intrinsic and extrinsic (neurohumoral) physiologic mechanisms that control the glomerular filtration rate (GFR) and sodium transport in the various segments of the nephron. No single nephron element by itself maintains sodium homeostasis, which is the result of the ex-

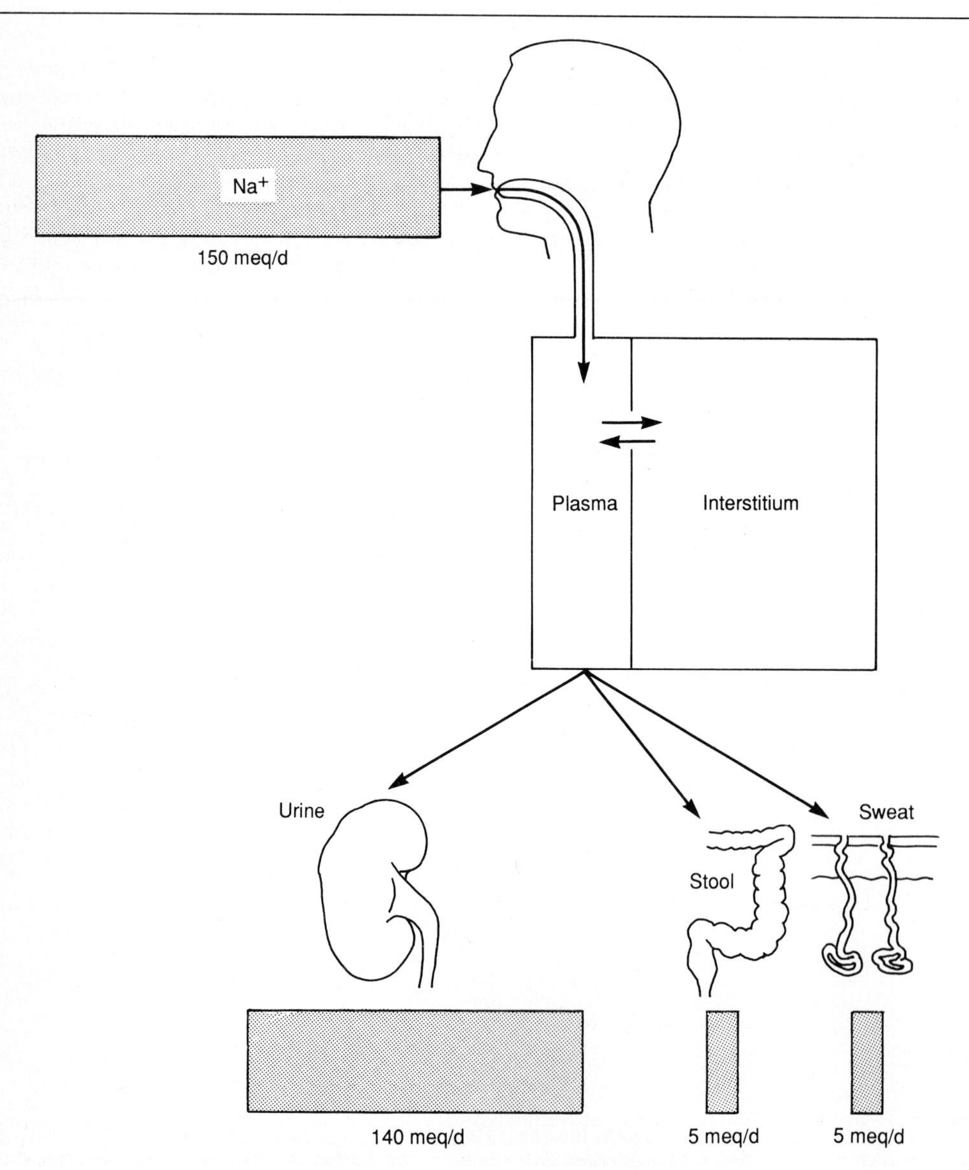

Figure 1-11. Normal sodium intake and routes of excretion.

quisitely coordinated and complex interaction of many factors.

OVERVIEW OF RENAL SODIUM HANDLING

Roles of GFR Versus Tubular Transport in Renal Sodium Excretion

Sodium is freely filtered and is not sieved by the glomerulus. The sodium concentration in the glomerular ultrafiltrate is thus virtually identical to that in plasma water.

The normal amount of sodium filtered at the glomerulus in humans is enormous. As shown in Fig 1-12, if the normal glomerular filtration rate (GFR) is taken as 100 mL/min (0.1 L/min) and the sodium concentration in the glomerular ultrafiltrate is 140 meq/L, then the amount of filtered sodium is over 20,000 meq/d (0.1 L/min × 140 meq/L × 1440 min/d). To put this quantity of filtered sodium in perspective, remember that the average urinary sodium excretion (equal to dietary consumption, neglecting nonrenal losses) is about 150 meq/d (Fig 1-11). The excreted sodium is then only 0.75% (150/20,000) of the filtered sodium load. Looked at in another way, renal sodium reabsorption is also huge (sodium reabsorption = sodium filtration − sodium excretion, or 20,000 − 150 = 19,850 meq/d) and indeed is 99.25% of the amount of sodium that is filtered.

Theoretically, urinary sodium excretion might be regulated by changes either in filtration or in reabsorption of sodium. For example, to double sodium excretion (eg, from 150 to 300 meq/d)—assuming that tubular transport of sodium is perfectly constant—it would be necessary for GFR to increase by only 0.75%, to 100.75 mL/min (original filtered sodium load of 20,000 meq/d plus additional 150 meq/d). Alternatively, if GFR were fixed, a small depression of tubular sodium reabsorption—from 19,850 to 19,700 meq/d (or 99.25% to 98.5% of the filtered load)—would produce the same increment in urinary sodium.

In fact, both mechanisms operate. While intrinsic transport mechanisms exist that adjust sodium reabsorption in response to changes in GFR and the filtered quantity of sodium (so-called glomerulotubular balance), these mechanisms are imperfect. A small increase or decrease in GFR therefore influences urinary sodium excretion.

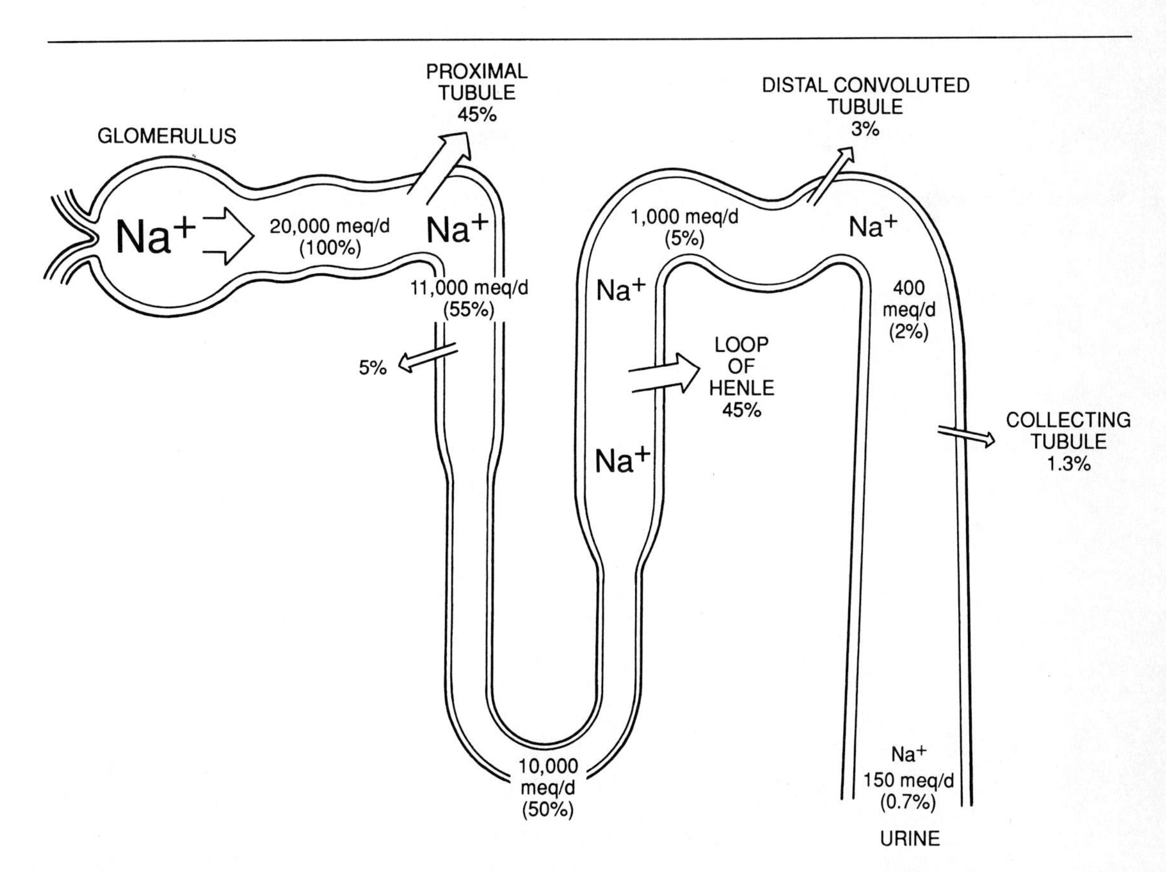

Figure 1-12. Normal amounts of sodium filtered, reabsorbed, and excreted.

Numerous neurohumoral factors also independently regulate sodium transport along the entire nephron, thereby modulating urinary sodium excretion at a given filtered sodium load. Renal sodium handling can be fully explained only by understanding the intricate and complexly regulated relationship between glomerular sodium filtration and tubular sodium reabsorption.

DETERMINANTS OF GFR & SODIUM FILTRATION

Forces Governing Filtration

All blood flowing into the kidney (Q_A) traverses the afferent arteriole and the glomerular capillary and exits via the efferent arteriole. The forces that govern fluid filtration within the glomerular capillaries are shown in Fig 1-13. Glomerular ultrafiltration is defined by Starling forces in a manner qualitatively similar to what occurs in systemic capillaries (Fig 1-5). An important quantitative difference is that the hydraulic pressure within the glomerular capillary is approximately two- to fourfold higher than that estimated within systemic capillaries (eg, 60 versus 15-35 mm Hg). Thus, the relative hydraulic force available to give glomerular ultrafiltration is extremely large.

Both the afferent and the efferent arterioles have adjustable resistances, termed R_A and R_E, respectively. R_A has been estimated to be approximately 0.3-0.6 the value of R_E in humans. The hydraulic pressure within the glomerulus, $\mathbf{P_{GC}}$, is affected by both of these resistances. R_A determines how much of the arterial pressure is transmitted into the glomerular capillary, whereas R_E determines how much of P_{GC} is dissipated into the efferent arteriole.

P_{GC} is the major force tending to push fluid across the capillary basement membrane. P_{GC} is opposed by the hydraulic pressure in Bowman's space (P_{BS}), which is relatively small—about 15 mm Hg, or approximately 25% of P_{GC}—and is not physiologically regulated.

A more important force opposing ultrafiltration is the oncotic pressure within the capillary, Π_{GC}. As in systemic capillaries, Π_{GC} is attributable principally to albumin, the oncotic pressure of which averages about 20 mm Hg in blood entering the glomerular capillary. Owing to the property

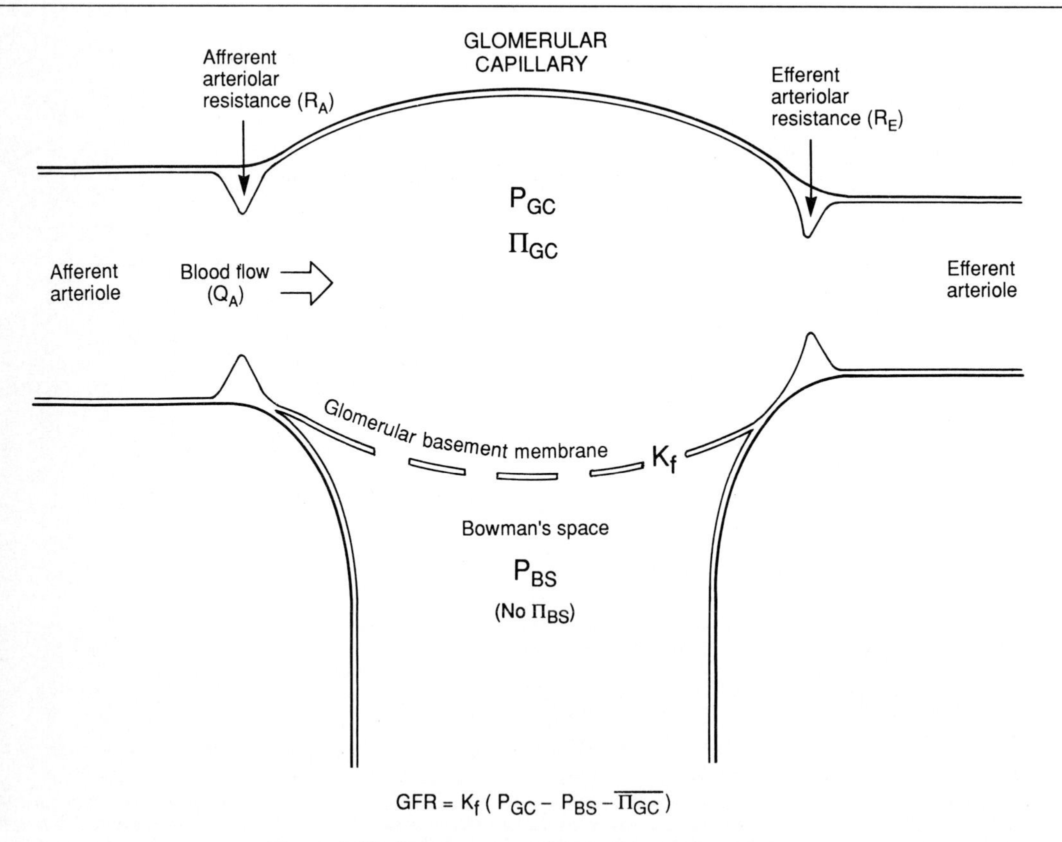

Figure 1-13. Factors responsible for glomerular filtration.

(called permselectivity) of the glomerular basement membrane, by which it efficiently excludes large anionic proteins, there is virtually no albumin and therefore no oncotic pressure in Bowman's space.

The permeability coefficient, $\mathbf{K_f}$, of the basement membrane is the hydraulic water permeability (water pores per surface area, L_p) multiplied by the total available glomerular membrane surface area (A); thus, K_f is also known as L_pA. K_f in a sense represents the total membrane "porosity" allowing for water flow. The glomerulus is quite permeable to water. Indeed, the glomerular K_f is the highest of any capillary system in the body.

The amount of fluid forced across the basement membrane at any point x (GFR_x) is defined by K_f multiplied by the sum of the glomerular Starling forces at that point: $P_{GC} - P_{BS} - \Pi_{GC}$ (Π_{BS} is trivial and is ignored):

$$GFR_x = K_f (P_{GC} - P_{BS} - \Pi_{GC})$$

Most of the determinants of GFR do not change as a function of glomerular capillary length. As illustrated in Fig 1-14, since there is little hydraulic resistance, P_{GC} is relatively constant along the glomerular capillary length, as are K_f and P_{BS} also. Therefore, at the beginning of the capillary, the net ultrafiltration pressure gradient (ΔP_{UF}) is as follows: P_{GC} (60 mm Hg) − P_{BS} (15 mm Hg) − Π_{GC} (20 mm Hg) = 25 mm Hg. However, Π_{GC} is not constant and predictably rises along the length of the glomerular capillary to a value estimated to be about 30 mm Hg by the end of the normal human glomerular capillary. The reason for this change is that albumin cannot cross the glomerular capillary basement membrane, so that its concentration increases as water is progressively filtered along the capillary length.

Because the oncotic pressure is higher at the end of the capillary than at the beginning, the net ultrafiltration pressure gradient (ΔP_{UF} at the end-capillary is reduced: P_{GC} (60 mm Hg) − P_{BS} (15 mm Hg) − Π_{GC} (30 mm Hg) = 15 mm Hg. The driving force is diminished but still greater than zero even by the end-capillary, so that the situation is termed glomerular pressure disequilibrium (ie, the hydraulic pressure still exceeds the oncotic

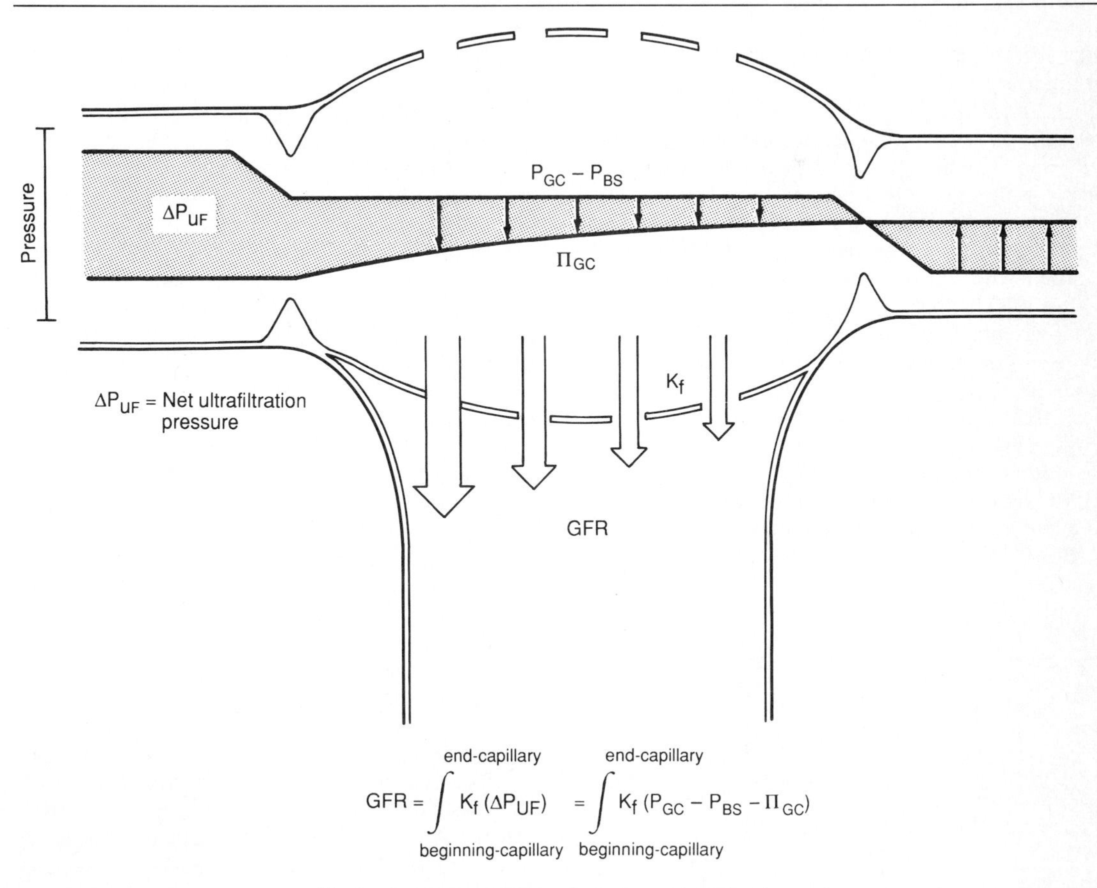

$$GFR = \int_{\text{beginning-capillary}}^{\text{end-capillary}} K_f (\Delta P_{UF}) = \int_{\text{beginning-capillary}}^{\text{end-capillary}} K_f (P_{GC} - P_{BS} - \Pi_{GC})$$

Figure 1-14. Determinants of glomerular filtration.

pressure). If the oncotic pressure had risen sufficiently to completely oppose the hydraulic pressure gradient and eliminate the driving force, glomerular pressure equilibrium would have been said to have occurred.

Because capillary oncotic pressure changes with length, total GFR along the entire capillary is defined by an integrated driving force:

$$GFR = \int_{\text{Beginning-capillary}}^{\text{End-capillary}} K_f (P_{GC} - P_{BS} - \Pi_{GC})$$

In the efferent arteriole, hydraulic pressure falls as a result of R_E to a level below that of oncotic pressure. The Starling forces in this peritubular capillary network are therefore in the absorptive direction (ie, ΔP_{uF} is reversed in the far right side of Fig 1–14).

REGULATION OF GLOMERULAR FILTRATION

Summarized in Table 1–1 are the various parameters capable of altering GFR. GFR increases if there is (1) an increase in K_f; (2) an increase in the hydraulic pressure gradient, primarily P_{GC} due to fall in R_A or rise in R_E (P_{BS} does not change very much under most physiologic conditions); or (3) a decrease in the mean oncotic pressure gradient, Π_{GC}, due to a fall in plasma albumin concentration or a rise in renal blood flow (Q_A).

As shown in Fig 1–15, renal plasma flow (Q_A) is a factor that can indirectly affect GFR because it alters the concentration profile of albumin within the capillary. Q_A is a function of the total renal vascular resistance, $R_A + R_E$. If the sum of $R_A + R_E$ decreases, Q_A rises. The increase in Q_A tends to attenuate the rise in albumin concentration that normally occurs along the capillary length (Fig 1–15B). Since there is little change in P_{GC} (due to the fall in both R_A and R_E), P_{BS}, and K_f, the reduction in the length-averaged Π_{GC} leads to a rise in the mean ΔP_{UF} (compare shaded areas in Fig 1–15A versus Fig 1–15B) and hence GFR

Table 1-1. Determinants of GFR.

	↑GFR	↓GFR
K_f	↑	↓
P_{GC}	↑	↓
R_A	↓	↑
R_E	↑	↓
Π_{GC}	↓	↑
[Albumin]	↓	↑
Q_A	↑	↓

rises. However, the increment in GFR is not proportionate to the increment in AP_A. Thus, the GFR:renal plasma flow (GFR/Q_A) ratio, called the filtration fraction, tends to decline as Q_A rises. As a corollary, there is a proportionate reduction in end-capillary protein concentration (Π_{GC}) as Q_A rises. This reduction in protein concentration that enters the peritubular circulation diminishes sodium chloride reabsorption in the proximal tubule (see below).

Intrinsic Regulation of R_A

As should be evident from the preceding discussion, the tones of the afferent and efferent arteriolar resistances are important both individually, for adjusting P_{GC} (Fig 1–14), and collectively, for regulating Q_A and hence Π_{GC} (Fig 1–15).

One way in which R_A is controlled is by **primary myogenic regulation,** mediated directly by the pressure or tension (or both) exerted on the wall of the afferent arteriole. As arteriolar pressure increases, afferent arteriolar resistance rises, and vice versa.

Afferent arteriolar resistance is also affected by delivery of sodium to and reabsorption of sodium by the macula densa region of the thick ascending limb of Henle. The afferent arteriole constricts (R_A increases) as sodium delivery increases. As shown in Fig 1–16, the thick ascending limb of Henle in reality—not the schematic version shown in most figures in this text—loops back to lie in close proximity to the afferent arteriole of its parent glomerulus. If GFR should rise for any reason, with increased sodium chloride delivery to the loop of Henle, sodium chloride reabsorption by the macula densa secondarily increases. A signal generated by increased sodium chloride transport by the macula densa cell—either electrical or chemical in nature (the signal is associated with a rise in intracellular calcium of these cells)—increases R_A. The rise in R_A then reduces P_{GC} and in that way returns GFR and sodium chloride delivery to the loop of Henle to preexisting normal values. This process, by which sodium chloride delivery to the macula densa region modulates GFR, is called **tubuloglomerular feedback.** Tubuloglomerular feedback operates normally to maintain stability of GFR.

The two factors that modulate R_A described above—afferent arteriolar myogenic tone and tubuloglomerular feedback—also affect renin release, as discussed previously (Fig 1–9). However, the changes are in opposite directions: An increase in afferent arteriolar pressure or macula densa sodium delivery increases R_A but suppresses renin release. A third factor, sympathetic nerve activity, also controls R_A and renin release, as discussed below.

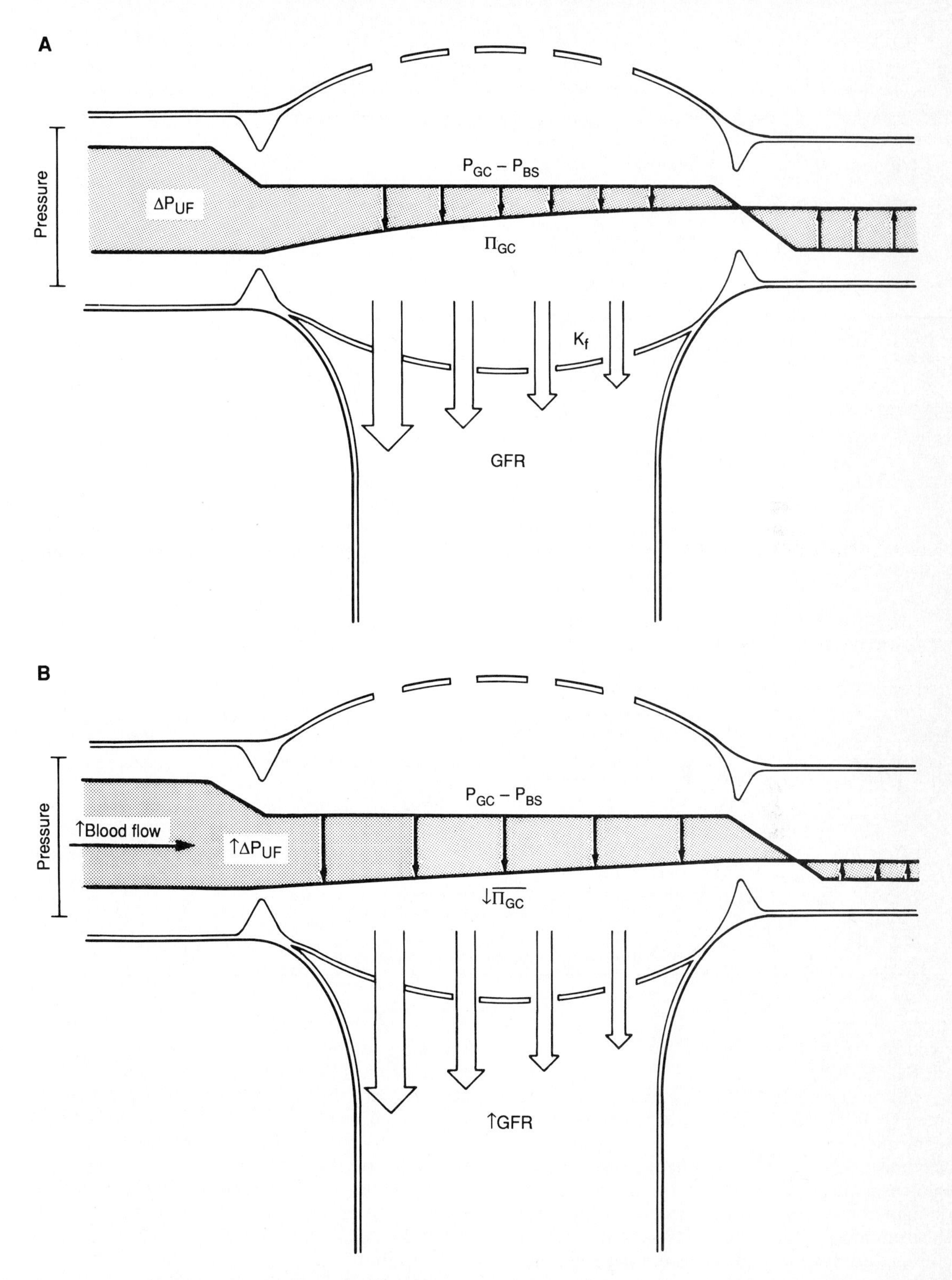

Figure 1-15. Determinants of glomerular filtration when renal blood flow is increased.

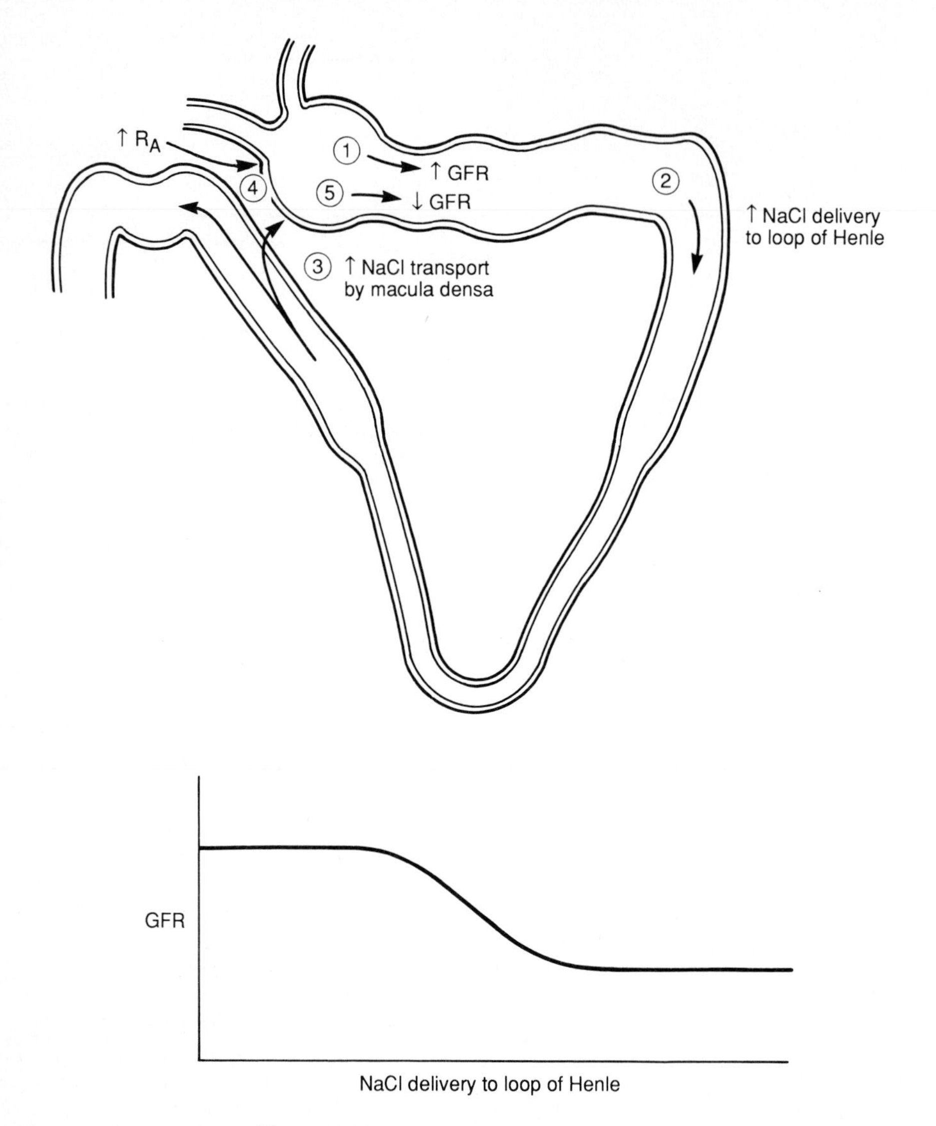

Figure 1–16. Tubuloglomerular feedback.

Neurohumoral Regulation of R_A, R_E, & K_f

Although a number of substances can alter glomerular function, the two major vasoconstrictive controllers of R_A, R_E, and K_f under physiologic circumstances are the **sympathetic nerves** and **angiotensin II,** as shown in Fig 1–17. Mutual local stimulation allows interconnection within this neurohumoral unit: The renin-secreting cells of the juxtaglomerular apparatus are densely innervated and under neurogenic (beta$_2$ receptor-mediated) control; angiotensin II affects sympathetic nerve output via presynaptic receptors.

As shown in Fig 1–17, in addition to previously described primary myogenic and macula densa-mediated mechanisms, R_A is controlled by the sympathetic nervous system. Although angiotensin II also has the ability to directly constrict R_A, such a change in R_A is not usually physiologically perceived because of functional antagonism by simultaneous induction by angiotensin II of vasodilatory prostaglandin production (notably PGE_2). However, R_A can be constricted by angiotensin II when the concentration is very high or when prostaglandin production is pharmacologically inhibited.

Under normal physiologic conditions, angiotensin II primarily affects R_E and is the principal controller of R_E. Both sympathetic nerves and angiotensin II can also greatly decrease K_f. Decrease of K_f by vasoconstrictors is thought to be mediated by mesangial cell contraction, which shrinks the

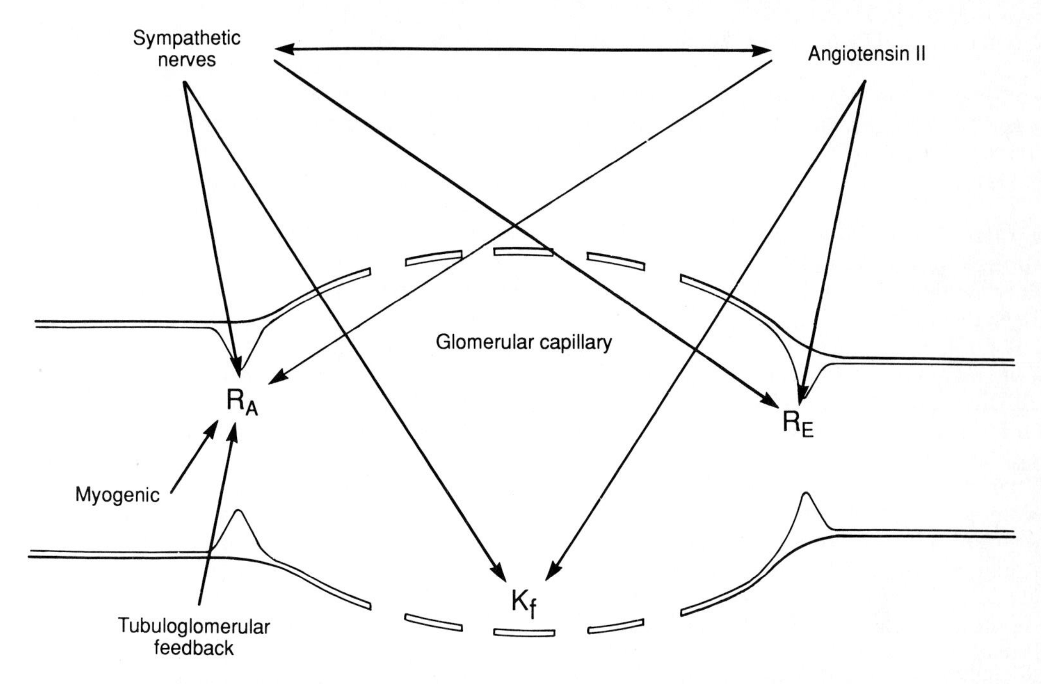

Figure 1-17. Regulation of glomerular capillary resistances and permeability.

glomerulus and diminishes the amount of glomerular basement membrane surface area available for filtration. Several substances also can reduce K_f—some indirectly, by activating angiotensin II (parathyroid hormone, certain vasoconstrictive prostaglandins, and cAMP); and some independently, without involving angiotensin II (ADH and endothelin).

Role of Vasodilators

The actions of the renal nerves and of angiotensin II on glomerular hemodynamics are physiologically opposed by vasodilator systems—notably the eicosanoids and ANF. As shown in Fig 1-18 and mentioned above, angiotensin II and norepinephrine induce prostaglandin E_2 (PGE_2 production, which acts in a negative feedback loop to antagonize the vasoconstrictor-mediated effects especially on R_A and sometimes on K_f but does not markedly affect R_E.

Under normal circumstances, prostaglandin levels are low and play only a minor role in controlling renal hemodynamics. Under hypovolemic conditions, however, prostaglandin production is stimulated by the high levels of angiotensin II and other vasoconstrictors and acts as a brake on the changes in R_A and K_f that would otherwise occur. Thus, neural- and angiotensin II-mediated afferent arteriolar and mesangial contraction is usually less than would occur if prostaglandins were not stimulated.

ANF can also be a physiologic antagonist of renal nerves and angiotensin II on R_A and K_f, shown in Fig 1-18. ANF causes afferent arteriolar vasodilation and mesangial relaxation, which tend to increase GFR. The importance of ANF in day-to-day adjustments of glomerular function in response to changes in dietary sodium intake is uncertain but is probably not great. ANF levels vary only minimally despite marked variations in sodium intake. However, when ANF levels are very high, due to overt atrial distention associated with acute extracellular volume expansion, congestive heart failure, or valvular heart disease or due to pharmacologic infusion of ANF, the influence on

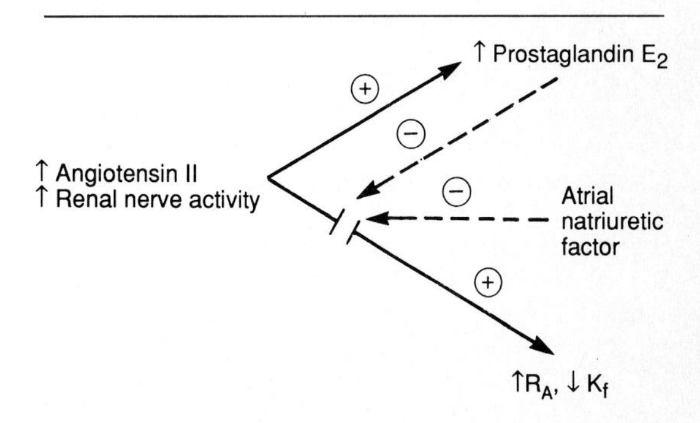

Figure 1-18. Feedback control of glomerular afferent arteriolar resistance and capillary membrane permeability.

glomerular hemodynamics can be quite significant.

Many other renal vasodilators exist, such as kinins (eg, bradykinin), endothelium-derived relaxing factor, and dopamine, but their role in physiologic regulation of GFR is still uncertain.

Integrated Control of GFR: Response to Alteration in Dietary Sodium

Given the many determinants and complex control of GFR, it may seem bewildering that filtration of sodium and water can be sensitively regulated. Having reviewed the physiologic controllers of GFR, we can now return to the response to ingestion of sodium.

When salt is ingested and the effective circulating volume is slightly expanded, there is a baroreceptor-mediated depression in sympathetic nerve and renin-angiotensin II activities, which coordinately decrease R_A and thus raise P_{GC} and increase K_f, as shown in Fig 1–19. These hemodynamic effects tend to increase GFR. A rise in ANF—to the extent that it occurs with this small increase in effective circulating volume, causes qualitatively similar glomerular microvascular changes. However, depression of sympathetic nerve and angiotensin II activities also decrease R_E, opposing the rise in P_{GC}. Finally, decreased total renal vascular resistance (R_A plus R_E) increases renal blood flow Q_A and tends to diminish the length-averaged Π_{GC}. Therefore, with only mild changes in effective circulating volume as induced by alterations in dietary sodium intake, GFR rises, though slightly. In fact, given the intrinsic error (about 5–10%) of conventional clearance techniques (see below), it is usually very difficult to measure any change in GFR in response to dietary sodium ingestion.

A greater change in effective circulating volume can have profound effects on GFR. For example, with hemorrhage, reduction in cardiac output and renal perfusion occurs with simultaneous baroreceptor-stimulated sympathetic nerve and angiotensin II activities. The high levels of neural activity and angiotensin II levels are united in increasing both R_A and R_E, which markedly diminish P_{GC} and Q_A, and in simultaneously contracting mesangial cells, which reduces K_f. Counteracting vasodilator levels (ANF and PGE_2) are low. The coordinated action of these factors causes GFR to fall markedly—in severe cases to nearly zero.

Stability of GFR in Response to Change in Arterial Pressure: Renal Autoregulation

The changes in GFR evoked by altered dietary sodium described above were a consequence of changes in effective circulating volume but not

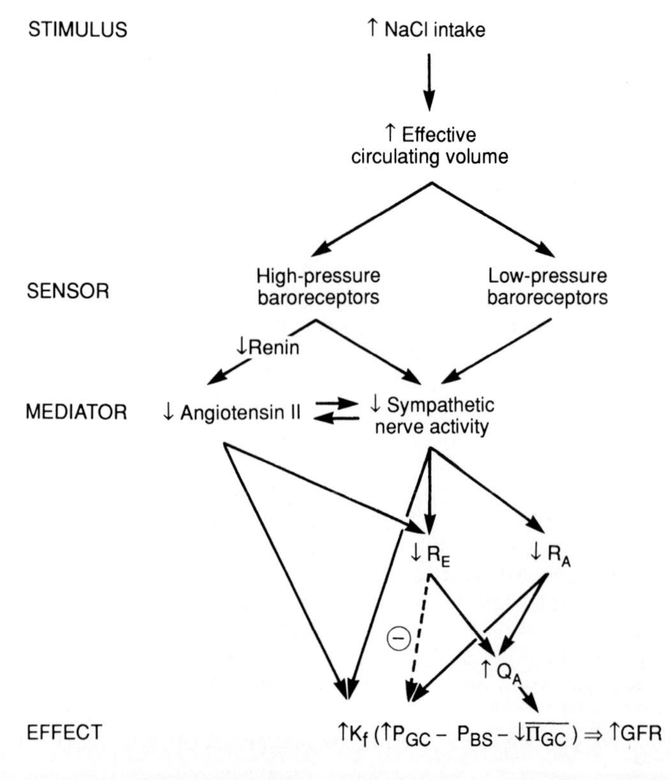

Figure 1–19. Glomerular response to an increase in sodium chloride intake.

blood pressure. With change in sodium intake, there is actually very little change in arterial blood pressure. However, even overt alterations in arterial blood pressure cause little change in GFR. This phenomenon, in which GFR remains stable despite primary acute changes in blood pressure, is termed renal autoregulation.

As shown in Fig 1-20, when blood pressure falls, afferent resistance also falls because of the primary myogenic vasodilatory response induced by the decrease in arteriolar wall pressure or tension (Fig 1-17). In addition, reduction in arterial pressure transiently decreases P_{GC} and thus GFR. The temporary decrease in sodium reabsorption by the macula densa cells changes the tubuloglomerular feedback signal to also decrease R_A (Fig 1-16).

Concurrently, reduction in arterial pressure and decrease in macula densa sodium delivery cause release of renin (Fig 1-9), resulting in local angiotensin II-mediated rise in R_E (Fig 1-20). The coordinated and reciprocal changes in R_A and R_E maintain P_{GC} and thus GFR constant (Table 1-1). Total renal vascular resistance (R_A plus R_E) is not substantially altered either, so renal blood flow also remains relatively unchanged. Thus, P_{GC}, GFR, and Q_A remain constant despite alterations in renal perfusion pressure (Fig 1-20). There are limits to these reciprocal changes in R_A and R_E, however, so that extreme changes in blood pressure in either direction cannot be tolerated and autoregulation breaks down.

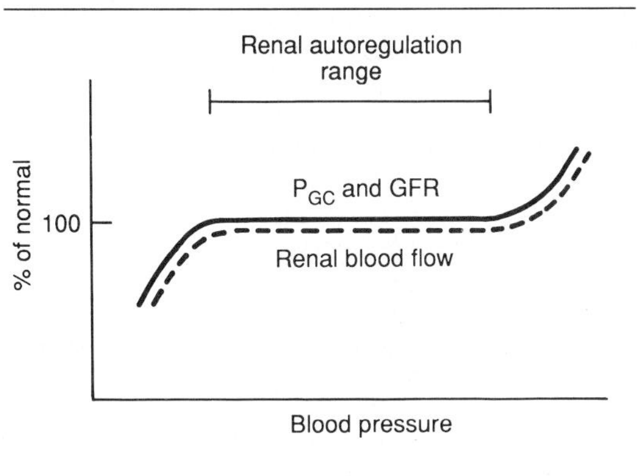

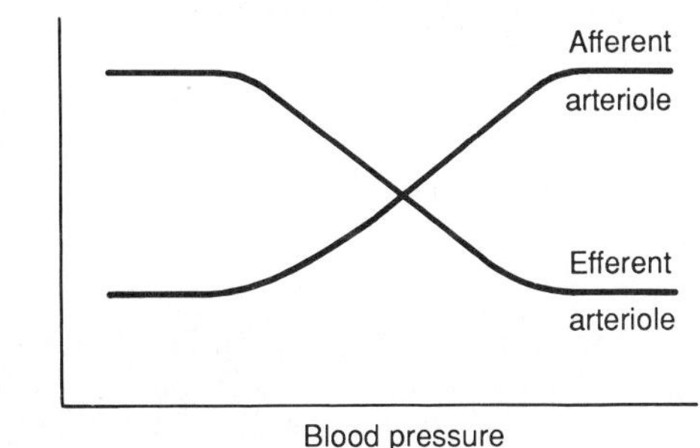

Figure 1-20. Renal autoregulation.

Some evidence suggests that a small subpopulation of nephrons in the juxtamedullary region of the kidney exhibit autoregulatory adjustments to changes in blood pressure less well than nephrons in the more superficial region of the cortex. If these nephrons lacked the autoregulatory response shown in Fig 1-20, they might be responsible for the phenomenon of **pressure natriuresis,** in which an acute rise in blood pressure causes the kidney to excrete sodium. According to this hypothesis, an increase in blood pressure would increase GFR in the juxtamedullary nephron subpopulation, causing them to increase sodium excretion. Since these nephrons represent but a small fraction of all nephrons and since efficient autoregulation by other nephrons occurs (Fig 1-20), a rise in GFR of the total kidney is usually undetectable when blood pressure rises acutely.

CLINICAL ASSESSMENT OF GFR: CREATININE CLEARANCE

Clinical measurement of GFR is important for estimating the contribution of sodium filtration to renal sodium homeostasis.

Marker of Glomerular Filtration

Any solute that is freely filtered at the glomerulus and neither reabsorbed nor secreted would serve as an ideal marker for glomerular filtration, as shown in Fig 1-21 (top). The total amount of glomerular filtration of such a solute *s*—plasma *s* concentration ($[P_s]$) × GFR—equals that excreted in the urine—urine *s* concentration ($[U_s]$) × urine flow rate (V):

$$P_S \times GFR = U_S \times V$$

Assuming the plasma concentration of *s* is constant during the period of the urine collection, GFR is calculated as follows:

$$GFR = \frac{U_S \times V}{P_S}$$

Creatinine Production & Creatinine Index

For research purposes, inulin and iothalamate serve as excellent glomerular filtration markers. Clinically, however, infusion of these substances is impractical, so that an endogenously produced marker, creatinine, is used. Creatinine is a muscle metabolite, and its production is proportionate to muscle mass. Creatinine generation and its input into plasma are quite constant during the day. Given a stable GFR and therefore a constant creatinine filtration (plasma creatinine concentration [P_{cr}] × GFR), it follows that creatinine excretion (urinary creatinine concentration [U_{cr}] × V) is also relatively constant and equal to creatinine generation:

$$\text{Creatinine generation} = P_{cr} \times GFR = U_{cr} \times V$$

Before considering the use of creatinine in estimating GFR, it is useful to recognize that this equality of creatinine generation and excretion is helpful in assessing the adequacy of a collected urine specimen. If a 24-hour urine collection is accurate, the total creatinine excreted should equal the individual's creatinine production during that period. In turn, creatinine production is predictable given a person's total muscle mass, which can be easily estimated by knowing body weight, gender, and age. Thus, total daily creatinine excretion, expressed as milligrams of creatinine excreted in the urine per kilogram of body weight, is known as the **creatinine index,** with normal values listed in Table 1-2. People with a large muscle mass (eg, athletes) or with active muscle breakdown (from muscle injury) may exceed the normal values, whereas lower values should be expected in individuals—especially old or chronically ill people—with muscle atrophy.

A urine collection with total creatinine content different from expected values in Table 1-2 usually represents a mistake in collection duration (ie, commonly the V term is in error, and the collection was really obtained over a shorter or longer time period than intended), though sometimes a creatinine analytical error occurs (ie, in the U_{cr} term). Such a collection cannot be relied upon for calculating creatinine clearance.

Creatinine Clearance

After the creatinine index is calculated to ensure that a urine collection is reliable, the urine collection values can be used to estimate creatinine clearance. Rearranging the equation above, the GFR is approximately equal to the clearance of creatinine, C_{cr}, as shown below:

$$GFR \approx C_{cr} = \frac{U_{cr} \times V}{P_{cr}}$$

Table 1-2. Creatinine index.

Creatinine index: Creatinine excretion ($U_{cr}V$) per day
Males: 20-26 mg/kg body weight per day
Females: 14-22 mg/kg body weight per day
(Values should be adjusted downward for advanced age or diminished muscle mass.)

Normal values for creatinine clearance are 75-145 mL/min per 1.73 m^2 in men and 70-130 mL/min per 1.73 m^2 in women. Creatinine clearance declines about 1 mL/min per 1.73 m^2 for each year of life.

Limitations of the Estimation of GFR by Creatinine Clearance

A major assumption used in the estimation of GFR by clearance methodology is that the urinary solute represents solely that which was filtered but did not undergo transport, depicted in Fig 1-21 (top). Unfortunately, as shown in Fig 1-21 (middle), creatinine is not a perfect marker for GFR because it undergoes tubular secretion by the organic acid secretory mechanisms resident in the proximal convoluted and straight tubules. Total urinary excretion of creatinine therefore represents that delivered by glomerular filtration plus that secreted by tubules after extraction from the remaining nonfiltered renal blood. Fortunately, the amount of tubular creatinine secretion is normally small, about 10% of GFR. Thus, creatinine clearance usually overestimates GFR by only about 10%. Drugs (such as histamine type 2 receptor antagonists) which block tubular creatinine secretion render creatinine a more accurate marker of GFR.

In some individuals with hypovolemia or primary renal disease (eg, glomerulonephritis), GFR is reduced but tubular secretory mechanisms remain intact. As shown in Fig 1-21 (bottom), tubular creatinine secretion may then be substantial, rendering creatinine clearance a marked overestimate of GFR. Creatinine secretion may occasionally exceed its filtration, such that the calculated creatinine clearance is more than double the true GFR.

Plasma Creatinine Concentration & Its Relationship to Creatinine Clearance

Even without a urine collection, C_{cr} can be estimated from the P_{cr} concentration. Since steady-state U_{cr} × V can be estimated from an individual's body weight, gender, and age (the creatinine index, Table 1-2), C_{cr} is proportionate to $1/P_{cr}$. A quantitative estimate of C_{cr} has been described as follows:

$$C_{cr} = \frac{(140 - \text{Age}) \times (\text{Wt}/72)}{P_{cr}}$$

For this equation, C_{cr} is given in mL/min, age in years, weight in kg, and P_{cr} in mg/dL. A 15% reduction is used for women. For example, the C_{cr} in a 40-year-old man weighing 72 kg with P_{cr} of 1 mg/dL would be (140 − 40) × (72/72)/(1) = 100

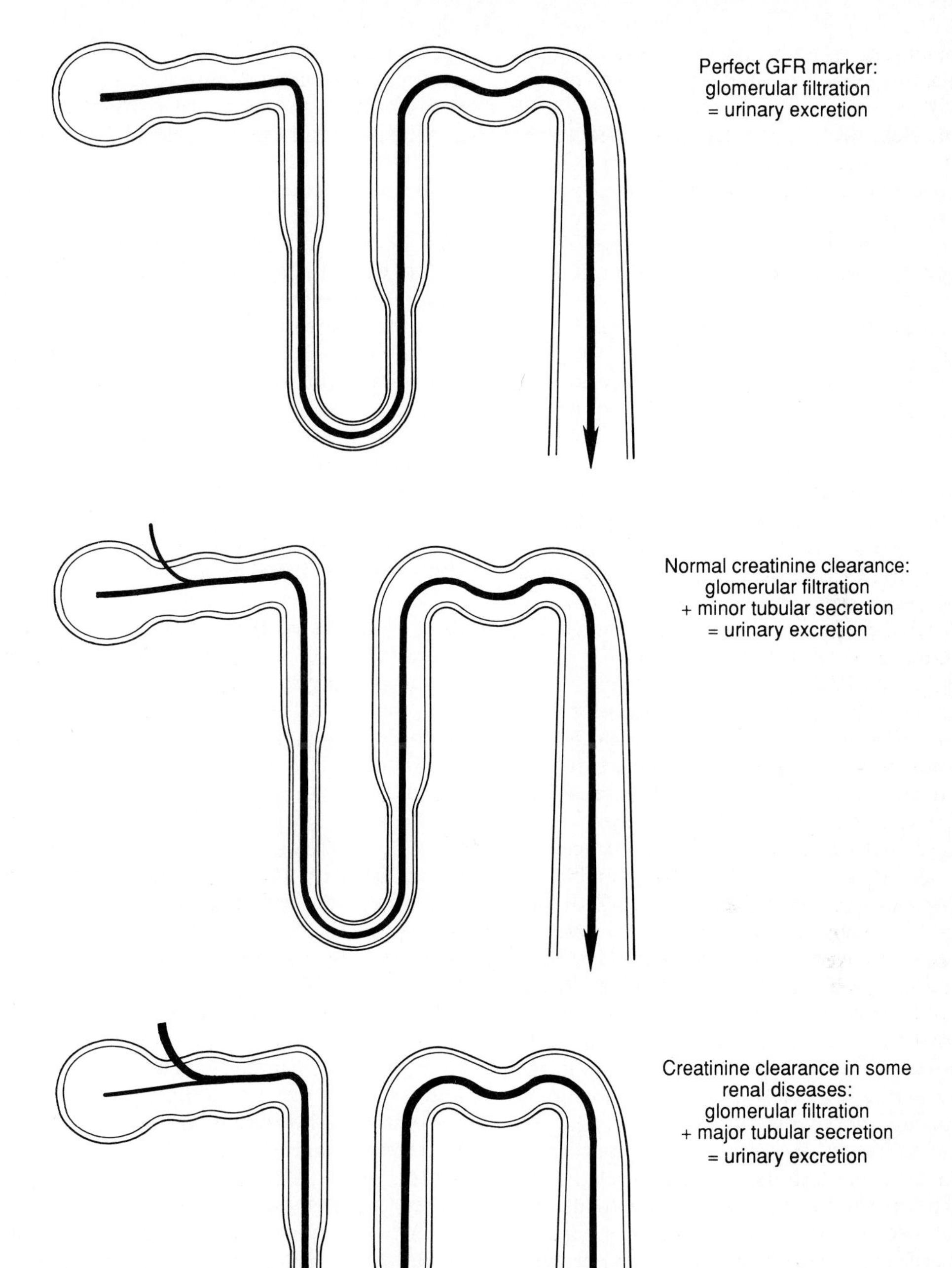

Figure 1-21. Creatinine clearance as a GFR marker.

mL/min. The C_{cr} in an 80-year-old woman weighing 36 kg with an identical P_{cr} of 1 mg/dL would be (0.85) × (140 − 80) × (36/72)/(1) = 25.5 mL/min. Note the 4-fold range in C_{cr} due to the difference in muscle mass associated with weight, age, and gender despite the similar P_{cr} levels in these two examples. The P_{cr} must therefore be carefully assessed with respect to factors affecting creatinine production before its relation to GFR is assumed.

Another caution concerning the use of P_{cr} relates to its inverse relationship to GFR. A marked reduction in renal function can occur without a large change in P_{cr}. For instance, halving of the GFR could cause the plasma creatinine concentration to change from 0.6 to 1.2 mg/dL, still within the normal range. As GFR falls, creatinine secretion becomes a larger percentage of filtration, further blunting the rise in P_{cr}. Thus, longitudinal collection of data is quite important when using the P_{cr} to assess renal function.

Urea Generation, Clearance, & Blood Level (BUN)

Urea represents the metabolic terminus for 90% of the nitrogen generated from protein catabolism, either from exogenous sources (diet or parenteral nutrition) or endogenous muscle breakdown. The kidneys excrete 90% of the urea; about 10% is excreted via the gastrointestinal tract.

Urea clearance is normally about 60% of GFR—ie, 40% of filtered urea is reabsorbed. When GFR declines or when tubular flow rate decreases (eg, due to hypovolemia or primary glomerular disease), a greater proportion of urea is reabsorbed. Thus, during hypovolemic or oliguric states, there is a decline in the urea clearance/GFR ratio and thus a disproportionate rise in the blood urea nitrogen (BUN). Conversely, as noted above during hypovolemic conditions, the P_{cr} tends to rise less than predicted owing to disproportionately more tubular secretion relative to GFR. Thus, as GFR falls during hypovolemia, there is a decrease in the ratio of urea clearance to creatinine clearance. If production of urea and creatinine remain constant, such changes in clearance cause a rise in the **BUN/plasma creatinine concentration ratio.** For example, the ratio of BUN/plasma creatinine might be (20 mg/dL)/(1 mg/dL), or 20, under normal conditions but (60 mg/dL)/(1.5 mg/dL), or 40, when hypovolemia supervenes.

SODIUM CHLORIDE REABSORPTION: OVERVIEW

Having dealt with the mechanisms, control, and assessment of GFR and the filtration of sodium, we now turn to the control of sodium reabsorption by the various nephron segments. As shown in Fig 1–12, reabsorption of sodium along the entire nephron is normally quite efficient, averaging about 99.3% of the filtered sodium load. The bulk of the reabsorption occurs in the proximal tubule and thick ascending limb of Henle. Fine control for reabsorbing the residual sodium is effected by the more distal nephron, notably the cortical collecting tubule.

In the following sections, the mechanisms and regulation of sodium reabsorption along the nephron will be reviewed, with an emphasis on sodium chloride. Control of reabsorption of sodium's other anion partner, bicarbonate, will be reviewed in Chapter 10.

SODIUM CHLORIDE REABSORPTION: PROXIMAL TUBULE

Active & Passive Transport Processes

The proximal convoluted tubule has two subsegments: S_1 and S_2. Most transport occurs in the S_1 subsegment, though the following discussion will describe transport by both subsegments combined.

The proximal tubule reabsorbs approximately 40–45% of the filtered load of sodium chloride. There are two modes of sodium chloride reabsorption in the proximal tubule, as illustrated in Fig 1–22. The first mode, accounting for about 50–65% of proximal sodium chloride reabsorption, is transcellular transport. Sodium and chloride are translocated across both luminal and contraluminal (basolateral) membranes of the proximal cell. The mechanisms by which sodium and chloride cross the luminal cell membrane are not known. They may enter the cell by means of parallel operation of Na^+/H^+ and Cl^-/OH^- antiporters. Nor is the process by which chloride leaves the cell understood for certain. Sodium exits the cell primarily via the Na^+/K^+-ATPase.

The other mode, accounting for 35–50% of proximal sodium chloride reabsorption, is passive paracellular transport. Sodium chloride moves from the lumen to the blood by diffusing between cells via the tight junction and the lateral intercellular space. The energy for this process derives from the fact that the chloride concentration in the luminal fluid of the proximal tubule is higher than in blood, as shown in Fig 1–23. In the early proximal tubule, water reabsorption due to isotonic sodium bicarbonate transport leads to concentration of the remaining solutes in the luminal fluid, principally sodium chloride. The fall in bicarbonate concentration (by about −20 meq/L) in the lumen therefore causes a reciprocal rise in

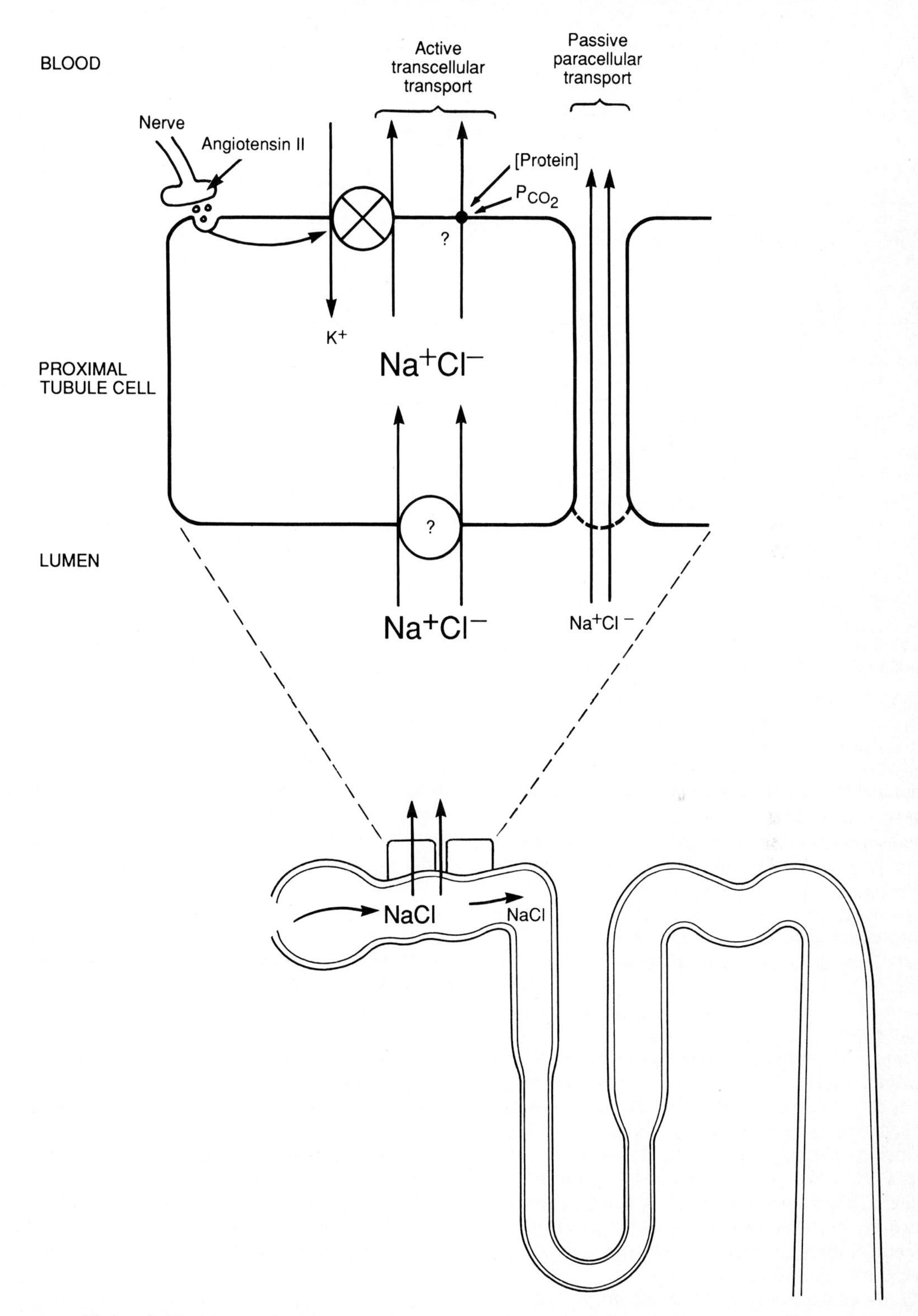

Figure 1-22. Mechanisms and regulation of sodium chloride transport in the proximal tubule.

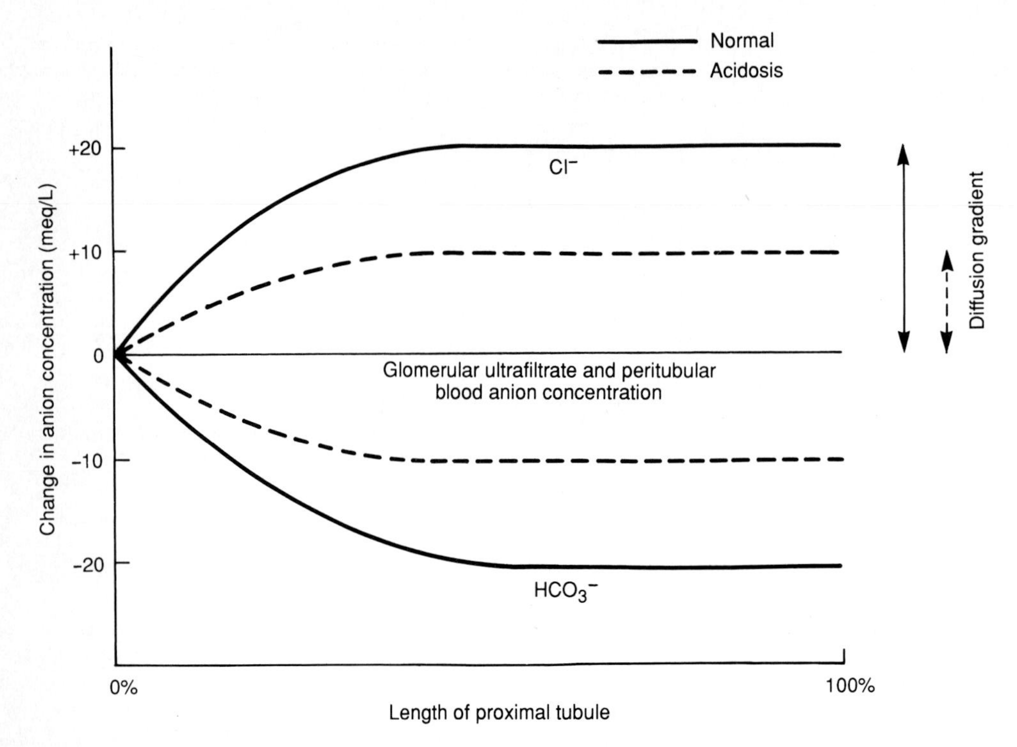

Figure 1-23. Axial change of Cl^- and HCO_3^- concentrations along proximal tubule during normal and acidotic conditions.

luminal chloride concentration. The 20 meq/L lumen-to-blood chloride concentration gradient provides the driving force for chloride diffusion along the rest of the tubule. Diffusion of the negatively charged chloride ion through the paracellular pathway creates a lumen-positive transepithelial potential difference (a "diffusion potential"), which then drives an equivalent amount of sodium.

Regulation

The active and passive transport systems in the proximal tubule are regulated in very different ways, summarized in Table 1-3. As shown in Fig 1-22, active sodium chloride is augmented by low-frequency sympathetic nerve stimulation, mediated by alpha$_1$ and perhaps alpha$_2$ receptors, which modulate Na^+/K^+-ATPase activity. Angiotensin II, by occupying presynaptic receptors or by central mechanisms, increases the gain of this adrenergic regulation.

Transcellular sodium chloride transport in the proximal tubule is also sensitively affected by peritubular protein concentration, but the mechanism is not understood. The absolute level of arterial albumin concentration and the extent to which it is concentrated in the glomerular capillary (filtration fraction, GFR/renal plasma flow ratio; see Fig 1-15) determine the protein concentration in the efferent arteriole and the postglomerular capillaries that surround the proximal tubule.

Finally, the prevailing arterial blood CO_2 tension and insulin concentration affect active sodium chloride transport. Hypercapnia potently inhibits whereas insulin stimulates proximal sodium chloride reabsorption.

The magnitude of passive sodium chloride reabsorption is inherently linked to the bicarbonate concentration gradient that is developed within the proximal tubule lumen (Table 1-3). Thus, as

Table 1-3. Physiologic determinants of NaCl reabsorption in the proximal tubule.

Transport	Factor	Impact
Active	↑Sympathetic nerve activity	Stimulates
	↑Peritubular [protein]	Stimulates
	↑P_{CO_2}	Inhibits
	↑Insulin	Stimulates
	↑GFR and NaCl delivery	Stimulates
Passive	↑[HCO_3^-] gradient	Stimulates
	↑Filtered [HCO_3^-]	Stimulates
	↑Angiotensin II	Stimulates
	↑GFR	Inhibits
	↑Nonreabsorbable solute	Inhibits

shown in Fig 1-23 (dotted lines), the chloride concentration gradient necessary for diffusion is diminished when bicarbonate transport in the early proximal tubule is reduced due to a low filtered bicarbonate concentration (in metabolic acidosis). A similar change occurs during inhibition of bicarbonate reabsorption (eg, by angiotensin II blockade or by a diuretic that inhibits carbonic anhydrase). An increase in GFR also tends to delay the fall in luminal bicarbonate concentration, thereby slowing the rise in luminal chloride concentration and hence sodium chloride diffusion.

The transepithelial chloride gradient can also be affected when there is a nonreabsorbable solute within the lumen. During osmotic diuresis, such a solute compels water movement into the lumen, diluting the chloride and thus preventing diffusional sodium chloride transport. A solute may be inherently nonreabsorbable, such as mannitol, or may be rendered nonreabsorbable by filtration in a quantity that exceeds proximal tubule transport capacity, such as glucose during hyperglycemia.

Load Dependence of Transport

The extent to which proximal reabsorption of a solute parallels a change in delivery of that solute has been termed **glomerulotubular balance.** In general, there is a very poor reabsorptive response in the proximal tubule when the filtered sodium chloride load increases owing to a rise in GFR—ie, proximal tubule glomerulotubular balance for sodium chloride is quite poor, as shown in Fig 1-24. As a result of unchanged reabsorption when the filtered sodium chloride load rises, delivery of sodium chloride out of the proximal tubule into the loop of Henle varies directly with GFR. This poor responsiveness for sodium chloride reabsorption when GFR changes is not a general transport characteristic of the proximal tubule: for other solutes, such as glucose and bicarbonate, glomerulotubular balance is excellent.

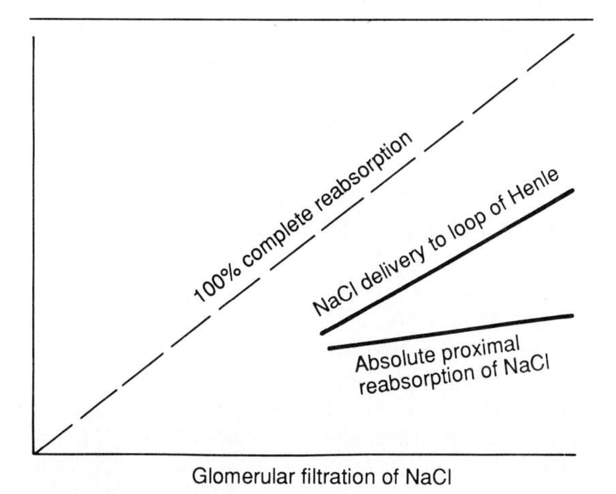

Figure 1-24. Load dependence of proximal tubule sodium chloride reabsorption.

Several offsetting factors may contribute to this poor load-dependent reabsorption for sodium chloride in the proximal tubule. Active transport of sodium chloride is inherently modestly flow-dependent, but this effect is counteracted by a decrease in peritubular protein concentration, which changes inversely with glomerular plasma flow and GFR (ie, filtration fraction generally falls as Q_A increases), as shown in Fig 1-15B. In addition, as luminal flow increases, the fall in luminal bicarbonate concentration and hence the rise in luminal chloride concentration are slowed, diminishing diffusional sodium chloride transport (similar to the situation with metabolic acidosis; dashed lines, Fig 1-23).

Response to Change in Effective Circulating Volume

The fraction of sodium chloride reabsorbed in the proximal tubule tends to diminish in response to an increase in effective circulating volume, and vice versa. An increase in effective circulating volume tends to increase GFR and diminish filtration fraction, so that most of the increment in filtered sodium chloride is unreabsorbed and transmitted to the loop of Henle (Fig 1-24). In addition, as part of the normal response to increased effective circulating volume, baroreceptor-mediated inhibition of sympathetic nerve and angiotensin II activities occurs, which independently decreases active sodium chloride reabsorption and further enhances sodium chloride delivery to the loop.

SODIUM CHLORIDE REABSORPTION: LOOP OF HENLE

The other major nephron segment responsible for sodium chloride reabsorption is the loop of Henle—notably, the thick ascending limb of Henle, as shown in Fig 1-25. Much of the sodium chloride escaping proximal reabsorption—representing 45% or more of all filtered sodium chloride—is reabsorbed in the loop of Henle (Fig 1-12).

Passive Transport in the Thin Ascending Limb of Henle

Little sodium transport occurs in the thin descending limb of Henle. In the thin ascending limb, there is a small amount of passive sodium chloride reabsorption, influenced by the efficiency of the countercurrent multiplication system and medullary hypertonicity (discussed more fully in Chapter 4). As summarized in Table 1-4, by

Table 1-4. Determinants of NaCl reabsorption in the loop of Henle.

Transport	Factor	Impact
Passive transport Thin ascending limb of Henle	Countercurrent multiplication system	
	↓Medullary blood flow	Stimulates
	↑Sympathetic nerve activity	Stimulates
	↑Angiotensin II	Stimulates
Active transport Thick ascending limb of Henle	↑NaCl delivery	Stimulates
	↑Sympathetic nerve activity	Stimulates
	↑ADH	Stimulates
	↑Prostaglandin E_2	Inhibits

regulating blood flow in the vasa recta of the medulla, angiotensin II and the renal sympathetic nerves probably affect the efficiency of the countercurrent system and thus passive sodium chloride reabsorption.

Active Transport in the Thick Ascending Limb of Henle

Most sodium chloride reabsorption in the loop of Henle occurs in the thick ascending limb. In cells of the thick ascending limb, as shown in Fig 1–25, the luminal transporter actually moves four ions at once: one sodium, one potassium, and two chlorides. Once inside the cell, the sodium exits the basolateral membrane via the Na^+/K^+-ATPase. The mechanism by which the chloride ions exit the cell has not yet been definitely elucidated but is probably via a chloride channel. Potassium recycles across the luminal membrane and creates a lumen-positive transepithelial potential difference that drives another sodium ion through the paracellular pathway. The sodium and chloride concentrations in the luminal fluid are lowered by this powerful transport system to about 30–50 meq/L each.

Regulation

Despite its quantitative importance in renal sodium chloride reabsorption, surprisingly little information is available about control of transport in the thick ascending limb of Henle. Stimulation of thick ascending limb transport by renal nerves and ADH has been demonstrated experimentally, though the physiologic import of such regulation is unresolved. A negative feedback system exists because antidiuretic hormone simultaneously induces prostaglandin E_2 production, an inhibitor of transport in this segment (Table 1–4). The Na^+-K^+-$2Cl^-$ luminal transporter can be pharmacologically inhibited by so-called loop diuretics such as furosemide, bumetanide, and ethacrynic acid.

Load Dependence of Transport

The thick ascending limb of Henle has an enormous capacity for sodium chloride reabsorption, and transport saturation is virtually impossible to achieve. Unlike the proximal tubule, the thick ascending limb of Henle demonstrates excellent, though not perfect, load dependence of transport (glomerulotubular balance), as shown in Fig 1–26. With an increment in sodium chloride delivery out of the proximal tubule, there is a nearly proportionate increase in reabsorption in the loop of Henle. Only about 5–15% of the increment in sodium chloride load escapes into the distal convoluted tubule.

Response to Change in Effective Circulating Volume

When GFR or proximal sodium chloride transport changes, most of the increment or decrement in solute delivery is greatly attenuated by the very high efficiency of the thick ascending limb of Henle. Since this transport "buffering" is intrinsically slightly imperfect, a small proportion of the sodium chloride diverted out of the proximal tubule enters the distal convoluted tubule (Fig 1–26). In addition, direct effects of renal nerve activity, angiotensin II, and ADH modulate the reabsorptive tone of the loop of Henle as a function of sodium chloride load (Table 1–4).

SODIUM CHLORIDE REABSORPTION: DISTAL CONVOLUTED TUBULE

Active Transport & Its Regulation

The distal convoluted tubule normally reabsorbs about 3% (range, 2–5%) of the filtered sodium chloride load (Fig 1–12). As shown in Fig 1–27, cells in the distal convoluted tubule have a unique luminal entry mechanism that involves the cotransport of sodium with chloride. The coupling ratio is 1:1, so that transport of sodium chloride is electroneutral. The Na^+/K^+-ATPase

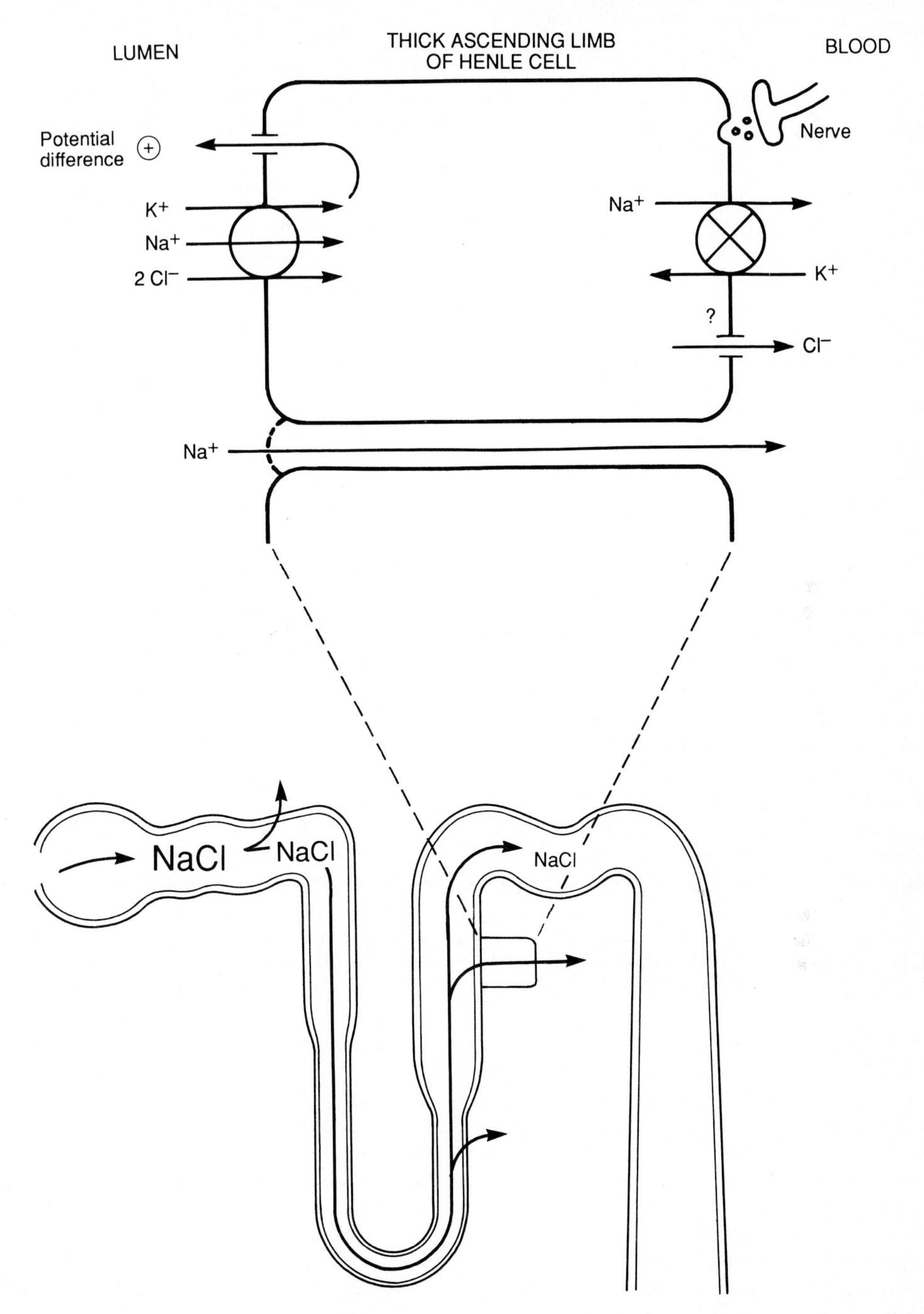

Figure 1-25. Mechanisms and regulation of sodium chloride transport in the thick ascending limb of Henle.

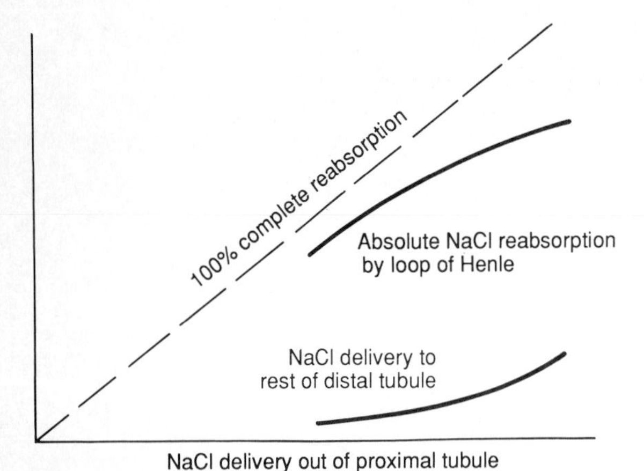

Figure 1-26. Load dependence of loop of Henle sodium chloride reabsorption.

pumps the sodium out of the cell, but the mechanism for chloride exit is unknown. Sodium and chloride concentrations in the luminal fluid decline slightly along the length of the distal convoluted tubule, approaching 25–30 meq/L.

Sodium chloride reabsorption in this nephron segment appears to be somewhat load-dependent, though the limits and mechanisms of this responsiveness are poorly understood. The sodium chloride cotransport mechanism on the luminal membrane is inhibitable by thiazide diuretics.

SODIUM CHLORIDE REABSORPTION: CORTICAL COLLECTING TUBULE

The remainder of the sodium reabsorbed in the nephron, roughly 1–2%, is reabsorbed in the collecting tubule. The collecting tubule is a relatively low-capacity sodium chloride transport site but has the ability to modulate the sodium concentration that ultimately reaches the urine. The best-understood segment is the cortical collecting tubule. Little sodium chloride reabsorption occurs in the outer medullary collecting tubule. Sodium chloride transport in the inner medullary collecting tubule appears to be similar in many respects to that in the cortical collecting tubule.

Active Transport

Sodium enters the principal cell of the cortical collecting tubule through ion-specific channels in the luminal membrane, as shown in Fig 1-28. The driving force for sodium entry is 2-fold. First, the sodium concentration is higher in the lumen than in the cell (> 25–30 meq/L versus < 10 meq/L, respectively). Second, the potential difference across the luminal membrane (about −90 mV in the cell and < −50 mV in the lumen) provides an electrical gradient favorable for sodium entry (described in more detail in Chapter 7). Rapid diffusion of the positively charged sodium ion creates a lumen-negative transepithelial potential difference, which is responsible for driving chloride through the paracellular pathway. Sodium is pumped out of the cell by means of the Na^+/K^+-ATPase. These collecting tubule transport mechanisms are capable of markedly reducing sodium and chloride concentrations in the luminal fluid to less than 1 meq/L each.

Regulation by Aldosterone

Sodium chloride transport in the cortical collecting tubule is most importantly controlled by the mineralocorticoid hormone aldosterone, as shown in Fig 1-28 and Table 1-5. Aldosterone from the blood enters the cell and combines with a high-affinity cytosolic mineralocorticoid receptor. The steroid-receptor complex migrates into the nucleus and binds to selective DNA sites, initiating RNA transcription and then translation of specific proteins. A cytosolic enzyme, 11β-hydroxysteroid dehydrogenase, metabolizes glucocorticoid hormones and prevents their occupancy of the mineralocorticoid receptor, which otherwise has an affinity for glucocorticoid hormones similar to its affinity for mineralocorticoid hormones.

After a lag period of one to several hours, the major physiologic changes induced by aldosterone begin to occur. A primary effect of aldosterone is to increase the number of sodium channels in the luminal membrane. By putting more "holes" for sodium entry through the luminal membrane, aldosterone enhances sodium reabsorption for any lumen-to-cell electrochemical gradient. The enhanced sodium flux increases the transepithelial lumen-negative potential difference and allows close coupling of chloride reabsorption. Aldosterone also stimulates the Na^+/K^+-ATPase. It is currently debated to what extent the change in Na^+/K^+-ATPase activity is primary or due to elevation of intracellular sodium concentration caused by increased influx of sodium through the luminal channels.

Aldosterone also augments luminal potassium

Table 1-5. Physiologic determinants of NaCl reabsorption in the distal nephron.

Transport	Factor	Impact
Distal convoluted tubule	↑NaCl delivery and flow rate	Stimulates
Collecting tubule	↑Aldosterone ↑NaCl delivery and flow rate ↑ANF	Stimulates Stimulates Inhibits

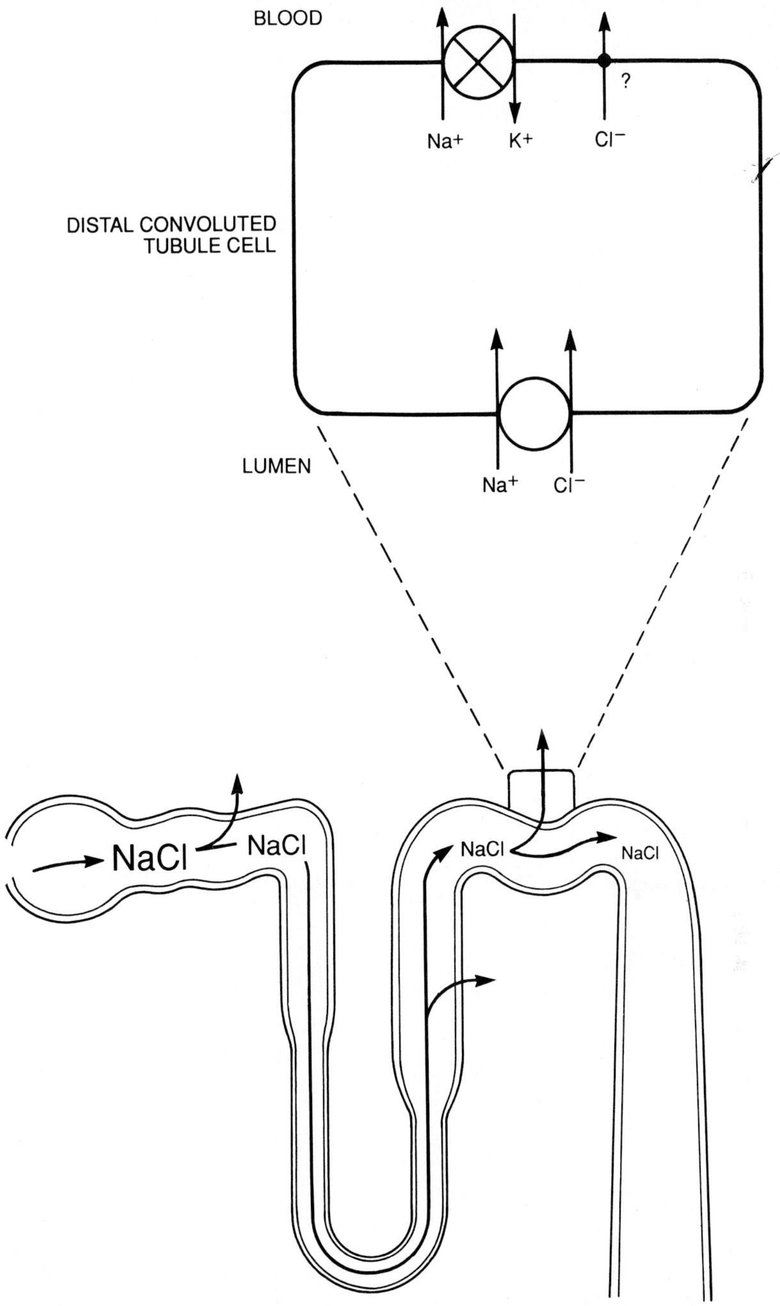

Figure 1–27. Mechanisms of sodium chloride transport in the distal convoluted tubule.

permeability (Fig 1–28) and a proton-translocating ATPase in the neighboring intercalated cells. Thus, in addition to exquisite control of sodium reabsorption, it also enhances potassium and hydrogen ion secretion (discussed in Chapters 7 and 10).

Pharmacologic inhibition of sodium transport in the cortical collecting duct can be achieved by direct blockade of the luminal sodium channel with so-called sodium channel blockers such as amiloride or triamterene. Inhibition of the renin-angiotensin II-aldosterone axis has the same effect, as with drugs that diminish renin release (eg, beta-blockers), angiotensin II formation (converting enzyme inhibitors), or binding of aldosterone to its receptor (eg, spironolactone).

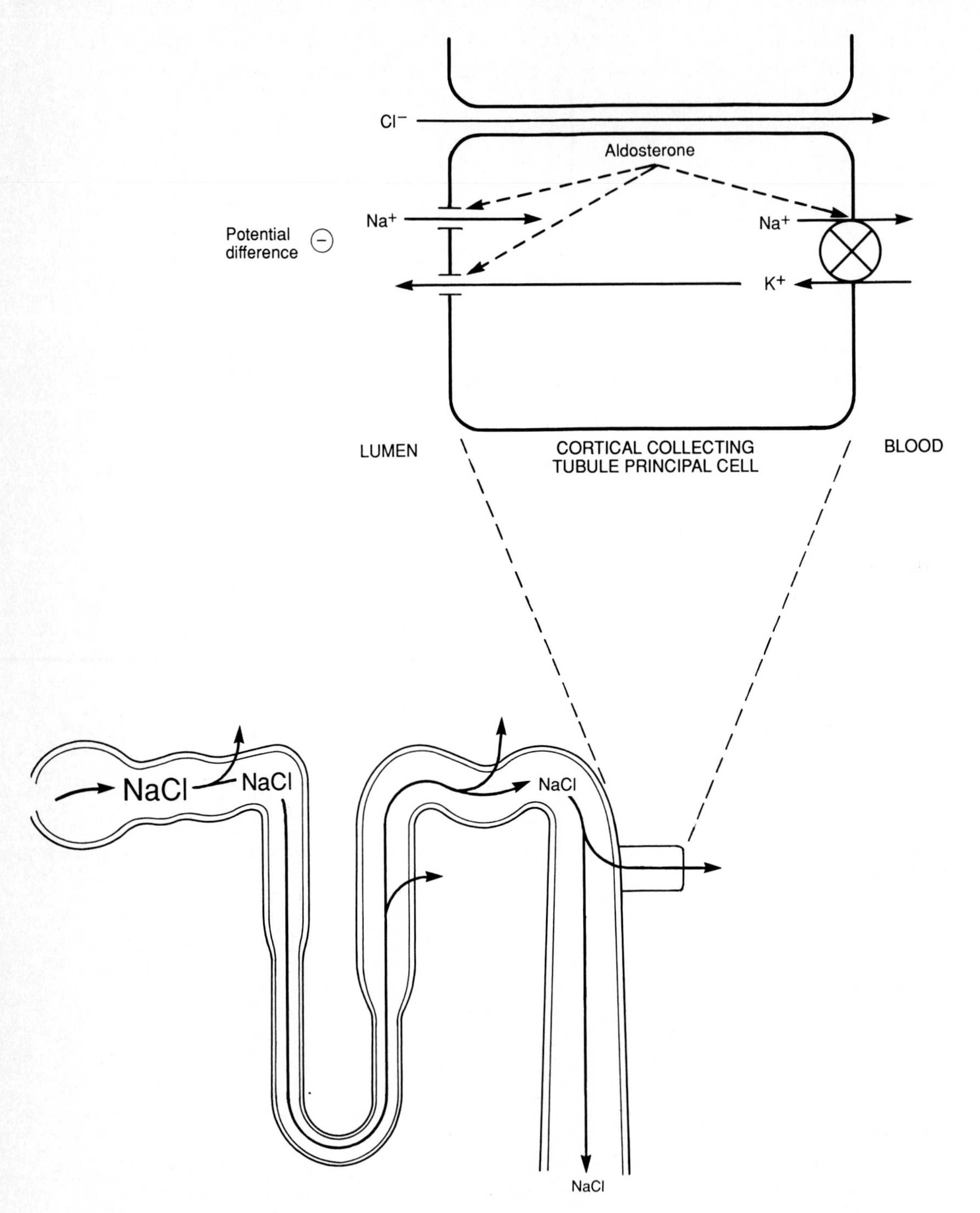

Figure 1-28. Mechanisms and regulation of sodium chloride transport in the cortical collecting tubule.

The physiologic significance of other hormonal regulators of cortical collecting duct sodium transport is not well documented. ANF affects sodium transport in the medullary collecting duct, though the in vivo physiologic significance of this effect is debated.

Load Dependence of Transport

The amount of sodium chloride delivered to the cortical collecting tubule can affect the amount reabsorbed, though systematic studies on this point have not been performed. Load dependence of sodium chloride reabsorption in the cortical collecting tubule is probably less efficient than in the thick ascending limb of Henle.

Response to Change in Effective Circulating Volume

Since intrinsic load-dependence transport in the collecting tubule is limited, a large percentage of an increment in sodium chloride load emerging from the distal convoluted tubule will be excreted when effective circulating volume is increased. In addition, increased effective circulating volume suppresses the renin-angiotensin II-aldosterone axis. The reduction in aldosterone serves to down-regulate sodium chloride reabsorption in the cortical collecting tubule. An increase in ANF level may also depress sodium transport in the medullary collecting tubule.

When effective circulating volume is decreased, the opposite occurs and the decrease in delivered sodium chloride load is effectively reabsorbed, due in large measure to the high plasma aldosterone level. It is important to recognize that the collecting tubule regulates the "floor" for urinary sodium and chloride concentrations, which can be extremely low (< 1 meq/L).

REGULATION OF RENAL SODIUM REABSORPTION

As should be obvious from the preceding discussion, the factors that affect the transit of sodium through the kidney are enormously complex. No one nephron segment controls sodium excretion. Rather, all nephrons contribute to an intricate control mechanism for modulating the urinary sodium excretion rate. Given the importance of careful sodium homeostasis for the viability of the organism, the complexity of this mechanism and the apparent redundancy of sites for renal sodium handling may be more easily appreciated.

At the risk of seeming reductionist and simplistic, we can say that two themes help to explain a great deal about renal sodium handling. The first is the intrinsic relationship between sodium filtration and reabsorption that exists in some parts of the nephron, and the second is the coordination of both glomerular and tubular function by the interrelated effects of the sympathetic nerves and the renin-angiotensin II-aldosterone and ANF hormonal systems.

Intrinsic Glomerulotubular Balance

An increment or decrement of glomerular sodium filtration is reflected, though at a much reduced amount, in urinary sodium excretion. The load dependence of sodium transport in the nephron as a whole is slightly imperfect (Figs 1–24 and 1–26). Given the high GFR and hence enormous daily filtered sodium load in normal individuals as well as this "slippage" in filtration-reabsorption coupling, relatively small changes in GFR can acutely have a significant impact on urinary sodium excretion.

Role of the Sympathetic Nerves, Renin-Angiotensin-Aldosterone, & ANF

Systemic vascular resistance, glomerular hemodynamics, and tubular transport function are regulated in coordinated fashion by sympathetic nerves, renin-angiotensin-aldosterone, and ANF as shown in Fig 1–29. These systems are all simultaneously regulated by baroreceptor signaling. In addition, each system directly impacts on the others: renin is under sympathetic nerve control; angiotensin II potentiates sympathetic nervous activity; and ANF inhibits sympathetic nerve activity as well as renin and aldosterone release. Finally, these systems coordinately modulate GFR at the same time as they change sodium transport directly in many nephron segments. The balance between sodium filtration and reabsorption is set by these neurohumoral systems; ie, they adjust the level of glomerulotubular balance.

Renal Response to Decrease in Effective Circulating Volume

As summarized in Fig 1–30, decreased effective circulating volume activates high- and low-pressure baroreceptors which in turn increase renin-mediated angiotensin II production and sympathetic nerve outflow while decreasing the release and plasma concentration of ANF. All of these neurohumoral changes tend to coordinately reduce GFR. Increased sympathetic tone and angiotensin II and decreased ANF act to decrease K_f and to increase afferent and efferent resistances, which lowers Q_A. The reduction in Q_A tends to increase the mean Π_{GC}, while the increase in afferent resistance has the opposite effect.

However, the effects of the sympathetic nervous system, angiotensin II, and ANF are not

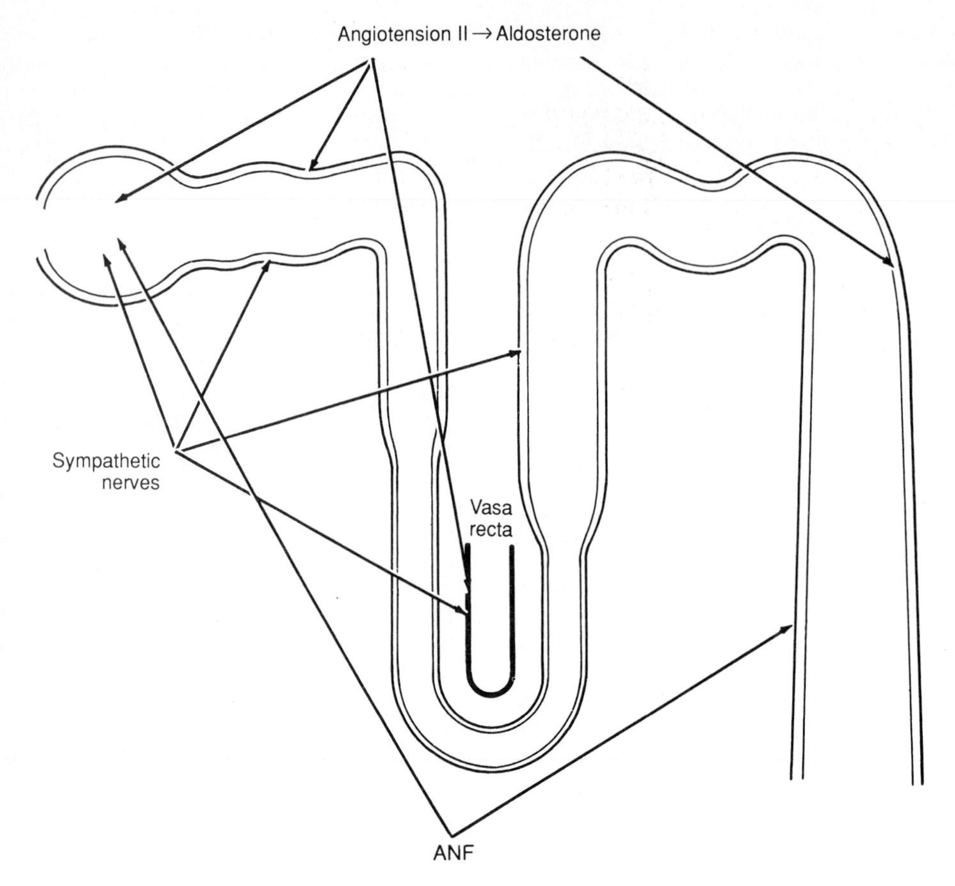

Figure 1-29. Renal sites of action of sympathetic nerves, angiotensin II, and aldosterone.

confined to the glomerular microcirculation. This potent neurohumoral unit has profound effects on sodium transport all along the nephron (Fig 1-29). As depicted in Fig 1-31, both angiotensin II and the renal sympathetic nerves directly increase sodium transport in the proximal tubule. In addition, an increase in filtration fraction consequent to the reduction in GFR and hence in the efferent arteriolar and peritubular capillary protein concentration directly augments active sodium transport in the proximal tubule.

Both angiotensin II and the sympathetic nervous system also decrease vasa recta capillary blood flow in the medulla, tending to enhance the efficiency of the countercurrent system and hence of passive sodium reabsorption from the thin ascending limb of Henle. The direct control of active sodium reabsorption in the thick ascending limb of Henle by sympathetic nerves as well as by ADH increases sodium reabsorption in this segment.

Finally, stimulation by angiotensin II of aldosterone release by the adrenal gland serves to increase sodium reabsorption in the cortical collecting tubule, while a decline in ANF causes increased sodium reabsorption in the inner medullary collecting tubule. in summary, a decrease in effective circulating volume is translated into a diminution of sodium filtration as well as augmentation of sodium reabsorption by all nephron segments, resulting in a fall in urinary sodium excretion (Fig 1-31).

Renal Response to an Increase in Effective Circulating Volume

We can finally summarize the final component of sodium homeostasis, the renal excretory response when sodium is ingested in excess. In general, the complex responses described above to a decrease in effective circulating volume are reversed when volume expands. Fig 1-32 summarizes the complete sequence of events leading to the sodium excretory response to an increase in sodium ingestion, including the following: (1) sodium input changes total exchangeable sodium; (2) a higher level of total exchangeable sodium, in

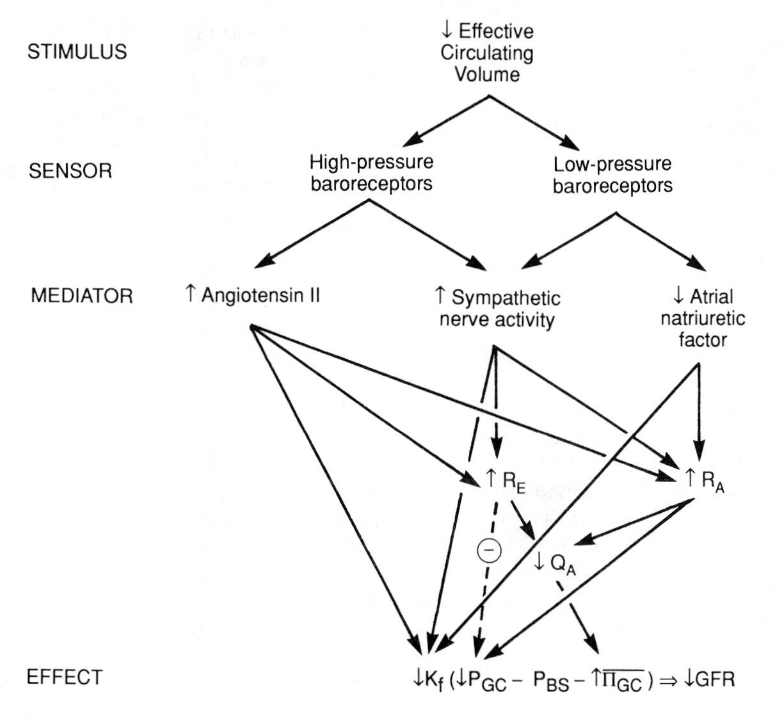

Figure 1-30. Glomerular response to a decrease in effective circulating volume.

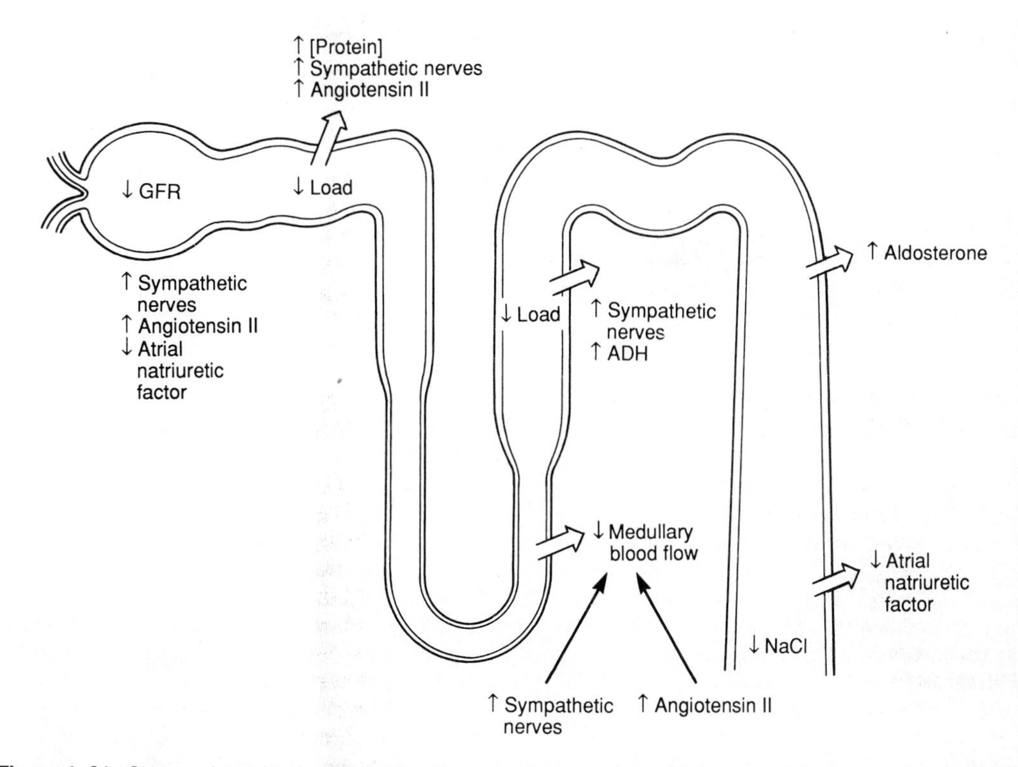

Figure 1-31. Sites and mechanisms of antinatriuretic response when effective circulating volume is decreased.

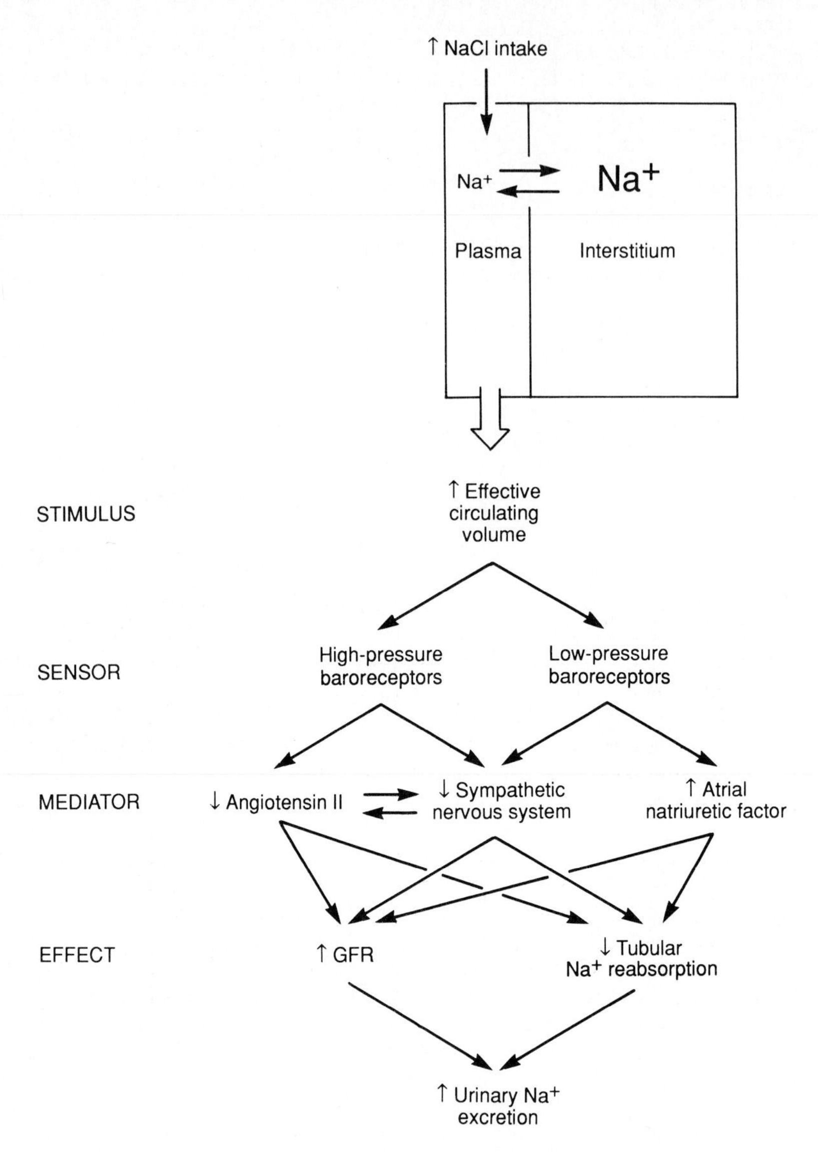

Figure 1-32. Renal response to an increase in sodium chloride intake.

equilibrium with the plasma volume as determined by the Starling forces, raises the effective circulating volume; (3) the increased effective circulating volume affects the tone of the high- and low-pressure baroreceptor monitors; (4) the baroreceptors depress the activity of the sympathetic nerve and the renin-angiotensin-aldosterone systems and enhance ANF release; and (5) these interrelated neurohumoral systems then control the determinants of GFR and the avidity of tubular sodium transport systems responsible for enhancing renal sodium excretion in a fashion opposite to what is shown in Figs 1-30 and 1-31.

LIMITS OF SODIUM CONSERVATION & EXCRETION

The fine control of GFR and tubular function by the baroreceptor-modulated neurohumoral systems in response to change in effective circulating volume allows for exquisite coupling of sodium excretion to sodium ingestion. As mentioned personally, urinary sodium excretion can match an extraordinary range of dietary sodium intake encountered in various societies, from less than 10 to more than 1000 meq/d, while incurring only modest changes in steady-state total body sodium content.

Relation of Extracellular Sodium Balance & Renal Sodium Excretion

The kidney has no direct sensing mechanism for the amount of sodium ingested. Renal sodium excretion under normal circumstances is the final expression of a sequence of signals that ultimately derive from the total exchangeable sodium content (Fig 1–32). This coupling of renal excretory function to total exchangeable sodium assumes that all components of the sodium homeostasis cascade are normal: cardiac function, capillary Starling forces, baroreceptor tone and afferent neural and central integrative functions, efferent sympathetic nervous system and hormonal activities, and renal function.

When all these conditions obtain, there appears to be a "zero set-point" of total body exchangeable sodium at which renal sodium excretion is negligible, as shown in Fig 1–33. Above this set-point, steady-state renal sodium excretion is a sensitive function of increasing extracellular volume or positive sodium balance. Below this set-point, renal sodium excretion remains virtually nil.

Sodium balance must exceed the normal zero set-point before the kidney responds by increasing sodium excretion. If a disease process were to render total body exchangeable sodium less than normal with negative sodium balance, renal sodium excretion would remain negligible even if sodium were being ingested. The kidney in this case does not "know" that sodium is being consumed—only that the extracellular volume and sodium balance are subnormal. If the sodium deficit were repaired to normalize extracellular volume, the kidney would then resume excretion of ingested sodium.

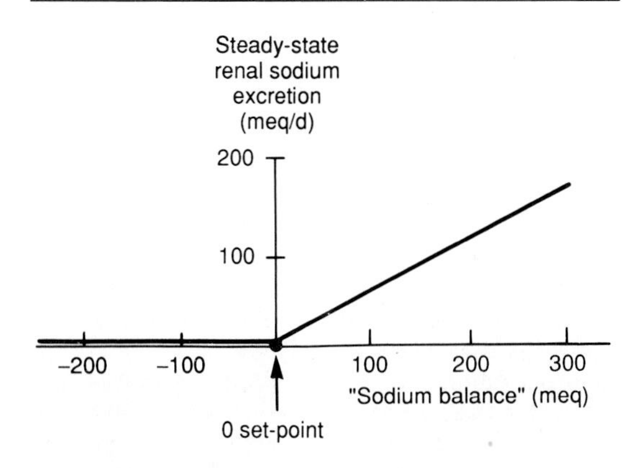

Figure 1–33. Renal sodium excretion as a function of sodium balance.

Renal Sodium Conservation in Response to Dietary Sodium Restriction

The response of a normal person to a reduction in sodium intake is shown in Fig 1–34. Sodium excretion is initially 150 meq/d (Fig 1–34A), equivalent to the normal sodium intake of 150 meq/d (ignoring nonrenal sodium excretion). The individual is therefore in sodium balance, and extracellular volume and body weight are stable (Fig 1–34B). When sodium intake is abruptly decreased to 10 meq/d (the vertical dotted line at day 0), renal sodium excretion initially is far in excess of intake. Negative sodium balance therefore is induced (shaded area in Fig 1–34A). Water loss parallels sodium loss, so that the plasma sodium concentration remains normal (discussed in Chapter 4). Thus, negative sodium balance is equivalent to the loss of isotonic extracellular volume, and body weight declines (Fig 1–34B). The depression in extracellular volume then initiates the series of events depicted in Fig 1–31 to reduce renal sodium excretion. Urinary sodium excretion gradually falls, halving approximately every day, so that in 3–5 days sodium excretion again equals intake at the new level of 10 meq/d. Sodium balance and stability of extracellular vollume are again achieved at this new steady-state level.

The cumulative change of the sodium body content during the 3- to 5-day period in which sodium excretion exceeds intake—ie, the net negative sodium balance—is about −140 meq (cumulative shaded area in Fig 1–34B). Since the concentration of sodium in extracellular fluid is about 140 meq/L, the isotonic loss of 140 meq of sodium is equivalent to about 1 L of extracellular volume or 1 kg of body weight loss. Put another way, this amount of negative sodium balance is necessary to activate the signaling mechanisms necessary to reduce renal sodium excretion by 140 meq/d, shown previously in Fig 1–33.

The 3- to 5-day time course for achieving a steady-state reduction in urine sodium excretion can be markedly shortened if the stimulus is stronger. If the extracellular volume is more rapidly depleted (eg, by hemorrhage, surgically induced volume losses, explosive diarrhea), the kidney responds much more quickly to reduce urinary sodium output—within an hour to several hours. Probably the best example of the rapidity with which the renal sodium excretory response can be achieved is the renal response to a change in posture. When an individual assumes the standing position, fluid is pooled in the lower extremities, the effective circulating volume is reduced, and urinary sodium excretion falls within an hour.

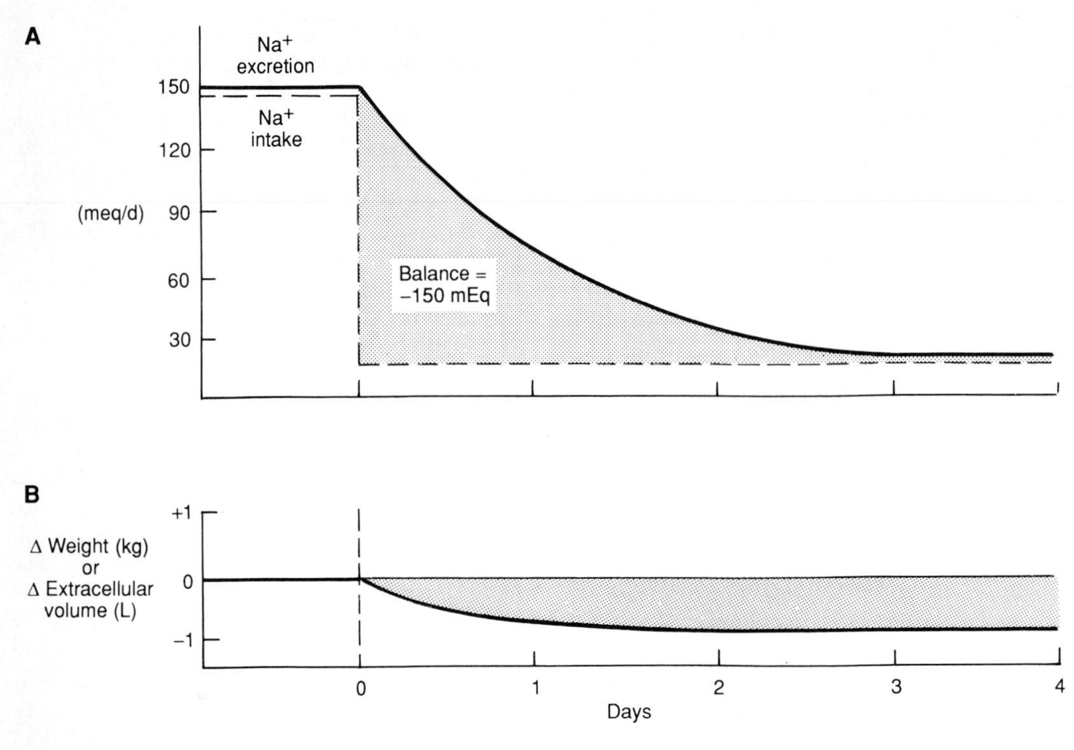

Figure 1-34. Time course of renal excretory response to a reduction in sodium chloride intake.

Renal Sodium Excretion in Response to Sodium Excess

The normal response to an increase in sodium intake is shown in Fig 1-35. In this figure, the basal sodium excretion is 150 meq/d, initially equivalent to dietary sodium intake. On day 0 (vertical dotted line), dietary sodium intake is shown to increase abruptly by an additional 150 meq/d. In the mirror image of the response to sodium deprivation, there is positive sodium balance of about 150 meq (shaded area in Fig 1-35A), resulting in an increase in extracellular volume and weight of about 1 L and 1 kg, respectively (shaded are in Fig 1-35B). A gradual increase in renal sodium excretion occurs over the ensuing 3–5 days until urinary sodium output equals sodium ingestion at the new steady-state level of 300 meq/d, so that extracellular volume is again stable.

As a rule of thumb, as shown in Fig 1-33, cumulative sodium retention—or amount of positive sodium balance—above the zero set-point is roughly equivalent to the steady-state sodium intake. That is, there is an excess of about 100–150 meq of sodium in the body (or about 1 L of extracellular volume) above the zero set-point when dietary sodium intake is 150 meq/d and about 300 meq of sodium (approximately 2 L of extracellular volume) when sodium intake is 200–300 meq/d.

The capacity of the kidneys to excrete sodium chloride excretion is enormous. In fact, the upper limit of sodium excretion is ill-defined. Sodium intake and urinary sodium excretion rates as high as 1500 meq/d are well described and remarkably well tolerated. At such an extremely high level of salt intake, there may develop mild change in blood pressure (unless the individual has a genetic predisposition to hypertension, in which case blood pressure may be somewhat more sensitive to dietary sodium chloride) and a trace amount of edema. While urinary sodium excretion in excess of 1500 meq/d is rare because oral sodium intake of this magnitude is not usually encountered, it can be provoked by brisk intravenous sodium chloride infusion. The urinary sodium concentration has been reported in such instances to exceed 500 meq/L.

Thus, urinary sodium excretion can vary over three orders of magnitude depending on the sodium chloride intake and physiologic conditions: from 1 to over 1000 meq/d, representing fractional sodium excretion rates (see below) from 0.01% to over 10%. An important lesson from the broad range of urinary sodium excretory rates achievable in response to altered sodium intake is that there is no "normal" sodium excretion. Sodium excretion is appropriate or inappropriate for an individual given his or her sodium intake and

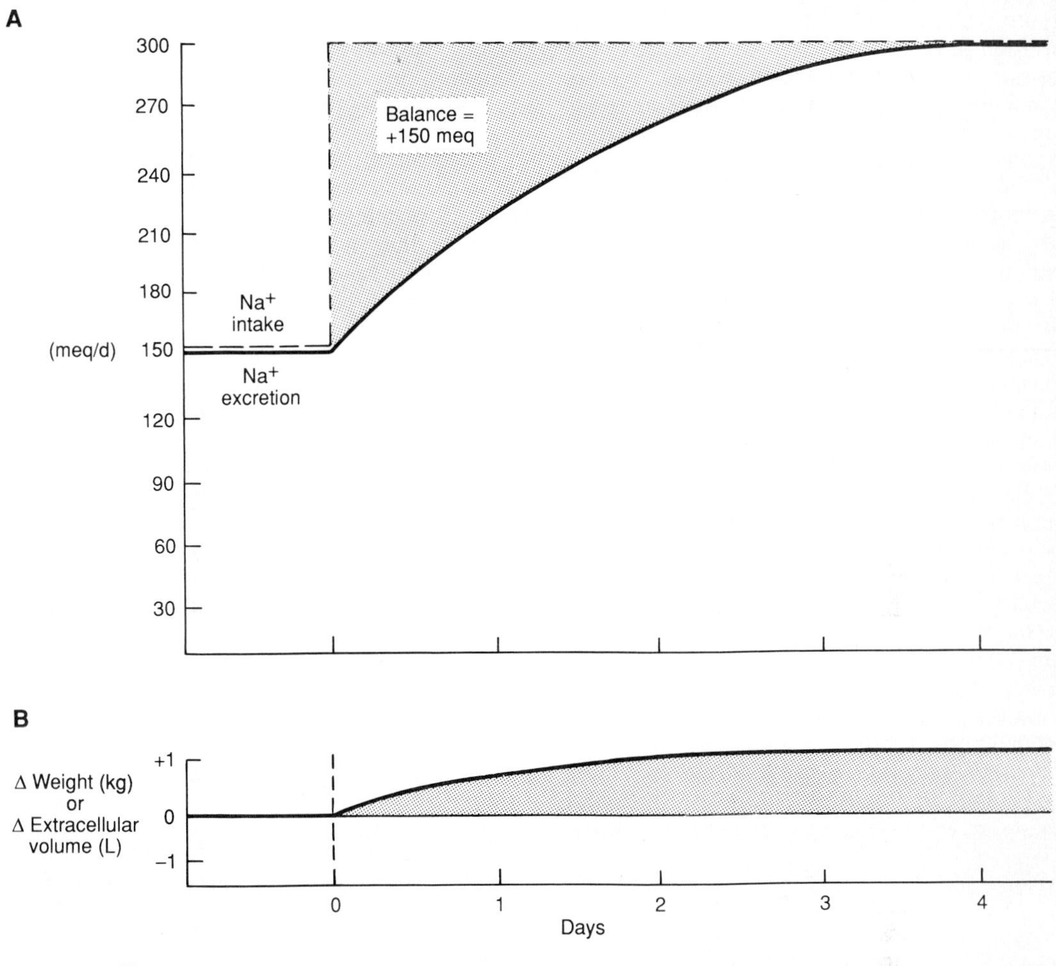

Figure 1-35. Time course of renal excretory response to a rise in sodium chloride intake.

physiologic state. For instance, a healthy person in a culture where little dietary salt is used may have a sodium intake and thus a urinary excretion rate of 10 meq/d. Conversely, another healthy individual who adds salt to all foods may have a sodium intake and excretion rate of 500 meq/d—50-fold higher. In both cases, sodium excretion matches intake and is physiologic and appropriate. By contrast, a sodium excretion rate of 10 meq/d in the setting of extracellular volume expansion or of 500 meq/d in the setting of extracellular volume contraction should both be considered grossly abnormal.

An important adaptive response occurs that allows sodium homeostasis to be maintained when the number of nephrons and GFR are reduced in the course of chronic renal disease. To retain absolute sodium excretion at a preexisting level when GFR and absolute sodium filtration are diminished, the fractional sodium excretion per nephron (see below) must be proportionately increased. The adaptive mechanisms that allow surviving nephrons to proportionately increase sodium excretion are not completely understood. An important factor is that the GFR in each residual nephron is quite high, which in itself leads to a relative augmentation of sodium excretion (Figs 1–24 and 1–26). Another factor is that some degree of chronic extracellular volume expansion usually initiates the cascade of events depicted in Fig 1–32 to diminish sodium excretion in the surviving nephrons.

CLINICAL ASSESSMENT OF RENAL SODIUM EXCRETION: URINE SODIUM CONCENTRATION & FRACTIONAL SODIUM EXCRETION

In the renal response to depletion or expansion of extracellular volume, urinary sodium and water outputs may vary independently. The sodium concentration in the urine is adjusted to achieve the necessary total sodium output in a given volume. In response to sodium deprivation, the urinary sodium concentration can be normally reduced to less than 10 meq/L, even less than 1 meq/L. In Fig 1–34, for example, if urine volume were 1 L/d, achieving the required urinary sodium output of 10 meq/d would necessitate a urine sodium concentration of 10 meq/L. But if the urinary output were 0.5 L/d or 2 L/d, urinary sodium concentration would stabilize at 20 meq/L or 5 meq/L, respectively. Minimum output of urinary sodium is not usually constrained by the minimal sodium concentration achievable in the urine.

The fact that sodium concentration varies inversely with urine volume to maintain constancy of total sodium output makes the urine sodium concentration itself somewhat unreliable for assessing either steady-state dietary sodium input or the extracellular volume status. Another way of approaching this problem is by calculating the amount of sodium excreted as a percentage of sodium filtered: the fractional sodium excretion. The sodium excreted is the urinary sodium concentration (U_{Na}) times the urinary volume excretion rate (V). The filtered sodium is the plasma sodium concentration (P_{Na^+}) times the creatinine clearance as an estimate of GFR. Thus, fractional sodium excretion (FE_{Na^+}) is expressed as follows:

$$FE_{Na^+} = \frac{U_{Na^+} \times V}{P_{Na^+} \times (U_{cr} \times V)/P_{cr}}$$

The V terms in the numerator and denominator cancel:

$$FE_{Na^+} = \frac{(U/P)_{Na^+}}{(U/P)_{cr}}$$

Sodium and creatinine concentrations measured in an untimed "spot" urine sample and in a simultaneously obtained plasma sample can be used to estimate the FE_{Na^+}. The advantage of FE_{Na^+} is that reciprocal changes in urine sodium concentration with urine volume are compensated, because urine creatinine concentration also varies inversely with urine volume.

For instance, in the example above in which urine volume is 1 L, if the urine sodium and creatinine concentrations were 10 meq/L and 140 mg/dL, respectively, with simultaneous plasma concentrations of 140 meq/L and 1 mg/dL, respectively, the FE_{Na^+} would be (10/140)/140/1) = 0.05%. If the urine volume halved—to 0.5 L/d—the urine sodium and creatinine concentrations would double, but the FE_{Na^+} would be unchanged: (20/140)/(280/1) = 0.05%. The lower limit of FE_{Na^+} normally achievable under stress is less than 0.01%. The upper limit is poorly defined but is greater than 10%.

While the emphasis has been on sodium, it should be recognized that urine chloride concentration and output usually closely match those of sodium. Under some circumstances, urinary excretion of sodium is obligated by a nonreabsorbable anion (eg, diuresis of bicarbonate or of ketoanions. The fractional chloride excretion rate is then a better reflection of effective circulating volume and extracellular volume than sodium (further discussed in Chapter 13).

Hypovolemic Disorders

2

INTRODUCTION

DEFINITION & CLASSIFICATION

Diminished Effective Circulating Volume

The approach to disorders of body volumes is dependent on one's frame of reference. As shown in Fig 2–1, the common factor in all hypovolemic disorders is diminution of the effective circulating volume. The arterial plasma volume is usually low, but the concomitant status of total body sodium and the extracellular volume may be decreased (eg, in hemorrhage, in disorders of skin, gastrointestinal or renal sodium loss), normal (eg, acute fluid translocation due to abnormal vascular permeability induced by sepsis, trauma, inflammation, or surgery), or increased (eg, in congestive heart failure, cirrhosis with ascites, or nephrotic syndrome).

Nosology Based on Extracellular Volume

Although it makes more sense to classify hypovolemic disorders pathophysiologically, according to the mechanism underlying the low effective circulating volume (Fig 2–1), it is customary to classify them according to total body sodium and extracellular volume status, chiefly because of therapeutic implications. The general approach to diminished effective circulating volume with low or normal total body sodium is to replenish the sodium deficit, while diuretics are often used in states of total body sodium excess.

Renal Sodium Retention

Renal sodium chloride retention is the final common pathway of all disorders associated with decreased effective circulating volume, no matter how much sodium is contained in the interstitial space (Fig 2–1). Thus, with respect to sodium excretion, the kidney "knows" and responds to the fullness of the circulation but is "ignorant" of the quantity of sodium within the body as a whole. Thus, even with overexpansion of body sodium, the kidney may respond to a "perceived" inadequacy of the arterial plasma volume because of transudation of fluid into the interstitium.

On the other hand, sodium retention should be considered physiologically inappropriate if the kidney retains sodium as an initiating event. Primary renal sodium chloride retention leads to expansion of the extracellular volume despite initial adequacy of the circulating volume, eventually causing hypertension and sometimes edema. The distinction between these primary and secondary antinatriuretic conditions will be further considered in Chapter 3.

Relation to Total Body Water Balance

Another simplification in the following discussion is that a deficit or surfeit of sodium will generally be considered to be isotonic; ie, a proportionate amount of water will be assumed to have been lost or gained. As discussed in Section II, this assumption is often invalid, since disproportionate changes in total body sodium and water occur in many hypovolemic conditions causing hyponatremia or hypernatremia. Nevertheless, the guidelines established to evaluate total body sodium and circulating volume status generally apply irrespective of the serum sodium concentration.

PATHOPHYSIOLOGY

Variable Total Body Sodium

Loss of sodium by hemorrhage or in fluids via the skin, gastrointestinal tract, or kidney leads to contraction of the plasma and interstitial volumes, shown in Fig 2–2B. Sodium can also be translocated from the vascular to the interstitial compartment when there is imbalance of the systemic capillary Starling forces, shown in Fig 2–2C. Interstitial fluid then increases at the expense of the plasma volume—but without change in total body sodium—as happens when vascular permeability increases acutely (eg, with sepsis, trauma, surgery, or inflammation) or when oncotic pressure is decreased or hydraulic pressure of the cap-

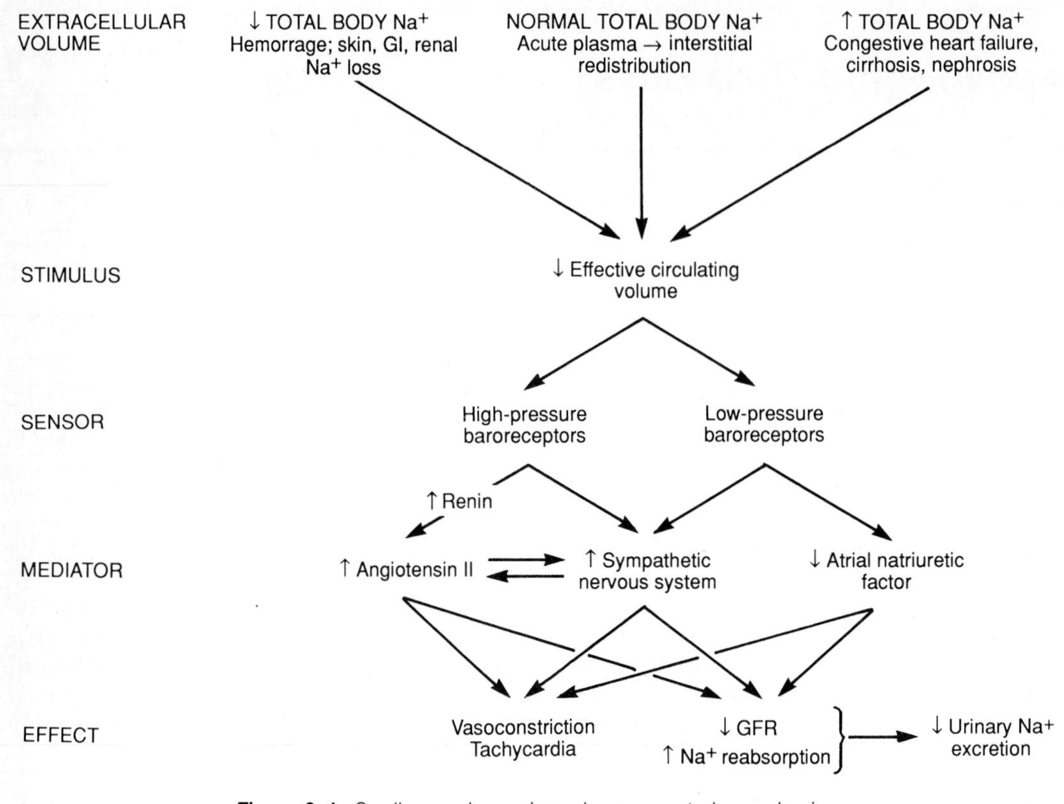

Figure 2-1. Cardiovascular and renal response to hypovolemia.

illary is increased (eg, with congestive heart failure, cirrhosis, or nephrotic syndrome).

As shown in Fig 2-1, whatever the total body sodium, a decrease in effective circulating volume sets into motion the cascade of events mediated by the neurohumoral effector systems previously reviewed in Chapter 1 that eventually lead to vasoconstriction, tachycardia, antinatriuresis (Figs 1-30 and 1-31), and antidiuresis (Chapter 4)—all designed to prevent circulatory collapse.

Absolute Versus Relative Deficiency of Plasma Volume

In some cases, the arterial plasma volume may be relatively low with respect to demand rather than being absolutely insufficient. For instance, when an arteriovenous fistula is created, a higher plasma volume is necessary to maintain circulatory sufficiency. Baroreceptors sense the relatively inadequate effective circulating volume and mediate sodium retention until a new, expanded plasma volume is sufficient to meet circulatory demand. An elevated absolute level of plasma volume is also appropriate in states in which neovascularity has occurred, such as pregnancy or cirrhosis. In the case of pregnancy, renal sodium retention also occurs to meet fetal sodium acquisition.

Evolution of States With Normal to Increased Total Body Sodium

When diminished effective circulating volume results from an abnormality of the capillary Starling forces, the renal response is to appropriately retain sodium (Fig 2-2C). As the individual ingests or is given sodium, much of the sodium retained enters the interstitium because of the perturbed Starling forces. If the disorder persists, total body sodium increases owing to expansion of the interstitial space, resulting in local or generalized edema. Thus, time distinguishes—somewhat arbitrarily from a pathophysiologic standpoint—hypovolemic states with normal versus increased total body sodium. Some of the latter edematous conditions will be considered subsequently under hypervolemic disorders in Chapter 3.

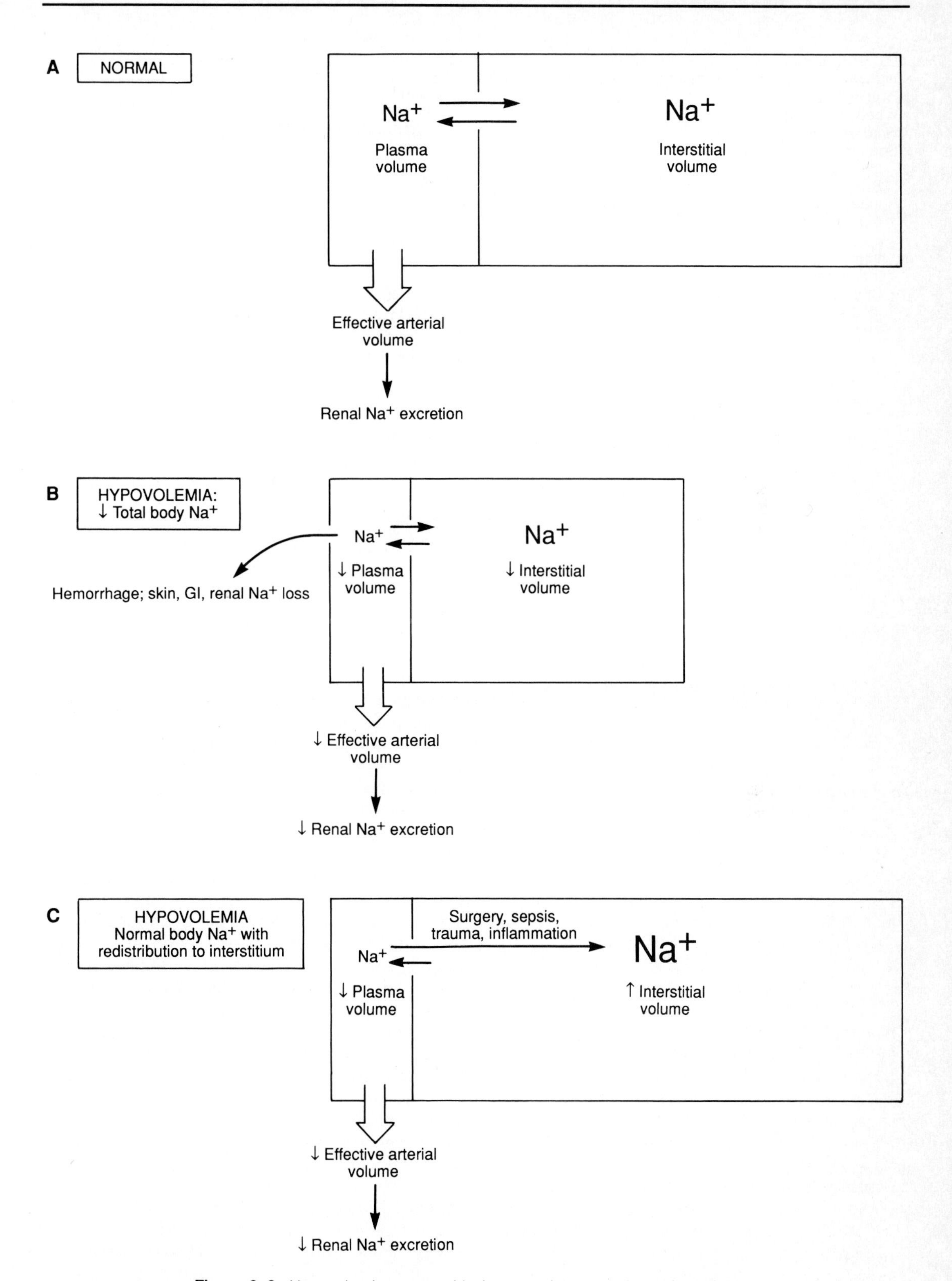

Figure 2-2. Hypovolemic states with decreased or normal total body Na^+.

Table 2-1. Hypovolemic disorders.

- ↓Total body sodium
  - Inadequate dietary sodium
  - Hemorrhage
  - Sodium loss via the skin
    - Excessive sweating
    - Burns
  - Sodium loss via the gastrointestinal tract
    - External drainage
    - Fistula
    - Vomiting or nasogastric suction
    - Intestinal inflammation
    - Diarrhea
  - Sodium loss via the kidney
    - ↑GFR
      - ↓Sympathetic nerve activity
      - ↓Angiotensin II
      - ↑ANF
      - Protein feeding
    - ↓Na^+ reabsorption
      - Proximal tubule
        - ↑Flow
        - Nonreabsorbable solute
        - ↓Bicarbonate concentration or reabsorption
        - ↓Peritubular [protein]
        - ↓Sympathetic nerve activity
        - ↓Angiotensin II
        - ↑PCO_2
        - ↓Insulin
        - Diuretics
          - Acetazolamide
      - Loop of Henle
        - ↑Flow
        - ↓ADH
        - ↓Sympathetic nerve activity
        - ↓Angiotensin II
        - ↑PGE_2
        - Bartter's syndrome
        - Diuretics
          - Furosemide
          - Ethacrynic acid
          - Bumetanide
      - Distal convoluted tubule
        - Diuretics
          - Thiazides
      - Cortical collecting tubule
        - ↓Renin
          - Diabetes
          - Hypertensive nephrosclerosis
          - Tubulointerstitial nephropathies
          - Drugs
            - Beta-blockers
            - Nonsteroidal anti-inflammatory drugs
        - ↓Angiotensin II
          - Converting enzyme inhibitors
            - Captopril
            - Enalapril
            - Lisinopril
          - Angiotensin II receptor antagonist
        - ↓Aldosterone activity
          - See Table 12-6.
- Normal body sodium
  - ↑Hydrostatic pressure gradient
    - Sepsis
    - Vasodilators
    - Heart failure
  - ↓Oncotic pressure gradient
    - Hypoalbuminemic states
    - Sepsis
    - Trauma
    - Surgery

continued

Table 2-1 (cont'd).

    - Burns
    - Lymphatic disease or obstruction
  - ↑Hydraulic permeability
    - Sepsis
    - Trauma
    - Surgery
    - Burns

DECREASED EFFECTIVE CIRCULATING VOLUME WITH DECREASED TOTAL BODY SODIUM

The causes of hypovolemic disorders are summarized in Table 2-1.

INADEQUATE DIETARY SODIUM CHLORIDE

Very low dietary sodium chloride intake, usually coupled with insufficient sodium retention by the kidney, is actually a rare cause of total body sodium chloride depletion. The kidney normally has a remarkably efficient ability to conserve sodium chloride—sodium chloride can be virtually eliminated from the urine—in response to reduction in dietary sodium (see Figs 1-33 and 1-34). Small obligate skin and gastrointestinal sodium losses still occur, but the quantities are very small (< 10 meq/d for both; Fig 1-11). As previously mentioned, several cultural groups have been observed in which the diet is adequate in calories but contains minimal sodium chloride—between 5 and 15 meq/d—with no ill effect on normal body volume homeostasis. With starvation- or dieting-induced sodium and calorie deprivation with ketosis, there is actual wasting of sodium into the urine, but this should be considered separately from the normal renal response when caloric intake is not curtailed.

Thus, dietary sodium chloride intake must be severely reduced for a prolonged time—often in conjunction with an impairment in renal sodium conservation—to cause any appreciable reduction in total body sodium and hypovolemia.

HEMORRHAGE

Even small amounts of blood loss—insufficient to cause hypotension—are sensed by the baroreceptor mechanisms, activating the neurohumoral effector systems shown in Fig 2-1 to cause renal sodium retention. The larger and faster the blood loss, the quicker the renal antinatriuretic response.

LOSS OF SODIUM VIA THE SKIN

Sweat is hypotonic to blood, with sodium and chloride concentrations of approximately 50 meq/L (range, 20–90 meq/L). While the usual volume of sweat is about 0.1 L/d, it can increase during heavy exercise in a hot climate to as much as 12 L/d (Chapter 4). Thus, while sweat is normally a trivial route for sodium chloride excretion (usually about 5 meq/d; Fig 1–11), it can be a major source of sodium loss—in excess of 600 meq/d—under some conditions.

The sweat volume and its concentration of sodium declines when sweating is prolonged from exertion or a warm environment. This acclimatization process is mediated by aldosterone and requires at least 2–5 days to become complete. Sweat sodium concentration can then be as low as 5–10 meq/L. Patients with mineralocorticoid deficiency (eg, primary adrenal disease) or hereditary electrolyte transport defects (eg, cystic fibrosis) cannot undergo this adaptation normally.

Severe sweating can cause other problems in addition to hypovolemia. Since sweat is hypotonic to blood, excess sweating with no replacement results in hypernatremia (Chapter 6). If exhaustion of sweating occurs, so that normal body temperature cannot be maintained during exercise or when environmental temperature is high, heat stroke with hyperpyrexia (core body temperature can sometimes exceed 41 °C) can supervene. The combination of hyperpyrexia and hypovolemia can cause severe neurologic impairment, including delirium, stupor, seizures, and coma. Severe potassium wastage in sweat and urine also is common, as discussed in Chapter 8, which can cause muscle injury, rhabdomyolysis, and acute renal failure.

Loss of sodium in the setting of burns can be quite large. Most of the functional sodium deficit is due to translocation from plasma into the interstitium and into cells due to thermal injury. However, large volumes of exudate from the burned area also account for sodium lost from the body. Fluid loss from skin can increase 20-fold following a full-thickness burn. One suggested estimate of daily external sodium loss is 200 meq/m^{2}of area burned.

LOSS OF SODIUM VIA THE GASTROINTESTINAL TRACT

Normally, only a small amount (~5 meq/d) of ingested sodium leaves the body via the stool (Fig 1–11). That there is net absorption of most dietary sodium chloride, however, should not conceal the fact that there are enormous bidirectional sodium fluxes within the intestinal tract. Similarly, water ingested averages 1.5 L/d, while stool water is only 0.1 L/d. Yet another 8.5 L of fluid enters the duodenum every day, primarily representing osmotically and hormone-induced fluid secretion in addition to contributions by pancreatic and biliary fluids. Thus, abnormal function or external drainage of segments of the gastrointestinal tract can result in marked electrolyte and volume shifts.

Pathophysiology

A brief review of the electrolyte composition in the various intestinal segments is depicted in Fig 2–3 and summarized in Table 2–2. In the stomach,

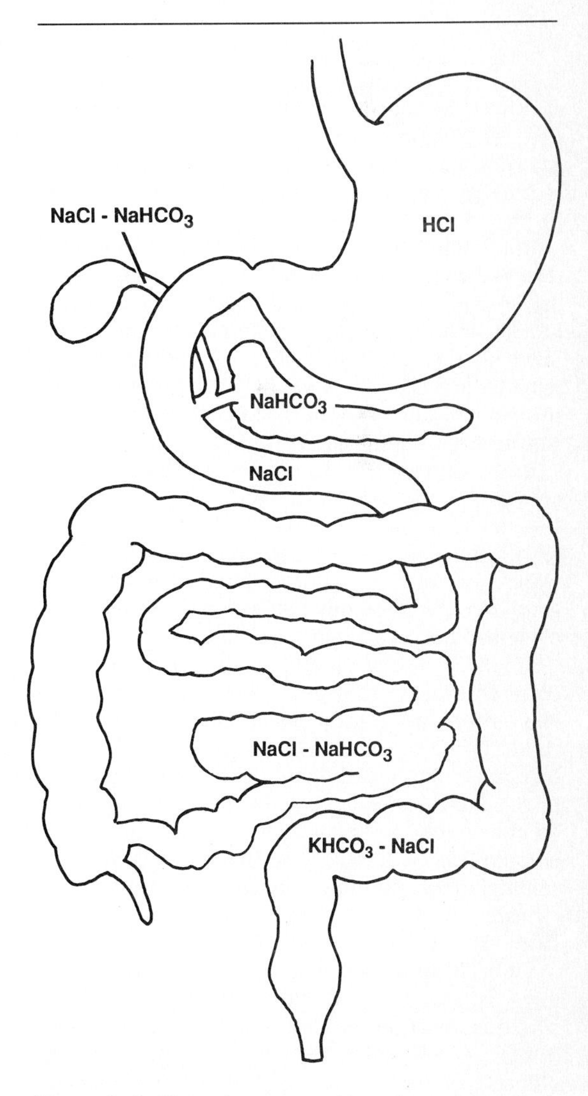

Figure 2–3. Electrolyte composition of gastrointestinal fluids.

Table 2-2. Representative gastrointestinal fluid volumes and electrolyte concentrations.

	Flow (mL/d)	$[Na^+]$ (meq/L)	$[Cl^-]$ (meq/L)	$[HCO_3^-]$ (meq/L)	$[K^+]$ (meq/L)
Gastric	1500	20 (acidic) 80 (nonacidic)	100	0	10
Duodenal	8500	60	100	10	15
Biliary	500	140	100	25	5
Pancreatic	500	140	40	120	5
Jejunal	3000	140	100	20	6
Ileal	600	80	60	75	8
Colonic	100	40 140 (↑ flow)	15	30	90 35–60 (↑ flow)

there is hydrochloric acid secretion, which can be enhanced by food ingestion, cholinergic stimulation, and various hormonal secretagogues. The gastric fluid can be very acidic, with pH 1.0–2.0. Normally, there is little sodium chloride in gastric fluid.

In the duodenum, epithelial electrolyte and water permeabilities are high, so that osmotically driven sodium chloride and water entry into the lumen occurs. There is also neutralization of much of the gastric acid by the bicarbonate-rich pancreatic juice. As a result, fluid entering the proximal small intestine becomes essentially an isotonic NaCl solution.

In the ileum, chloride and water absorption occurs simultaneously with bicarbonate secretion (via parallel Na^+/H^+ and Cl^-/HCO_3^- antiporters), which raise the bicarbonate concentration while lowering the chloride concentration. The fluid becomes a roughly half-and-half sodium chloride-sodium bicarbonate isotonic solution.

Finally, in the colon, aldosterone stimulates sodium chloride reabsorption over bicarbonate reabsorption and induces potassium secretion. The resultant major electrolyte in stool water is therefore potassium bicarbonate, with a lesser amount of sodium chloride. In diarrheal states, the ability of aldosterone to change the composition of the colonic fluid is lessened, so that there is proportionately more sodium compared to potassium.

Etiology

External drainage or diversion of a fistula of any gastrointestinal segment predictably leads to sodium depletion in an amount estimated from Table 2–1. Whether sodium is accompanied by chloride or bicarbonate depends on the segment drained (Fig 2–3). In general, as sodium traverses the gastrointestinal tract, the accompanying anion changes from chloride to bicarbonate. Bile electrolyte composition is similar to that of blood, while pancreatic juice is very rich in bicarbonate. Potassium is lost in large amounts only when volume flow through the colon increases. Urinary potassium loss can occur simultaneously owing to hypovolemia-induced hyperreninemic hyperaldosteronism (Chapter 8).

With vomiting or gastric drainage, only a small amount of sodium (about 20 meq/L) is lost in the gastric fluid (Table 2–1). Additional sodium is lost into the urine, however, because of an obligate sodium bicarbonate diuresis (see Chapter 13).

Electrolyte and volume transport within the intestine can be disrupted by abnormal bowel motility, intestinal inflammation, or ileus. As shown in Table 2–1, large amounts of sodium (about 500 meq) and water (about 8.5 L) normally enter the small intestine daily. To the extent that this sodium and water are functionally "outside" the body, sequestration and abnormal reabsorption of this fluid can potentially markedly deplete the effective circulating volume.

Any of the many causes of diarrhea can result in profound fluid and electrolyte imbalance. While the colon has a large capacity to augment fluid absorption—from its normal level of about 0.5 L/d to as much as 7 L/d—this site can be overwhelmed by large solute and water delivery when small intestinal transport is markedly abnormal. More commonly, however, diarrhea is caused by an intrinsic abnormality in colonic transport, induced by an osmotic agent, a secretagogue, abnormal motility, or inflammation. Stool volume in acute, cAMP-mediated secretory diarrheal illnesses such as cholera can be as much as 1 L/h. The normally low colonic sodium concentration (40 meq/L) rises as stool volume increases and approaches that of plasma (140 meq/L).

LOSS OF SODIUM VIA THE KIDNEY

The normal renal response to reduction in effective circulating volume is to diminish urinary sodium excretion and prevent any further volume reduction (Fig 2–1). As summarized in Fig 2–4, several factors can disrupt this normal antinatriuretic response to hypovolemia. The kidney may then be the primary cause of sodium loss from the body. Actions of diuretics in this regard will be discussed more fully in the next chapter.

Primary Increase in GFR

An increase in GFR is often part of the mechanism by which a sodium load is excreted (Fig 1–19). But if GFR acutely rises without a primary stimulus, normal sodium transport mechanisms along the nephron can be overwhelmed. The resulting natriuresis can potentially induce hypovolemia. If hypovolemia develops, it tends to counteract the primary stimulus for glomerular hyperfiltration, and GFR becomes normal again. In addition, compensatory changes in tubular sodium reabsorption are induced with prolonged glomerular hyperfiltration, so the natriuresis in such conditions is usually transient.

Primary glomerular hyperfiltration can occur with imbalance of the pressures and flow which determine GFR. These determinants are coordinately regulated by vasoconstrictors (sympathetic nerves and angiotensin II) and vasodilators (ANF). Thus, disorders associated with primary reduced sympathetic nerve tone (eg, autonomic insufficiency or blockade) or hypoangiotensinemia (hyporeninemia or converting enzyme inhibition) or with a primary acute elevation of ANF (during paroxysmal atrial tachycardia or pharmacologic administration of ANF) can potentially increase GFR inappropriate to volume status, cause natriuresis, and deplete body sodium (Fig 2–4).

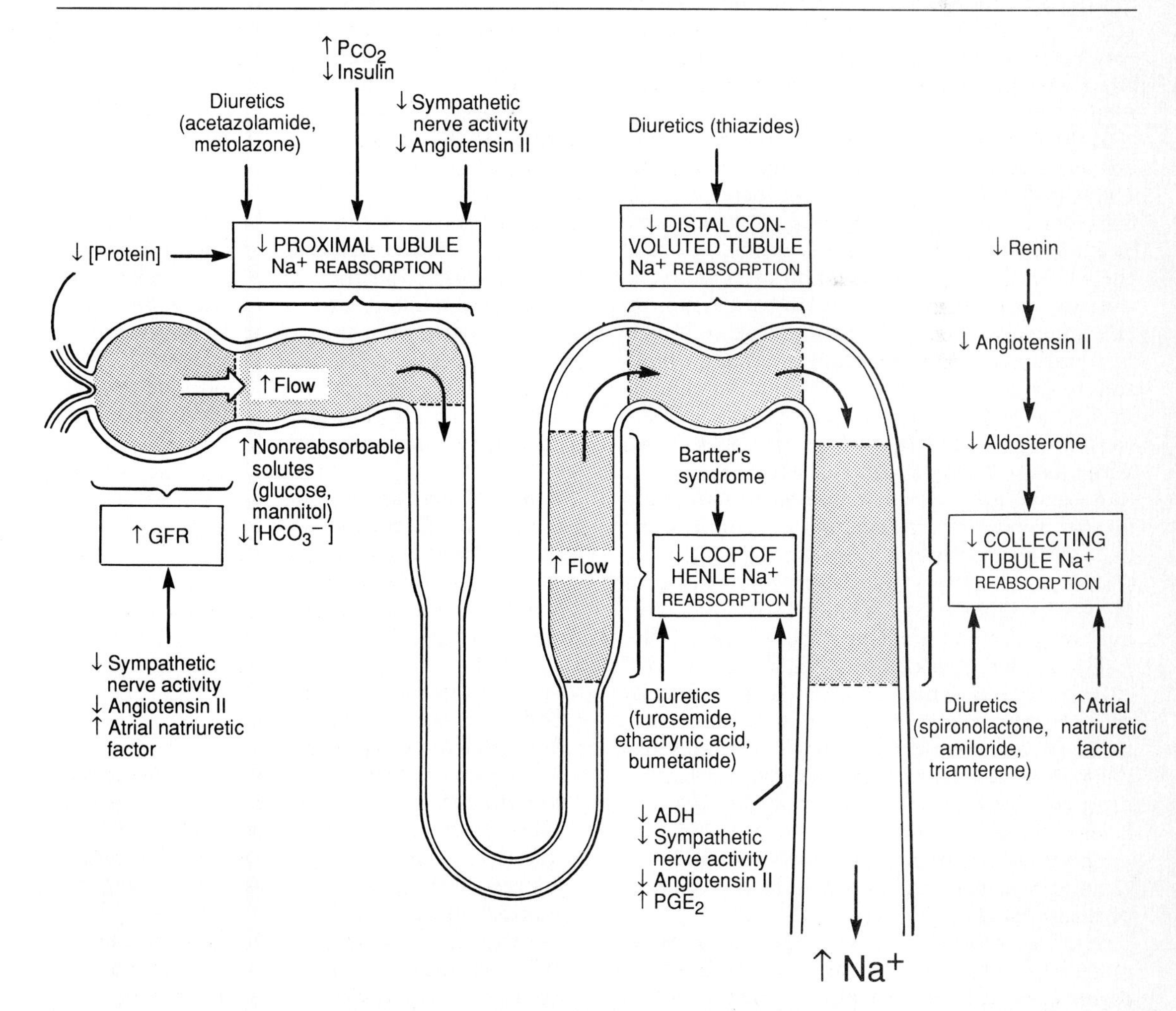

Figure 2–4. Nephron sites and mechanisms responsible for natriuresis.

Protein ingestion during a meal also increases GFR acutely and causes a brief natriuresis.

The effect of primary reduction in sympathetic activity or angiotensin II to augment GFR is not only transient but usually of modest degree. An expected increase in GFR due to the rise in P_{GC} (due to decreased R_A), increased Q_A (due to decreased sum of R_A and R_E), and increased K_f is functionally counteracted by the tendency for P_{GC} to diminish (due to decreased R_E) (Table 1-1). Nevertheless, because both glomerular and tubular elements are coordinately regulated by sympathetic tone and angiotensin II (Fig 1-29) and peritubular protein concentration (defined by the glomerular filtration fraction), even the slight increase in GFR causes natriuresis because it is coupled with a simultaneous fall in sodium chloride transport, especially in the proximal tubule.

Primary Decrease in Proximal Tubular Sodium Reabsorption

Transport of sodium chloride in the proximal tubule can be affected by alteration in luminal flow rate and composition, intrinsic passive or active transport mechanisms, or neurohumoral regulatory determinants (Fig 2-4).

If there is an increase in GFR and luminal flow for any reason, the proximal tubule has little intrinsic ability to respond with an increase in sodium chloride transport (Fig 1-24). The increment in filtered sodium is therefore transmitted out of the proximal tubule into the loop of Henle.

Restraint of the passive, diffusive component of sodium chloride reabsorption can occur if there is filtration of an intrinsically nonreabsorbable solute, as occurs during mannitol's administration, or a solute in a concentration that exceeds the normal proximal transport capacity, as occurs with glucose during hyperglycemia.

A decrease in filtered bicarbonate concentration or tubular bicarbonate reabsorption can reduce the driving force responsible for passive diffusion of sodium chloride. For instance, in metabolic acidosis there is a diminished bicarbonate concentration gradient (dotted lines in Fig 1-23), which causes a blunted rise in luminal chloride concentration and therefore sodium chloride reabsorption. Decreased sodium bicarbonate and hence sodium chloride reabsorption in the proximal tubule also can be induced by carbonic anhydrase inhibition, as with acetazolamide (the thiazide metolazone also has this property), or inhibition of angiotensin II activity (due to hyporeninemia, converting enzyme inhibition, or receptor blockade).

With regard to the active component of proximal reabsorption of sodium chloride, a potent regulator is peritubular capillary protein concentration, in turn defined by the end-glomerular capillary protein concentration and glomerular filtration fraction (Table 1-3). Proximal sodium chloride transport is therefore diminished when filtration fraction is reduced by a rise in renal blood flow (Fig 1-15), as with vasodilating drugs or hormones, or with reduction in renal sympathetic nerve or angiotensin II activity.

Finally, active sodium chloride transport in the proximal tubule is controlled by sympathetic nerves, P_{CO_2}, and insulin. Disruption of these regulators of proximal transport may be responsible at least in part for the natriuresis and chloruresis seen during autonomic insufficiency or sympathetic blockade, acute hypercapnia and hypoinsulinemic states (eg, diabetic or starvation ketoacidosis).

Primary Decrease in Sodium Reabsorption in the Loop of Henle

A marked increase in luminal flow causes more sodium chloride to escape from the thick ascending limb (Fig 1-26). Loop sodium chloride transport can also be inhibited be reduced directly or indirectly (due to changes in medullary blood flow) by diminished ADH, sympathetic nerve, or angiotensin II activity (Table 1-4). Factors that increase PGE_2, which inhibits transport, have the same effect.

A diuretic that inhibits the Na^+-K^+-$2Cl^-$ luminal transporter, such as furosemide, bumetanide, and ethacrynic acid, potently decreases sodium chloride transport in the loop. Finally, there is a rare hereditary disorder characterized by an intrinsic defect in sodium chloride transport in the loop of Henle, called Bartter's syndrome. The clinical characteristics of this syndrome resemble in many respects those associated with chronic administration of a loop diuretic (Chapter 13).

Primary Decrease in Sodium Reabsorption in the Distal Convoluted Tubule

The luminal Na^+-Cl^- cotransporter in this nephron segment is specifically and potently inhibited by thiazide diuretics.

Primary Decrease in Sodium Reabsorption in the Cortical Collecting Tubule

Sodium chloride reabsorption in the cortical collecting tubule is under the sensitive control of aldosterone. Sodium transport can therefore be reduced by decreased activity of one or more components of the renin-angiotensin-aldosterone-cortical collecting tubule axis.

Hypovolemic disorders caused by aldosterone deficiency or inactivity are usually associated with hyperkalemia and metabolic acidosis, and the detailed discussion of etiology is contained in Chapter 9 and Chapter 12 (Table 12-6). Briefly, diabetes, hypertensive nephrosclerosis, and intrinsic tubulointerstitial disease can affect renin biosynthesis. Other factors that inhibit renin release include beta-blockers and prostacyclin synthesis inhibitors, such as nonsteroidal anti-inflammatory drugs. Primary hypoangiotensinemia can be induced pharmacologically by converting enzyme inhibition (eg, enalapril, captopril, or lisinopril) or by specific angiotensin II receptor blockers. Primary hypoaldosteronism can arise in the course of adrenal disease. Finally, there can be unresponsiveness of the target end-organ, the cortical collecting tubule. Such end-organ unresponsiveness can occur during blockade of the mineralocorticoid receptor within the cell by the diuretic spironolactone or when tubulointerstitial disease has physically destroyed the responding cells. Finally, the diuretics amiloride and triamterene and the hormone ANF serve to shut down the luminal aldosterone-regulated channels by which sodium enters the cell.

Renal Salt Wasting

As discussed above, inhibition of sodium reabsorption at any nephron site can potentially cause natriuresis and hypovolemia. Transport inhibition at a single site may be functionally transient; however, if hypovolemia occurs, it can secondarily stimulate sodium transport in the other nephron sites to compensate for the primary change. However, this sequence may not hold when the primary change in sodium transport has occurred terminal segment of the nephron, the cortical collecting tubule.

As previously shown in Fig 1–34 and reviewed in Fig 2–5A (top panel, solid line), in response to reduction in dietary sodium restriction to 10 meq/d (vertical dotted line), urinary sodium excretion is normally lowered in 3–5 days to the same amount, to reachieve sodium balance, associated with cumulative weight loss of only 1 kg (Fig 2–5A, bottom panel, solid line). But if the cortical collecting tubule is dysfunctional—intrinsically or as a result of diminished renin, angiotensin II, or aldosterone activities—sodium cannot be excluded from the urine. A much higher minimal threshold (typically > 50 meq/L) occurs, as shown by the dotted line in Fig 2–5A (top panel) with negative sodium balance (shaded area). Relentless renal sodium excretion in excess of sodium intake causes progressive loss of extracellular volume and weight (Fig 2–5A, lower panel, dotted line) despite increasing hypovolemia. This condition is a specific syndrome known as renal salt wasting.

It is important to differentiate clinical renal salt wasting due to inhibition of sodium transport in the cortical collecting tubule from the consequences of sodium transport inhibition in upstream segments such as the loop of Henle, as shown in Fig 2–5B. With initiation of furosemide administration, which potently reduces loop of Henle sodium transport, there is a transient increase in sodium excretion (top panel, dotted line), negative sodium balance, and weight loss (bottom panel, dotted line). This volume contraction activates the renin-angiotensin-aldosterone system, so that enhanced sodium transport in the proximal and cortical collecting tubules compensates for the increased sodium load emerging from the loop. A new steady state then obtains in which sodium excretion is equivalent to intake despite continued diuretic administration, though at the expense of diminished extracellular volume, sustained weight loss, and mild hypovolemia. At this point, if dietary sodium is reduced to 10 meq/d (vertical dotted line), the further stimulus to sodium retention in the proximal and cortical collecting tubules can reduce urinary sodium excretion to reestablish sodium balance (top panel). There is a normal further reduction in weight of about 1 kg (bottom panel), even with the persistent inhibition of loop sodium reabsorption by furosemide. Progressive renal salt wasting with further loss of weight and hypovolemia does not occur.

In summary, renal salt wasting is a specific syndrome diagnosed by inability to conserve sodium and maximally lower the urinary sodium concentration even when provoked by dietary sodium deprivation or other cause of hypovolemia, leading to progressive volume contraction and weight loss. The syndrome of renal salt wasting is unique to dysfunction of the cortical collecting tubule—it is usually also associated with significant dysfunction of other transport processes within this nephron segment and hence hyperkalemia (see Chapter 9) and hyperchloremic metabolic acidosis (type IV generalized distal renal tubular acidosis; see Chapter 12). In contrast, inhibition of sodium reabsorption in the proximal, loop, and distal convoluted tubule segments can be self-limiting—overcome by significant volume contraction. In such cases, elevation in urinary sodium concentration is transient and negative sodium balance and the degree of hypovolemia are then stable and nonprogressive.

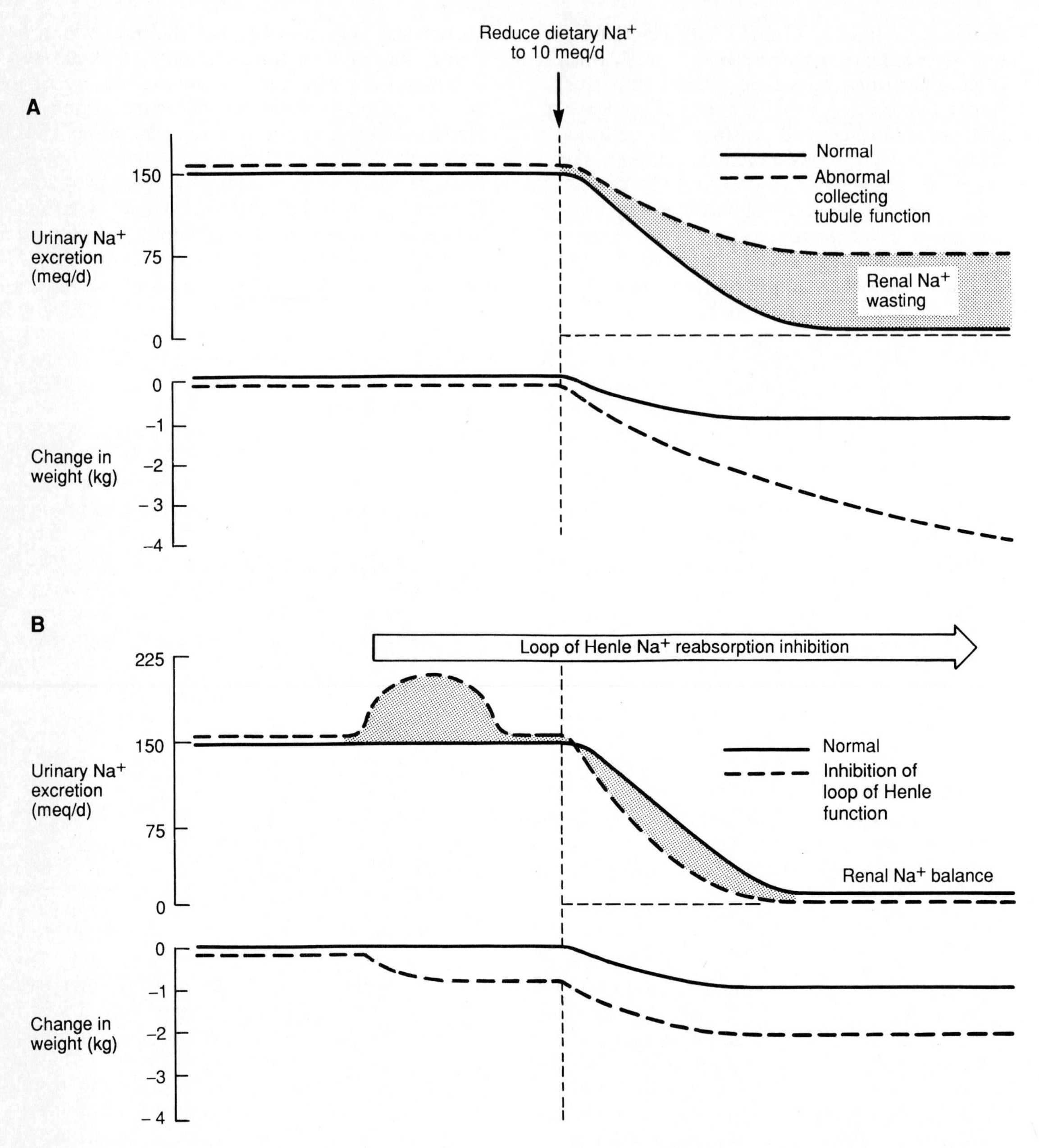

Figure 2-5. Renal Na^+ excretory response to reduction in dietary Na^+ in setting of abnormal function of collecting tubule ***(A)*** or loop of Henle ***(B)***.

DECREASED EFFECTIVE CIRCULATING VOLUME WITH NORMAL TOTAL BODY SODIUM

ABNORMAL CAPILLARY STARLING FORCES OR PERMEABILITY

Hypovolemia may occur as a result of capillary leakage. Such translocation of fluid from the plasma to the interstitial compartment can arise if there is imbalance in the Starling pressures that govern the normal distribution of volumes or if there is a change in the capillary permselectivity to albumin as summarized in Table 2–3 and in Fig 2–6.

First, increased hydrostatic pressure gradient can arise from decreased precapillary resistance (eg, from sepsis or vasodilators) or increased postcapillary resistance (eg, heart failure). A decrease in interstitial pressure could theoretically have the same effect, but this does not occur under clinical circumstances.

Second, decreased oncotic pressure gradient can occur if plasma albumin concentration decreases (ie, hypoalbuminemic states) or if albumin traffic from plasma to interstitium increases. Injury or functional change mediated by cytokines in the capillary endothelial cell can alter the permselective barrier to albumin (eg, during sepsis, trauma, surgery, or burns). Sodium and water follow albumin into the interstitium or to another compartment in functional contiguity—often called a "third space". Typical third spaces include a body cavity (eg, peritoneum in peritonitis or bowel in intestinal inflammation), a newly created compartment (a burned or traumatized area or the retroperitoneum during pancreatitis), or even cells (due to anoxic, traumatic, or thermal injury). Increased albumin accumulation in the interstitium can also occur when albumin leakage from the capillary is normal but there is impairment of lymphatic drainage (eg, lymphatic diseases or obstruction).

Third, increased hydraulic permeability can give rise to enhanced solute and water flow for any given capillary pressure gradient. Such a permeability change can also occur during sepsis, trauma, surgery, and burns.

Table 2-3. ↑Plasma-to-interstitial fluid movement.

Pressure	Examples
↑Hydrostatic pressure gradient	
↓Precapillary resistance	Sepsis, vasodilators
↑Postcapillary resistance	Heart failure
↓Interstitial pressure	—
↓Oncotic pressure gradient	Hypoalbuminemic states
↓Plasma oncotic pressure	
↑Interstitial oncotic pressure	
↑Protein permeability	Sepsis, trauma, surgery, burns
↓Lymph flow	Lymphatic disease or obstruction,
↑Hydraulic permeability	Sepsis, trauma, surgery, burns

SYMPTOMS

CARDIOVASCULAR

The symptoms of hypovolemia relate to diminution of regional tissue perfusion, as summarized in Fig 2–7. Symptoms are sometimes rather vague. Circulatory signs include diminished central venous pressure and progression from orthostatic tachycardia to resting tachycardia and from orthostatic hypotension to resting hypotension and shock.

Peripheral vasoconstriction usually accompanies hypovolemia, causing cool, clammy skin. However, when hypovolemia occurs with sepsis or in the setting of acidemia, peripheral vasodilation with warm skin can be the predominant manifestation. There is typically decreased salivary and other mucous membrane secretions, decreased sweating, and poor skin turgor.

Increased angiotensin II and ADH levels induced by hypovolemia stimulate thirst, and the patient may be aware of oliguria and an increase in urinary concentration with relatively dark urine. Ironically, despite the depletion of total body sodium chloride, salt craving does not occur in humans as in some other species.

NEUROMUSCULAR

With mild volume depletion, there may be symptoms of lassitude, anorexia, and nausea. As the hypovolemia becomes more severe, dizziness or syncope may accompany the orthostatic change in blood pressure and the patient may sense palpitations. With severe hypovolemia and hypoperfusion of the brain, stupor or coma may supervene.

Vasoconstriction and decrease in muscle perfusion (and potassium deficiency when present) can also lead to muscular weakness and often cramps.

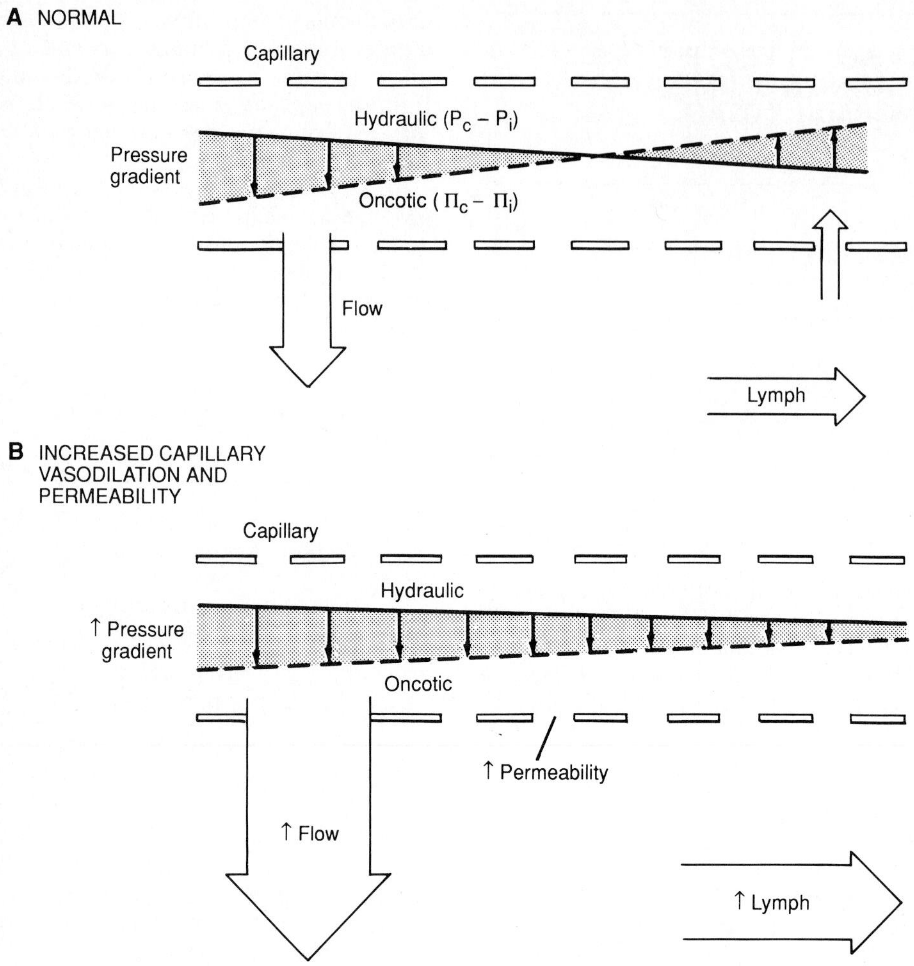

Figure 2-6. Change in transcapillary fluid movement in response to vasodilation and increased vascular permeability.

DIAGNOSIS

APPROACH TO ETIOLOGIC DIAGNOSIS

History & Physical Examination

The history and physical examination are the mainstays of evaluating the cause of hypovolemia. Since hypovolemia often occurs because sodium output exceeds input, both should be assessed very carefully. A careful analysis of sodium input should include the diet as well as supplemental sources of sodium in drugs or intravenous fluids. Equally important though not always feasible is an accurate assessment and quantitation of external sodium loss in urine and gastrointestinal fluids. Unfortunately, the amount of circulating volume lost externally during hemorrhage or lost into the interstitium during trauma, surgery, burns or sepsis is usually difficult to quantitate directly.

Previous Illness & Drugs

History and physical signs of sepsis (previous infection, fever), gastrointestinal disease (diarrhea, vomiting, abdominal tenderness, hepatic enlargement or tenderness, ascites), heart disease (chest pain, shortness of breath with exertion or posture, murmurs, and gallops), and kidney disease should be sought. A careful drug history is essential, especially for cathartics, diuretics, inhibitors of the renin-angiotensin-aldosterone

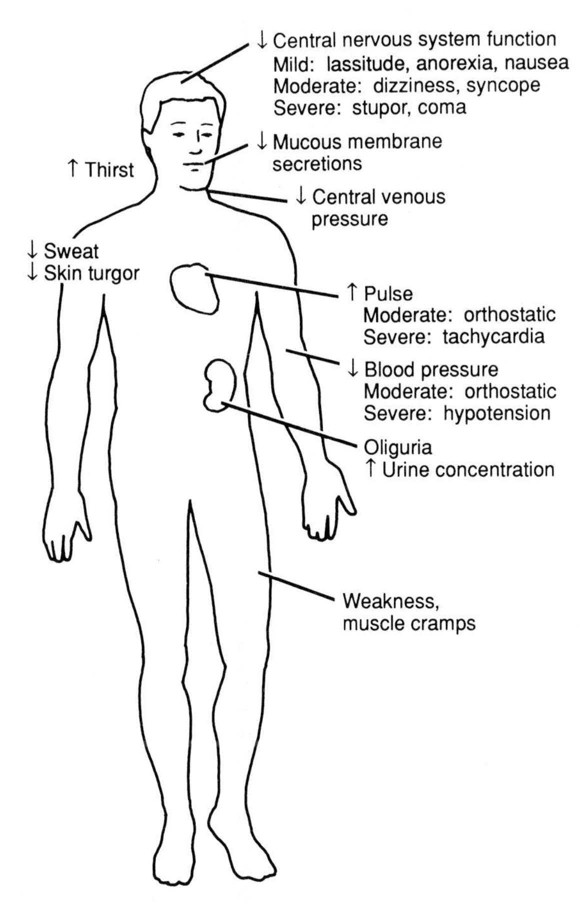

Figure 2-7. Symptoms and signs of hypovolemia.

axis, nonsteroidal anti-inflammatory agents, and cardiovascular drugs.

CARDIOVASCULAR RESPONSE

Degree of Hypovolemia

The assessment of volume status is a clinical one—there is no laboratory test for this purpose that is invariably reliable. Mild hypovolemia can be appreciated if there is a pulse rate increment greater than normal (> 10 beats/min) when assuming the upright posture. Such orthostatic tachycardia indicates that circulating volume is insufficient to maintain normal cardiac output. The central venous pressure is low (ie, < 3 cm H_2O), reflected by flat neck veins in the supine position.

Further decrease in effective circulating volume associated with moderate hypovolemia causes resting tachycardia, orthostatic hypotension, and peripheral vasoconstriction. With more severe hypovolemia, cardiac performance and vasoconstriction no longer are sufficient to maintain blood pressure, and resting hypotension, hypoxemia, and shock supervene.

The physical examination is usually sufficient for the diagnosis of central hypovolemia. Occasionally, however, coexistent neurologic, cardiac, or pulmonary disease makes the usual hemodynamic functional assessment difficult. To ensure that left ventricular filling pressure is optimal in a patient with preexisting cardiopulmonary disease, direct measurement of pulmonary capillary wedge pressure—as an index of left atrial diastolic pressure—is not only useful but sometimes mandatory.

Rule Out Primary Myocardial & Neural Dysfunction

As shown in Fig 2–8, signs of hemodynamic instability or tissue perfusion usually indicate insufficient priming of the pump (ie, diminished cardiac preload) due to hypovolemia but may point to a problem with the pump itself. Primary causes of diminished cardiac performance must be eliminated before one can assume that hypovolemia has caused inadequate cardiac output. A problem of the central or peripheral sympathetic nervous system can also masquerade as hypovolemia. Autonomic dysfunction prevents appropriate vasoconstriction in response to standing despite normal cardiac preload and intrinsic cardiac function and thereby induces orthostatic hypotension.

RENAL RESPONSE

Decrease in GFR

With hypovolemia, the same glomerular hemodynamic changes shown in Fig 1–30 due to mild volume depletion evoked by dietary sodium restriction occur, but they are more exaggerated. GFR falls as a result of increase in R_A and R_E and decrease in K_f, all due to the coordinated and synergistic actions of the sympathetic nerves and angiotensin II.

As GFR falls, creatinine and urea clearances also decline, so that the BUN and serum creatinine concentrations rise (assuming relatively constant rates of urea and creatinine production). These parameters are relatively insensitive markers and may be within or only slightly above normal with mild-to-moderate reduction in GFR, but they are invariably elevated (> 20 mg/dL and > 1.4 mg/dL, respectively) when GFR is markedly reduced, as summarized in Table 2–4.

The BUN:plasma creatinine concentration ratio is also helpful when assessing hypovolemic states. This ratio, normally about 10–20:1, increases during hypovolemia, sometimes up to 40:1. The ratio increases both because the numerator increases and the denominator decreases: as GFR declines

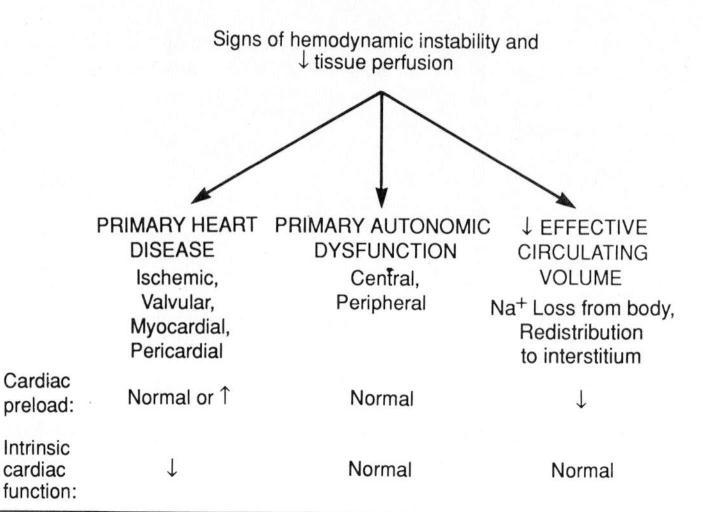

Figure 2-8. Diagnosis of hypovolemia.

and oliguria develops, there is a disproportionate increase in urea reabsorption and creatinine secretion. Elevation in this ratio is also is useful in determining that an oliguric state is due to hypovolemia rather than to intrinsic tubular disease (eg, acute tubular necrosis), in which the ratio is unaltered because urea reabsorptive and creatinine secretory changes as a function of GFR do not occur.

Antinatriuresis

During hypovolemia, urinary sodium concentration should be almost unmeasurable (< 1 meq/L) with fractional sodium excretion (FE_{Na^+}) much less than 0.1%, sometimes close to zero. There may be factors that mitigate this maximal response, however.

The time course of the stimulus is a major factor defining the renal antinatriuretic response. As shown in Fig 1–34, when change in effective circulating volume is slow and mild following reduction in dietary sodium, 3–5 days are required for the urinary sodium concentration to reach an appropriate level. When the stimulus is of greater severity, as in hemorrhage or copious vomiting or diarrhea, the renal response is much more prompt—within minutes to hours.

The magnitude of the renal response may be reduced for several other reasons. First, if there is dysfunction of the renin-angiotensin-aldosterone-cortical collecting tubule axis, there may be impairment in reduction of urinary sodium concentration to very low levels—a mild or overt form of renal salt wasting (Fig 2–5A). Second, if there is osmotic diuresis (eg, in uncontrolled diabetes), pharmacologic diuresis (eg, with acute diuretic administration), or an acid-base disorder (eg, dieting- or starvation-induced ketoacidosis), sodium may be obligated into the urine despite hypovolemia (Fig 2–4). Finally, a secondary, intercurrent intervention may obscure the primary response to hypovolemia. For instance, the simple maneuver

Table 2-4. Laboratory parameters associated with hypovolemia.

	Values During Hypovolemia[1]	Typical Value on Usual Diet[2]
BUN	> 20 mg/dL	10 mg/dL
Serum creatinine	> 1.4 mg/dL	1 mg/dL
BUN/creatinine	> 20	10
Urine [Na^+]	< 20 meq/L	100 meq/L
Urine Na^+ excretion	< 8 meq/d	150 meq/d
FE_{Na^+}[3]	< 0.2%	0.7%
Urine volume	< 400 mL/d	1500 mL/d
Urine osmolality	> 1000 mosm/kg	600 mosm/kg
Urine specific gravity	> 1.040	1.020

[1]Unless kidney disorder is responsible for the hypovolemic state.
[2]Note that these are not "normal values."
[3]$FE_{Na^+} = (U/P)_{Na^+}/(U/P)_{cr}$

of putting an edematous patient at bed rest may convert an antinatriuretic state to a natriuretic one by allowing mobilization of dependent interstitial fluid into the vascular compartment.

Given the constraints mentioned above, commonly used and clinically generous guidelines for the physiologic renal response associated with diminished effective circulating volume are summarized in Table 2–4. Urine sodium concentration should be less than 20 meq/L, representing an absolute sodium excretion rate of less than 8 meq/d (assuming a low urine output of 400 mL/d). In general, dietary sodium intake is not that low.

Fractional excretion of sodium (FE_{Na^+}) is also predictably reduced during hypovolemia. With a normal GFR of 100 mL/min and sodium filtration of 20,000 meq/d, a sodium output of 8 mg/d would represent a FE_{Na^+} of 0.04% (8 meq/d divided by 20,000 meq/d). However, GFR is usually decreased under these conditions. If GFR were reduced fivefold, to 20 mL/min, then the comparable FE_{Na^+} should be 0.2%.

Antidiuresis

As will be discussed in greater detail in Chapter 4, 10% or more reduction in effective circulating volume causes baroreceptor-mediated activation of ADH and thirst. The renal effects of ADH include increased efficiency of the countercurrent multiplication system and increased hydro-osmotic permeability of the medullary collecting tubule. Optimally during hypovolemia, as shown in Table 2–4, urine volume is reduced to a minimal level, less than 400 mL/d, along with a maximal increase in urinary osmolality, to greater than 1000 mosm/kg, reflected by a urinary specific gravity of greater than 1.040.

However, as discussed more fully in Chapter 6, several factors can abrogate this normal oliguric response: abnormality in the hypothalamic-posterior pituitary-ADH release mechanism; abnormality in the renal countercurrent system (eg, by a loop or osmotic diuretic); or abnormality in the hydro-osmotic response to ADH (eg, potassium deficiency or hypercalcemia).

Variability of the Serum Sodium Concentration

It is very important to reemphasize that disorders of sodium metabolism and of water metabolism are not necessarily coexistent. In particular, as explained more fully in Chapter 4, the serum sodium concentration is ***not*** a guide to extracellular or total body sodium stores. The serum sodium concentration is merely a reflection of total body solute divided by total body water. As a ratio, therefore, it reflects proportionate—not absolute—changes in its two components.

While hypovolemia is often associated with stimulation of ADH and hyponatremia, hyponatremia may also occur when total body sodium is normal (eg, syndrome of inappropriate ADH) or high (eg, edema states). Thus, the diagnosis of circulating volume depletion remains a clinical one, based on history and physical examination and not the serum sodium concentration alone.

Coincident Acid-Base Disorders

Disturbances of hydrogen ion metabolism frequently accompany states of sodium depletion, as summarized in Table 2–5. Serum bicarbonate concentration, arterial blood gases, and serum electrolytes (anion gap) are very helpful in the diagnosis and necessary for the treatment of acid-base disorders (Chapters 11–13).

Disorders associated with translocation of fluid from plasma to interstitium do not directly cause acid-base derangement. Secondary stimulation of respiratory centers (during fever, sepsis, pain, or burns) can lead to respiratory alkalosis, and sequestration of fluid in the interstitium may eventually induce metabolic alkalosis. The acid-base disorder accompanying gastrointestinal fluid loss can usually be predicted by whether the fluid lost (Fig 2–3 and Table 2–2) is bicarbonate-rich (leading to metabolic acidosis) or chloride-rich (leading to metabolic alkalosis).

Renal sodium loss is usually associated with acid-base derangement. Since most bicarbonate reabsorption occurs in the proximal tubule and net acid formation occurs in the cortical collecting duct, disruption of sodium reabsorption and of hydrogen ion secretion in these two sites predictably causes hypovolemia with metabolic acidosis. On the other hand, when sodium chloride transport is disrupted at other sites, in the thick ascending limb of Henle or distal convoluted tubule, there is enhanced sodium delivery to the cortical collecting tubule, which accelerates hydrogen ion secretion so that metabolic alkalosis accompanies hypovolemia.

Coincident Potassium Disorders

Disorders of potassium homeostasis also occur frequently with sodium loss, summarized in Table 2–5. Since aldosterone is an important stimulant of potassium excretion and because hyperaldosteronism is part of the reflex defense of extracellular volume depletion, urinary potassium wasting frequently accompanies hypokalemia (Chapter 8).

Although negative potassium balance and hypokalemia are the rule with hypovolemia, there are several exceptions (see Chapter 7). First, sufficient time may not have elapsed to permit potassium wastage. Second, severe hyperosmolality and acidemia (eg, in diabetic ketoacidosis) can mask the drop in plasma potassium that would otherwise result from potassium depletion. Third,

Table 2-5. Hypovolemic states.

	Concurrent Acid-Base Disorder	Concurrent $[K^+]p$
Hemorrhage (acute)	—	—
Translocation: Plasma → interstitium		
Acute	±Respiratory alkalosis	—
Chronic	±Metabolic alkalosis	↓
Gastrointestinal Na^+ loss		
Gastric	Metabolic alkalosis	↓
Duodenal	Metabolic alkalosis	↓
Biliary	—	—
Pancreatic	Metabolic acidosis	↓
Ileal	Metabolic acidosis	↓
Diarrheal	Metabolic acidosis	↓
Renal Na^+ loss		
↑GFR	—	↓
↓Proximal Na^+ reabsorption		
↓$NaHCO_3$ plus ↓NaCl reabsorption	Metabolic acidosis	↓
↓NaCl reabsorption	—	↓
↓Loop of Henle Na^+ reabsorption	Metabolic alkalosis	↓
↓DCT Na^+ reabsorption	Metabolic alkalosis	↓
↓CCT Na^+ reabsorption	Metabolic acidosis	↑

DCT = distal convoluted tubule; CCT = cortical collecting tubule.

potassium loss may be limited by low sodium delivery to the cortical collecting tubule when GFR is excessively compromised in the hypovolemic state.

Finally, potassium wasting does not occur—in fact, potassium retention and hyperkalemia may result—when hypovolemia results from renal sodium chloride wasting due to impaired function of the cortical collecting tubule, the primary site for potassium secretion. The combination of hypovolemia and hyperkalemia is therefore a clue that adrenal disease (eg, Addison's disease) may be present that requires emergent glucocorticoid therapy (hydrocortisone sodium succinate, 100 mg intravenously every 8 hours, plus normal saline, 100–200 mL/h intravenously) for correction.

Other Laboratory Parameters

In many forms of hypovolemia, loss of isotonic sodium chloride from the vascular space concentrates the residual protein and red cells, resulting in hyperproteinemia and polycythemia. A reduction in GFR and effective circulating volume also decreases renal uric acid clearance, so hyperuricemia ensues. Hypovolemia brings into play many acute stress hormones (increased cortisol, epinephrine) and hematologic responses (leukocytosis, thrombocytosis, and acute phase serum proteins).

THERAPY

DECISION TO TREAT

The critical dependence of normal cardiac function and tissue perfusion on the effective circulating volume makes treatment of symptomatic hypovolemia necessary in virtually all cases. Of course, many hypovolemic states are transient and will correct spontaneously if sodium loss abates and the individual has access to sodium chloride. Even if hypovolemia is reasonably well tolerated at a certain point in time, volume repletion should be considered to guard against further loss of the depleted reserve with risk of irreversible organ damage. Treatment is modified depending on the cause of the hypovolemia and on assessment of ongoing loss. The specific treatment of aldosterone deficiency is discussed in Chapters 9 and 12.

ROUTE, RATE, & TYPE OF REPLACEMENT

Assessment of Amount of Volume Depletion

The degree of depletion of the effective circulating volume is a clinical judgment. In some respects, the renal response is probably the most sensitive index of mild hypovolemia, and the laboratory parameters listed in Table 2–4 can be of great help in assessment. Remembering the response to dietary sodium restriction (Figs 1-33 and 1-34), negative sodium balance of only 150 meq—equivalent to losing only 1 L or 6% of total extracellular fluid—is sufficient to stimulate the renal sodium reabsorptive systems maximally to completely remove sodium from the urine (Fig 1-33). When about 10% of extracellular fluid is lost, ADH is stimulated by volume baroreceptors, and oliguria and a concentrated urine result. Rather vague neuromuscular symptoms such as anorexia, lassitude, apathy, and weakness also herald the onset of mild hypovolemia (Fig 2-7).

The presence of moderate hypovolemia, signified by symptoms and signs of orthostatic hypotension, tachycardia, mild peripheral vasoconstriction, and diminished central venous pressure in addition to intense antinatriuresis and antidiuresis, is associated with loss of about 10-20% of effective circulating plasma volume. Hypovolemia of this magnitude may be due to loss of 0.5-1 L of blood (with hemorrhage) or, equivalently, by extracellular volume loss of 2-4 L (due to skin, gastrointestinal, or renal sodium loss).

More severe hypovolemia is signaled by signs of shock with overt circulatory insufficiency and hypotension even in the supine position, severe peripheral vasoconstriction, and inadequate tissue perfusion evidenced by hypoxemia and significant neurologic impairment, including stupor and coma. Rapid infusion of large amounts of blood or isotonic fluid is required to resuscitate the circulation.

If known, acute weight loss can be the most precise and quantitative index of the deficit in extracellular volume. A change in weight in a short period of time is usually equivalent to the change in body water. With isotonic losses, as from gastrointestinal or renal routes, weight change in kilograms is equivalent to extracellular volume change in liters.

The use of cardiac monitoring can be very helpful to assess the adequacy of preload. Pulmonary capillary wedge pressure less than 10 cm H_2O and central venous pressure less than 3 cm H_2O usually indicate inadequate filling of the left and right heart, respectively.

Therapy in the Acute Situation

For hypovolemic shock, a large-bore catheter (> 16-gauge) should be used to gain access to the circulation. The fluid administered to resuscitate the circulation should be cross-matched whole blood if obtainable when hemorrhage has occurred; otherwise, isotonic crystalloid solutions should be given as listed in Table 2-6. The rate of administration depends on the severity of clinical symptoms, varying from 1 to 5 L/h.

Use of colloid-containing solutions (eg, plasma or albumin solutions) has both potential benefit and a risk of harm. They have the theoretical advantage of causing water translocation from the interstitial fluid to the blood volume. On the other hand, if there is abnormal capillary permeability and capillary leakage, the administered protein leaves the vascular space, enters the interstitial compartment—bringing additional water along—and exacerbates the hypovolemia. Thus, unless capillary integrity is assured, colloid-containing solutions should generally be avoided.

Whether to use additives such as bicarbonate (Chapter 12) and potassium (Chapter 8) depends on the electrolytic composition of the fluid lost. For gastrointestinal fluid loss, Table 2-7 provides guidelines for replacement therapy (based on the data of Table 2-2).

The volume and rate of administration of therapeutic fluids depends on the clinical response. There are no absolute guidelines except that the goal of therapy is to normalize hemodynamic indices and tissue perfusion. Thus, fluid therapy should be curtailed when there is normalization of blood pressure (eg, > 90 mm Hg), heart rate, and

Table 2-6. Electrolyte concentrations of commonly used intravenous solutions.

	Osmolality (mosm/kg H_2O)	$[Na^+]$ (meq/L)	$[Cl^-]$ (meq/L)	Glucose (g/L)
Isotonic				
Normal saline	308	154	154	0
D_5 normal saline	560	154	154	50
Ringer's lactate[1]	273	130	109	0
Hypotonic				
D_5 water (D_5W)	252	0	0	50
D_5 0.25-normal saline	320	34	34	50
0.5-normal saline	154	77	77	0
D_5 0.5-normal saline	405	77	77	50

[1]Also contains 28 meq/L HCO_3, 4 meq/L K^1, 2 meq/L Ca^{2+}.

Table 2-7. Replacement guidelines for gastrointestinal fluid loss.

	Volume Replacement (per L lost)	Alternative Volume Replacement (per L lost)		KCl[1] (meq/L)	$NaHCO_3$ (meq/L)
		Normal Saline	D_5W		
Salivary	0.25-normal saline	250	750	20	45 (1 amp)
Gastric	0.25-normal saline	250	750	20	—
Small bowel	0.75-normal saline	750	250	5	22 (0.5 amp)
Pancreatic	0.5-normal saline	500	—	5	90 (2 amps)
Biliary	0.75-normal saline	750	250	5	45 (1 amp)
Diarrheal	0.5-normal saline	500	500	40	45 (1 amp)

[1]Increase supplement if K^+ depleted; add urinary K^+ loss.

atrial filling pressure (eg, central venous pressure > 5 cm H_2O or pulmonary capillary wedge pressure > 10 mm Hg), with improvement of level of consciousness, skin color and temperature, and urine sodium and water excretion rates. Frequent check of vital signs, the physical and neurologic status, and urine output should be performed to determine when therapy should be slowed to avoid volume overload. Cardiac pressure monitoring is also especially helpful in patients with limited cardiopulmonary reserve, in whom excessive volume repletion incurs a risk of pulmonary edema. As a rule of thumb, roughly half the estimated sodium and volume deficit should be repleted in the first 24 hours.

Therapy of the Nonacute Situation

In most nonemergent cases of mild to moderate hypovolemia, the repletion solution should contain sodium chloride, the major constituent of the extracellular space. The basic fluid for expansion of the extracellular fluid is therefore isotonic sodium chloride, also known as normal saline. As was the case in emergent treatment, depending on concurrent acid-base and potassium status, bicarbonate and potassium also may be needed in the volume resuscitation, but these should be considered adjuncts and will be considered separately in Chapters 8 and 12.

Whereas the volume of distribution of isotonic saline is the extracellular volume, the volume of distribution of solutions that have water without sodium chloride (such as 5% dextrose in water, or D_5W), is total body water. Considering that the extracellular volume is less than half of total body water, a given volume of isotonic saline will expand the extracellular volume more effectively than a comparable volume of D_5W. By similar reasoning, colloid-containing solutions are used in emergent situations because a given volume of plasma or blood expands the plasma volume 3–5 times more effectively than an equal volume of saline given the relative sizes of the plasma and extracellular volumes (ie, 4.5% versus 20% of body weight). Hypotonic intravenous solutions are sometimes required in hypernatremic states—after the extracellular volume is repleted by isotonic solutions—to dilute the body tonicity back to normal (Chapter 6). Again, as with the acute situation, it is essential to closely monitor the cardiovascular and renal responses to guide the rate and amount of sodium repletion.

Less vigorous intravenous or even enteral or oral repletion can be initiated when symptoms and signs of extracellular volume depletion mandate even less urgent treatment. The sodium content of common enteral solutions is shown in Table 2–8.

Prevention of Extracellular Volume Depletion

While the limit for sodium conservation from the body is ideally about 10 meq/d (5 meq/d from sweat, 5 meq/d from stool, and 0–1 meq/d from urine), a more generous estimate of obligate sodium loss is usually 50–100 meq/d. This takes into account that renal sodium conservation may not be perfect—for a variety or reasons—and provides a small "buffer" against extracellular volume depletion. This is the basis for the recommended minimal daily allowance of sodium often listed as 50 meq/d, or about 1150 mg. Fifty to 100 meq/d of sodium is therefore administered to an individual without overt extracellular volume depletion but who is not eating (eg, after uncomplicated surgery), or added to the estimated sodium repletion in disorders associated with sodium redistribution or external sodium loss.

If the patient cannot take nourishment orally or must receive intravenous fluids, given a reasonable need for sodium of 50–100 meq/d, for water of about 1–1.5 L/d (Chapter 4), and for potassium of 40–60 meq/d (Chapter 7), the usual basal

Table 2-8. Electrolyte composition of enteral formulas.

Name	Vol/d Needed to Meet 100% RDA (mL)	Osmolality (mosm/kg)	Na^+ (meq/L)	K^+ (meq/L)	Ca^{2+} (mg/L)	P (mg/L)	Mg^{2+} (mg/L)
Whole protein							
Osmolite	1887	300	28	26	528	528	211
Ensure	1887	470	37	40	528	528	211
Ensure HN	1321	470	40	40	757	757	302
Sustacal HC	1180	650	37	38	847	847	340
Magnacal	1000	590	44	32	1000	1000	400
Enrich	1391	480	37	40	719	719	287
Blenderized							
Vitaneed	2000	310	22	32	500	500	200
Defined formulas							
Vivonex	1800	550	20	30	556	556	222
Vital	1500	500	20	34	667	667	267
Other							
Portagen	960	320	21	21	938	703	208
Amin-Aid	...	700	<15	<6	...	...	...

intravenous "maintenance" infusion is D_5 0.5-normal saline with 20–40 meq/L potassium chloride administered at a rate of 50–75 mL/h.

When a patient has a known propensity to develop extracellular volume depletion, anticipating and prophylactically replacing the loss can prevent the symptoms and signs of volume depletion. The composition of the fluid should mimic that of the fluid lost. For instance, with gastrointestinal fluid loss, the sodium content of the repletion solution should be equal to that measured in the loss, if possible, or estimated from Table 2-7. The route of sodium therapy should be oral if possible, or enteral or intravenous if the patient cannot eat.

3 Hypervolemic Disorders

INTRODUCTION

DEFINITION

The hypervolemic disorders are so named because the extracellular volume is expanded. Since the volume of this space is determined principally by sodium (Figs 1–1 and 1–3), an equivalent statement is that total body sodium is increased.

In general, sodium retention and rise in extracellular tonicity obligates proportionate water retention due to osmoregulation provided by ADH. If there is decreased effective circulating volume to act as a nonosmotic stimulus for ADH, excessive water retention can occur, with dilution of body solute and hyponatremia (discussed in more detail in Chapter 5).

CLASSIFICATION & PATHOPHYSIOLOGY

Renal Sodium Retention Due to Decreased Effective Circulating Volume

As shown in Fig 3–1B, there are disorders—congestive heart failure, cirrhosis, and the nephrotic syndrome—in which effective circulating volume is "sensed" by the baroreceptor systems as being inadequate. The consequence is that sodium and water intake are in excess of renal excretion for a prolonged period. Categorization of these disorders as hypervolemic is a misnomer because from a pathophysiologic viewpoint they are really hypovolemic disorders, as shown in Fig 3–2. Cardiovascular-renal afferent and efferent systems are all physiologically responding to a perceived deficiency in effective circulating volume.

Although an oversimplification, Fig 3–3A portrays the final common pathway whereby edema formation occurs in these hypervolemic conditions. There is baroreceptor-mediated activation of the sympathetic nervous system and the renin-angiotensin-aldosterone and vasopressin systems, which in turn tend to decrease GFR and cause antinatriuresis and antidiuresis. The positive sodium balance restores the plasma volume and effective circulating volume toward normal in a compensated state, but at the expense of transudation of fluid into the interstitial space, eventually perceived as edema. The difference between a compensated and decompensated state is to a large extent a matter of time, sodium availability, and the degree to which sodium retention alone can compensate for the underlying disorder.

Increased Total Body Sodium Due to Primary Renal Sodium Retention

There is a distinct group of disorders in which the kidney is the "prime mover" for the increase in total body sodium and water, as shown in Fig 3–1C. The kidney in these cases is not responding to a systemically originated, baroreceptor-mediated signal of diminished effective arterial volume, but rather retains sodium *despite* an initially normal extracellular volume (Fig 3–2).

As shown in Fig 3–3B, primary renal sodium retention may be due to physiologically unprovoked decrease in GFR, such as structural diseases of the glomerulus or imbalance of the forces that regulate glomerular filtration, or increased tubular sodium transport, as occurs in hyperreninemic or hyperaldosteronemic states and in some cases of cirrhosis. The inappropriate sodium retention causes an increase in the plasma volume, circulatory overload, and hypertension. This form of hypertension is known as volume-dependent hypertension since it is caused by excessive intravascular volume rather than vasoconstriction. Increased plasma volume results in transudation of sodium and water into the interstitial space, and edema can occur.

Variability in Degree of Sodium Retention

The afferent limb of sodium homeostasis remains intact. As shown in Figs 3–2 and 3–3B, increased effective circulating volume due to renal sodium retention is sensed appropriately by baroreceptors, resulting in physiologic inhibition of

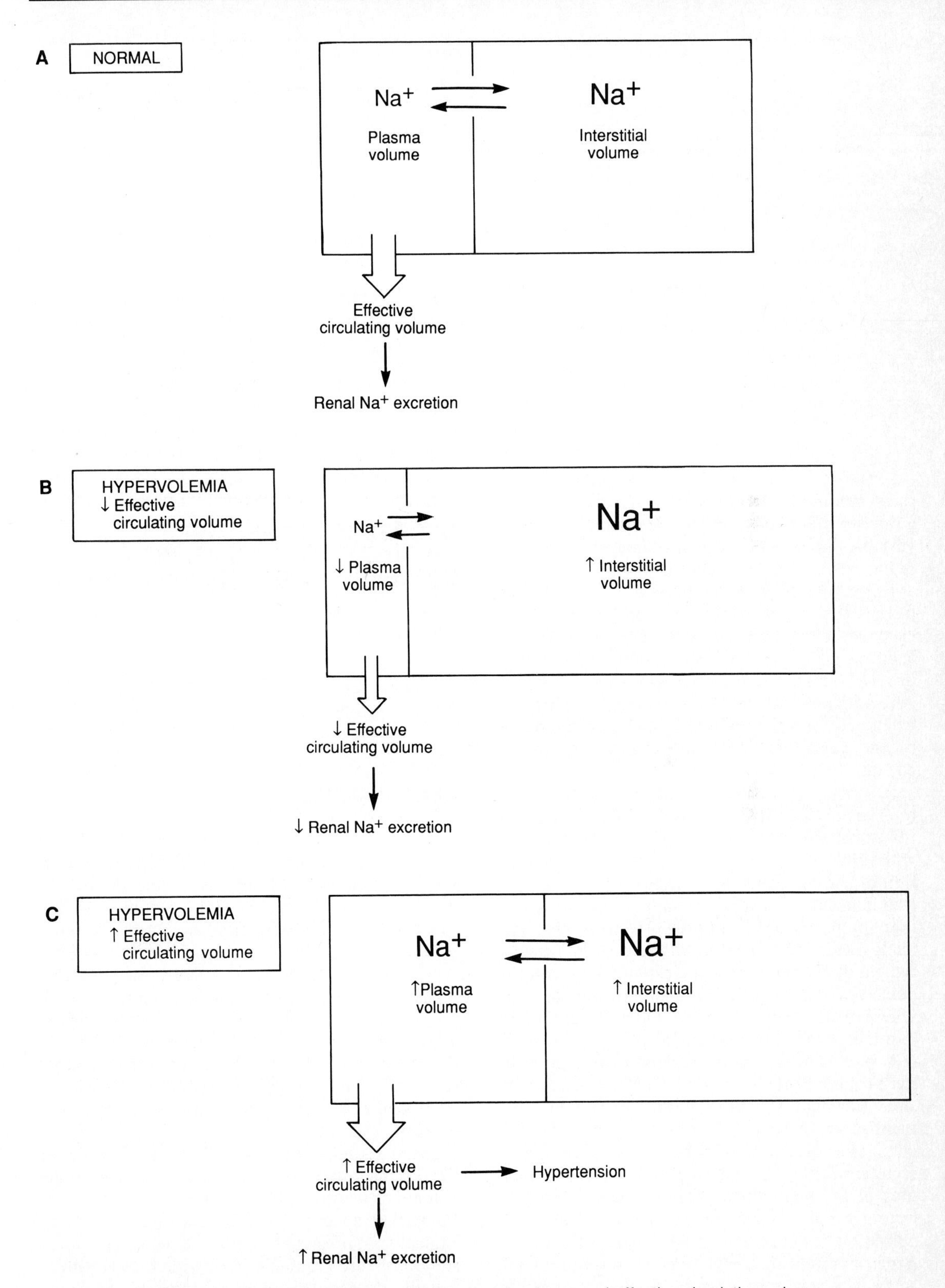

Figure 3–1. Hypervolemic states with decreased or increased effective circulating volume.

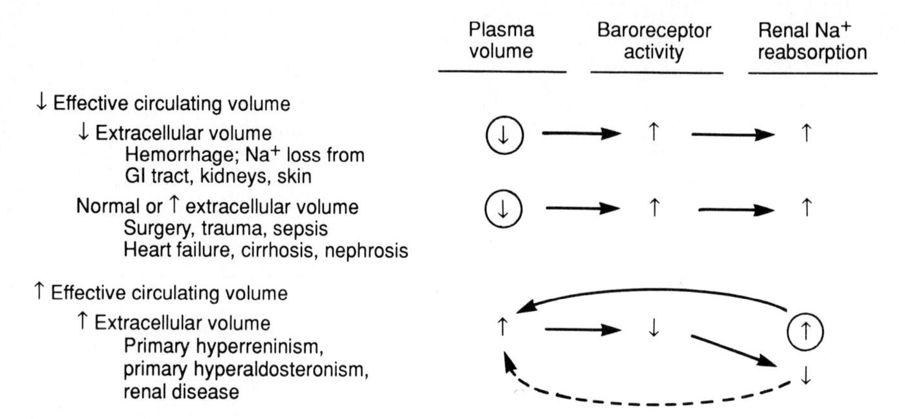

Figure 3-2. Differences in plasma volume and renal sodium reabsorption in hypervolemic states.

the sympathetic nervous system and the renin-angiotensin-aldosterone system. However, there is a variable impact of these neurohumoral changes to reduce renal sodium retention.

Primary renal sodium retention may sometimes be relentless—completely unresponsive to the high effective circulating volume. For instance, in acute renal failure or acute glomerulonephritis, urinary sodium excretion remains low and fixed. Any administered sodium is retained, resulting in progressive positive sodium balance, hypertension to the degree of being malignant, and severe edema.

In other cases, the kidney retains responsiveness to external signals, though with a higher set-point for extracellular volume regulation. For instance, with primary hypermineralocorticoidism, stimulation of sodium transport in the cortical collecting tubule occurs. For a while, sodium output is less than input, which expands the plasma volume and leads to hypertension and an increase in interstitial fluid, though not to the extent that edema is perceived. Consequent to this extracellular volume expansion, however, the kidney responds in a qualitatively appropriate physiologic fashion (Chapter 1) by decreasing sodium reabsorption in the proximal tubule or loop of Henle (the exact site is debated). Urinary sodium excretion then returns to a level equivalent to intake—the adaptive changes in the proximal tubule or loop serving to counterbalance the continued antinatriuretic stimulus of the mineralocorticoid in the cortical collecting tubule. When this compensation has occurred, it is called the **escape** phase of primary hypermineralocorticoidism. Thus, resumption of sodium balance occurs but at the expense of extracellular volume expansion and hypertension.

DECREASED EFFECTIVE CIRCULATING VOLUME WITH INCREASED TOTAL BODY SODIUM

The causes of hypervolemic disorders are summarized in Table 3-1.

HEART FAILURE

Fig 3-3A summarizes the pathophysiology of sodium retention in congestive heart failure. In brief, a decrease in myocardial function from any cause diminishes the effective circulating volume, sensed especially by high-pressure baroreceptors. As myocardial performance fails, progressively higher cardiac filling pressures are required which are transmitted to the capillary circulation, causing volume to be transudated from the plasma to the interstitial space. Stimulated renal sympathetic nerve activity and the renin-angiotensin-aldosterone axis cause vasoconstriction and antinatriuresis.

Renal sodium retention allows reexpansion of the plasma volume toward normal, increasing the volume of transudation into the interstitial space. Interstitial space expansion eventually leads to clinically appreciated edema. Owing to hydrostatic pressure gradients, right-sided heart failure usually creates dependent edema—of the lower legs in ambulatory patients or of the back and sacral areas in supine patients. Sodium and water retention in left-sided heart failure causes elevation

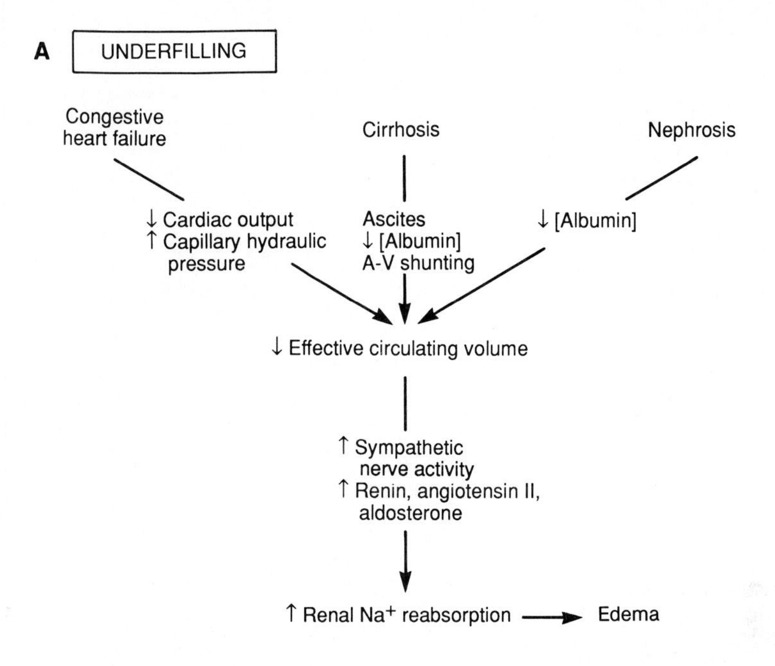

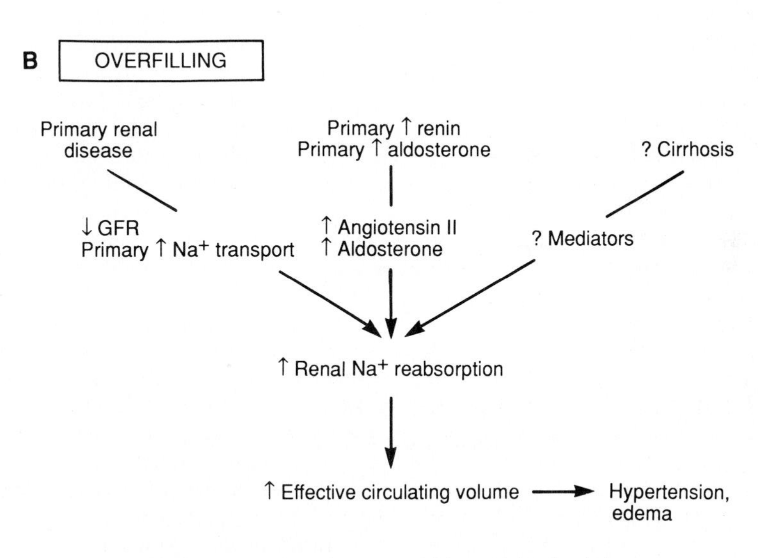

Figure 3–3. Pathophysiology of hypervolemic states.

of the pulmonary vascular pressure, resulting in pulmonary edema.

When the atria are dilated in some forms of congestive heart failure, cardiomyopathies, or valvular heart disease, an increased ANF level is observed. Although clearly the antinatriuretic stimuli are dominant, it is currently unclear what modulatory role ANF plays in this condition. It is reasonable to propose that antinatriuresis would be even more intense if ANF were not elevated.

CIRRHOSIS

The classic explanation for the sodium retention in cirrhosis is that there is a diminished effective circulating volume, similar to congestive heart failure, as shown in Fig 3–3A. Several factors may contribute to the decrease in effective circulating volume. First, abnormal Starling forces in the hepatosplanchnic circulation cause transudation of solute and volume into the abdominal cavity—

Table 3–1. Hypervolemic disorders.

- $\downarrow$Effective circulating volume
  - Heart failure
  - Cirrhosis
  - Nephrotic syndrome
- $\uparrow$Effective circulating volume
  - Rapid NaCl administration
    - Salt water near-drowning
    - Oral ingestion of salt
    - Intravenous saline
  - Primary $\downarrow$GFR
    - $\uparrow$Vasoconstrictor activity
      - $\uparrow$Sympathetic nerve activity
      - $\uparrow$Angiotensin II
        - Potassium deficiency
      - Acute tubular necrosis
    - $\downarrow$Vasodilatory activity
      - $\downarrow PGE_2$
        - Nonsteroidal anti-inflammatory drugs
      - $\downarrow$ANF
    - Impaired autoregulation
      - Renal artery stenosis plus converting enzyme inhibition
    - Primary glomerular disease
      - Renal artery disease
      - Renal vasculitis
      - Glomerulonephritis
      - Tubulointerstitial nephropathies
    - Urinary tract obstruction
  - $\uparrow$Tubular sodium transport
    - Proximal tubule
      - $\uparrow$Sympathetic nerve activity
      - $\uparrow$Angiotensin II
      - $\uparrow$Insulin
    - Loop of Henle
      - $\uparrow$ADH
      - $\uparrow$Sympathetic nerve activity
      - $\uparrow$Angiotensin II
      - $\downarrow PGE_2$
    - Cortical collecting tubule
      - Hypermineralocorticoidism
        - $\uparrow$Renin
          - Renal artery stenosis
          - Renin-secreting tumors
          - Magnesium deficiency
        - $\uparrow$Aldosterone
          - Adrenal adenoma
          - Cushing's disease
          - Adrenogenital syndromes
        - 11β-Hydroxysteroid dehydrogenase inhibition
          - Licorice
          - Chewing tobacco
    - Unknown site
      - Potassium deficiency
      - Cyclic edema
      - Essential hypertension

ie, ascites formation—which depletes the plasma volume. Second, if albumin synthesis is impaired, hypoalbuminemia causes transudation of solute and water into the interstitium. Third, as arteriovenous fistulae develop in the microcirculation, the existing plasma volume is rendered relatively inadequate for the increased vascular capacitance. Thus, both absolute and relative **underfilling** of the effective circulating volume contribute to activation of baroreceptors, enhancement of neurohumoral effector mechanisms, and an appropriate antinatriuretic response.

It has been increasingly appreciated, however, that sodium retention in cirrhosis may not be due solely to a diminished effective circulating volume. Increased renal sodium retention has been documented before any cardiovascular compromise can be detected. There appears to be a signal—perhaps an increase in afferent sympathetic neural traffic associated with hepatic venous outflow obstruction—that causes the kidney to increase sodium reabsorption. According to this pathophysiologic sequence, as shown in Fig 3–3B, some cirrhotics may have **overfilling** of the effective circulating volume. As such, cirrhosis might be better considered in the group of disorders described below with primary renal sodium retention. Sodium retention during cirrhosis may well be due to both mechanisms, each occurring to varying degrees in a given patient and at different times in the progression of the cirrhotic state.

NEPHROTIC SYNDROME

In the classic explanation for sodium retention in the nephrotic syndrome, hypoalbuminemia due to urinary loss with insufficient hepatic synthesis allows translocation of fluid from the plasma to interstitial spaces (Fig 3–3A). The decrease in effective circulating volume causes appropriate baroreceptor-mediated activation of the neurohumoral effector system to cause sodium retention with progressive expansion of the interstitial space. Edema formation occurs when the plasma albumin is < 2 g/dL and is often generalized (anasarca) rather than dependent.

The explanation given above is sufficient to explain edema formation in only some cases of the nephrotic syndrome. In other nephrotic patients, however, findings inconsistent with this sequence occur—notably, hypertension and suppression of the renin-angiotensin-aldosterone system. In these cases, primary glomerulonephritis-like sodium retention appears to occur (Fig 3–3B). Conflicting evidence exists regarding whether the predominant nephron site of the enhanced renal sodium transport is the proximal tubule or collecting tubule, though most evidence favors the latter. Thus, sodium retention and edema formation in the nephrotic syndrome, as was the case with cirrhosis, is often a result of underfilling or overfilling of the circulation.

INCREASED EFFECTIVE CIRCULATING VOLUME WITH INCREASED TOTAL BODY SODIUM

Sodium input into the body can occasionally be sufficiently rapid to overwhelm renal sodium excretory capacity. More commonly, renal sodium excretory mechanisms may be uncoupled from control by the effective circulating volume as a result of a primary decrease in GFR or a primary increase in tubular sodium reabsorption, a summary of which is shown in Fig 3–4.

EXCESSIVE SODIUM CHLORIDE INTAKE

The kidney normally has an extraordinary ability to rid the body of excess sodium, at least 1 meq/min (ie, ≥ 1500 meq/d). Under very unusual circumstances, enough sodium can be ingested or administered to overwhelm renal saluretic capacity. This has been described during salt water near-drowning (when several liters of markedly hypertonic sodium—0.5 mol/L—combined in sea water can be ingested rapidly), with rapid ingestion of large amounts of very salty fluids, or with too-rapid intravenous infusion of normal saline solution. Hypernatremia and acute hypertension usually occur in these situations.

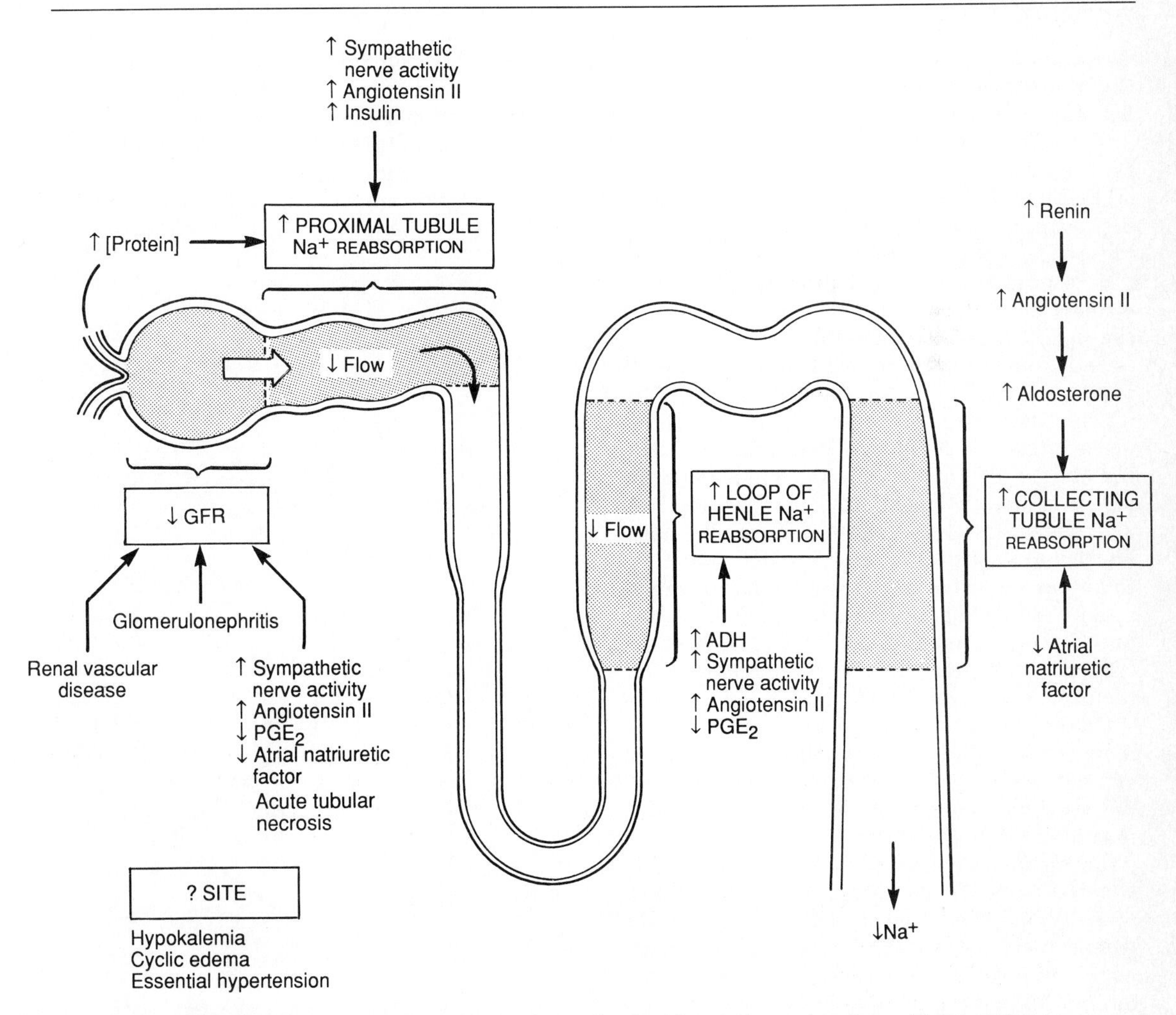

Figure 3-4. Nephron sites and mechanisms of antinatriuresis in renal sodium-retaining states.

PRIMARY DECREASE IN GFR

Primary glomerulopathic conditions, glomerulonephritis, renal vascular diseases, and urinary tract obstructive disorders that lower GFR may cause reduction in sodium excretion (Fig 3–4). If the new level of sodium excretion is less than intake, positive sodium balance can ensue, causing progressive expansion of the plasma volume and hypertension. Transudation of fluid into the interstitial space also occurs, sometimes called nephritic edema because of the renal pathogenetic origin.

Glomerular hypofiltration can be generally ascribed to two major functional causes. There may be a depression of the driving force for glomerular ultrafiltration by imbalance of the renal nerve and angiotensin II activities, causing an increase in R_A, a decrease in R_E, or both. Alternatively, K_f can be diminished either physically or functionally by vasoconstrictive substances released by the diseased glomerulus.

Overactivity of the Glomerular Vasoconstrictive Systems

A modest increase in nerve activity or angiotensin II level often causes only a small reduction in GFR. The rise in R_E which elevates P_{GC} serves to oppose the concurrent tendency of GFR to fall because of a rise in R_A and a fall in both Q_B (with rise in mean Π_{GC} and K_f. However, more severe increase in renal nerve activity or angiotensin II can significantly decrease GFR.

The vasoconstrictor-induced antinatriuresis is opposed over time by two factors: hypertension, which itself causes natriuresis (the pressure natriuresis effect; *see p 19*); and induction of prostaglandin synthesis (Fig 1–18), which moderates the effects of catecholamines and angiotensin II on determinants of glomerular hemodynamics, notably R_A and K_f. Thus, the level of sodium excretion following a primary increase in sympathetic nerve activity or circulating catecholamine level (eg, pheochromocytoma) or angiotensin II concentration (eg, renal artery stenosis or a renin-secreting tumor) is not entirely predictable.

On the other hand, functional effects of overactivity of the angiotensin II system can be more obvious if one of its major physiologic "brakes," the vasodilatory prostaglandins (especially PGE_2), is removed. For this to occur, the angiotensin II level must first be elevated by a decrease in effective circulating volume. Should there then be concurrent inhibition of prostaglandin synthesis, the hemodynamic and antinatriuretic effects of angiotensin II are unopposed, causing further decrease in GFR and exacerbating the relative stimulation of tubular sodium transport. A deficiency of ANF, the other major antagonist to vasoconstriction, could theoretically lead to similar consequences, but this has not been described.

Primary renal vasoconstriction occurs during processes of renal ischemia or toxicity collectively known as acute tubular necrosis. The causes and mediators of this renal vasoconstriction are not currently known, but severe reduction in GFR and acute renal failure can result. Well established in animals but not as yet in humans is the fact that **potassium deficiency** decreases GFR due to enhanced synthesis of angiotensin II as well as another potent vasoconstrictor, thromboxane.

Lack of Angiotensin II Activity During Renal Hypoperfusion

Paradoxically, absence of a normally functioning renal angiotensin II system can also diminish glomerular function under special circumstances. Recall—as reviewed in Fig 3–5 (solid lines)—that angiotensin II is essential for the increase in R_E to maintain P_{GC} and GFR constant when renal perfusion pressure is lowered: the autoregulatory response. Blockade of angiotensin II formation when renal perfusion pressure is simultaneously reduced prevents appropriate rise in R_E—in fact, R_E actually falls—so P_{GC} and GFR decrease (dashed lines in Fig 3–5).

This scenario occurs clinically when a converting enzyme inhibitor is used in the setting of renal

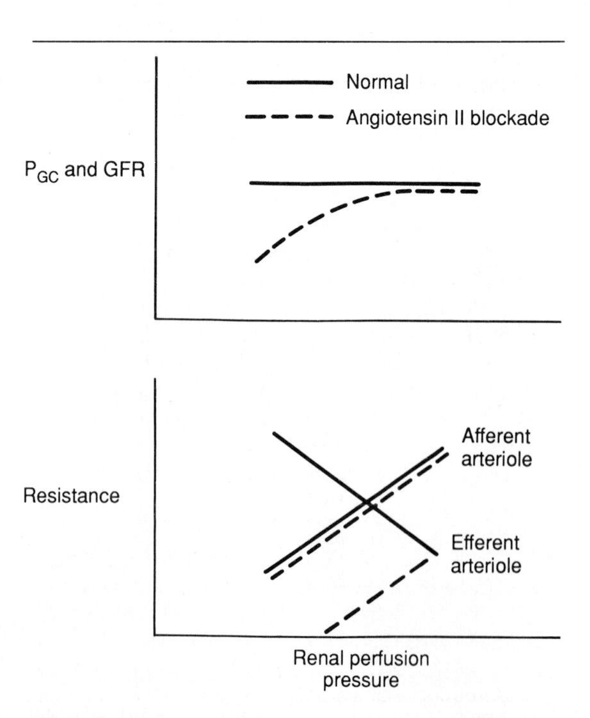

Figure 3–5. Disruption of glomerular autoregulation during angiotensin II blockade.

artery stenosis. The stenosis itself creates a situation where renal perfusion pressure is less than systemic arterial pressure, often at a critically low autoregulatory level. The administration of a converting enzyme inhibitor further lowers blood pressure and therefore renal perfusion pressure falls to a very low value. Since there is simultaneous antagonism of angiotensin II action, P_{GC} and hence GFR cannot be maintained. If the stenosis is unilateral, the ipsilateral decrease in GFR is usually not appreciated clinically because the perfusion pressure and GFR in the unstenosed contralateral kidney is not reduced or may even be slightly increased. If there is renal artery stenosis of a solitary kidney or bilateral stenosis, however, a fall in GFR will be clinically apparent during angiotensin II blockade.

Primary Glomerular Disease

It is easy to appreciate that any process which disrupts glomerular capillary blood flow or which physically or immunologically injures the glomerulus can impair GFR to cause sodium retention. Glomerular hypoperfusion states or primary diseases of the glomerulus include bilateral renal artery obstruction, renal vasculitis, and glomerulonephritides,including those which cause the nephrotic syndrome (Fig 3–3B). Glomerular inflammation also may induce synthesis of cytokines and vasoconstrictors, including angiotensin II, thromboxane, and leukotrienes, which exacerbates the reduction in GFR. GFR is sometimes reduced in tubulointerstitial nephritis, though the dominant tendency for disruption of tubular transport mechanisms more usually renders the kidney prone to sodium wasting rather than retention (Fig 2–4).

Urinary Tract Obstruction

When tubular or urinary flow is obstructed, the rise in tubular pressure can be sufficient to oppose P_{GC} and hence reduce GFR. Obstruction also activates afferent renal nerve activity and stimulates renin, both of which contribute to the antinatriuresis. If the process is unilateral, normal sodium excretion is assumed by the contralateral kidney. If the process is bilateral, however, partial or complete cessation of GFR predictably leads to retention of all sodium input. The obstruction can be intranephronal, due to casts (multiple myeloma or acute tubular necrosis), crystal deposition (acute urate nephropathy), or tissue inflammation and edema (tubulointerstitial nephropathies). Obstruction can also be extrarenal, due to urethral obstruction from stones, tumors, or retroperitoneal fibrosis, or due to bladder, urethral, or prostatic diseases.

PRIMARY INCREASE IN TUBULAR SODIUM TRANSPORT

Compared with the number of causes of diminished sodium transport, there are relatively few examples of sustained primary increased transport of sodium by the kidney. Increased transport by one nephron segment is often compensated by a reciprocal change in a subsequent segment, so that little net change in sodium reabsorption occurs. Hypermineralocorticoidism, which increases sodium transport by the collecting tubule, represents the best-described instance of a primary enhancement of sodium transport. Even in this instance, however, the resulting extracellular volume expansion and hypertension lead to escape from sodium retention by upstream nephron sites, as described above.

Increased Sodium Transport in Proximal Tubule

Marked, primary increases in sympathetic nerve activity and angiotensin II tend to decrease GFR, increase the filtration fraction, and directly increase proximal sodium transport, so that synergistic effects on renal sodium reabsorption are observed (Fig 1–24). Hyperinsulinemia increases proximal sodium reabsorption and may be important in mediating the edema formation sometimes observed during carbohydrate refeeding following starvation.

Increased Sodium Transport in the Loop of Henle

The upstream events that decrease luminal flow plus the direct effects of increased sympathetic nerve, angiotensin II, and ADH activity can augment loop of Henle sodium transport. Prostaglandin inhibition, by releasing the physiologic brake on transport, can also enhance sodium reabsorption in the loop of Henle.

Increased Sodium Transport in the Cortical Collecting Duct

Hypermineralocorticoid states predictably lead to enhanced sodium transport in the cortical collecting tubule, extracellular volume expansion, and hypertension. The interstitial volume expansion is usually not extensive enough to be appreciated as edema owing to the mineralocorticoid escape phenomenon discussed above.

As discussed more fully in Chapters 8 and 12, because of concurrent increases in potassium secretion and hydrogen ion secretion, hypokalemia and metabolic alkalosis accompany the sodium retention of hypermineralocorticoid states. Causes include primary hyperreninemia (renal artery stenosis, renin-secreting tumors, magnesium deficiency), adrenal overproduction of mineralocorticoids (adrenal adenomas, Cushing's disease, adrenogenital syndromes), and exogenous substances (in some kinds of licorice or chewing tobacco) that inhibit the enzyme (11β-hydroxysteroid dehydrogenase) which normally prevents endogenous glucocorticoids from having potent mineralocorticoid effects.

Increased Sodium Transport at Unknown Nephron Site

Potassium deficiency increases sodium transport. The nephron site of action is uncertain, but the loop of Henle is a likely candidate. The concomitant reduction in GFR mentioned above probably contributes to the potent antinatriuretic effect of potassium deficiency, which can lead to edema formation.

Cyclic edema is a disorder frequent in women that is characterized by intermittent bouts of sodium retention and lower extremity edema. The pathophysiologic mechanism and the site of enhanced sodium reabsorption are unknown. Abnormal vascular permeability has been suggested as a possible mechanism for the sodium accumulation, since fluid retention is most prominent in the upright posture. Such a capillary permeability abnormality may be idiopathic, associated with a family history of diabetes, or induced by a hypothalamic-endocrine disorder, perhaps involving prolactin or dopamine. Another factor that may contribute to the sodium retention may be an exaggerated insulin response to ingestion of carbohydrate.

Diuretics (especially thiazides), which are frequently used by individuals afflicted with idiopathic cyclic edema, have been suggested to aggravate and in some cases even to cause the syndrome. According to this hypothesis, a real or perceived increase in body volume prompts the patient to take a diuretic, which causes resolution of the edema but induces potassium depletion. When the diuretic is stopped, the potassium-deficient state causes rebound sodium retention and the edema recurs, prompting reinstitution of the diuretic.

Finally, abnormal renal sodium retention has been suggested to be a critically important contributor to the pathophysiology of some patients with essential hypertension. A subset of patients have a blood pressure response when sodium intake is altered (ie, "salt-sensitive" hypertension), but the mediators and sites of this well-documented effect are unknown. Indeed, some theorists have proposed that in virtually all patients with essential hypertension, there is intrinsically enhanced renal sodium transport, serving to re-set the blood pressure-renal sodium excretion balance that is necessary to maintain the hypertension.

SYMPTOMS

The common symptoms of the hypervolemic disorders relate to edema and, in the entities associated with increased plasma volume, to circulatory overload, as shown in Fig 3–6.

CARDIOPULMONARY

With primary renal sodium retention in the setting of normal intrinsic cardiac function, excessive preload causes hypertension. The hypertension may be severe, even malignant, with retinal changes; central nervous system symptoms, including seizures; and renal damage, manifested by proteinuria and reduction in GFR.

With progressive circulatory congestion and increase in cardiac filling pressures, pulmonary edema eventually develops, manifested by symptoms of dyspnea on exertion, orthopnea, paroxysmal nocturnal dyspnea, bronchospasm, and finally resting shortness of breath.

SKIN & CONNECTIVE TISSUE

Depending on the duration and severity of the process, the degree of peripheral edema associated with all hypervolemic disorders can vary from slight and of merely cosmetic concern to severe and debilitating. The systemic capillary Starling forces determine the distribution of the excess sodium and water in the interstitial areas. In hypoproteinemic and renal failure disorders, the distribution tends to be more diffuse, as anasarca, whereas in heart failure, interstitial fluid initially accumulates in dependent areas.

DIAGNOSIS

APPROACH TO ETIOLOGIC DIAGNOSIS

History & Physical Examination

The first step in approaching the patient with peripheral edema is to ascertain whether there is simply venous or lymphatic obstruction, with local rather systemic imbalance of the capillary Starling forces.

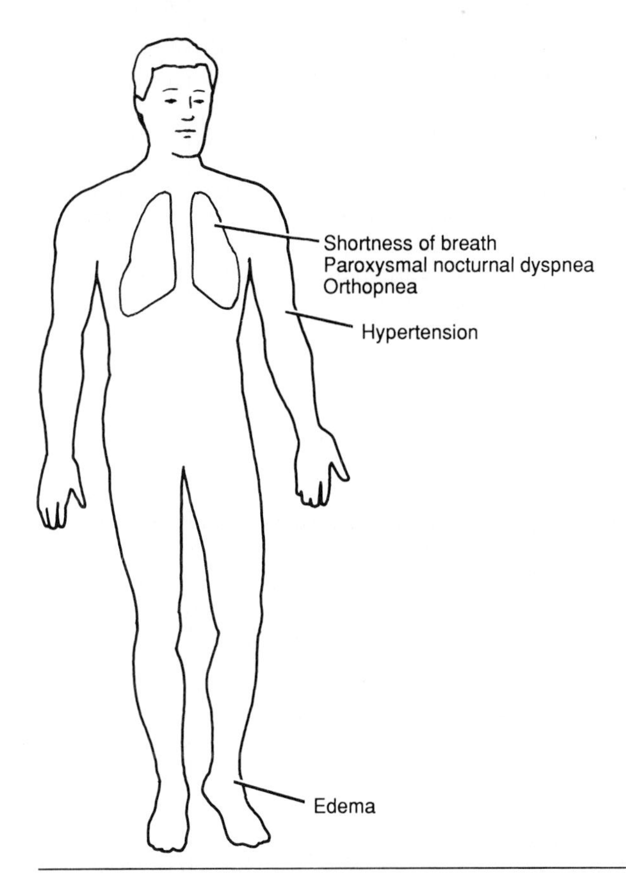

Figure 3-6. Symptoms and signs of hypervolemia.

The major functional differentiation of the edematous disorders revolves around whether the renal sodium retention is physiologically appropriate, ie, responsive to a reduced effective circulating volume (congestive heart failure, cirrhosis, or nephrotic syndrome); or inappropriate, due to a primary renal disorder responsible for expanding the effective circulating volume (Fig 3–1).

Historical clues and symptoms and signs of the following may be present to distinguish between these two groups: heart failure (chest pain, orthopnea, paroxysmal nocturnal dyspnea, cardiomegaly, arrhythmias, gallops, central venous pressure elevation, elevated cardiac filling pressures, etc); cirrhosis (history of alcoholism, hepatitis, jaundice, hepatomegaly, splenomegaly, ascites, spider angiomas, palmar erythema, gynecomastia, testicular atrophy, etc); nephrotic syndrome (history of marked proteinuria, hematuria, etc); or primary renal disease (history of hypertension, proteinuria, hematuria, abnormal urinalysis, stones, renal infections, familial renal disease, etc).

A careful drug history is essential in evaluating all hypervolemic states, especially for drugs that affect the renin-angiotensin system (converting enzyme inhibitors, prostaglandin synthesis inhibitors, mineralocorticoids, nephrotoxic agents) as well as cardiovascular acting drugs and diuretics.

Renal Response

The urine sodium concentration may vary substantially in the hypervolemic states depending on the phase of the process, as summarized in Table 3-2. As discussed previously (Table 2–4), the urine sodium should be very low (concentration < 20 meq/L) in physiologic response to diminished effective circulating volume unless the kidney itself is responsible for depleting extracellular volume. In the hypervolemic disorders, renal sodium reab-

Table 3-2. Urine sodium excretion in hypervolemic states.

	Urine Na^+ Excretion
↓Effective arterial volume	
Decompensated	< Intake[1]
↑Effective arterial volume	
Decompensated	< Intake[1]
Compensated	= Intake

[1]Severe: Urine $[Na^+]$ < 20 meq/L.

sorption must be less than intake at least for a short period of time to allow the edema to develop: the so-called **decompensated phase.**

The stimulus for renal sodium conservation may not be as profound as with overt hypovolemia, and urine sodium may not be maximally reduced. Furthermore, in some cases, sodium retention eventually returns the effective circulating volume sufficiently close to normal that urine sodium excretion again approximates sodium intake. In this **compensated phase,** the patient resumes external sodium balance (sodium output = sodium input), though at the expense of marked extracellular volume expansion and edema. Thus, the urine sodium concentration and output can be helpful in the edematous disorders for defining whether a compensated state has resumed.

A mild reduction in GFR is part of the renal response of hypervolemic states with diminished effective circulating volume, as occurs in the hypovolemic states (Table 2–4). Elevation of the BUN:plasma creatinine ratio is then expected in these hypervolemic states.

Coincident Disorders of Acid-Base, Serum Sodium, & Serum Potassium

As summarized in Table 3–3, various abnormalities in arterial pH, plasma osmolality, and potassium homeostasis frequently accompany hypervolemic conditions. Metabolic acidosis is common when decreased sodium filtration or enhanced proximal and loop sodium reabsorption occurs in association with diminished effective circulating volume. Acidosis may occur because acidification in the cortical collecting tubule is dependent, at least in part, on the rate of sodium delivery (Chapter 10). Any abnormality in the renin-angiotensin-aldosterone axis aggravates the tendency to acidosis. Finally, since potassium secretion is also dependent on sodium delivery to the cortical collecting tubule and on aldosterone, hyperkalemia may also occur (type IV generalized distal renal tubular acidosis).

The opposite electrolyte changes occur—metabolic alkalosis and hypokalemia—when hypervolemia is due to primary stimulation by mineralocorticoid of cortical collecting tubule sodium transport.

Cirrhosis is often associated with stimulation of ventilation and respiratory alkalosis. Although hyperkalemia can occur in cirrhosis (especially when GFR is profoundly reduced), hypokalemia is more commonly observed as a consequence of inadequate potassium intake, gastrointestinal potassium loss, use of diuretics, and severe hyperaldosteronism.

In general, renal water retention parallels sodium retention in the hypervolemic states. In the decompensated phase, the patient may be relatively oliguric and the urine highly concentrated. In the compensated phase, however, urinary volume may be normal. In states with markedly reduced effective circulating volume, hyponatremia may occur as a result of nonosmotic release of ADH (Chapter 5). Sodium chloride intoxication and primary renal sodium-retaining disorders may be accompanied by mild hypernatremia.

THERAPY

INDICATIONS FOR TREATMENT

Importance of Treating Underlying Disease

In most cases, treatment of edema is palliative but does not address the underlying problem.

Table 3–3. Hypervolemic states.

	Common Concurrent Acid-Base Disorder	Common Concurrent $[Na^+]_p$	Common Concurrent $[K^+]_p$
↓Effective circulating volume			
Congestive heart failure	Metabolic acidosis	↓	—(↑)
Cirrhosis	Respiratory alkalosis; metabolic acidosis	↓	↓(↑)
Nephrosis	Metabolic acidosis	↓	—
↑Effective circulating volume			
↑ECV solute			
NaCl intoxication	Metabolic acidosis	↑	—
↓Renal Na⁺ excretion			
↓GFR	Metabolic acidosis	—	↑
↑Tubular Na⁺ reabsorption			
↑Mineralocorticoid	Metabolic alkalosis	Slight ↑	↓

Whenever possible, the primary disease should be treated. Every attempt should, of course, be made to normalize cardiac dynamics in heart failure (eg, valve replacement or pericardiocentesis if indicated, inotropes, afterload reduction), improve hepatic function in liver failure (eg, remove toxins, administer prednisone in some cases), treat glomerulonephritis in nephrotic syndrome (eg, administer prednisone, cytotoxic agents, or anticoagulants, depending on the type of glomerulonephritis), relieve postrenal obstruction (eg, urethral or ureteral catheterization), and so on. It is unwise to attempt to reduce extracellular volume in states in which very high diastolic filling pressures are needed for normal cardiac performance, as in constrictive pericarditis, hypertrophic cardiomyopathy, or aortic stenosis.

Indications for treatment to reduce extracellular sodium content include pulmonary compromise and overt pulmonary edema, impaired cardiovascular function due to excessive vascular congestion, abdominal pain or abdominal herniation due to severe ascites, and impaired mobility due to large amounts of peripheral edema. Given the potential hazards of diuretics, pharmacologic treatment of edema simply for cosmetic purposes is not a solid indication. Conservative care, such as maintaining a supine posture and using supportive stockings, can help mobilize edema.

RESTRICTION OF SODIUM INTAKE

Evaluation of Dietary Sodium

As indicated in Table 3–2, the difference between a compensated and decompensated state—betweeen neutral or positive sodium balance—may simply be in the amount of ingested sodium. The kidneys may be "set" at a certain level of sodium excretion by the prevailing effective circulating volume or underlying renal disease. A good dietary history is therefore essential when considering an edematous patient for dietary counseling. Occasionally, sodium chloride is a component of drugs, and this source should be calculated in addition to that in diet. The data regarding sodium content of common foods shown in Table 3–4 are useful for estimating intake and for guiding dietary sodium restriction.

Type of Sodium Salt Needing Restriction

Sodium as its **chloride** salt is what is reabsorbed by the kidney and contributes to the extracellular volume. Sodium with any other anion, such as bicarbonate or glutamate, need not be excluded, since these sodium salts are filtered easily, cannot be reabsorbed by tubular transport mechanisms, and are excreted. Thus, the only sodium of concern in foods listed in Table 3–4 that potentially contributes to edema formation is sodium chloride.

Amount of Sodium Restriction

The degree of sodium chloride restriction depends on the severity of the edematous state, the degree of remaining renal sodium excretion, the potency of the diuretics to be employed, and the compliance of the patient. In some cases, sodium restriction is the sole therapeutic modality required, but should be employed in all cases as an adjunct to diuretic therapy so as to achieve and maintain the negative sodium balance needed for edema reduction. It should be remembered there is little physiologic need for sodium—usually less than 10 meq/d. Assuming no concurrent redistribution of plasma volume (eg, due to surgery, trauma, or sepsis), sodium intake should be severely curtailed in patients with acute renal failure.

As shown in Table 3–5, sodium intake may be halved simply by omitting very salty foods (Table 3–4) and by not adding salt when eating. Greater reductions can be achieved by not adding salt during cooking and by eating only low-sodium foods.

DIURETICS

If the extracellular volume expansion has caused an acute clinical indication for therapy or if dietary sodium restriction alone is insufficient to reverse positive sodium balance, diuretics are needed. Clinical considerations govern the route, type, and dosage regimen of the diuretic. Unfortunately, only a few generalizations are possible. Dosing information for commonly used oral diuretics is given in Table 3–6. Dosage ranges of intravenous and oral forms of these diuretics are generally similar. Concurrent effects of diuretics on the renal handling of K^+, H^+, Ca^{2+}, and Mg^{2+} are shown in Fig 3–7.

Route, Type, & Amount of Diuretic Administration

Many diuretics can be given intravenously or orally; the choice depends on the acuteness of the clinical situation. Acute symptomatic cerebral or pulmonary edema obviously requires prompt intravenous therapy, whereas less emergent conditions can be treated with oral agents.

The choice of a diuretic requires consideration of the severity of the hypervolemic state, the degree of impairment of renal function, and the potency of the diuretic. Specific indications are discussed in detail below, but in general thiazides are

Table 3-4. Sodium content of foods.[1]

	Very high (> 190 mg, > 30 meq)	High (230-690 mg, 10-30 meq)	Moderate (69-230 mg, 10-30 meq)	Low (< 69 mg, < 3 meq)
Milk (1/2 cup or as stated)		Processed cheese (1 oz)	Cheese (1 oz) Puddings Yogurt	Milk Ice cream Cream cheese (1 oz) Sherbet Sour cream (1 tbsp)
Beans and nuts (1/2 cup or as stated)	Salted cashews Fermented soy beans	Baked beans Salted peanuts	Almonds	Kidney and other dry beans Chestnuts Lentils Peanut butter (1 tbsp) Peas Soy beans (cooked) Walnuts
Meat (2 oz or as stated)	Ham Beef (dried, chipped) Prepared dinners (1) Canned meats (1)	Fish (canned) Beef (corned) Chicken (canned) Hamburger (fast food) (1) Pizza (1 slice) Taco (1)	Fish baked with butter Shellfish Salami, bologna (1 slice)	Egg (1) Fish Beef (lean) Chicken Pork Veal Lamb
Fruit (1/2 cup or as stated)				All fresh fruits
Vegetables (1/2 cup or as stated)	Tomato sauce	Vegetables (canned) Vegetables with butter sauce Tomato juice	Chard	All fresh vegetables
Grains		Danish pastry English muffin	Biscuit (1) Bread (1 slice) Oatmeal (instant) Cereal (1/2 cup) Doughnut (1) Coffee cake (1/8) Sweet roll (1) Cake (1/12) Pie (1/8) Pancake (1) Roll (1) Popcorn (1/2 cup) Potato chips (10) Corn chips (1 oz)	Oatmeal (regular
Other	Soups (1 cup) Salt (1 tsp) (84 mg) Pickle (1)	Olives (4)	Butter (1 tbsp) Margarine (1 tbsp) Salad dressing (1 tbsp) Sauces (1 tbsp) Ketchup (1 tbsp) Horseradish (1)	Coffee (1/2 cup) Alcoholic beverages Tea (1 cup) Fruit juice (1 cup) Soft drinks (1 cup) Candy (1 oz) Jam (1 tbsp) Syrup (1 tbsp) Jelly (1 tbsp) Sugar (1/2 cup) Oil (1 tbsp) Pepper (1 tsp)

[1]United States Department of Agriculture Home and Garden Bulletin No. 233.

Table 3-5. Dietary sodium restriction.

	Daily Na$^+$		Omit Added NaCl	Omit NaCl in Diet Prep.	Restrict Foods Based on NaCl Content
	meq/d	g/d			
Average American diet	150+	3.5+	—	—	—
Mild restriction	100	2.3	+	—	+
Moderate restriction	50	1.2	+	+	++
Severe restriction	25	0.6	+	+	+++

used when there is mild to moderate hypervolemia or GFR reduction (eg, 50–100 mL/min) and loop diuretics are used with more severe hypervolemia or GFR impairment (eg, < 50 mL/min).

The dose of diuretic and duration of therapy is guided by clinical criteria. Reduction or elimination of cerebral, pulmonary, or peripheral edema or of hypertension is usually the goal. The dose of diuretic is adjusted to achieve and maintain these therapeutic objectives.

Diuretics That Increase GFR

There is no clinically available agent that consistently increases GFR in order to overwhelm tubular transport mechanisms and cause natriuresis. Experimentally, intravenous infusion of **ANF** in pharmacologic concentrations can act in this fashion, with direct and indirect tubular effects contributing to its natriuretic potency (Fig 1–8).

ANF has been shown to be effective acutely in increasing cardiac output in congestive heart fail-

Table 3-6. Diuretics.

	Single Oral Dosage Range (mg)	Dosing Frequency (per day)
Proximal tubule		
Acetazolamide (Diamox)	125-250	1-6
Thick ascending limb of Henle		
Furosemide (Lasix)	20–120	1–4
Ethacrynic acid	25–200	1–4
Bumetanide (Bumex)	0.5–2	1–4
Distal convoluted tubule		
Thiazides		
Chlorothiazide (Diuril)	500–1000	1–2
Hydrochlorothiazide (Esidrix, Hydro-Diuril, Oretic)	25–100	1–2
Trichlormethiazide (Metahydrin, Naqua)	1–4	1–2
Hydroflumethiazide (Diucardin, Salurom)	25–100	1–2
Bendroflumethiazide (Naturetin)	5–10	1–2
Quinethazone (Aquamox)	50–100	1–2
Metolazone (Diulo, Zaroxolyn)	5–20	1
Chlorthalidone (Hygroton)	25–100	1
Polythiazide (Renese)	1–4	1
Methyclothiazide (Aquatensen, Enduron)	2.5–10	1
Indapamide (Lozol)	2.5–5	1
Cortical collecting tubule		
Spironolactone (Aldactone)	25–100	1–4
Amiloride (Midamor)	5–10	1–2
Triamterene (Dyrenium)	50–100	2–3

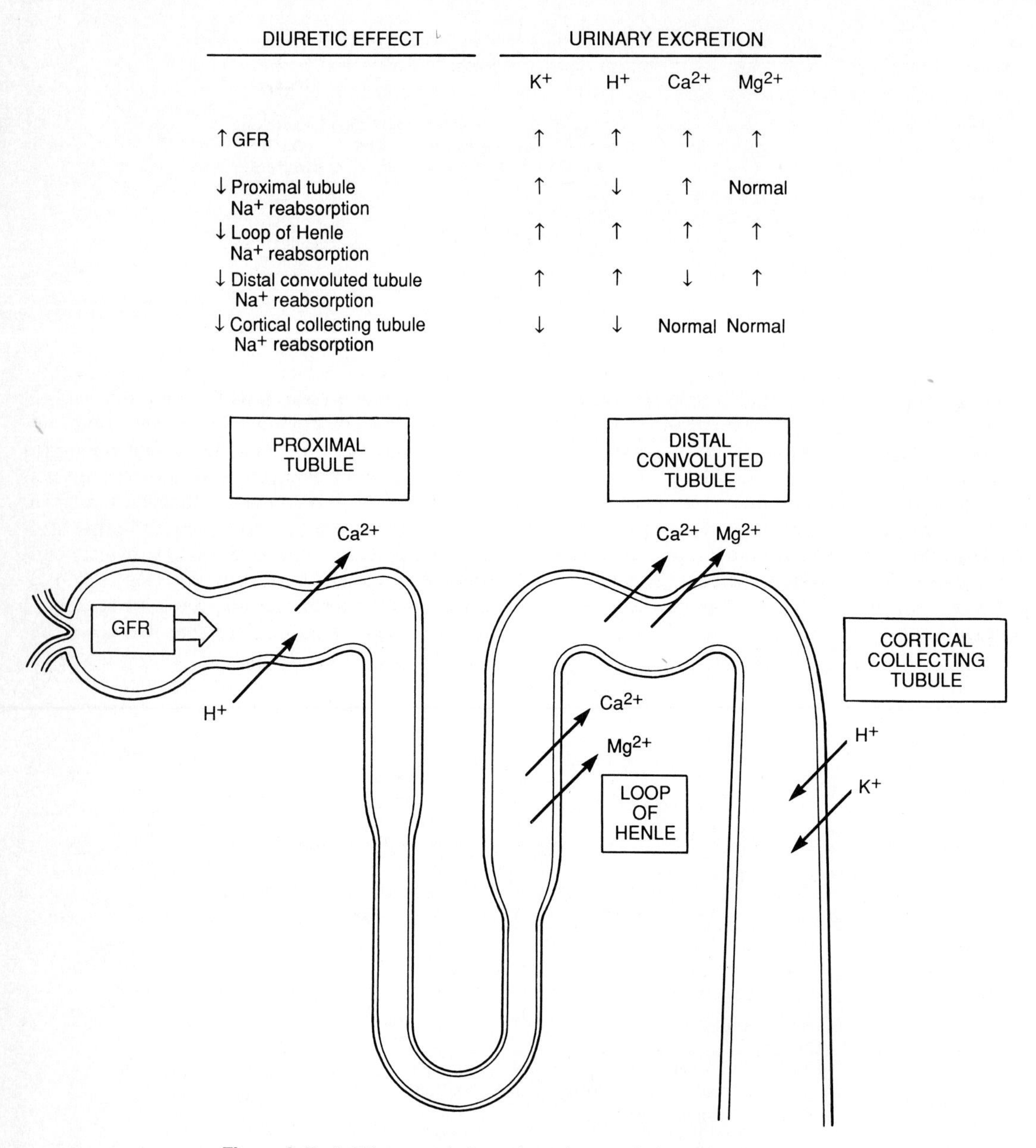

DIURETIC EFFECT	URINARY EXCRETION			
	K^+	H^+	Ca^{2+}	Mg^{2+}
↑ GFR	↑	↑	↑	↑
↓ Proximal tubule Na^+ reabsorption	↑	↓	↑	Normal
↓ Loop of Henle Na^+ reabsorption	↑	↑	↑	↑
↓ Distal convoluted tubule Na^+ reabsorption	↑	↑	↓	↑
↓ Cortical collecting tubule Na^+ reabsorption	↓	↓	Normal	Normal

Figure 3-7. Acid-base and electrolyte changes induced by diuretics.

ure, by decreasing preload and afterload. The natriuretic potency of ANF is limited in congestive heart failure, nephrosis, and cirrhosis, but there is synergistic efficacy when the drug is combined with a loop diuretic. Hypotension is the principal side effect of this potent vasodilator, treated simply by decreasing the intravenous infusion rate. Whether the effects of ANF are sustained over prolonged intervals has not yet been demonstrated. ANF is not yet available for clinical use, although trials are currently in progress.

Diuretics That Inhibit Sodium Transport in the Proximal Tubule

There are two classes of proximally acting diuretics: osmotic agents, which interrupt sodium chloride reabsorption; and carbonic anhydrase inhibitors, which interrupt sodium bicarbonate reabsorption. These diuretics are not generally used for hypervolemic disorders and have few specific indications.

The prototype osmotic diuretic is **mannitol.** Besides inhibiting proximal sodium chloride reabsorption, mannitol increases inner medullary blood flow, which reduces medullary hypertonicity and sodium reabsorption in the thin ascending limb of Henle. As an impermeant solute and vasodilator, its principal indications are for abrupt reduction in acute cellular swelling, as in cerebral edema (caused by tumor, posthypoxia, metabolic conditions, or following acute dialysis—the dialysis disequilibrium syndrome) or to acutely increase the flow of tubular fluid and medullary blood, as in incipient acute renal failure (eg, caused by myoglobinuria or radiographic contrast dye). It is used only acutely and not for prolonged periods. A dose of 1–2 g/kg of mannitol (15–20% solution) intravenously achieves its peak effects on cerebral edema within 60–90 minutes. The acute increase in plasma osmolality causes plasma volume expansion, with risk of circulatory overload. The hyperosmolality is associated with hyponatremia (Chapter 5) and hyperkalemia (Chapter 9). Mannitol is excreted by the kidney and causes a brisk natriuresis. Urinary potassium, calcium, and magnesium excretion are also increased—usually of little importance.

The prototype of the carbonic anhydrase inhibitors is **acetazolamide.** This agent inhibits the cellular and luminal isoenzymes of carbonic anhydrase in the proximal tubule and results in marked disruption of sodium bicarbonate reabsorption (Chapter 12). It also inhibits the carbonic anhydrase enzyme responsible for aqueous humor formation and is used for treating glaucoma. Acetazolamide is available for oral as well as intravenous administration. It is well absorbed orally ($>$ 90%), is tightly protein-bound (90%), and is secreted in the proximal tubule to yield a sodium bicarbonate diuresis that can reach 20% or more of the filtered bicarbonate load. Phosphate excretion is enhanced, and the large delivery of sodium bicarbonate to the collecting tubule also increases potassium secretion.

Besides glaucoma, therapeutic indications for acetazolamide include some drug overdoses requiring urinary alkalinization (eg, aspirin), and symptomatic metabolic alkalosis in the setting of a cardiopulmonary status intolerant of sodium chloride. Other indications include prevention of acute mountain sickness and hypokalemic periodic paralysis, though the mechanism of action in prevention of these disorders is unknown. Complications of acetazolamide include type I proximal renal tubular acidosis (Chapter 12), hypokalemia (Chapter 8), and nephrolithiasis. Acetazolamide is not very effective for eliminating sodium chloride from the body, however, and has limited utility in most hypervolemic disorders.

Diuretics That Inhibit Sodium Transport in the Thick Ascending Limb of Henle

The major loop diuretics are the sulfonamide derivatives **furosemide** and **bumetanide** and a phenoxyacetic acid derivative, **ethacrynic acid.** Loop diuretics are well absorbed (bioavailability is about 80%), tightly protein-bound (96–97%), freely filtered as well as secreted by the proximal tubule, and act when in the tubular fluid on the luminal Na^+ /K^+ /2Cl^- transporter of the thick ascending limb of Henle (Fig 1–25). They have no effect from the peritubular (basolateral, or blood) side. Potassium, calcium, and magnesium transport in the loop are diminished along with sodium chloride absorption, and the enhanced delivery of sodium chloride to the cortical collecting tubule augments distal potassium and hydrogen ion secretion. Prostaglandin-mediated increase in renal blood flow also occurs with loop diuretics.

Loop diuretics can be given intravenously or orally (Table 3–6). They are the most potent diuretics available and can achieve acutely a maximal fractional sodium excretion rate of 25–30%. Indications for a loop diuretic include conditions of markedly increased extracellular volume or circumstances in which an acute brisk natriuretic response is necessary, as in pulmonary edema. A loop diuretic is also used in chronic, moderate to severe hypervolemic conditions which are unresponsive to dietary sodium chloride restriction or to less potent diuretics or when renal function is significantly depressed. Loop diuretics induce venodilation and a decrease in systemic vascular resistance that precede the natriuretic effects, which makes them very useful in the treatment of congestive heart failure. Other indications include hypercalcemia (Chapter 17) and hypermagnesemia (Chapter 23) and conversion of oliguric to nonoliguric acute renal failure.

Complications of loop diuretics are generally predictable based on their physiologic action: hypovolemia leading to hyponatremia and decreased GFR if the natriuresis is excessive, hypokalemia, hypocalcemia, hypomagnesemia, hyperuricemia, metabolic alkalosis, and an inability to concentrate the urine. Ototoxicity is a dose-related, usually reversible complication due to effects of loop diuretics on endolymph ion transport and is exacerbated by concurrent aminoglycoside administration. The loop diuretics that are sulfonamide derivatives may cause allergic responses in susceptible individuals.

Diuretics That Inhibit Sodium Transport in the Distal Convoluted Tubule

The **thiazides** are sulfonamide-derived agents that inhibit the electroneutral sodium chloride transporter on the luminal membrane of the distal convoluted tubule (Fig 1-27). Most thiazides are generally well absorbed (bioavailability is 60–70%, except chlorothiazide, which is 10%); are moderately protein-bound (50–90%); are freely filtered; and, like the loop diuretics, are secreted by the proximal tubule. The decreased sodium absorption in the distal convoluted tubule by thiazides simultaneously enhances calcium reabsorption (Chapter 17), decreases magnesium reabsorption (Chapter 22), stimulates potassium and hydrogen ion secretion downstream in the cortical collecting tubule (Chapters 8 and 13), and reduces the ability to dilute the urine (Chapter 4).

Thiazides are available for oral (Table 3-6) as well as for intravenous administration. They are moderately natriuretic and can achieve a maximal fractional sodium excretion rate of 5–10%—substantially less than those of loop diuretics. Thiazides are used so long as acute, life-threatening symptoms are not present. Indications for a thiazide include essential hypertension, as the initial treatment of mild to moderate hypervolemic disorders, and in miscellaneous conditions such as idiopathic hypercalciuria (Chapter 17), diabetes insipidus (Chapter 6), and type I proximal RTA (Chapter 12). With the exception of metolazone, thiazides are ineffective when the GFR is significantly reduced (ie, 30 mL/min or less).

Toxicity of thiazides can be generally predicted from their physiologic actions: hypovolemia with hyponatremia and decreased GFR if natriuresis is excessive, hypokalemia, and hyperuricemia and occasionally gout in susceptible individuals. Zinc deficiency, hypercholesterolemia (associated with low-density lipoprotein), and hypertriglyceridemia also occur. As sulfonamide derivatives, thiazides have been associated in susceptible individuals with allergic and idiosyncratic reactions, including hemolytic anemia, neutropenia, thrombocytopenia, pancreatitis, cholestatic jaundice, and others.

Diuretics That Inhibit Sodium Transport in the Cortical Collecting Tubule

Two classes of oral agents block sodium transport in the cortical collecting tubule: competitive antagonists of the aldosterone receptor, the prototype being **spironolactone;** and direct inhibitors of the luminal sodium channel, the prototypes being **amiloride** and **triamterene.** In both cases, the functional effects of aldosterone are inhibited, so that sodium reabsorption is diminished and potassium and hydrogen ion secretion are simultaneously depressed. These drugs have reasonable oral bioavailability (50–70%) and variable protein binding (spironolactone 90%, triamterene 50%, amiloride nil). Spironolactone requires at least 2–3 days of administration for its active metabolite (camrenone) to reach a maximally effective concentration.

Spironolactone, amiloride, and triamterene are all weak diuretics, able to achieve a maximal fractional sodium excretion rate of 2–3%, and thus are less potent than thiazides. They are useful in states with a high aldosterone level and relatively well-preserved GFR, such as mild to moderate cirrhosis. These agents are also widely used in conjunction with thiazides to spare the potassium wasting that would otherwise occur (Chapter 8).

The side effects of spironolactone, amiloride, and triamterene are also largely attributable to their physiologic actions and include the potential for inducing hyperkalemia and type IV generalized distal RTA (Chapter 12). Oral potassium supplements should not be used with this class of diuretics. When the GFR is reduced, potassium homeostasis is critically dependent on efficient cortical collecting tubule function, so these diuretics should not be used in patients with renal insufficiency or those in whom potassium homeostasis is compromised for any other reason (eg, adrenal insufficiency, converting enzyme inhibition, diabetes). Other side effects include gynecomastia, impotence, and irregular menses (spironolactone), and nephrolithiasis and pancytopenia (triamterene).

APPROACH TO DIURETIC RESISTANCE

Diuretics are sometimes ineffective in reducing edema or lose potency with time. The approach to this problem is summarized in Table 3-7.

Reassess Underlying Antinatriuretic Stimuli

Excessively rapid diuresis can cause loss of response to a diuretic. External sodium loss faster than can be replenished from extravascular stores may exacerbate the diminished extracellular volume already present in congestive heart failure and many cases of cirrhosis or nephrosis. If cardiac, renal, and hepatic function are normal, generalized edema can be mobilized at a rate as high as 2 L/d without impairing plasma volume. However, removal of fluid, such as ascites, from a sequestered space generally cannot exceed 0.9 L/d (usually only 0.2–0.4 L/d) without jeopardizing the plasma volume.

Decreased GFR can also change the steady-state response to a diuretic. For instance, nephrotoxins, including drugs that inhibit prostaglandins, such as aspirin and nonsteroidal anti-inflammatory agents, can aggravate sodium retention.

Table 3-7. Management of diuretic resistance.

1. Decrease intensity of diuretic therapy if effective circulating volume is further compromised.
2. Discontinue nephrotoxins and nonsteroidal anti-inflammatory drugs.
3. Decrease sodium intake.
4. Decrease dosing interval.
5. Administer diuretic intravenously.
6. Change to more potent class of diuretic.
7. Add another class of diuretic.
8. Ultrafiltration
    - Hemodialysis
    - Peritoneal dialysis
    - Pure ultrafiltration
    - Continuous arteriovenous hemofiltration

Reduce Sodium Intake

If a reasonable natriuresis is attained (ie, > 50–100 meq/d) and the patient is still not losing extracellular volume and weight, the sodium intake may be too high. Further reduction in dietary sodium guided by the information in Table 3–4 to a level below output will allow weight loss to recur.

Increase the Frequency of Dosing or Change the Route

When thiazides and the loop diuretics are given once a day, much of the diuretic action has been completed within 6 hours and there is a compensatory antinatriuresis over the remaining 18 hours. To provide a more sustained natriuretic state, the diuretic should be given more frequently: twice or even four times daily. Occasionally, bioavailability is hindered by intestinal edema, especially in right-sided heart failure. Efficacy can be regained by changing to intravenous administration of a loop diuretic even at the same dosage.

Change to a More Potent Class of Diuretic

In general, the (ascending) rank order of potency of diuretics are those that inhibit sodium transport in the cortical collecting tubule (spironolactone, amiloride, and triamterene) → in the distal convoluted tubule (thiazides) → in the loop of Henle (furosemide, bumetanide, ethacrynic acid).

The maximal effective dosages per day can be calculated from Table 3–6 (highest single dose × most frequent dosing interval). For instance, for hydrochlorothiazide, the highest single dose is 100 mg and the highest dosing frequency is twice daily, so the maximal daily dose is 200 mg/d. If the therapeutic end point has not been attained when the maximal daily dose of an agent is reached, it may be stopped and a drug from the next more potent class initiated. It is not rational to add another agent in the same class (eg, spironolactone to amiloride, one thiazide to another, or furosemide to ethacrynic acid).

Add a Diuretic From Another Class

If a potent diuretic has not proved sufficiently efficacious, an option is to add another agent from a different class. For instance, increased efficacy of furosemide is rarely achieved with doses higher than about 400 mg/d (and ototoxicity is risked). If a patient's edema were refractory to furosemide alone, a response might be achieved by adding metolazone, a thiazide diuretic with some carbonic anhydrase inhibitory activity. Thus, these two diuretics would inhibit sodium transport in proximal, loop, and distal convoluted tubule sites. Often a good response can be seen with the combination of these two diuretics even when the dose of each is below a maximal level.

ULTRAFILTRATION TECHNIQUES

When renal function is severely compromised or nonexistent such that efficient pharmacologically induced natriuresis is prevented, other modalities are available to remove sodium from the body.

Ultrafiltration During Dialysis

In the setting of renal failure, dialysis is necessary to remove nitrogenous wastes and control potassium and acid-base balance. Dialysis may be effected using an extracorporeal system, **hemodialysis,** or using the patient's own peritoneal membrane, **peritoneal dilysis.** In both of these modalities, small solutes like urea are removed by diffusion, the driving forces being the blood-to-dialysate concentration gradient. Since the dialysate is essentially isotonic, there is no net sodium or water movement (ie, no ultrafiltration) during the dialysis procedure itself.

However, net sodium and water removal from the body can be effected by imposing hydraulic pressure to convectively force fluid across the dialyzer or peritoneal membrane. In the case of hemodialysis, this driving force is the blood-to-dialysate hydrostatic pressure gradient, called the transmembrane pressure. By imposing a given pressure gradient across a membrane with a preselected hydraulic permeability, a predictable isotonic volume can be ultrafiltered from blood. For instance, consider a membrane with hydraulic permeability of 3 mL/h per mm Hg of pressure. If the mean extracorporeal blood pressure were 100 mm Hg and the dialysate pressure were −100 mm Hg, the transmembrane pressure would be 200 mm Hg (100 − [−100]) and the fluid removal would be 3 mL/h per mm Hg × 200 mm Hg, or 600 mL/h. Rates of isotonic fluid of 2 L/h or more can be achieved depending on the membrane hydraulic permeability and the transmembrane pressure gradient prescribed. Hemodialysis requires percutaneous venous access (femoral or subclavian cannulation) or an arteriovenous fistula or graft.

In the case of peritoneal dialysis, fluid ultrafiltration is driven by an osmotic pressure gradient achieved by adding glucose to the dialysate. Assuming a 1-hour dwell time per cycle of dialysate instillation during acute peritoneal dialysis, the mean fluid removed with 1.5% glucose in the dialysate is 75 mL/cycle, with 2.5% it is 150 mL/cycle, and with 4.25% it is 350 mL/cycle. Ultrafiltration rate peaks at a 2-hour dwell time per cycle and then declines at longer dwell times. For example, assuming a 4-hour dwell time per cycle characteristic of peritoneal dialysis (eg, continuous ambulatory peritoneal dialysis), the mean fluid removed with 1.5% glucose in the dialysate is 130 mL/cycle, with 2.5% it is 250 mL/cycle, and with 4.25% it is 750 mL/cycle. Peritoneal dialysis requires a temporary or permanent (Tenckhoff) catheter inserted through the abdominal wall.

Ultrafiltration Without Dialysis

Dialysis with small solute removal by diffusion need not occur in order to have ultrafiltration. A conventional artificial kidney without dialysate flowing through it is capable of performing **pure or isolated ultrafiltration** (PUF, IUF). A conventional delivery machine used for hemodialysis is used—but without the dialysate flow—to impose a hydrostatic pressure gradient across the dialyzer membrane. Isotonic fluid removal is calculated and controlled in the same way as when dialysis is simultaneously occurring (see above).

A similar principal is used with **continuous arteriovenous hemofiltration (CAVH),** in which fluid is constantly ultrafiltered across a semipermeable membrane using the patient's own circulatory pressure as the driving force. This modality is capable of low rates of fluid removal but can be used continuously, so that large amounts of edema can be removed on a daily basis. Arteriovenous access and systemic heparinization are required.

SECTION I REFERENCES: SODIUM HOMEOSTASIS & DISORDERS: HYPOVOLEMIA & HYPERVOLEMIA

Berry CA: Heterogeneity of tubular transport processes in the nephron. Ann Rev Physiol 1982;44:181–201.

Brenner BM: Control of glomerular function by intrinsic contractile elements Fed Proc 1983;42:3045–3079.

Cogan MG: Atrial natriuretic peptide. Kidney Int 1990;37:1148–1160.

Cogan MG: Angiotensin II: A powerful controller of sodium transport in the early proximal tubule. Hypertension 1990;15:451–458.

DiBona GF: The functions of the renal nerves. Rev Physiol Biochem Pharmacol 1982;94:75–181.

Dworkin LD, Ichikawa I, Brenner BM: Hormonal modulation of glomerular function. Am J Physiol 1983;244:F95–F104.

Edelman IS, Leibman J: Anatomy of body water and electrolytes. Am J Med 1959;27:256–277.

Narins RG, Chusid P: Diuretic use in critical care. Am J Cardiol 1986;57:26A–32A.

Navar LG: Renal autoregulation: Perspectives from whole kidney and single nephron studies. Am J Physiol 1978;234:F357–F370.

Rector FC Jr: Sodium, bicarbonate and chloride absorption by the proximal tubule. Am J Physiol 1983;244:F461–F471.

Schrier RW: Body fluid volume regulation in health and disease: A unifying hypothesis. Ann Intern Med 1990;113:155–159.

Schrier RW: Pathogenesis of sodium and water retention in high-output and low-output cardiac failure, nephrotic syndrome, cirrhosis, and pregnancy. N Engl J Med 1988;319:1065–1072 and 1127–1134.

Skorecki KL, Brenner BM: Body fluid homeostasis in man. Am J Med 1981;70:77–88.

Strauss M et al: Surfeit and deficit of sodium: A kinetic concept of sodium excretion. Arch Intern Med 1958;102:527–536.

Taylor AE: Capillary fluid filtration, Starling forces, and lymph flow. Circ Res 1981;49:557–575.

Section II: Water Homeostasis & Disorders: Hyponatremia & Hypernatremia

Normal Water Homeostasis 4

INTRODUCTION

DISTRIBUTION OF WATER

Total Body Water

The most widely accepted estimate of total body water is **60% of body weight** in young men. Total body water is somewhat less, about 50% in young women. Water content declines in both sexes with age, reaching 50% and 45% of body weight in men and women over 60 years old, respectively. Children in the first year of life have a higher body water content (65–75%).

Intracellular & Extracellular Volumes

As previously reviewed in Chapter 1 (Fig 1-1) and shown again in Fig 4–1, total body water is divided into two large components—the intracellular and extracellular volumes—with a distribution ratio of about 55:45. Thus, in an "average" 70-kg young man, total body water is 42 L (60% of body weight), comprising an intracellular volume of 23 L and an extracellular volume of 19 L. The extracellular volume is further divided into plasma (4–5% of body weight) and interstitial fluid compartments, as well as less readily accessible transcellular fluid compartments.

As illustrated in Fig 4–1, **the primary intracellular cation is potassium.** There is relatively little potassium in the extracellular volume (about 2% of total body potassium). Conversely, the **predominant extracellular cation is sodium,** and there is relatively little sodium in the intracellular volume (about 3% of total body sodium). This distributional polarity of solute concentrations is maintained by the powerful activity of the Na^+/K^+-ATPase resident on all cell membranes. The major negatively charged solutes within cells—organic anions and phosphates—also differ from those in the extracellular volume, mainly chloride and bicarbonate.

PLASMA SODIUM CONCENTRATION

Ratio of Total Body Sodium & Potassium to Total Body Water

Water flows without obvious restriction across cell membranes, driven by osmotic and hydrostatic pressure gradients so that the concentrations of the intracellular and extracellular solutes are maintained at the same level. For any given total amount of body water, the proportion in the intracellular and extracellular volumes is governed by the relative amounts of potassium and sodium, respectively.

In the simple model of Fig 4–1, the extracellular sodium concentration ($[Na^+]_e$) at any time equals the intracellular potassium concentration ($[K^+]_i$).

$$[Na^+]_e = [K^+]_i$$

The actual concentration of $[Na^+]_e$ and $[K^+]_i$ are then defined by the ratio of total body sodium and potassium to water content. Since the plasma is part of the extracellular compartment and has the same composition, plasma sodium concentration is also equal to the amount of total body sodium plus potassium in relation to total body water:

$$\text{Plasma }[Na^+] = [Na^+]_e = [K^+]_i = \frac{\text{Total body sodium} + \text{Total body potassium}}{\text{Total body water}}$$

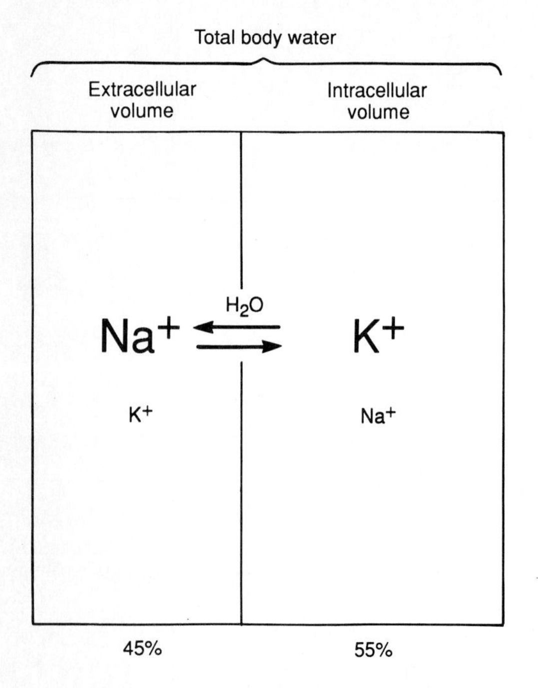

Figure 4–1. Distribution of body water.

Physiologic Control of Plasma Sodium Concentration

Under normal circumstances, renal mechanisms maintain plasma sodium concentration within very fine limits of **140 ± 5 meq/L.**

A decline in plasma sodium concentration could arise in three ways: (1) a decrease in total body sodium; (2) a decrease in total body potassium; or (3) an increase in total body water. Most commonly, plasma sodium is altered by change in total body sodium or water (although similar changes could be induced by change in total body potassium).

As shown in Fig 4–2A, consider the case in which there is pure sodium loss from the body, initially without change in total body potassium or water. Sodium loss first leads to a reduction in $[Na^+]_e$, resulting in imbalance of $[Na^+]_e$ and $[K^+]_i$. Water then rushes from the extracellular to the intracellular space to restore osmotic equality of $[Na^+]_e$ and $[K^+]_i$, though at a level less than their previous value. In addition, because the total body content of potassium remains unchanged but total body sodium has been reduced, the proportionate shift of water causes relative expansion of the intracellular volume and contraction of the extracellular volume. At this stage, total body water—the sum of the two compartments—is still normal.

As discussed in detail subsequently in this chapter, the kidney's physiologic compensation in response to the decline in $[Na^+]_e$ and $[K^+]_i$ is to excrete water. The reduction in total body water causes a proportionate rise in both $[Na^+]_e$ and $[K^+]_i$, back to a normal level. Total extracellular volume is then further reduced, while the intracellular volume shrinks back to its preexisting normal value.

Alternatively, as shown in Fig 4–2B, consider the consequences of water gain to the body, which expands the intracellular and extracellular volumes proportionately, diluting the contents of both compartments. A reduction in $[Na^+]_e$ and $[K^+]_i$ thereby occurs. As in the previous example, the body's subsequent physiologic response to a low $[Na^+]_e$ is to excrete the excess water, which normalizes the intracellular and extracellular electrolyte concentrations and volumes.

In both cases, changes in $[Na^+]_e$ and plasma sodium concentration were sensitively detected and total body water quickly altered by renal mechanisms to restore plasma sodium concentration to normal. There are physiologic conditions—notably a disturbance in the effective circulating volume—in which this fine control of plasma sodium concentration can be disrupted and overridden.

PLASMA OSMOLALITY

Relation of Plasma Osmolality to Plasma Sodium Concentration

Up to this point in our discussion, the simplification has been employed that the principal cation in the extracellular volume was considered to be solely sodium. This simplified model has greater than 95% accuracy compared with the actual situation: extracellular and plasma osmolality can be reasonably considered to represent sodium salts. Thus, plasma osmolality is estimated well by 2 × plasma sodium concentration.

However, to be even more exact, the contributions of two other solutes, glucose and urea, should be included to more closely approximate the plasma osmolality:

$$\text{Plasma osmolality} = 2[Na^+] + [\text{Glucose}] + [\text{Urea}]$$

The molecular weight of glucose is 180, and that of the two nitrogens in urea is 28. Plasma contents of both are usually expressed as mg/dL (instead of mg/L), so molecular weights must be divided by 10. Thus, [glucose] can be estimated by the plasma glucose content (in mg/dL)/18 and [urea] can be estimated by the BUN (in mg/dL)/2.8:

$$\text{Plasma osmolality (mosm/kg)} = 2[Na^+] + \frac{\text{Glucose}}{18} + \frac{\text{BUN}}{2.8}$$

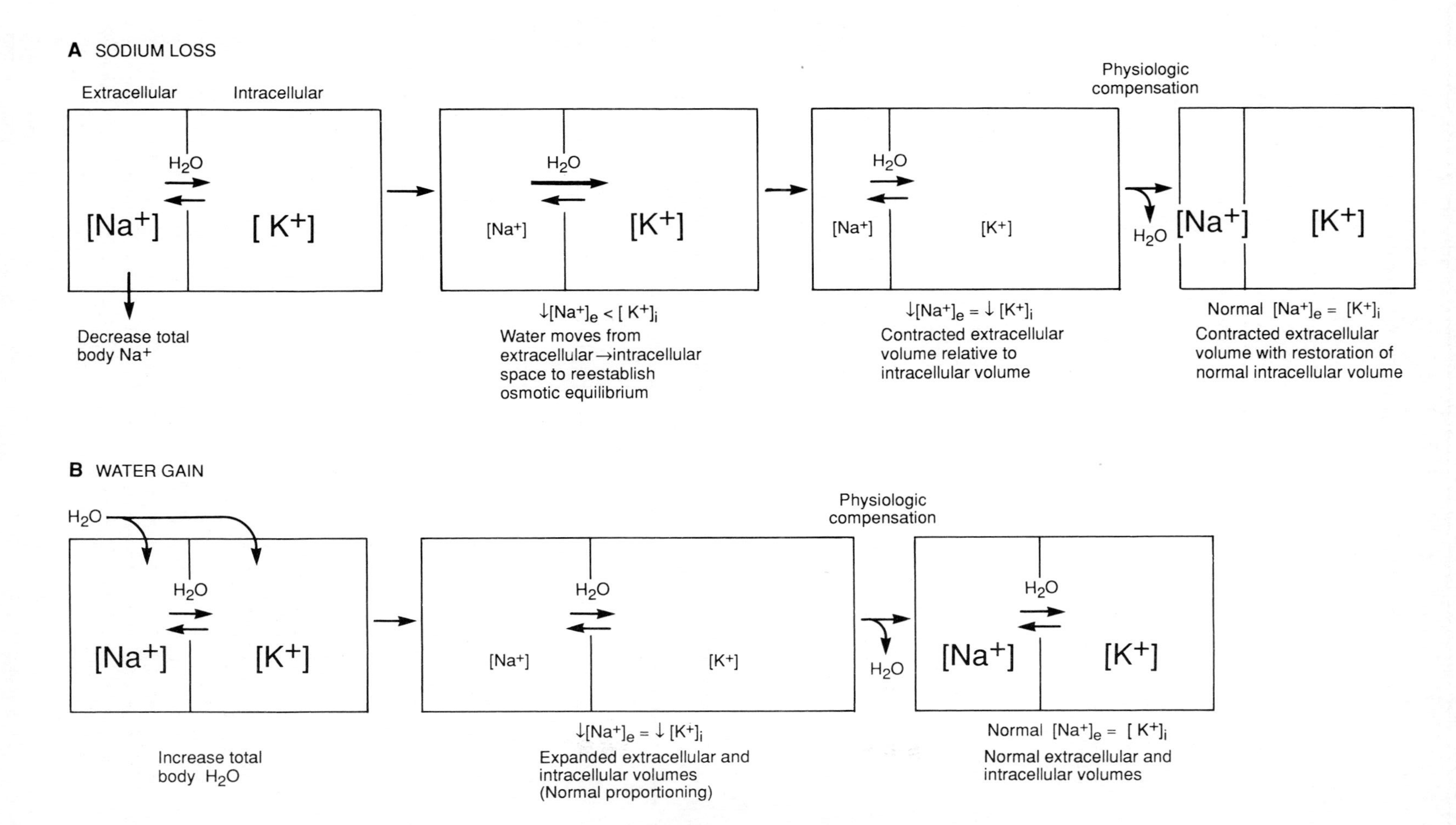

Figure 4-2. Changes in water distribution in response to sodium loss or water gain.

This estimation of plasma osmolality is usually accurate to within 1–3% (ie, within 9 mosm/kg) of what is obtained directly by osmometry. To verify this calculation, plasma osmolality can be assessed directly and is usually measured by its colligative properties: freezing point depression or vapor pressure.

Physiologic Control of Plasma Osmolality

Although control of the plasma sodium concentration is generally considered to be equivalent to control of plasma osmolality, it is the latter which is actually regulated. As mentioned above, sodium and its associated anions usually account for 97–98% of the plasma osmolality. Therefore, in practice, a change in plasma osmolality is due to a proportionate change in sodium concentration, but occasionally it can be due to a marked alteration in glucose, BUN, or another exogenously administered osmole. Very sensitive mechanisms maintain plasma osmolality within 1–2% of normal, **287 ± 7 mosm/kg.**

WATER INTAKE, ABSORPTION, DISTRIBUTION, & ROUTES OF EXCRETION

WATER INTAKE

Water intake is, to a large extent, culturally determined. About 85% of water ingested derives from food itself or the metabolism of fat. As shown in Fig 4–3, the average amount of water ingested by an urban adult in a temperate climate is about 1.5 L/d.

WATER ABSORPTION & DISTRIBUTION

Water is rapidly and almost completely absorbed by the intestine. Once absorbed, water can easily cross cell membranes, which are freely permeable to water, and cells therefore usually behave as near-perfect osmometers. Water distributes according the amount of sodium and potassium in the extracellular and intracellular compartments, diluting the constituents proportionately (Fig 4–2B). The time course of distribution is on the order of minutes—much less than the 2–3 hours needed for the kidneys to dispose of an acute water load, as shown in Fig 4–4.

ROUTES OF WATER EXCRETION

As shown in Fig 4–3, about one-third of the usual amount of ingested water compensates for fixed insensible water losses averaging about 0.5 L/d that occur in stool (0.1 L/d), in air exhaled after humidification by the lungs (0.3 L/d), and in sweat (0.1 L/d). The remaining water is excreted in the urine, the volume of which is variable (average, 1 L/d). Water balance is usually sensitively adjusted by variations in rates of water ingestion—governed by thirst mechanisms—and of renal water excretion.

ANTIDIURETIC HORMONE (ADH) & THE AFFERENT LIMB OF WATER HOMEOSTASIS

OVERVIEW OF ADH CONTROL

As shown in Fig 4–5, the key player in mediating balance between water ingestion and excretion is antidiuretic hormone (ADH), also known as arginine vasopressin (AVP). A very small change in plasma sodium concentration consequent to imbalance of total body sodium and water is sensed by specialized osmoregulatory cells in the hypothalamus, which sensitively control ADH release from the posterior pituitary. A large change in effective circulating volume, sensed by the high- or low-pressure baroreceptors described in Chapter 1 (Fig 1–6), can also affect ADH release. In general, thirst is also controlled, with less sensitivity, by the same stimuli that regulate ADH. ADH is the principal hormonal regulator of urine volume and concentration and controller of total body water homeostasis under normal conditions.

REGULATION OF ADH SECRETION

Structure, Synthesis, Neuronal Transport, & Release

ADH is a nonapeptide produced by magnocellular neurons in the **supraoptic and paraventricular nuclei of the hypothalamus,** which have axonal extensions to the posterior pituitary. Oxytocin is synthesized in neighboring cells. The precursor of ADH, a prohormone of molecular weight 20,000, contains three peptides in series: ADH, neurophysin, and a glycopeptide. The prohormone is packaged into 120- to 180-nm granules. As the granules travel from the hypothalamus to the **posterior**

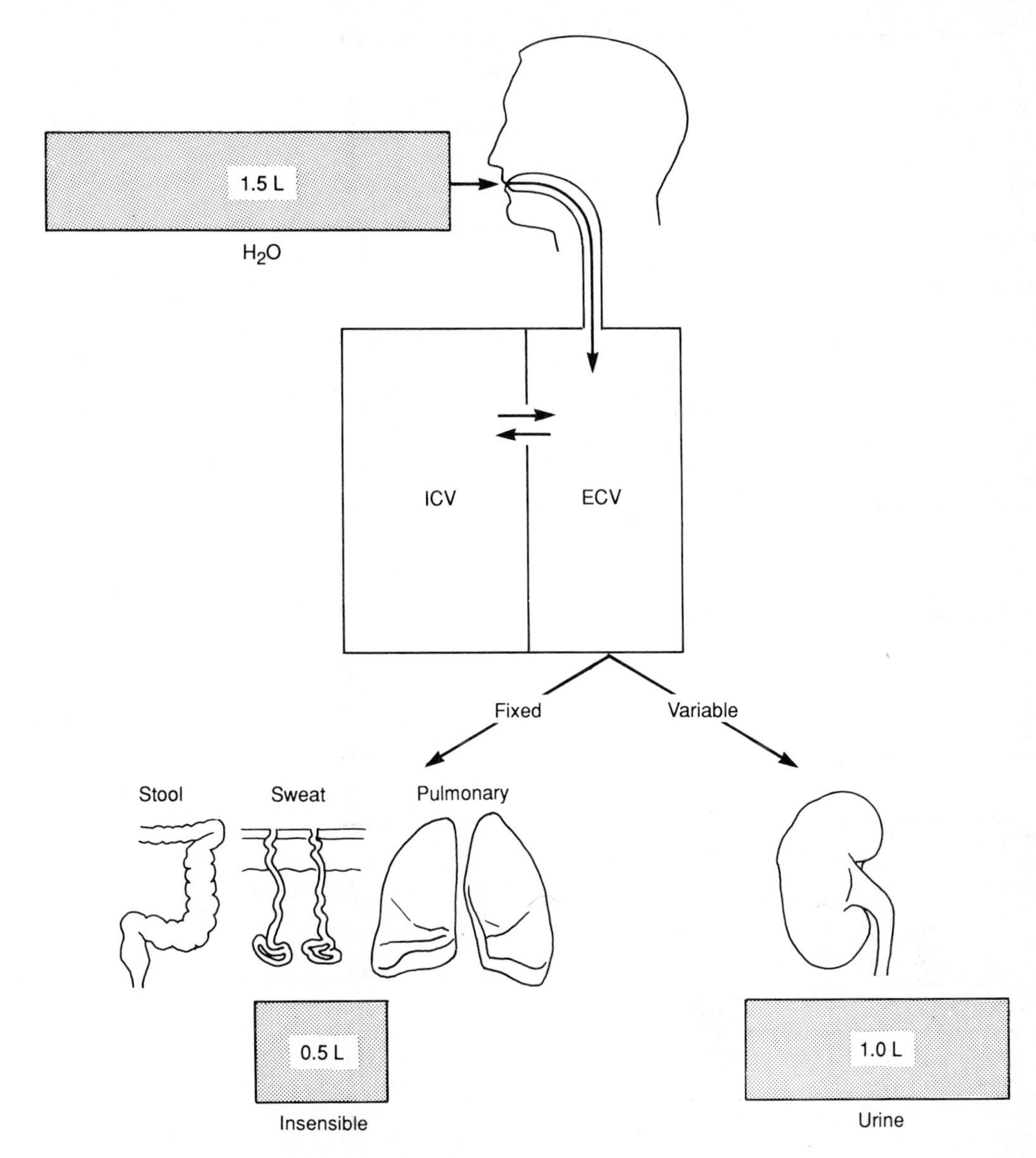

Figure 4–3. Normal water intake and routes of excretion.

pituitary, the precursor is progressively cleaved to yield ADH and its protein partner, neurophysin.

Vasopressin and its associated neurophysin are released from the posterior pituitary by calcium-dependent exocytosis when the axon is depolarized by osmoreceptor or baroreceptor stimuli. ADH has a short plasma half-life—about 10–20 minutes—due to rapid hepatic and renal clearance and degradation by a specific vasopressinase in plasma.

Osmoreceptor Control of ADH Secretion

Maintenance of water balance begins with sensing of plasma osmolality, which in practical terms is the plasma sodium concentration. This sensing mechanism is performed by specialized neurons in the anterolateral hypothalamus, near—but not in—the supraoptic nucleus. Some evidence suggests that these neurons respond to osmolality changes primarily outside the blood-brain barrier, but the issue is not settled. These specialized neural osmoreceptors control the neurons responsible for ADH production and secretion.

As shown in Fig 4–6 , below a threshold plasma osmolality (~280 mosm/kg), osmoreceptors are quiescent and ADH secretion is minimal. When plasma osmolality rises above this threshold value, osmoreceptor cells are progressively stimulated to cause ADH release. When plasma osmo-

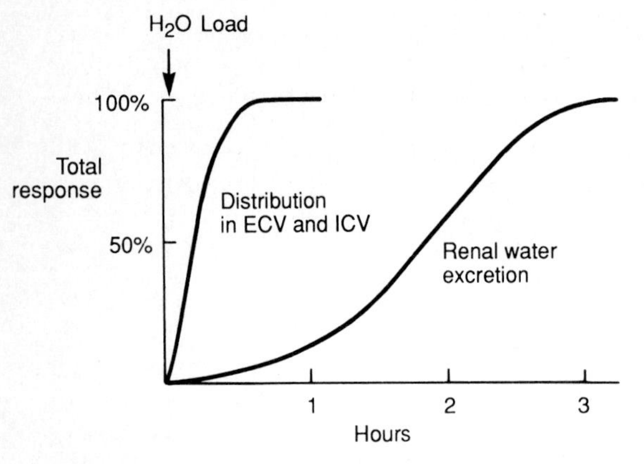

Figure 4-4. Time course of distribution and renal excretion following a water load.

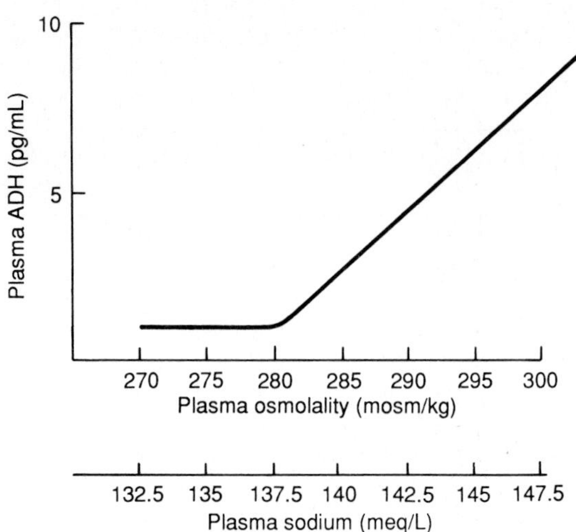

Figure 4-6. Plasma ADH as a function of plasma osmolality and sodium concentration.

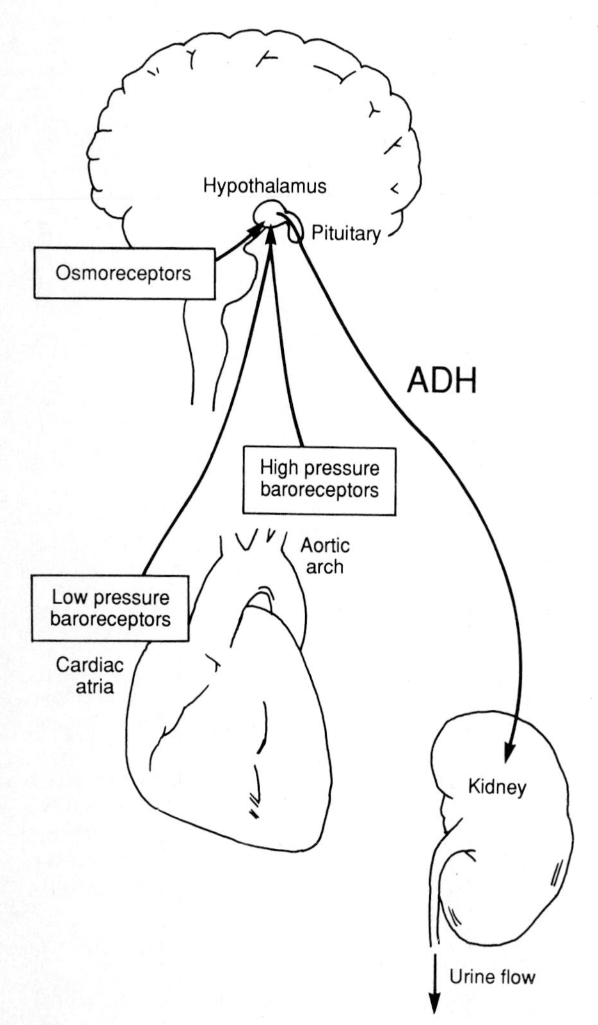

Figure 4-5. Regulation and action of antidiuretic hormone (ADH).

lality reaches about 290–292 mosm/kg, an ADH level (5 pg/mL) is obtained that causes the maximal renal antidiuretic response. Thus, a change of about 10–12 mosm/kg—only a 3–4% change in plasma osmolality—causes ADH to span its entire physiologically active concentration range.

Although the relationship between ADH and osmolality is shown as two linear functions in Fig 4-6, available data do not exclude the possibility of a simple exponential relationship. The point at which the two lines meet—the threshold for ADH release—has been shown to vary from 275 to 290 mosm/kg in different individuals. Similarly, the slope of the ADH response as a function of plasma osmolality in the suprathreshold range can vary by tenfold in different individuals under varying physiologic conditions.

The solute specificity of osmoregulation is important. Sodium as its chloride salt is the usual solute influencing osmoregulatory stimulation of ADH. The corresponding plasma sodium concentration responsible for the exquisite control of ADH secretion is 137–145 meq/L (Fig 4-6). Some plasma solutes such as mannitol or glucose (without insulin) have a similar osmoregulatory action.

However, a few solutes, such as urea and sometimes glucose, when added to blood to increase plasma osmolality, do not stimulate ADH release. The explanation for this finding is that stimulation of ADH occurs when osmoregulatory cells shrink in response to increased extracellular osmolality so long as the external solute cannot enter the cell, as is the case with sodium, mannitol, or glucose (without insulin). However, if an extracellular solute such as urea or glucose (when insulin is present) is permeant and can enter the cell rapidly, no change in cell volume is induced, so that ADH secretion is unaltered. These observations have led to the concept of **effective osmotic pressure,** which discounts any change in plasma osmolality due to substances that easily permeate cells.

Osmoreceptors also regulate **thirst.** The neurons that regulate thirst are similar to—but not identical with—those that control ADH release. A higher plasma osmolality (295 mosm/kg) and sodium concentration (145 meq/L) than are needed to induce ADH release are required to trigger these osmoreceptor neurons to cause a conscious desire for drinking—the so-called dipsogenic threshold—as shown in Fig 4-7. Thus, thirst is not stimulated until plasma osmolality is 2-3% above normal, but at that point even a small increment in osmolality strongly stimulates thirst. A maximum thirst response is observed when plasma osmolality approaches 300 mosm/kg and plasma sodium concentration 148 meq/L. Maximal desire for water can lead to water consumption rates of 20-25 L/d.

Nonosmotic, Baroreceptor Control of ADH Secretion

ADH release is also under the control of the **effective circulating volume.** The same baroreceptor systems mentioned in Chapter 1 (Fig 1-6) responsible for controlling the sympathetic nervous system also control ADH secretion. These neural pathways originate from both the low-pressure sensors, predominantly in the cardiac atria, and the high-pressure sensors, predominantly in the aortic arch and carotid sinus (Fig 4-5). Afferent stimuli are carried by the vagal and glossopharyngeal nerves, with primary synapses in the nuclei of the tractus solitarius. Secondary projections then relay signals regarding circulatory fullness to the neurons in the hypothalamic nuclei that control ADH synthesis and secretion.

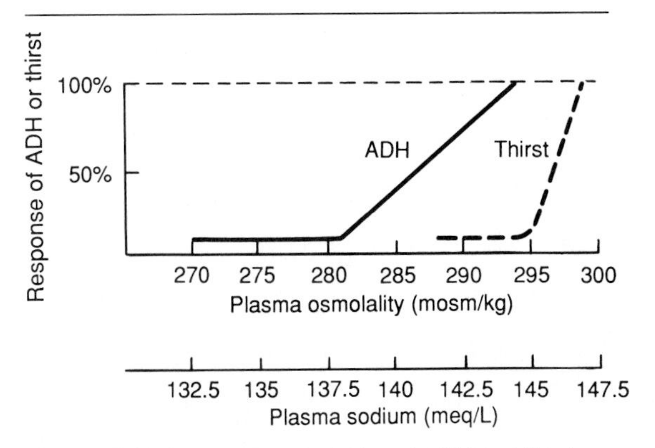

Figure 4-7. Comparative activities of ADH and thirst as a function of plasma osmolality.

From a qualitative point of view, ADH release is less sensitive to small changes in effective circulating volume. As noted in Fig 4-6, a change in plasma osmolality of less than 1% is sufficient to demonstrably affect the circulating ADH level. In contrast, plasma ADH level is unaltered by reduction in effective circulating volume by less than 10%. However, with further hypovolemia, there is a potent, exponential stimulation of ADH secretion, as shown in Fig 4-8.

When present, baroreceptor control of ADH secretion can override osmoreceptor control. With severe hypovolemia, ADH secretion is stimulated even if there is hyponatremia and hypo-osmolality, which would otherwise suppress ADH release. As shown in Fig 4-9, effective circulating volume contraction causes ADH to be released at a subnormal osmolality threshold. In the range above the new threshold, the slope of the ADH secretion as a function of plasma osmolality is relatively normal, though it may be increased somewhat. The converse changes occur with volume overload. Thus, the effective circulating volume can affect the set-point of ADH release and hence the level at which plasma sodium concentration and osmolality are regulated.

The baroreceptors also control the threshold for thirst. As with osmoreceptor control (Fig 4-7), the stimulus needed for inducing thirst appears to be greater than that needed to release

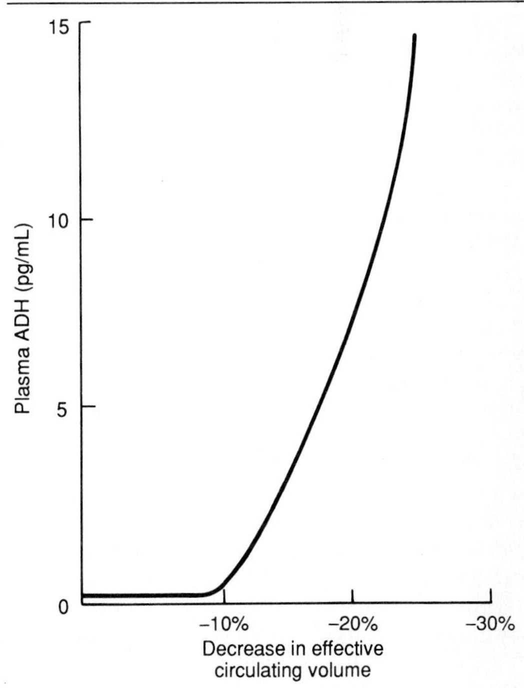

Figure 4-8. Control of ADH by circulating volume.

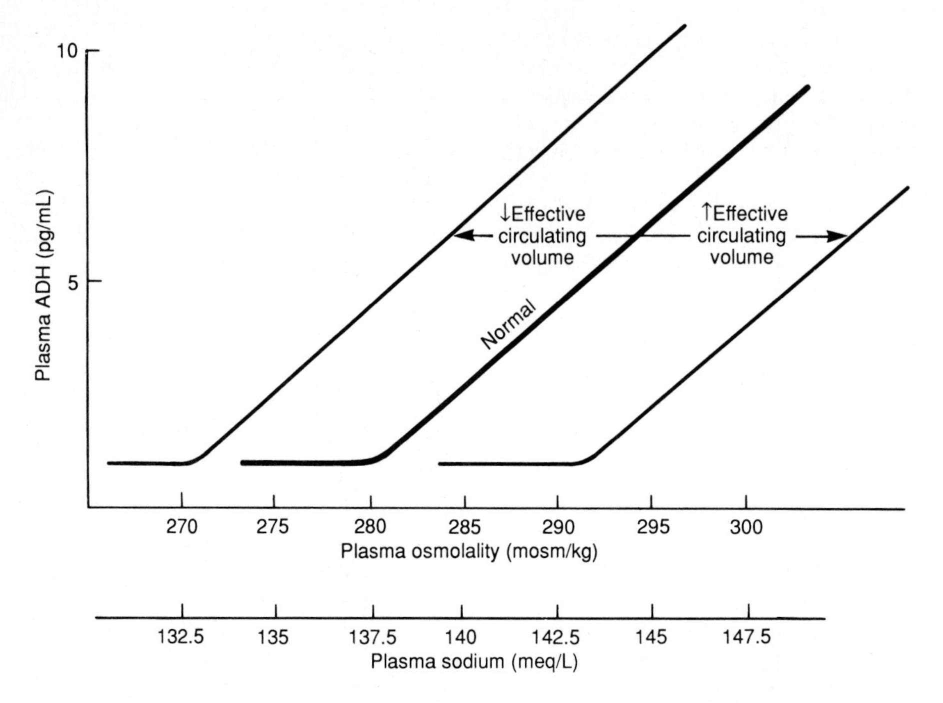

Figure 4–9. Effect of circulating volume on ADH osmoregulation.

ADH. Thus, a greater degree of hypovolemia is required to stimulate thirst than for ADH secretion.

Angiotensin II may at least partially mediate dipsogenic control induced by extracellular volume depletion. When plasma angiotensin II exceeds twice the normal concentration, it activates thirst-controlling paraventricular nuclei (subfornical organ and organum vasculosum of the lamina terminalis, part of the A_3V_3 region), which lack a blood-brain barrier.

Other Controllers of ADH Secretion

As shown in Table 4–1, many other factors besides osmolality and volume status have been demonstrated to affect ADH release, usually transiently. A very potent stimulus for ADH secretion is nausea. ADH levels as high as 500 pg/mL—100 times the level needed for maximal antidiuresis—can be acutely achieved by an emetic stimulus. The origin of the neural pathways affecting ADH secretion is probably in the brain stem, the chemoreceptor zone in the area postrema of the medulla. Many drugs and conditions also activate this dopaminergic pathway to enhance ADH release (eg, morphine, vincristine, cyclophosphamide, apomorphine, nicotine), while blockade of dopaminergic pathways (eg, with haloperidol, fluphenazine, or promethazine) suppresses ADH release.

An interesting reflex occurs when water is ingested, in which the physical act of swallowing suppresses ADH. This oropharyngeal-neuroendocrine reflex occurs before the ingested water has a chance to dilute the body solutes and occurs even when drinking a hypertonic solution. The reflex is transient, lasting no more than about 30 minutes.

Another stimulus to ADH secretion is hypoglycemia, though a greater than 50% fall in blood glucose is required. The physiologic roles of angiotensin II, pain and stress, and hypoxia and hypercapnia to cause sustained and physiologically meaningful increases in ADH secretion are debated. Finally, many drugs act through central mechanisms, generally poorly understood, to stimulate or inhibit ADH secretion. Drugs that block opioid pathways (low-dose morphine or alcohol) tend to inhibit ADH secretion. ADH itself might inhibit its own release by a short-loop negative feedback system.

Table 4-1. Control of ADH secretion.

↑ADH	↓ADH
↑Plasma osmolality	↓Plasma osmolality
↓Effective circulating volume	↑Effective circulating volume
Neurotransmitters, neuropeptides	Neurotransmitters
Angiotensin II	Norepinephrine
Acetylcholine	?ADH
Epinephrine, β_2 agonists	
Histamine	
Bradykinin	
Prostaglandins	
β-Endorphin	
Drugs	Drugs
Morphine	Morphine (low dose)
Vincristine	Haloperidol
Cyclophosphamide	Fluphenazine
Apomorphine	Promethazine
Nicotine	Oxilorphan
	Butorphanol
	Alcohol
	Carbamazepine
	Glucocorticoids
	Clonidine
Other	Other
Nausea	Swallowing
Hypoglycemia	
Hypoxia or hypercapnia	

RENAL WATER EXCRETION & THE EFFERENT LIMB OF WATER HOMEOSTASIS

OVERVIEW OF RENAL WATER EXCRETION

Usual Independence of Urinary Solute & Water Excretion Rates

Under normal conditions, renal water excretion is regulated to compensate for water ingestion minus insensible nonrenal water losses (Fig 4-3). The appropriate control of water excretion requires the coordinate action of the renal countercurrent multiplication system and ADH. Excretion rates of electrolytes such as sodium, potassium, hydrogen ion, calcium, phosphate, and magnesium are regulated by different factors (see Chapters 1, 7, 10, 15, 18, and 21), often responsive to different physiologic needs than those which control water excretion. Thus, within limits discussed in detail subsequently, the urine may be concentrated or dilute irrespective of and independent of electrolyte excretory rates.

Primary Roles of the Countercurrent System & ADH

Renal water excretion is primarily regulated by altering the reabsorption of water in the collecting tubule. As shown in Fig 4-10, water is freely filtered at the glomerulus (GRF), and about 45% is reabsorbed in a relatively fixed manner in the proximal nephron. Another 40% of the filtered water is reabsorbed in a relatively unregulated fashion in the loop of Henle, by means of the countercurrent multiplication system, and 5% is reabsorbed in the distal convoluted tubule. The remaining amount of water—about 10% of the quantity filtered—either is unreabsorbed when ADH is absent, causing diuresis, or is reabsorbed by the collecting tubule when ADH is present during antidiuresis. In both cases, the medullary interstitium is hypertonic, maintained by active and passive transport processes in the loop of Henle and collecting tubule—known as the countercurrent system.

GLOMERULAR FILTRATION & PROXIMAL REABSORPTION OF WATER

GFR

Water is filtered unrestricted in the glomerulus. The determinants of GFR were discussed in detail in Chapter 1. In a normal individual, GFR is quite large ($\sim$150 L/d) compared to the usual water excretory demands ($\sim$1 L/d). Thus, only very severe reduction in GFR is rate-limiting with respect to normal water excretion.

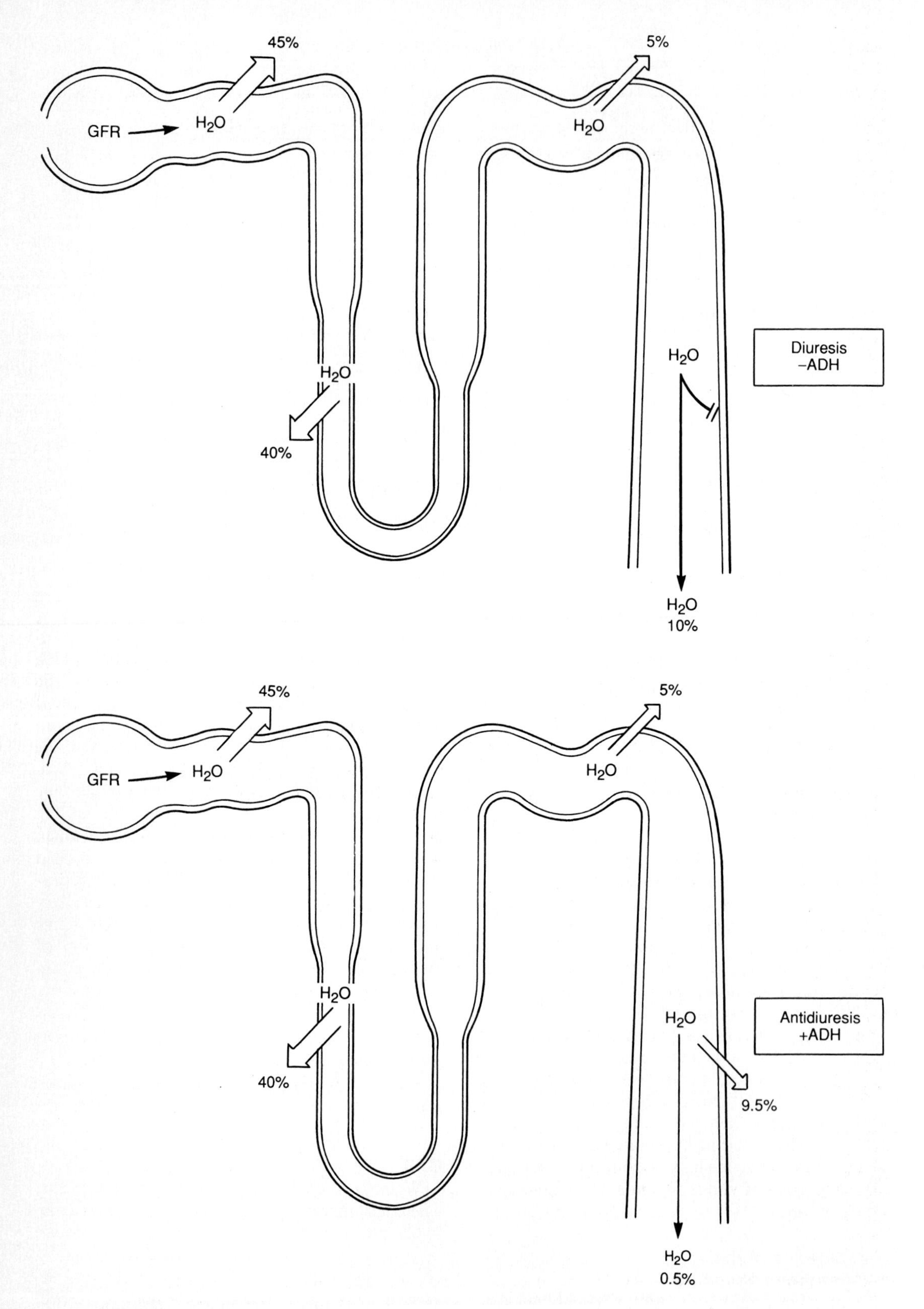

Figure 4–10. Sites and amounts of water reabsorbed and excreted in diuresis and antidiuresis.

Proximal Tubule

Water permeability in the proximal convoluted tubule is quite high. Only a very small transepithelial osmotic gradient is required to drive the large water flux that occurs in this nephron segment. Luminal fluid osmolality gradually falls along the length of the proximal tubule, becoming about 5 mosm/kg hypotonic compared with blood. The reabsorbate osmolality is therefore about 5 mosm/kg hypertonic to blood.

About 40–50% of the filtered water is reabsorbed by the proximal convoluted tubule. If GFR changes, this fraction of filtered water that is proximally reabsorbed stays relatively constant. The proximal tubule is not a target for ADH action and responds insensitively to physiologic demands for water conservation or excretion.

WATER REABSORPTION IN THE LOOP OF HENLE & COLLECTING TUBULE: THE COUNTERCURRENT SYSTEM

Thin Descending Limb

The thin descending limb of Henle is water-permeant but sodium- and urea-impermeant. There is therefore water reabsorption but no sodium or urea transport in this segment.

The medullary interstitial fluid in contact with the thin descending limb becomes progressively hypertonic, changing from close to isotonicity (290 mosm/kg) near the cortex to very concentrated (1200 mosm/kg) in the papilla (see below). As a result, as shown in the left side of Fig 4–11, the slightly hypotonic fluid emerging from the proximal tubule loses water continuously as it flows in the thin descending limb during its descent through the medulla.

Thin Ascending Limb

The thin ascending limb of Henle has very different permeability properties, being water-impermeable but sodium- and urea-permeable. As shown in the middle portion of Fig 4–11, sodium, which has been concentrated by the removal of water in the descending limb, is reabsorbed by diffusion down its concentration gradient. On the other hand, the urea concentration is higher in the medulla than in the lumen, so urea enters the tubular fluid by passive diffusion. Thus, the total volume of fluid in the thin ascending limb remains constant, but it becomes progressively lower in sodium concentration and higher in urea concentration.

Thick Ascending Limb

The motive force–the so-called single effect—initially responsible for establishing the solute concentration gradients in the medulla is active sodium chloride transport in the thick ascending limb of Henle. In this segment, sodium, urea, and water permeabilities are low but an active process reabsorbs sodium (Fig 1–25) via a luminal, furosemide-sensitive Na^+-K^+-$2Cl^-$ transporter. The high sodium concentration in the medulla is generated to a large extent by this process. Owing to the powerful sodium reabsorptive force of the medullary portion of the thick ascending limb of Henle plus further sodium abstraction in the cortical thick ascending limb of Henle, the distal convoluted tubule and cortical collecting tubule, luminal fluid eventually becomes quite dilute, with sodium concentration of less than 25 meq/L and osmolality of about 50 mosm/kg.

Note that this is the only active, metabolic energy-requiring step in the countercurrent multiplication system. Further medullary hypertonicity is established passively, by the differential solute and water permeabilities and gradients in the thin limbs (see above) and collecting tubule (see below).

Medullary Collecting Tubule

As shown in the right side of Fig 4–11, cells of the medullary collecting tubule become permeable to water when ADH is present. The hypertonic medullary interstitium then drives water reabsorption by osmosis from the hypo-osmolar luminal fluid and progressively concentrates its remaining major solute, urea. Since urea permeability in the inner medullary collecting tubule is also increased by ADH, urea is reabsorbed by passive diffusion to reestablish the medullary urea lost into the thin ascending limb of Henle. Thus, urea recycles from the medullary collecting tubule into the medullary interstitium and back into the thin ascending limb of Henle. Urea may also retroflux into the medullary interstitium from the pelvic urine. Thus, with ADH present, a concentrated urine of low volume occurs as a result of high water permeability in the medullary collecting tubule and well-maintained medullary interstitial hypertonicity.

The medullary collecting tubule becomes water- and solute-impermeable in the absence of ADH. The hypo-osmolar fluid emerging from the loop of Henle and distal convoluted tubule is passed unchanged through the collecting tubule and forms the large volume of dilute urine. Under these conditions of ADH absence, the high luminal flow rate and inability of urea to recycle depletes the medulla of urea and thereby reduces medullary hypertonicity.

Medullary Hypertonicity

Both sodium and urea contribute in roughly equal proportions to medullary hypertonicity. The osmolality increases from the corticomedullary

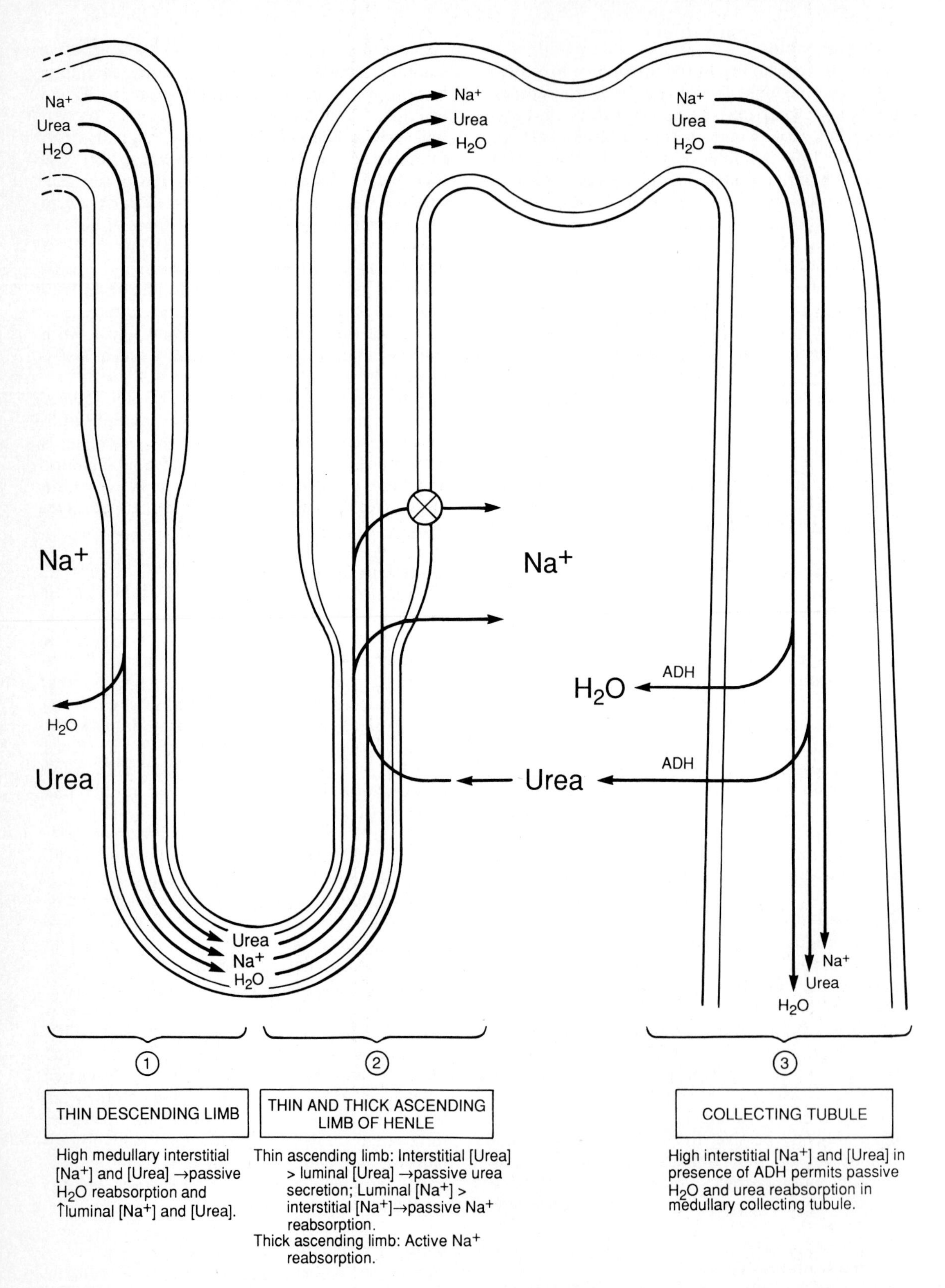

Figure 4-11. Mechanisms of water reabsorption by the renal countercurrent system.

junction to the papilla, where a level as high as 1200 mosm/kg can be achieved. In the presence of ADH, water equilibrates between the medullary collecting tubule and the innermost papilla. Thus, 1200 mosm/kg is also the most hypertonic the urine can be under maximal antidiuretic conditions. Cells in the renal medulla avoid shrinkage and protect themselves from this high extracellular osmolality by generating intracellular osmolytes—principally polyols, amino acids, and methylamine—including sorbitol, myoinositol, betaine, glycerophosphocholine, and others.

Under water diuretic conditions, when ADH is absent, urea is excreted and urea recycling is interrupted. Medullary interstitial hyperosmolality is diminished, often to levels as low as 400–800 mosm/kg. Note, however, that under these conditions—ie, lacking ADH—the urine can be far less concentrated, with an osmolality as low as 50 mosm/kg, reflecting that of the luminal fluid emerging from the cortex. Thus, the medullary collecting tubule must stay water-impermeable to prevent water reabsorption.

Vasa Recta

The descending and ascending vasa recta maintain the steep medullary concentration gradient by exchanging solutes and water along their lengths. This countercurrent arrangement prevents dissipation of solute gradients. The descending vasa recta pick up solutes while losing water as it enters the hypertonic medulla, while the ascending vasa recta does the opposite.

Mechanism of Action of ADH

ADH binds to high-affinity receptors on the medullary collecting tubule, as shown in Fig 4–12. The renal ADH receptor is called a V_2 receptor, which is functionally different from the ADH receptor on vascular smooth muscle cells, called the V_1 receptor. The renal V_2 receptor on the medullary collecting tubule cell is coupled via a GTP-requiring stimulatory G protein (G_s) to the enzyme adenylate cyclase to form cAMP. Before it is metabolized by phosphodiesterase, cAMP activates the cAMP-dependent protein kinase A, which enhances insertion of so-called aggrephores into the

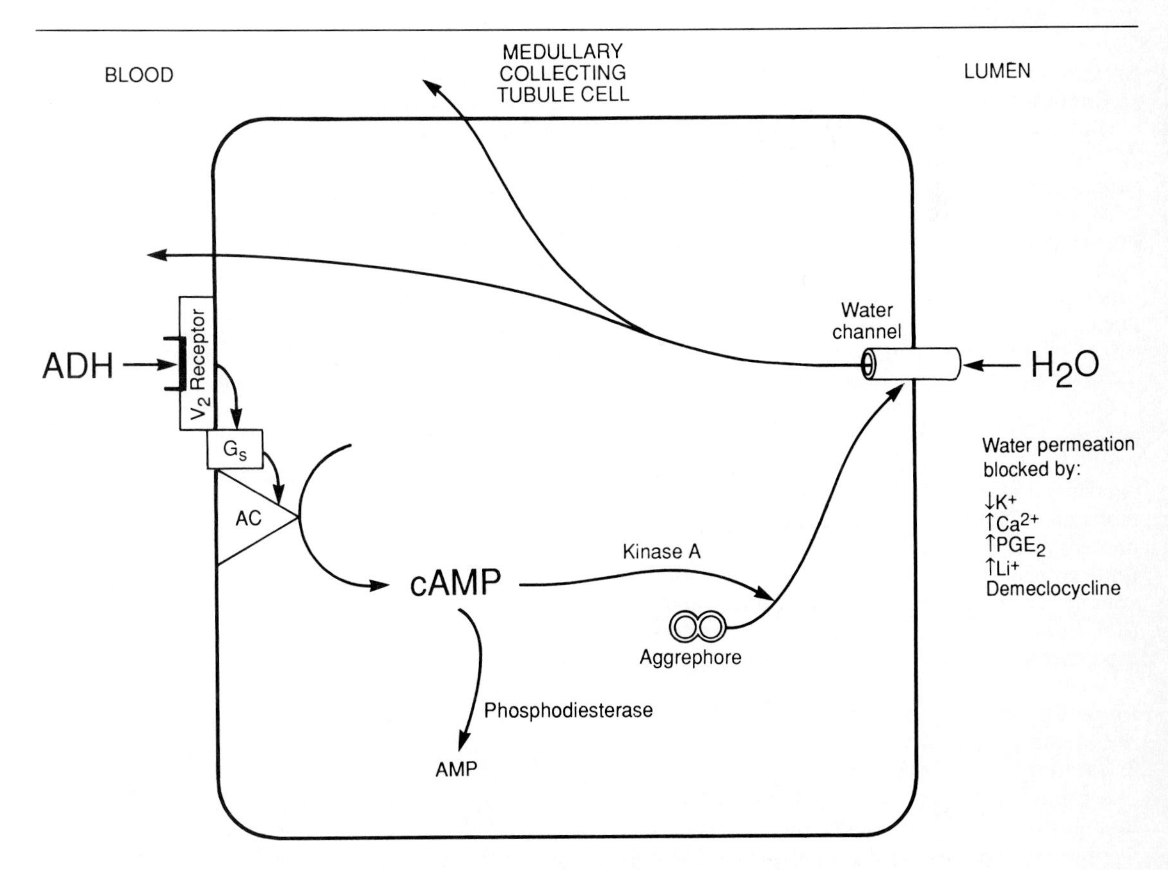

Figure 4–12. Cellular mechanisms of water transport in the medullary collecting tubule.

luminal membrane. The aggrephores are believed to be water channels, which allow water to cross the luminal cell membrane, which is otherwise water-impermeable. Water exits the cell across the always permeable, unregulated basolateral membrane.

Several factors can block this sequence of events in the medullary collecting tubule of ADH-receptor-G_s-adenylate cyclase-cAMP-kinase A-aggrephore insertion-water permeation. Among these are hypokalemia, hypercalcemia, and prostaglandin E_2 and several drugs, such as lithium and demeclocycline, though the intracellular steps affected have not been definitively identified.

Evidence in animals supports an additional role of ADH in regulating sodium reabsorption by the thick ascending limb of Henle, thereby helping to "prime the pump" of the countercurrent system. However, in humans, receptors for ADH probably do not exist on cells of the medullary thick ascending limb of Henle, so the role of ADH is confined to augmenting water permeability in the collecting tubule.

REGULATION OF RENAL WATER REABSORPTION

Relation of Water Homeostasis to Sodium Homeostasis

Under day-to-day conditions, thirst and the ADH-renal system responsible for excretion of water are regulated to maintain the plasma osmolality and sodium concentration within ± 1% of normal values (Fig 4-6). ADH-mediated water excretion by the kidney is usually independent of the total amount of total body sodium. Thus, ADH normally controls the *ratio* of total body sodium to total body water and, within limits, does not "know" the quantity of total body sodium.

Only if total body sodium becomes significantly depressed or elevated (< 10% or > 10% of normal) is the osmoregulatory function of ADH overridden (Fig 4-9). ADH then becomes a hormone participating in effective circulating volume homeostasis. A common theme in electrolyte and fluid homeostasis is that the effective circulating volume is of preeminent importance and, if threatened, is defended by renal mechanisms at the expense of body solute concentration.

Renal Response to Change in Plasma Osmolality & Sodium Concentration

A primary increase in total body sodium (or potassium) or loss in total body water causes a rise in plasma osmolality and sodium concentration. As summarized in Fig 4-13, the hypothalamic osmoreceptor neurons sense this change in plasma tonicity and stimulate the adjacent neural systems responsible for the sensation of thirst and for ADH release. Subsequent ingestion of water and diminished renal water excretion serve to produce positive water balance, which normalizes the plasma osmolality and sodium concentration.

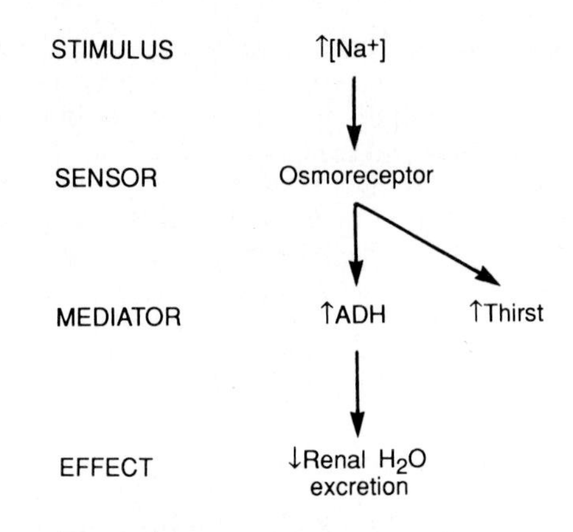

Figure 4-13. Response to hypernatremia.

As shown in Fig 4-14, the opposite changes occur when there is a primary decrease in total body sodium (or potassium) or increase in total body water to decrease the plasma osmolality and sodium concentration. Osmoreceptor-mediated decline in thirst and ADH serve to diminish water ingestion and increase renal water excretion to return plasma osmolality and sodium concentration to normal.

Renal Response to Change in Effective Arterial Volume

With a greater than 10% depression of effective circulating volume, baroreceptor-mediated augmentation of ADH secretion and thirst (Fig 4-9). As shown in Fig 4-15, subsequent increase in wa-

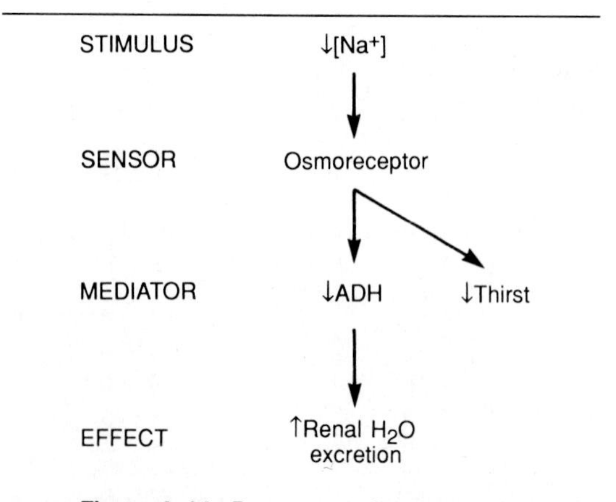

Figure 4-14. Response to hyponatremia.

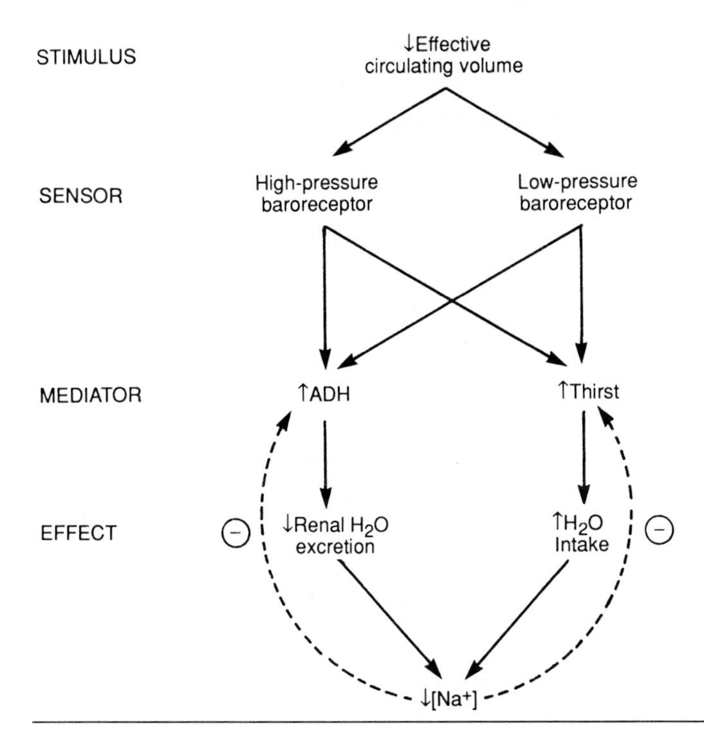

Figure 4–15. Response to hypovolemia.

ter ingestion and decrease in renal water excretion causes net water retention and dilution of plasma sodium. The resulting hypo-osmolality and hyponatremia feedback to cause inhibition of ADH and thirst. When sufficient water retention has occurred, there is a new steady-state set-point for regulating plasma osmolality and sodium concentration (Fig 4–9). The opposite changes occur in response to hypervolemia.

When effective circulating volume is diminished, note that the baroreceptor-neural signaling pathways to increase ADH are virtually the same as those described in Chapter 1 which increase the activity of the adrenergic nervous system and activate the renin-angiotensin-aldosterone system (Figs 1–7 and 1–9). Thus, in graded response to decreased effective circulating volume, there is a coordinated, integrated, and progressive antidiuresis and antinatriuresis. With sufficient sodium chloride and water retention, circulating volume may be restored to nearly normal. Then, a compensated state can develop with return of renal sodium and water excretion rates equivalent to intake. But this compensated state ensues only at the expense of a high level of neurohumoral vasoconstrictor activity and hyponatremia.

LIMITS OF WATER TRANSPORT REGULATION

The osmoreceptor-ADH-medullary collecting tubule system normally exhibits extraordinary sensitivity and efficiency in water conservation or excretion. The normal range of urine osmolality is from **50 to 1200 mosm/kg.** This 24-fold range in urine osmolality translates into renal water excretory rates of 0.4 to over 10 L/d to preserve normality of plasma osmolality and sodium concentration.

Urine Osmolality as a Function of ADH

As shown in Fig 4–16, urine osmolality is exquisitely sensitive to ADH. ADH has a rapid mode of onset, with full hydro-osmotic effect occurring in 20–30 minutes. The full range of urine osmolality achievable, from 50 to 1200 mosm/kg, is a linear function of plasma ADH concentration from 0 to 5 pg/mL. An ADH level greater than 5 pg/mL elicits no further urine concentration. Note that the normal set-points of plasma osmolality (287 mosm/kg) and sodium concentration (140 meq/L) correspond to a plasma ADH of about 3 pg/mL, a concentration roughly midway in the range functionally controlling urine concentration.

Urine Specific Gravity as an Estimate of Osmolality

The urine specific gravity—the urine "density" compared to water—is a first approximation of urine osmolality. The range of urine specific gravity—1.000–1.040—corresponds in a roughly linear fashion with osmolality from 0 to 1200 mosm/kg (ie, a change in specific gravity of 0.010 per

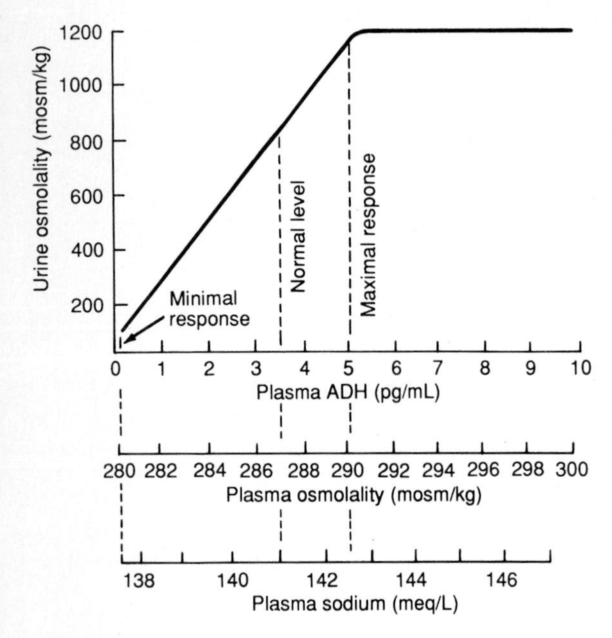

Figure 4-16. Physiologic relationships of urinary concentration to plasma levels of ADH, osmolality, and sodium concentration.

300 mosm/kg change in osmolality). Isosthenuria—urine osmolality equal to that of blood (287 mosm/kg)—is about 1.010. The utility of the specific gravity measurement holds only for low-molecular-weight electrolytes and urea and not for larger molecules such as sugars and radiocontrast dyes.

RENAL WATER CONSERVATION: CONCENTRATION OF THE URINE

Response to Dehydration

As depicted in Fig 4-16, a rise in plasma osmolality and sodium concentration of less than 1% (eg, 2 mosm/kg and 1 meq/L, respectively) causes an increase in ADH level of 1 pg/mL, which is sufficient to increase urine osmolality by about 200 mosm/kg. Maximal urinary concentration of 1200 mosm/kg—achieved when ADH is 5 pg/mL—corresponds to a plasma osmolality of 290 mosm/kg or greater, or a plasma sodium concentration of 142 meq/L or greater.

The time course response of renal water conservation depends on the rate of water loss and preexisting hydration status. A typical antidiuretic response to simple water deprivation is shown in Fig 4-17. When water ingestion is restricted, insensible water loss causes a progressive rise in plasma osmolality. Maximal urinary concentration and minimal urinary volume in response to the hyperosmolality is achieved in about 8 hours.

Factors Affecting Maximal Urinary Concentration

Maximal urinary concentration declines when the medullary countercurrent system is functionally or structurally impaired, as summarized in Fig 4-18 and Table 4-2.

The number of nephrons and overall GFR is important. The countercurrent multiplication system becomes progressively nonfunctional when the nephron number is reduced in the course of chronic renal disease, as shown in Fig 4-19. The abnormality in urinary concentration applies to any renal disease of glomerular but especially of tubulointerstitial origin. With severe, chronic reduction in GFR, maximum achievable urine osmolality is no higher than 300-500 mosm/kg and is sometimes even lower than plasma osmolality. The decrease in concentrating ability associated with nephron loss may be due to the solute load per residual surviving nephron, to single-nephron solute diuresis (see below), to intrinsic tubular abnormalities, or to other unidentified causes. A decrease in urinary concentrating ability also occurs when nephron number is reduced during the normal process of aging.

Sodium reabsorption in the thick ascending limb of Henle is required for the single effect of the countercurrent multiplication system, and this function is inhibited by loop diuretics. As shown in Fig 4-20, maximum urinary concentrating ability is also reduced during osmotic diuresis. At very high solute clearances, the maximal urine concentration approaches that of plasma—ie, is isosthenuric, about 290 mosm/kg.

A preexisting water diuresis can wash out the medullary interstitial urea and diminish concentrating capacity. As basal urine flow increases, subsequent maximal response to ADH is lower than normal as a result of the reduction in medullary tonicity—sometimes as low as 400 mosm/kg, as shown in Fig 4-21. By similar mechanisms, a very low protein intake that lowers urea generation reduces medullary interstitial osmolality. Reduced protein intake for only 3 days is sufficient to reduce maximal urine concentration by 200 mosm/kg. In malnourished patients, maximal urine osmolality may be substantially reduced—to as low as 400-600 mosm/kg.

Finally, as shown in Fig 4-18, the requisite action of ADH can be impaired as a result of deficient hormone release, as in various hypothalamic-pituitary disorders; can be functionally antagonized by hypokalemia, hypercalcemia, increased prostaglandin (PGE_2) production, or drugs such as lithium or demeclocycline; or can be is rendered ineffective by end-organ unresponsive-

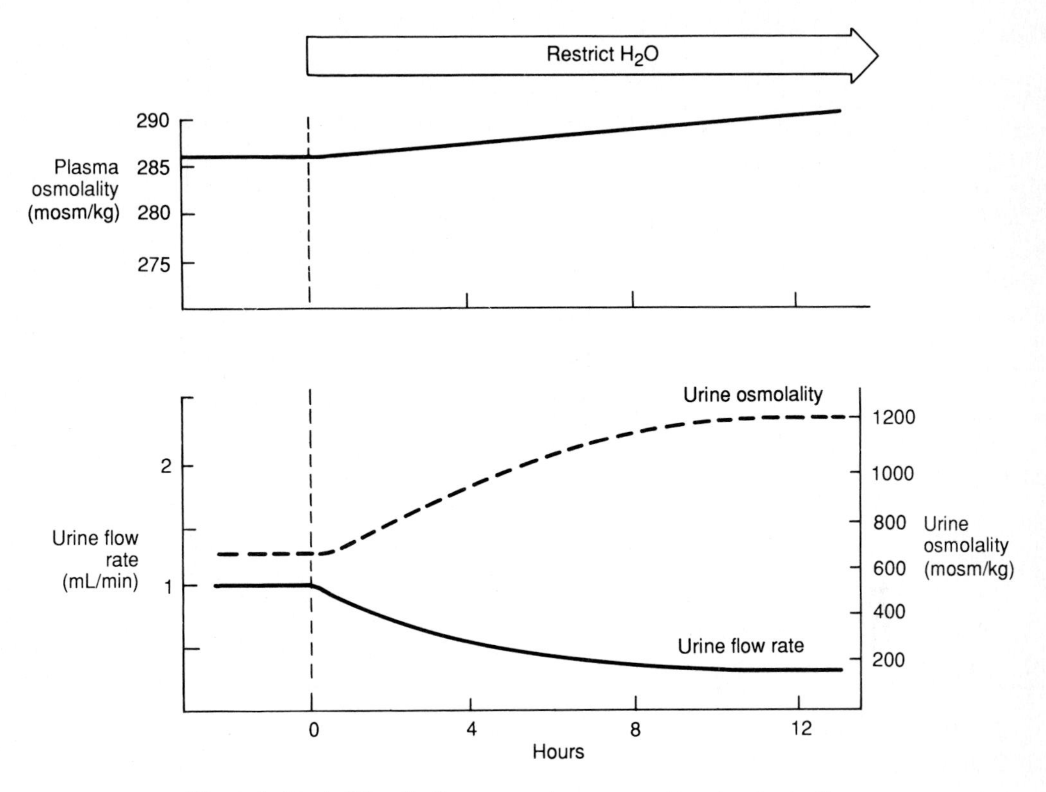

Figure 4-17. Antidiuretic time course in response to water deprivation.

ness when damage to the medullary collecting tubule occurs, as in some tubulointerstitial diseases.

Similar to the escape from the antinatriuresis induced by sustained mineralocorticoid exposure (Chapter 3), there is an escape from the antidiuresis due to prolonged ADH administration. When ADH levels are rendered consistently high, the initial high urine osmolality and low urine volume last only about a week, at which time urine osmolality begins to progressively decline and urine volume to increase to levels approaching those before ADH was given. The cellular mechanism responsible for ADH escape is currently unknown.

Minimal Urine Volume During Antidiuresis

While the preceding discussion has assumed that solute excretion rates are regulated independently of water flow and ADH, they in fact can affect the obligate urine volume under severe antidiuretic conditions. When maximal urinary osmolality is achieved, the corresponding urine volume depends on the solute load simultaneously requiring excretion.

An average adult generates about 800 mosm/d of total solute, with a range of 500–1500 mosm/d. Slightly more than half of these solutes derive

Table 4-2. Requirements for maximal water conservation.

	Altered By:
GFR	Chronic renal disease Age
NaCl reabsorption in thick ascending limb of Henle	Loop diuretic Osmotic diuresis
Medullary hypertonicity	Preexisting water diuresis ↓Protein intake
↑ADH	↓Hypothalamic-pituitary ADH release ↓K^+, ↑Ca^{2+}, ↑PGE_2, ↑Li^+, demeclocycline Medullary collecting tubule abnormality

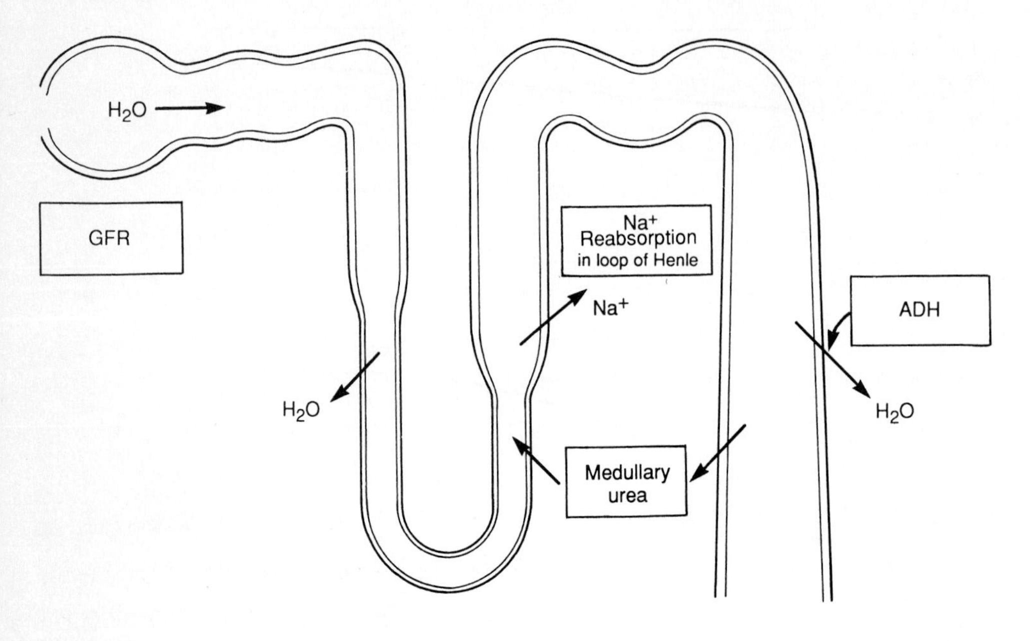

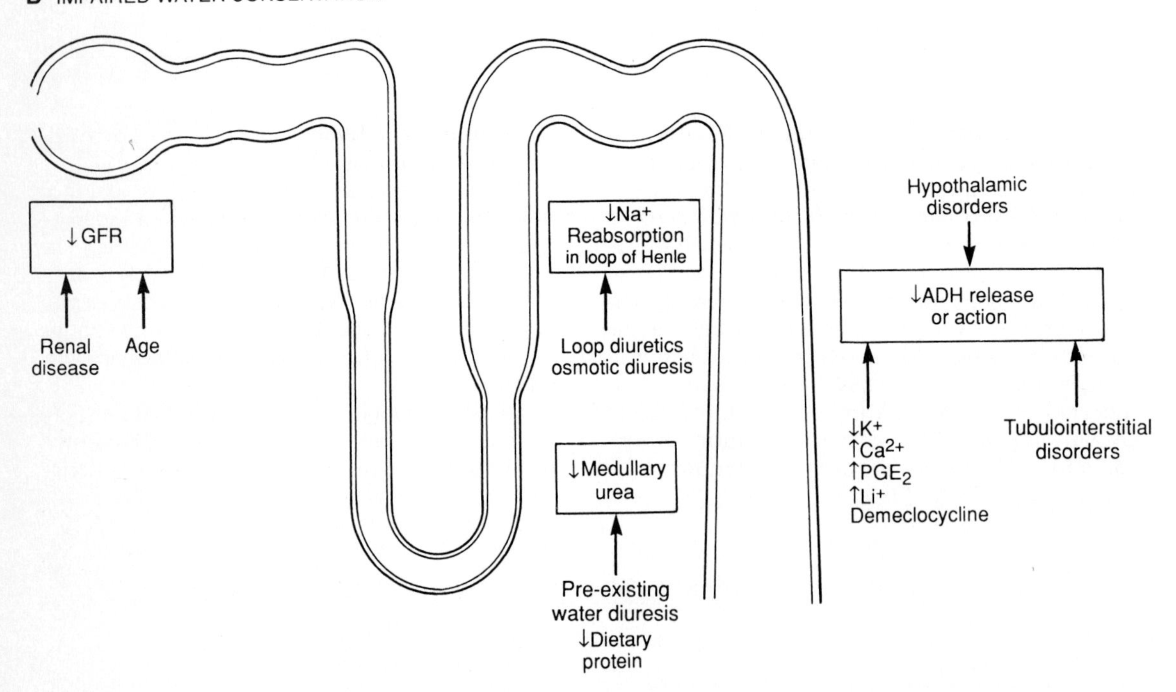

Figure 4–18. Pathophysiology of urinary concentrating disorders.

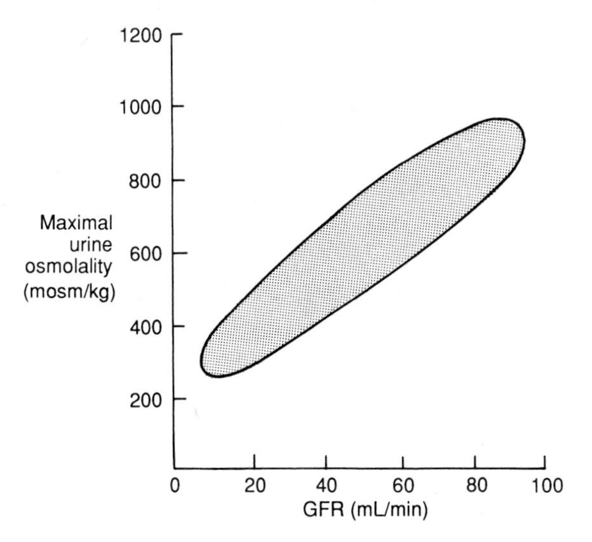

Figure 4-19. Maximal urinary concentrating ability in chronic renal disease

from dietary sodium (100–200 meq/d), potassium (50–100 meq/d), chloride (100–200 meq/d), and other electrolytes (25–50 meq/d) and the rest from urea (250–400 mosm/d) produced by exogenous or endogenous protein catabolism. As shown in Fig 4-22, if obligate solute excretion is 800 mosm/d and the average basal urine osmolality is 600–800 mosm/kg, the urine volume required to excrete this solute load is about 1–1.3 L.

If ADH were to increase and render urine osmolality maximal at 1200 mosm/kg, the minimal urine volume achievable with a solute excretion rate of 800 mosm/kg would be 0.67 L/d (800

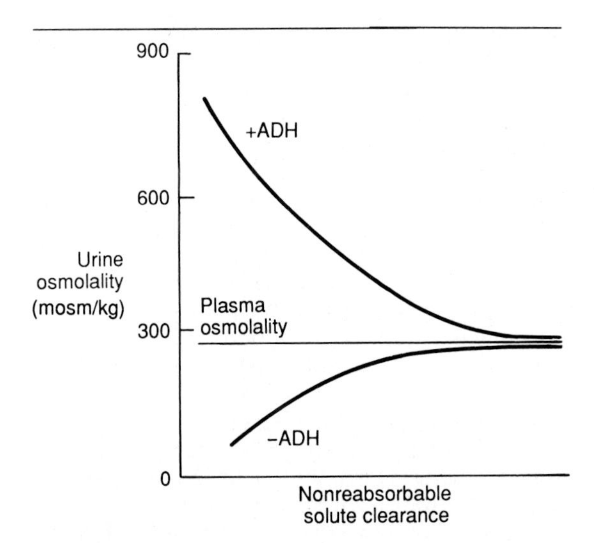

Figure 4-20. Impact of solute diuresis on concentrating and diluting capacity.

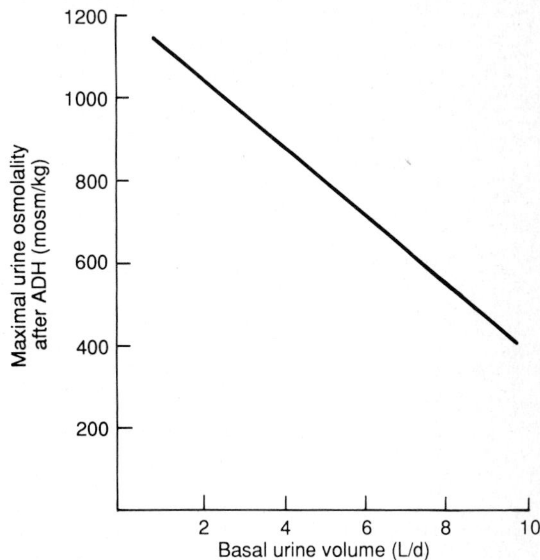

Figure 4-21. Impact of previous water diuresis on maximal concentrating ability.

mosm/d divided by 1200 mosm/kg), as shown in the insert in Fig 4-22. As also illustrated, even if solute generation were halved to 400 mosm/d, the minimal urine volume under these maximal antidiuretic conditions would be 0.33 L/d. In usual practice, the lower limit of renal water excretion is usually considered to be **0.4 L/d,** but it actually varies according to the obligate solute load that requires excretion.

RENAL WATER ELIMINATION: DILUTION OF THE URINE

Response to Overhydration

As shown in Fig 4-16, when plasma osmolality falls to reach a threshold value of 280 mosm/kg, or a plasma sodium concentration reaches 137 meq/L, circulating ADH is virtually undetectable. When ADH is absent, water permeability in the medullary collecting tubule is negligible, so the minimal urine osmolality of 50 mosm/kg is then obtained. With severe sodium chloride and protein restriction, even lower urine osmolality may be achieved (30 mosm/kg).

A typical time course of diuretic response to a 1-L water load is shown in Fig 4-23. About half an hour is required for the water to be absorbed and to dilute the plasma (Fig 4-4). In this case, the plasma osmolality is reduced by about 2.5%. The maximal diuretic response of the osmoreceptor-ADH-medullary collecting tubule axis is then observed after another half-hour. Urine flow rate is sustained at maximal levels for 2 hours. In this

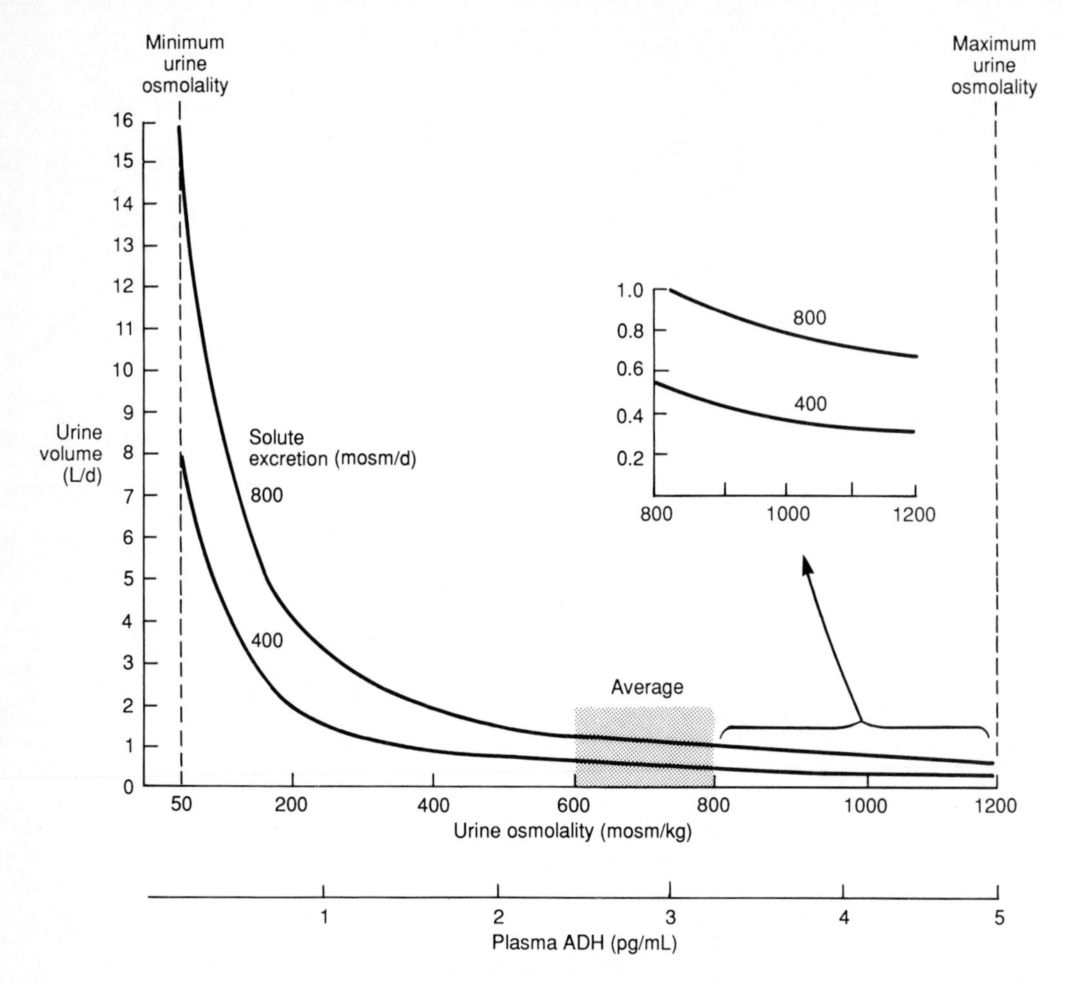

Figure 4–22. Relationships of urinary volume, solute excretion rate, urine osmolality, and ADH.

time, over 90% of the ingested water load is excreted and the plasma osmolality is restored to normal. Note that water excretion by the kidney is much faster than sodium excretion. Whereas most of a water load is excreted within 2 hours, 3–5 days are required to fully excrete a sodium load (Fig 1–35).

Factors Affecting Maximal Water Excretion

As shown in Fig 4–24 and Table 4–3, factors that affect GFR, solute reabsorption required for creation of luminal fluid hypotonicity, and depression of ADH secretion can affect the magnitude of water diuresis.

An acute decline in GFR and tubular fluid flow rate can disrupt diluting capacity. Minimal urine osmolality then approaches and can even slightly exceed blood osmolality. However, when GFR is chronically reduced in the course of renal disease, the ability to dilute the urine is well preserved, although normal minimal urine osmolality (50 mosm/kg) is usually not achieved. As shown in Fig 4–25, as the GFR falls in chronic renal disease or in the process of aging, the maximal water clearance in absolute amount falls (bottom), but the proportion that can be excreted as a fraction of GFR remains fairly constant at a normal value—about 10% (top). Thus, even at a GFR of 10 mL/min, maximal water elimination can be as high as 10% of GFR—1 mL/min or 1.4 L/d.

The cortical diluting segment, especially the distal convoluted tubule, is important in abstracting solute to dilute the urine. Thiazides are particularly potent in abrogating maximal urinary dilution by impairing transport in this nephron segment. This capacity can also be affected by a large obligate solute load, especially if the solute is completely unreabsorbable during an osmotic diuresis, as shown in Fig 4–20 (bottom curve).

Finally, maximal water diuresis is dependent upon ADH activity being shut off. Hypovolemia

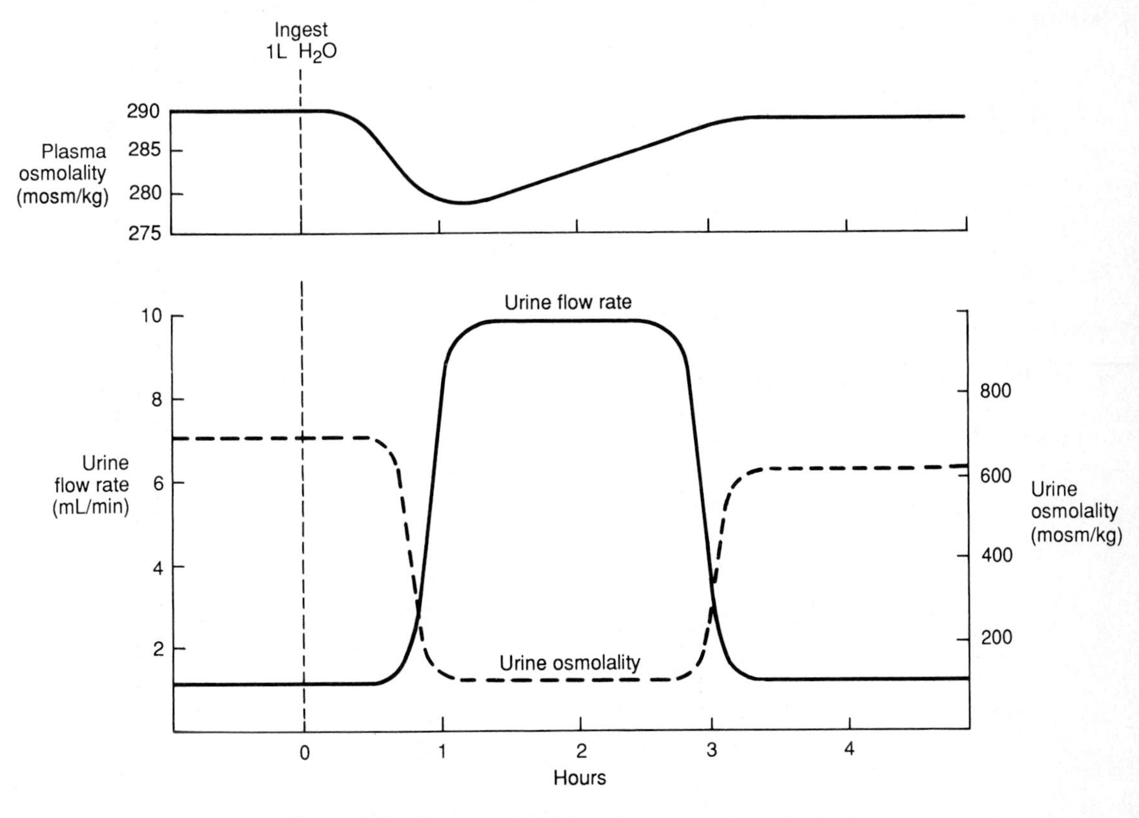

Figure 4–23. Time course of diuretic response to water load.

causes ADH to increase despite hypo-osmolality (Fig 4–15). Nonsteroidal anti-inflammatory agents also serve to potentiate ADH activity by depressing production of prostaglandin E_2, a functional antagonist of ADH.

Maximal Urine Volume During Diuresis

When all renal transport functions are optimally functioning and ADH is at a minimal level, as shown in Fig 4–22, the minimum urine volume depends on the obligate solute load requiring excretion. With 800 mosm/d to be excreted, the maximal urine volume is 16 L/d (50 osm/kg × 800 mosm/d). With 400 mosm/d solute load, the maximal urine volume is 8 L/d. Thus, the solute load affects the urine volume under diuretic conditions as it does under antidiuretic conditions.

With maximal diuresis when ADH is absent, urine volume is approximately equal to luminal fluid volume delivered out to the collecting tubule, which is about 10% of the GFR (Fig 4–10). Under these conditions, free water clearance (see below for definition) is about the same as urine volume. Thus, if GFR were 100 mL/min (144 L/d), maximal water excretion could be as high as 10 mL/min (10% of GFR), or 14.4 L/d. This re-

Table 4–3. Requirements for maximal water excretion.

	Altered By:
GFR	Chronic renal disease Age
Solute reabsorption in the cortical diluting segment	Obligate solute excretion Osmotic diuresis Thiazides
↓ADH	↓Effective circulating volume Nonsteroidal anti-inflammatory drugs

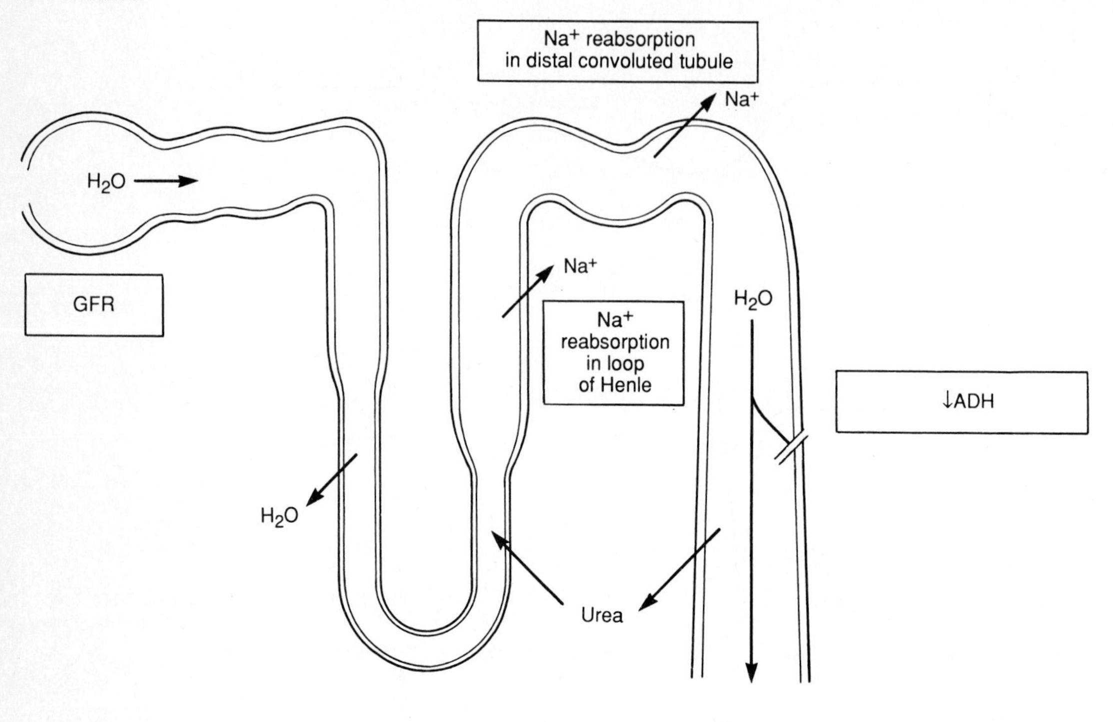

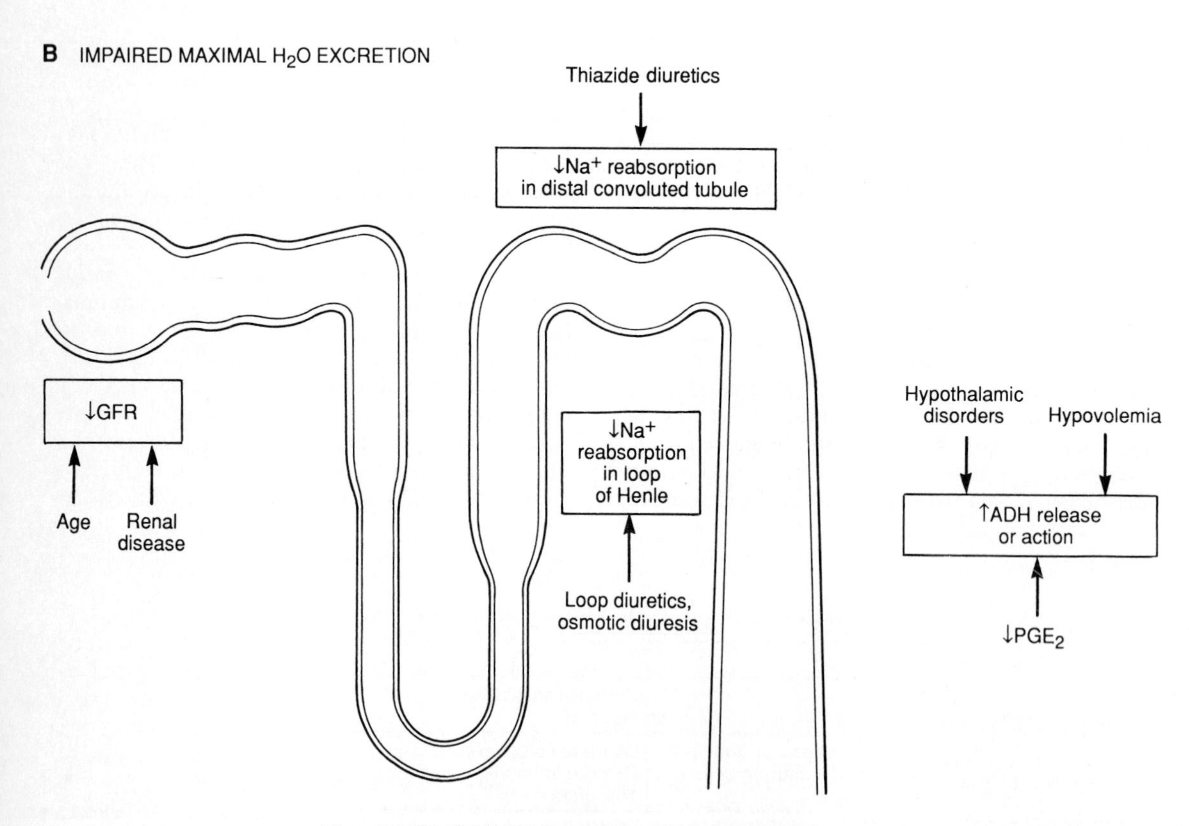

Figure 4-24. Pathophysiology of urinary diluting disorders.

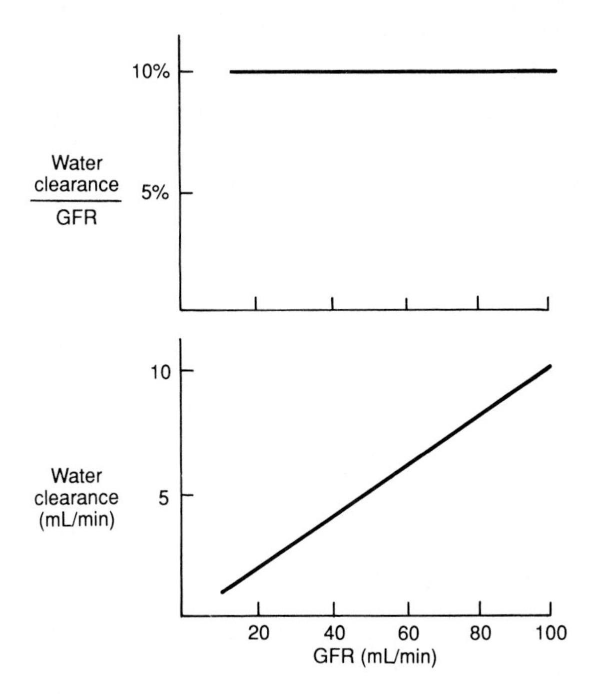

Figure 4–25. Diluting capacity in chronic renal disease.

markable diluting capacity and maximal diuretic ability by the kidney are rarely needed for long periods of time, since massive water ingestion (ie, approaching 14.4 L/d) is very unusual.

Electrolyte-Free Water Clearance & Reabsorption

Up to this point, it has been assumed that urinary concentration or dilution was responsible for water conservation or excretion, respectively. It has been assumed that if the urine had an osmolality greater than that of plasma, net water retention would result, tending to reduce the plasma osmolality, and vice versa.

This formulation, however, takes into account only the measured, cryoscopic urine osmolality. It is important from a physiologic perspective to consider also the identity of the urinary solutes. Remembering that the body may be simply viewed as containing only sodium, potassium, and water, plasma sodium concentration (P_{Na^+}) is derived as follows:

$$\text{Plasma } [Na^+] = \frac{\text{Total body sodium} + \text{Total body potassium}}{\text{Total body water}}$$

In this context, the impact on plasma sodium concentration would be derived from the relative excretion in the urine of sodium and potassium versus water: (urine sodium + potassium)/urine water compared with (total body sodium + potassium)/total body water. Formally, then, the loss of free water in the urine should be calculated from the urine volume, urine sodium (U_{Na^+}, and potassium (U_{K^+}) concentrations and plasma sodium concentration as follows:

$$\text{Free water clearance} = \text{Urine volume} \times \left(1 - \left[\frac{(U_{Na^+} + U_{K^+})}{P_{Na^+}}\right]\right)$$

Urinary free water clearance may be positive, with loss of water that tends to concentrate the body solutes; neutral, with no impact on body tonicity; or negative, with free water reabsorption that tends to dilute the body solutes.

For example, consider the excretion of urine at 1 L/d containing sodium or potassium (or both) totaling 140 meq/L, equal to plasma sodium concentration. The osmolality of the urine (280 mosm/kg) is then approximately equal to that of plasma. The excretion of urine with this composition would of course not alter the plasma sodium concentration. Free water clearance is 0 L/d: 1 L/d × [1 − (140/140)].

In contrast, consider the consequence of excreting 1 L/d of urine with the same osmolality (280 mosm/kg) but containing only urea, with no sodium or potassium (total of 0 meq/L). In this case, urinary volume excretion is equivalent to losing free water; water is being lost without either of the electrolytes (sodium or potassium) that comprise the body's solutes. In this case, urinary loss of water without sodium or potassium concentrates the remaining solutes in the body and increases plasma osmolality. Free water clearance is thus 1 L/d: 1 L/d × [1 − (0/140)].

There is an apparent paradox here because the measured osmolality of the urea-containing urine in the second case is equal to that of the sodium- and potassium-containing body fluids, yet the urine is considered physiologically to be free water. Urea should be viewed as a substance that is produced in large quantities and is rapidly excreted, without remaining in the body to substantially contribute to osmolality of total body water. It is a major constituent of the urine, however, and the urinary urea concentration is inversely proportionate to urine volume as defined by the ADH level. From a diagnostic standpoint, the osmolality of the urine—whether electrolytes or urea—reflects the ADH level (Fig 4–22). The presence of urea, concentrated in the urine under the influence of ADH, serves to minimize the excreted urine volume. But from a physiologic standpoint, a urine concentrated with urea but not sodium and potassium implies loss of free water.

5 Hyponatremic Disorders

INTRODUCTION

DEFINITION & CLASSIFICATION

Remembering that sodium and its associated anions account for 97–98% of the plasma osmolality, the plasma sodium concentration and osmolality generally vary in parallel and are sensitively maintained in a given individual within a narrow range (2–3% of normal). Hyponatremia is defined as a sodium concentration less than **135 meq/L**, with a corresponding plasma osmolality of less than **280 mosm/kg.**

A normal physiologic reduction in plasma sodium concentration and osmolality occurs during pregnancy. Plasma sodium concentration falls by an average of 5 meq/L) and osmolality by 10 mosm/kg within 5–8 weeks of gestation, and these values then remain stable for the duration of pregnancy. The osmotic threshold for ADH release is reset at a lower level without change in the slope relating plasma ADH concentration to plasma osmolality (Fig 4–6).

RELATIONSHIP OF WATER & SODIUM METABOLISM

Plasma sodium concentration represents the ratio of total body sodium (and potassium) to total body water. A reduction in this ratio might conceivably arise from (1) a low quantity of total body sodium with a lesser reduction in total body water; (2) normal total body sodium content with excess total body water; or (3) an excess of total body sodium with an even greater excess of total body water. As will be discussed below, all three causes of hyponatremia exist in practice.

The plasma sodium concentration therefore gives no insight into the state of total body sodium. Independent clinical assessment is mandatory for judging the absolute amount of total body sodium, as discussed in Chapter 1. Hyponatremia signifies only that there is a *relative* excess of water compared to sodium. Thus, hyponatremia implies that normal osmoregulation is disordered and there is a primary abnormality of water metabolism.

CONDITIONS IN WHICH HYPONATREMIA & HYPO-OSMOLALITY ARE NOT SYNONYMOUS

Under a few conditions, changes in plasma sodium concentration may be disjoined from changes in osmolality, as summarized in Table 5-1.

Hyponatremia With Hyperosmolality

Glucose (without insulin) cannot permeate cell membranes. During hyperglycemia, the initial osmotic imbalance between the extracellular and intracellular compartments causes water to flow out of cells, diluting the extracellular sodium including that in plasma. When osmotic equilibrium is reestablished, the plasma sodium concentration is depressed even while plasma osmolality is elevated because of the presence of glucose. For each 100 mg/dL increment in plasma glucose concentration, the corresponding decrement in plasma sodium concentration is approximately −1.6 meq/L.

For instance, if the plasma glucose rose from 100 to 600 mg/dL—an increment of 500 mg/dL—the plasma sodium concentration would be diminished by 8 meq/L (140 − [−1.6 meq/L × 5]), from 140 to 132 meq/L. However, the plasma osmolality would initially be increased by 28 mosm/kg, due to the addition of glucose. The plasma sodium concentration in this condition is no longer a valid index of the more important physiologic parameter, the plasma osmolality. The same situation applies following administration of other impermeant solutes such as mannitol, glycerol, or radiocontrast dye, which are often intravenously administered in very hypertonic solutions of 500–2000 mosm/kg. In all these cases, hyponatremia may occur despite plasma hyperosmolality.

Table 5-1. Conditions causing dissociation of normal relationship between plasma sodium concentration and plasma osmolality.

	Plasma $[Na^+]$	Plasma Osmolality
Addition to plasma of impermeable solute	↓	↑
Glucose		
Mannitol		
Glycerol		
Radiocontrast dye		
Addition to plasma of permeable solute	Normal	↑
Urea		
Ethanol		
Methanol		
Ethylene glycol		
Laboratory artifact	↓	Normal
↑Lipids		
↑Proteins		

Normonatremia With Hyperosmolality

During azotemic states, the plasma sodium concentration does not fall. A rise in extracellular urea concentration quickly causes an elevation in intracellular urea concentration owing to the high membrane permeability for urea. Because no osmotic gradient is established, water does not move out of the cells and there is no dilution of the plasma sodium concentration. However, the increased urea concentration contributes to an increase in plasma osmolality. Thus, in the case of azotemia, normonatremia is observed with hyperosmolality.

Ingestion of certain toxins, such as ethanol, methanol, and ethylene glycol, also leads to an increase in plasma osmolality. These substances are readily permeable across cell membranes and so, like urea, do not alter the plasma sodium concentration. Since they are sometimes ingested in large quantity and each has a low molecular weight, significant elevation in their plasma concentration can occur, with measurable contribution to the plasma osmolality. For instance, at the potentially lethal plasma level of ethanol (350 mg/dL), methanol (80 mg/dL), and ethylene glycol (21 mg/dL), with molecular weights of 46, 32, and 62, respectively, the measured plasma osmolality is increased by 76, 25, and 3 mosm/kg, respectively.

The usual convenient calculation to estimate plasma osmolality as described in Chapter 4—$2[Na^+] + glucose/18 + BUN/2.8$—normally correlates well with the measured osmolality ($\pm$ 9 mosm/kg). However, this calculation neglects all other potential osmoles except sodium, glucose, and urea. Thus, when toxins contribute to an increase in the actual measured plasma osmolality, there is a discrepancy with the calculated osmolality (which neglects these toxins), creating an **osmolar gap.** An osmolar gap greater than 9 mosm/kg usually indicates the presence in blood of a substantial amount of a low-molecular-weight, nonsodium, nonglucose, nonurea substance. A high anion gap metabolic acidosis also ensues following metabolism of these toxins, as discussed in Chapter 12.

Artifactual Hyponatremia: Pseudohyponatremia

Normally, the plasma is about 93% water and 7% solid, the latter representing principally lipids and proteins. Sodium is dissolved only in the aqueous phase. With modern autoanalyzer equipment in most clinical laboratories, measurement of serum sodium is obtained from a fixed-volume sample, which is normally 93% aqueous.

If the solid phase increases—in hyperlipidemic or hyperproteinemic states—the aqueous portion of the fixed serum volume declines commensurately. Although the sodium concentration in the aqueous phase remains normal, the aqueous contribution to the total sample and hence the total sodium content is less than normal when assessed by the autoanalyzer, resulting in reporting of an artifactually low sodium concentration. If a laboratory were to measure sodium using a sodium-selective electrode, which accurately assesses sodium concentration in the aqueous phase, a normal value would be obtained. Large lipid or protein molecules contribute little to osmolality—and do not affect the methods for measuring osmolality—so that reported plasma osmolality in such conditions is normal.

Autoanalyzer measurement of plasma sodium concentration decreases about 1 meq/L for a rise in lipid concentration of 500 mg/dL, so that an erroneous serum sodium value below 135 meq/L reported by the laboratory due to hyperlipidemia is unusual. A protein concentration greater than 10 g/dL is required to impact significantly on the measurement of plasma sodium concentration, as sometimes occurs in dysproteinemic states such as multiple myeloma. When these conditions occur, hyponatremia should be considered artifactual, since the sodium concentration in the aqueous phase of serum and the plasma osmolality are actually normal.

PATHOPHYSIOLOGY

Water Intoxication

One way in which hyponatremia can arise, as shown in Fig 5–1, is by ingestion or administration of more water than the kidney can excrete. Despite normal osmoregulation such that ADH is appropriately absent and the urine osmolality is as low as feasible (50 mosm/kg), maximal renal water excretion is insufficient to keep up with water ingestion. Water ingestion may be so massive that it exceeds diluting capacity and maximal free water excretion (10% of GFR) of the normal kidney, or it may merely be large enough to exceed renal water excretory mechanisms that are diminished by low GFR (Fig 4–25), low obligate solute load (Fig 4–22), or thiazides (Fig 4–24).

ADH Stimulation by Diminished Effective Circulating Volume

More commonly, hyponatremia is due to renal water retention resulting from ADH release (Fig 5–1). the enhanced ADH secretion may be physiologically appropriate for diminished effective circulating volume (Figs 4–9 and 4–15). In these hyponatremic cases, total body sodium may be low, normal, or high depending on the cause (Chapters 2 and 3).

ADH Secretion Despite Normal Effective Circulating Volume & Hypo-osmolality

A syndrome exists in which ADH is released without osmolality-dependent or volume-dependent physiologic stimulation to cause renal water retention and hyponatremia (Fig 5–1). Since ADH is high despite low plasma osmolality and euvolemia, urine is nondilute (ie, > 50 mosm/kg). This is called the **syndrome of inappropriate ADH (SIADH).**

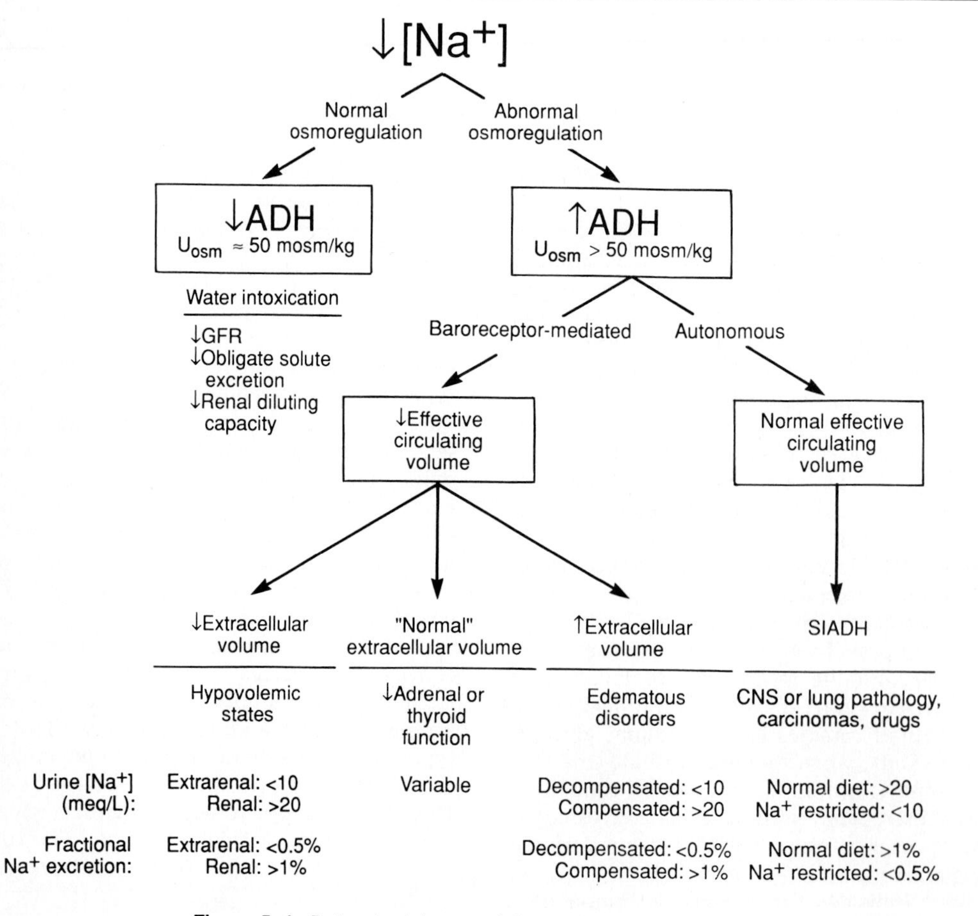

Figure 5–1. Pathophysiology and diagnosis of hyponatremia.

WATER INTOXICATION

Causes of hyponatremia are listed in Table 5-2.

EXCESS WATER INGESTION COMPARED TO RENAL DILUTING CAPACITY

Normal GFR

The kidney has an enormous capacity to excrete water (Fig 4-23). When GFR and tubular sodium transport mechanisms are normal and ADH is absent, a normal individual with GFR of 100 mL/min or 144.4 L/d can excrete an amount of urinary volume equivalent to 10% of GFR, or 14.4 L/d (Fig 4-15). Some psychotic individuals or other patients with abnormal thirst have a very high rate of water ingestion but rarely exceeding 14.4 L/d, so that overt hyponatremia does not ensue. These individuals with psychogenic polydipsia are markedly polyuric, of course.

Subnormal GFR

Free water excretion depends on the tubular fluid volume delivered out of the distal convoluted tubule into the collecting tubule, which in turn is a relatively fixed fraction (about 10%) of the GFR (Fig 4-10). As shown in Fig 4-25, absolute free water clearance capacity declines as GFR falls. Thus, if GFR is low, renal water excretory mechanisms can be overwhelmed at a much lower level of water ingestion than is required normally.

Low Urinary Obligate Solute Excretion Rate

Another determinant besides GFR of the maximal volume of free water in a diuretic state is the obligate solute load requiring urinary excretion. In individuals who are protein-malnourished (with minimal urea generation) and ingesting diminished amount of electrolytes, obligate solute excretion can be low. As shown in Fig 4-22, if obligate solute load is half-normal—400 mosm/d—the minimal urine osmolality of 50 mosm/d allows excretion of only 8 L/d despite normal GFR and ADH suppression. While unusual, such poorly nourished individuals who consume very large volumes of solute-free beverages—over 8 L/d—may develop hyponatremia. This condition has been termed "beer potomania" when beer is the beverage consumed. Even less ingested volume is required to induce hyponatremia if GFR is also reduced for any reason.

Table 5-2. Causes of hyponatremia.

Causes
Water intoxication
Markedly ↑H_2O intake: primary polydipsia
↓GFR
↓Obligate urinary solute excretion
Thiazides
↓Effective circulating volume
Hypovolemia (see Chapter 2)
Hemorrhage
Skin
Plasma → interstitial sodium translocation
Normovolemia
Adrenal or thyroid insufficiency
Hypervolemia (see Chapter 3)
Congestive heart failure
Cirrhosis
Nephrosis
SIADH
Nausea, pain
Psychosis
Carcinomas
Bronchogenic
Duodenal
Pancreatic
Thymoma
Ureteral
Lymphoma
Ewing's
Mesothelioma
Bladder
Prostatic
Pulmonary disorders
Pneumonia (viral or bacterial)
Pulmonary abscess
Tuberculosis
Aspergillosis
Positive-pressure breathing
Asthma
Pneumothorax
Cystic fibrosis
Central nervous system disorders
Encephalitis (viral or bacterial)
Meningitis (viral, bacterial, tuberculous, fungal)
Head trauma
Brain abscess
Guillain-Barré syndrome
Acute intermittent porphyria
Subarachnoid hemorrhage or subdural hematoma
Cerebellar and cerebral atrophy
Cavernous sinus thrombosis
Neonatal hypoxia
Hydrocephalus
Shy-Drager syndrome
Rocky Mountain spotted fever
Delirium tremens
Cerebrovascular accident (cerebral thrombosis or hemorrhage)
Multiple sclerosis
Acute psychosis

[1]Adapted in part from Robertson GL, Berl T: Water metabolism. Chap 11, pp. 385-432, in: *The Kidney,* 3rd ed. Brenner BM, Rector, FC Jr (editors). Saunders, 1986.

Impaired Diluting Mechanism: Thiazides

Finally, besides GFR, obligate solute load, and ADH level, the ability of the kidneys to maximally excrete free water depends on the normal extraction of solutes, principally sodium, from the tubular fluid. The terminal portion of the cortical nephron, the cortical thick ascending limb of Henle, the distal convoluted tubule, and the cortical collecting tubule are responsible for avid sodium reabsorption, so luminal fluid attains an appropriately low sodium concentration (< 25 meq/L) and osmolality (50 mosm/kg).

Interruption of sodium transport in the distal convoluted tubule is especially damaging to the ability to dilute the urine. Thiazide diuretics, which inhibit sodium transport in this segment (see Chapter 3), diminish renal diluting capacity—often by as much as 50% below normal. Therefore, an otherwise normal individual ingesting a thiazide diuretic would be able to excrete only about half the free water normally possible, or about 5% of the GFR (ie, 7.2 L/d). If water ingestion were to exceed 7.2 L/d, dilutional hyponatremia would occur.

More commonly, hyponatremia develops during thiazide administration only when several other mechanisms operate in concert to disrupt the renal diluting capacity. A frequent example is an elderly, somewhat poorly nourished individual being treated with a thiazide diuretic in whom slight hypovolemia develops. In this setting, thirst and water ingestion may be high as a result of the hypovolemia (Fig 4–15). In addition, all determinants of renal diluting capacity are perturbed: GFR is reduced due to age and hypovolemia (Fig 1-30); intrinsic diluting capacity is lowered by the thiazide and the reduced solute excretory requirement (Fig 4–22); and ADH secretion is stimulated by hypovolemia (Fig 4–15). As a result, urinary water excretion can be is substantially reduced—to even less than intake—resulting in dilutional hyponatremia.

DECREASED EFFECTIVE CIRCULATING VOLUME

As discussed in detail in Chapter 2, effective circulating volume may be reduced in association with a diminished, normal, or increased extracellular volume and total body sodium. The pathophysiology of disordered water metabolism in all these disorders is similar: baroreceptor-mediated, nonosmotic enhancement of ADH secretion causing antidiuresis, water retention, and hyponatremia (Fig 5–1).

DECREASED EXTRACELLULAR VOLUME

A reduction of 10% or more of the plasma volume (equivalent to 0.5% or more of body weight, or 350 mL in a 70-kg individual) or an equivalent amount of extracellular sodium lost in sweat or in gastrointestinal fluids by vomiting, diarrhea, or external drainage or diversion of an intestinal fistula—induces an appropriate antidiuretic (Fig 4–15) and antinatriuretic response (Fig 1–31).

When the kidney is itself the cause of the sodium loss, most commonly due to osmotic diuresis, diuretics, or functional lack of angiotensin II or aldosterone activities, the urine may be less concentrated, have a normal or even high volume (owing to the solute diuresis; Fig 4–20), and contain sodium.

NORMAL EXTRACELLULAR VOLUME

Adrenal or Thyroid Insufficiency

Both glucocorticoid and thyroid hormones are necessary for normal cardiac function. An impairment of systemic hemodynamic integrity has been generally held to be the most important factor responsible for the baroreceptor-mediated nonosmotic stimulation of ADH release in glucocorticoid deficiency (due either to hypopituitarism or to hypoadrenalism) or hypothyroidism (due to hypothalamic or thyroid disease). Independent direct effects on renal function and water transport due to glucocorticoid or thyroid hormone deficiencies may also contribute to the hyponatremia of these disorders.

Redistribution of Sodium

Translocation of fluid from the plasma to the interstitial compartment can arise with imbalance of the Starling forces or change in the hydraulic or protein permeabilities of the capillary (Chapter 2). Hypovolemia due to such changes occurs during hypoalbuminemia, sepsis, trauma, surgery, or inflammation (eg, peritonitis, pancreatitis). Again, a urine with low volume, high osmolality, and low sodium concentration reflects the baroreceptor-mediated integrated activation of ADH, the sympathetic nervous system, and the renin-angiotensin-aldosterone system.

INCREASED EXTRACELLULAR VOLUME

Congestive Heart Failure

As discussed in Chapter 3, diminished myocardial function is sensed by the high-pressure baroreceptors, causing the kidney to retain sodium and expand the extracellular volume. Initially, water is retained in proportion to sodium—ie, there is normal osmoregulation. However, as heart failure worsens, baroreceptor stimulation of ADH secretion occurs, causing water to be retained in relative excess of sodium, resulting in hyponatremia. Since it signifies fairly advanced myocardial dysfunction, the development of hyponatremia in congestive heart failure is an ominous sign.

The initial renal response during decompensated, moderate to severe congestive heart failure is antinatriuresis (urine sodium concentration < 20 meq/L) and antidiuresis due to the high ADH level (Fig 5–1). With sufficient sodium and water retention to improve myocardial function, a compensated phase can occur, with resumption of sodium and water balance (Table 3–2).

Cirrhosis

In the early stages of cirrhosis in some patients, primary renal sodium retention may cause relative overfilling of the effective circulating volume (Fig 3–3). At this stage, water metabolism is relatively normal and hyponatremia does not occur. Later, overt underfilling of the effective circulating volume may occur, when hypoalbuminemia, arteriovenous fistulas, and ascites have developed. In this stage, nonosmotic stimulation of ADH and hyponatremia may supervene. Poor nutrition and depression of GFR often contribute to the difficulty in excreting free water in cirrhosis. Urinary sodium excretion is usually low but—as in congestive heart failure—can be equivalent to intake in the compensated state (Fig 5–1).

Nephrotic Syndrome

As with cirrhosis, primary renal sodium retention and circulatory overfilling appears to occur in some patients with nephrotic syndrome. Water homeostasis is usually normal in such patients. However, in those nephrotics with low plasma volume and underfilling of the circulation due to hypoalbuminemia, water retention and hyponatremia can arise as a result of baroreceptor stimulation of ADH. The hyperlipidemia that is commonly seen in nephrotics should be excluded as causing a factitious reduction of serum sodium concentration (Table 5–1).

NORMAL EFFECTIVE CIRCULATING VOLUME: THE SYNDROME OF INAPPROPRIATE ADH (SIADH)

OVERVIEW

Disorders exist in which the release of ADH and water retention are not attributable to the two physiologic stimuli for ADH release: hyperosmolality and decreased effective circulating volume. When ADH release is unrelated to these normal pathways, the subsequent dilution of body solutes and hyponatremia is aptly termed the syndrome of inappropriate ADH secretion (SIADH).

The neurophysiologic inputs for ADH release arise both from the central nervous system and from the chest via baroreceptors and neural input (Fig 4–5). It is not surprising, therefore, that the causes of SIADH are in general related to disorders affecting the central nervous system—structural, metabolic, psychotic, or pharmacologic—or the lungs. Also, some carcinomas independently synthesize and release ADH.

PATTERNS OF ADH RELEASE

By definition, hyponatremia arises because ADH is inappropriately released at a subnormal plasma osmolality. In some cases, however, a degree of osmoregulatory control of ADH secretion persists. In other cases, ADH secretion is completely divorced from osmoregulation. There are three major patterns in SIADH of plasma ADH concentration expressed as a function of plasma osmolality and sodium concentration, as shown in Fig 5–2:

(1) Leak of ADH: In this pattern, ADH is low but measurable in the blood even at plasma osmolality below the normal threshold (280 mosm/kg). If plasma osmolality progressively rises, ADH secretion begins to increase at the normal plasma osmolality threshold and then rises normally. The defective urinary dilution owing to inability to completely suppress ADH at low plasma osmolalities has been ascribed to persistent leakage of ADH from the hypothalamus-posterior pituitary neurons that synthesize, transport, and secrete ADH. Such an ADH leak is sometimes seen in basilar skull fractures and other conditions.

(2) Reset threshold for ADH release: In this pattern of SIADH, ADH secretion is appropriately suppressed at very low plasma osmolalities but begins to increase at a subnormal threshold. As plasma osmolality increases further, the slope

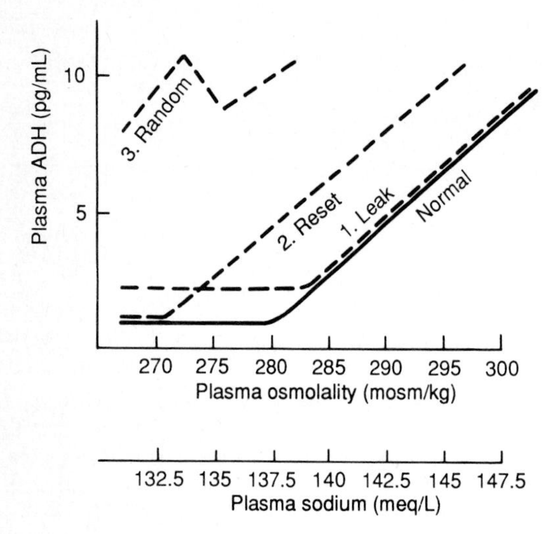

Figure 5-2. Patterns of ADH release in SIADH.

of the ADH rise is usually normal. Thus, ADH osmoregulation is qualitatively normal but shifted to a lower level. Normal limits of urinary dilution and concentration are attainable, but at subnormal plasma osmolalities. Such reduction in osmotic threshold for ADH release is reminiscent of that induced by hypovolemia (Fig 4-9) or during pregnancy. This pattern is seen with some neurologic, psychiatric, and pulmonary disorders and sometimes as a result of drug action or toxicity.

(3) Random secretion: In some cases of SIADH, ADH secretion is chaotic, totally divorced from osmoregulation. This pattern is sometimes observed with carcinomas, central nervous system diseases, and other conditions.

Finally, there are a small number of hyponatremic patients with an ADH secretory pattern indistinguishable from normal. In these patients, the pathophysiologic mechanism of the hyponatremia is unclear. There may be tubular supersensitivity to ADH or an ADH-like molecule, undetected by the radioimmunoassay, responsible for water retention.

MAINTENANCE OF SODIUM BALANCE

Sodium Excretion Equivalent to Intake

SIADH is a good example of the independent control of sodium and water metabolism. In this condition, ADH secretion is disordered, causing excessive water retention. But the factors that control sodium excretion (Chapter 1), notably the sympathetic nervous system and the renin-angiotensin-aldosterone axis, are relatively normal, so that sodium balance is unperturbed. Despite hyponatremia, sodium excretion remains equal to intake and normovolemia obtains. On a usual diet, sodium excretion is 100–200 meq/d, so that urine sodium concentration is virtually always over 20 meq/d in SIADH (Fig 5-1).

However, if sodium intake were to decrease for any reason, the appropriate antinatriuretic response would occur. For instance, if hyponatremia in a patient with SIADH became severe and caused anorexia or a seizure and cessation of sodium intake, urinary sodium concentration would appropriately fall to less than 20 meq/L and fractional sodium excretion to less than 1%. Thus, because the individual is in sodium balance, varying levels of urinary sodium excretion can be observed, depending on dietary sodium and volume status.

Consequences of Slight Extracellular Volume Expansion

With normal sodium homeostasis but excess water retention, the extracellular volume is actually slightly expanded in SIADH. The renal effects of this mildly volume-expanded state include relatively increased urea (see Chapter 1) and uric acid clearances, frequently resulting in low BUN (< 10 mg/dL) and hypouricemia (plasma uric acid level < 4 mg/dL). Plasma renin activity is generally low, but the plasma aldosterone level tends to be normal (due to an independent effect of hyponatremia itself), serving to maintain normal acid-base and potassium homeostasis. The slight expansion of the extracellular volume is insufficient to cause overt hypervolemia, hypertension, or edema.

SYMPTOMS

Though the spectrum of symptoms is similar, severity is usually greater at any given serum sodium concentration with acute hyponatremia than with chronic hyponatremia.

NERVOUS SYSTEM

Central Nervous System

As shown in Fig 5-3, symptoms of hyponatremia relate chiefly to the central nervous system. Although the brain cells partially compensate for extracellular hypo-osmolality by losing intracellular solute, primary water retention is nevertheless associated with brain swelling.

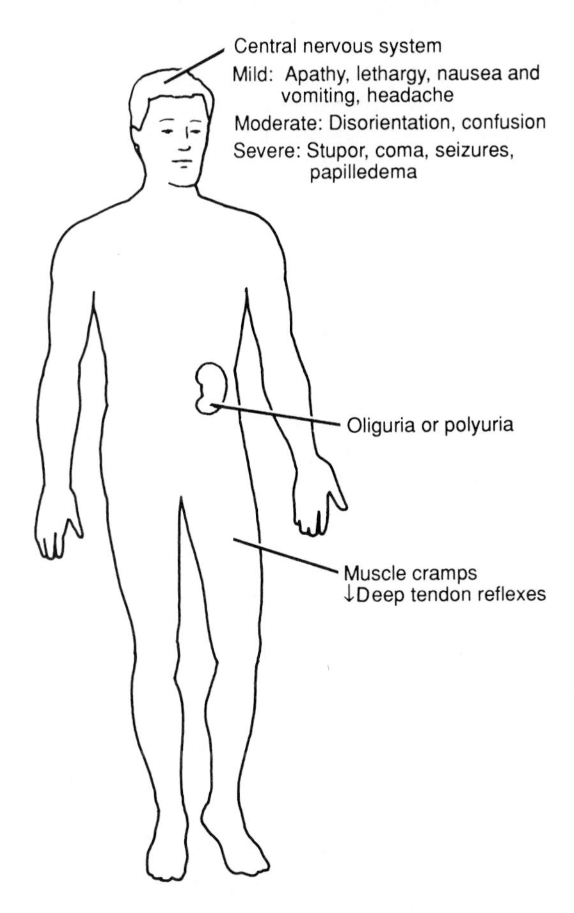

Figure 5-3. Signs and symptoms of hyponatremia.

Both the absolute magnitude and the rate of fall of serum sodium concentration correlate with cerebral edema and symptomatology. In general, mild chronic hyponatremia—serum sodium concentration above 125 meq/L—is fairly well tolerated. There are progressive neurologic symptoms below this level. Symptoms include apathy and lethargy, nausea and vomiting, and headache, which can progress to symptoms of disorientation and confusion with chronic hyponatremia of moderate severity (serum sodium 110–125 meq/L).

Severe symptoms such as stupor, coma, or seizures can occur with profound chronic hyponatremia (serum sodium < 110 meq/L) or with acute hyponatremia even when serum sodium has quickly fallen to a value that is still above 120 meq/L. These severe symptoms due to brain swelling may be reflected in signs of increased intracranial pressure, and papilledema may be observed. Neurologic impairment is sometimes irreversible, especially in women. The mortality rate can exceed 50% with chronic hyponatremia of 110 meq/L or less, especially if alcoholism or cachexia is an underlying condition.

Peripheral Nervous System

Rapid changes in plasma osmolality are frequently associated with muscle cramps and depression of deep tendon reflexes.

CARDIOVASCULAR

There are no consistent hemodynamic or electrocardiographic effects of hyponatremia per se. Concomitant disorders of total body sodium and effective circulating volume are more important in this regard. As mentioned previously, water retention and extracellular volume expansion in SIADH is insufficient to be manifested as circulatory overload, hypertension, or edema.

DIAGNOSIS

APPROACH TO ETIOLOGIC DIAGNOSIS

History & Physical Examination

A careful assessment of water intake, both oral and parenteral, is a necessary first step in evaluating the patient with hyponatremia. The patient may mention polydipsia, but frequently this symptom is elicited only upon careful questioning. Estimates of external water loss, including sweat and gastrointestinal losses, as well as urine are required as well. Solute and protein intakes should also be assessed, since these affect water excretion.

SIADH is diagnosed to some extent by exclusion, when water intoxication, adrenal or thyroid, and hypovolemic disorders with or without edema have been excluded (Table 5-2). SIADH is usually associated with nausea, pain, or diseases of the chest or central nervous system. In this context, neurologic examination is essential, as is chest examination and chest x-ray.

Assessment of Effective Circulating Volume

As shown in Fig 5-1, the major differential diagnosis of ADH-mediated hyponatremia is between disorders having decreased effective circulating volume with appropriate secretion of ADH and disorders causing SIADH. The approach to determining whether there is diminished effective circulating volume has been outlined in Chapter

2. Total body sodium and extracellular volume may be low (hypovolemic disorders), normal (adrenal or thyroid insufficiency), or increased (edematous disorders).

Drug History

A great many drugs may cause ADH release or potentiate its renal response or impair renal diluting capacity, such as thiazides (Table 4–3). Note should also be taken of other drugs that impact on the effective circulating volume, such as diuretics other than thiazides, nonsteroidal anti-inflammatory drugs, and antihypertensive agents (Chapter 1).

RENAL RESPONSE

Urine Volume & Osmolality

In hyponatremic patients due to water intoxication, urine is appropriately dilute: 50 mosm/kg (unless GFR is reduced or thiazides are ingested), corresponding to a urine specific gravity of about 1.002 (Fig 5–1). Polyuria is also present.

All other patients with chronic hyponatremia have an elevated ADH level, driven by physiologic baroreceptor-mediated stimuli or in the setting of SIADH (Fig 5–1). Since ADH is not suppressed by the hypo-osmolality, urine osmolality of these patients is not very dilute; ie, it is substantially in excess of the normal minimal value of 50 mosm/kg. The urine osmolality may not be maximally high (ie, is < 1200 mosm/kg) and often is in the range of 500–800 mosm/kg because escape from the urine concentrating action of ADH occurs when ADH is chronically elevated. Thus, steady-state urine volume in states of chronic ADH excess is usually relatively normal.

Sodium Excretion

Euvolemic hyponatremic patients, with water intoxication or SIADH, are in sodium balance and have sodium excretion rates equal to intake (Fig 5–1). Urine sodium concentration is therefore usually over 20 meq/L, and fractional sodium excretion is over 1%. However, as mentioned previously, if such a patient has low sodium ingestion for any reason or has supervening hypovolemia, he or she would have an appropriately low urine sodium concentration (< 10 meq/L).

The diminished effective circulating volume in the remainder of patients with hyponatremia—ie, those with hypovolemia, adrenal or thyroid insufficiency, or an edematous disorder—is associated with antinatriuresis, urine sodium concentration usually less than 20 meq/L, and fractional sodium excretion below 0.5% in the decompensated state. If sodium and water retention returns the patient to a compensated state, urine sodium can then approximate intake (Fig 5–1). Thus, in all hyponatremic conditions, urine sodium must be evaluated in clinical context and can vary depending on various factors including diet, volume status, and stage of underlying disease process.

Potassium & Acid-Base Homeostasis

Except as induced by the underlying disorder, hyponatremia itself and ADH excess do not affect renal potassium handling or acid-base homeostasis.

THERAPY

SPECIFIC APPROACHES TO THERAPY

Hyponatremia Due to Water Intoxication

As shown in Fig 5–4, if hyponatremia is due to water intoxication, simply denying the patient access to water or discontinuing the thiazide diuretic will allow the normal diuretic response to rapidly correct the hyponatremia. Psychiatric intervention is very important in attempting to curtail subsequent hyponatremic episodes in patients with primary psychogenic polydipsia.

Hyponatremia Due to Hypovolemia

When water retention and hyponatremia are consequent to physiologic baroreceptor stimulation of ADH associated with hypovolemia, appropriate therapy is to repair the extracellular volume deficit (Fig 5–4). Administration of isotonic saline solution by the intravenous route is the usual therapy, though blood or oncotic agents may be used acutely (Chapter 2).

When normal saline repletes the volume deficit, baroreceptor stimulation of ADH ceases and normal osmoregulation of ADH resumes. At this point, ADH secretion is suppressed appropriately for the degree of hyponatremia, allowing the urine to become dilute and the excess water to be excreted. Hypertonic (3%) saline solution is unnecessary for two reasons: First, isotonic saline used in the initial therapy (308 mosm/kg) is actually somewhat hypertonic to the patient. Second—and more importantly—the hypovolemia is more efficiently and safely corrected by normal saline. The normal kidney freed of ADH action has a tremendous ability to excrete free water (> 0.5 L/h) and will correct the hyponatremia rapidly.

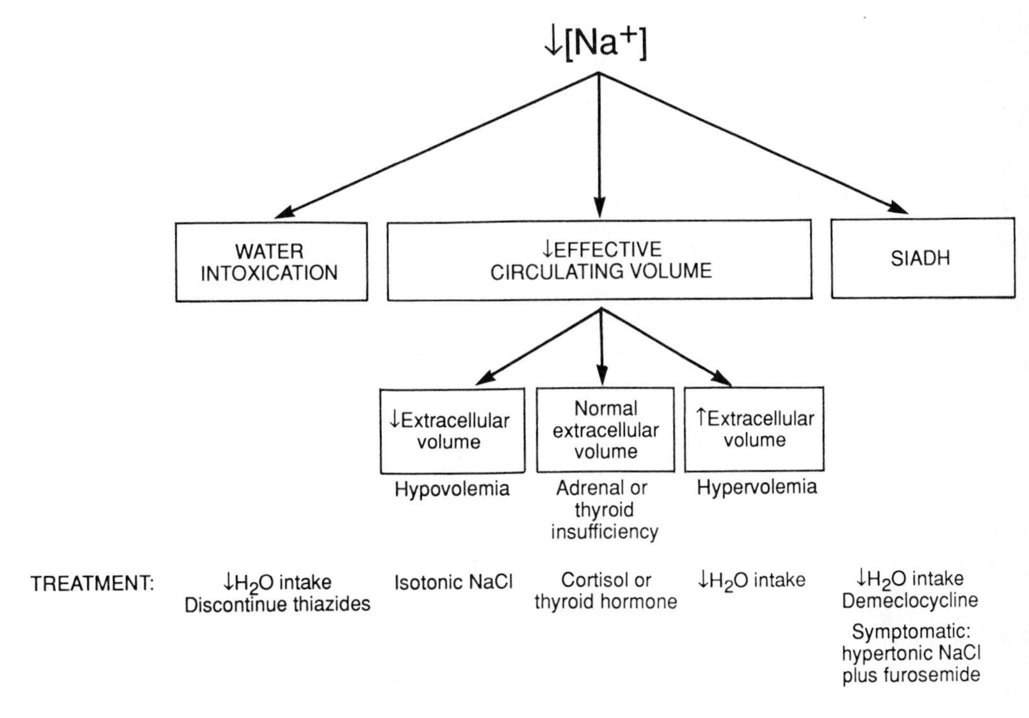

Figure 5-4. Treatment of hyponatremia.

Hyponatremia Due to Adrenal or Thyroid Disease

It is essential to rule out adrenal and thyroid insufficiency in a patient with hyponatremia, either clinically or by formal measurement. If present, appropriate hormonal replacement is of course indicated and the hyponatremia will then be corrected.

Hyponatremia Due to Hypervolemia

The specific treatment of cardiac output in congestive heart failure, of underlying liver disease in cirrhosis, and of the renal disease responsible for the nephrotic syndrome should be the first approach to treating these edematous hyponatremic states (Chapter 3). If the underlying condition can be improved, the nonosmotic, baroreceptor-mediated stimulation of ADH will be lessened, allowing for better free water excretion and at least partial repair of the hyponatremia. Otherwise, dietary water restriction is necessary to manage the hyponatremia (Fig 5–4).

Hyponatremia Due to SIADH

In patients with SIADH, the cause of the ADH secretion should be removed if possible (Table 5-2). Drugs that stimulate ADH release (Table 4-1) or impair renal diluting capacity or enhance ADH responsiveness should be stopped, including thiazides, nonsteroidal anti-inflammatory drugs, and chlorpropamide. If SIADH is due to central nervous system or pulmonary infection, the hyponatremia usually resolves with treatment of the infection. When specific approaches fail, hyponatremia is treated by water restriction, demeclocycline, or hypertonic saline plus furosemide (Fig 5–4), as discussed below.

AMOUNT OF WATER EXCESS

Acute Hyponatremia

Especially in the hospital setting, acute hyponatremia is commonly due to rapid hypotonic fluid administration to a patient with a high ADH level resulting from nonosmotic stimulation or SIADH. If total body solutes and mass are assumed to be constant, short-term water excess may be estimated from free water balance (solute-free water intake minus solute-free urine output) or simply weight gain.

Chronic Hyponatremia

The development of hyponatremia is usually a slow, insidious process in which weight gain is unknown. Assuming that the low plasma sodium concentration is due solely to excess water and that total body sodium and potassium are constant, the water excess may be calculated by the following equation:

$$\text{Excess water} = \text{Total body water} \times \text{Fractional change in plasma [Na}^+]$$

Since total body water in a man can be estimated as 0.6 × body weight and the fractional change in plasma sodium concentration (P_{Na^+}) can be approximated by (140 − P_{Na^+}/140, the equation above can be reduced as follows:

$$\text{Excess water} = (0.6 \times \text{Body weight}) \times \frac{(140 - P_{Na^+})}{140}$$

In practice, the goal of therapy is not to correct plasma sodium concentration to normal but rather to a value slightly less—usually 130 meq/L—because of the risk of overcorrection. The total amount of water to be removed can be quickly estimated by substituting 130 for 140 meq/L in the above equation.

TREATMENT OF SEVERE, SYMPTOMATIC HYPONATREMIA

If there are seizures or disturbance in consciousness, hyponatremia should be rapidly treated at any level of plasma sodium concentration, as outlined below. This problem is usually confined to patients with SIADH.

Rate & Degree of Correction

As a guideline, therapy should raise plasma sodium concentration above the critical level of 120 meq/L. Significant neurologic morbidity results from overcorrection, so a reasonable and safe goal for plasma sodium concentration is 130 meq/L but no higher.

Controversy exists about whether too-rapid correction of hyponatremia—by more than 2 meq/L per hour—is hazardous and can provoke a potentially fatal neurologic disorder called central pontine myelinolysis. An opposing opinion is that hyponatremia itself causes central pontine myelinolysis, not its correction, and therefore requires as prompt resolution as possible. A reasonable compromise recommendation would be that plasma sodium concentration should be increased by at least 0.5–1 meq/L per hour but not by more than 1.5 meq/L per hour. Thus, a symptomatic patient with a plasma sodium concentration of 100 meq/L should be managed in such a way as to increase the sodium concentration to 130 meq/L over the course of about 30 hours.

Hypertonic Saline Plus Furosemide

Although the pathogenesis of hyponatremia is water retention, there is no simple way of ridding the body of pure water. Since ADH is elevated in most hyponatremic conditions, logical therapy would be the to give an ADH V_2 receptor antagonist, but these agents are not yet clinically available.

Since a patient with SIADH is euvolemic and in sodium balance, administration of hypertonic sodium alone is ineffective for changing plasma sodium concentration. The excess sodium would be rapidly excreted in a highly concentrated urine. Thus, if hypertonic fluid alone were infused, a comparably hypertonic urine would be excreted, without achieving a change in water balance. If, however, furosemide is simultaneously administered, the kidney's ability to concentrate the urine is abolished despite the presence of ADH (Fig 4–18). With furosemide present, hypertonic saline is infused while isotonic urine is excreted. If the total amounts of sodium in the infused and excreted volumes are equal, there is a net loss of free water.

Using this strategy, plasma sodium concentration can be rapidly corrected with hypertonic sodium plus furosemide (1 mg/kg intravenously) given as necessary (every 4–6 hours) to clamp urine osmolality at an isotonic level (ie, achieve isosthenuria of 290 mosm/kg, specific gravity 1.010). Total urine sodium loss is measured hourly and replaced quantitatively in absolute amount using hypertonic (3%) saline (which has a sodium concentration of 513 meq/L and osmolality of 1026 mosm/kg). The necessity for measuring urine volume and sodium excretion frequently as well as systemic hemodynamics usually require that the patient be closely observed in an intensive care facility. This method can achieve maximal net free water loss of about 0.5 L/h with a concomitant rise in plasma sodium concentration of about 1.5 meq/L per hour.

The difference in isotonic urine volume excreted and hypertonic saline volume infused is essentially the free water loss. For instance, consider the following case: Urine volume excretion using furosemide is 300 mL/h with sodium concentration of 150 meq/L, representing a total urinary sodium loss of 45 meq/h. Replacement of 45 meq/h using hypertonic saline would require 88 mL/h (45 meq/h divided by 513 meq/L), resulting in net water loss of 212 mL (300 − 88 mL).

TREATMENT OF ASYMPTOMATIC HYPONATREMIA

Asymptomatic, stable, mild hyponatremia in the range 130–135 meq/L, during the course of disorders associated with chronically decreased effective circulating volume or SIADH is usually well tolerated and not treated. As a rule of thumb, a serum sodium concentration of 120–130 meq/L is usually treated by dietary water restriction to prevent symptoms in case progression of the hyponatremia should occur. Since there is a high prevalence of developing symptoms, a plasma sodium concentration of less than 120 meq/L should serve as an indication for treatment even if the patient is asymptomatic.

Water Restriction

Given insensible free water loss via the skin, lungs, and gastrointestinal tract, restriction of oral water intake inevitably leads to a reduction in total body water content. Restriction of oral water intake and hypotonic fluid administration to less than 1000 mL/d leads to a gradual increase (over the course of days) in serum sodium concentration, with more rapid correction achieved by severe water restriction (to $<$ 500 mL/d). If a diuretic is necessary for a hyponatremic patient with edema, a loop diuretic may be preferable to a thiazide since the latter impairs diluting capacity.

Prevention of Chronic Hyponatremia

To prevent recurrence of hyponatremia once it has been corrected by water restriction, the patient should be instructed to continue to curtail water intake.

In patients with SIADH in whom water restriction proves difficult, lithium or democlocycline has had some utility in antagonizing the hydro-osmotic effect of ADH on the medullary collecting tubule. Of the two drugs, democlocycline, 600–1200 mg/d orally, is more effective and can effectively reduce urine osmolality. It has been reported to decrease GFR, especially in hyponatremic cirrhotics, and should be used cautiously in any patient with preexisting renal insufficiency.

6 Hypernatremia

INTRODUCTION

DEFINITION

All hypernatremic states are states of hyperosmolality. The reverse is not necessarily true, however, in that some hyperosmolal states have a normal plasma sodium concentration (Table 5–1). Hyperosmolality with normonatremia may be due to high plasma concentration of glucose, urea, or a low-molecular-weight toxin such as ethanol, methanol, or ethylene glycol. Hypernatremia is diagnosed when the plasma sodium concentration is greater than **145 meq/L** and hyperosmolality when the plasma osmolality is greater than **295 mosm/kg.**

PATHOPHYSIOLOGY

Hypernatremia represents an imbalance of total body sodium and water and can arise if there is (1) rapid hypertonic sodium ingestion with insufficient time or opportunity for water ingestion; (2) a defect in thirst or in ability to drink water; or (3) water ingestion insufficient to cope with renal or extrarenal water loss. In practice, the latter two scenarios are almost always responsible for hypernatremia. Thus, as with hyponatremia, hypernatremia should be considered a disorder of water metabolism.

In all these cases, physiologic osmoregulatory responses are intact, so that ADH is appropriately high (Fig 6–1) and the urine is maximally concentrated and there is a strong stimulation of thirst.

Hypernatremia may also arise if thirst and water ingestion are unable to keep up with a primary polyuric disorder, called diabetes insipidus, caused by failure of ADH to be secreted or inability of ADH to exert its functional hydro-osmotic effect on the medullary collecting tubule. In these cases, the urine remains inappropriately dilute despite hyperosmolality (Fig 6–1). If thirst and water ingestion do not compensate for severe renal water wastage, hypernatremia and hypovolemia may occur.

In all these disorders of nonrenal or renal water loss, normal osmoregulatory mechanisms sensitively and efficiently regulate thirst. Therefore, water loss by sweating, burns, diarrhea, or osmotic diuresis (glycosuria) does not predictably lead to hypernatremia unless water access or ingestion is limited for another reason. Even with severe diabetes insipidus—with dilute urine output of 10–15 L/d—equivalent water ingestion permits the hypernatremia to be very mild (145–150 meq/L) if present at all.

STATES WITH APPROPRIATE ELEVATION & ACTIVITY OF ADH

Causes of hypernatremia are listed in Table 6–1.

INCREASED EXTRACELLULAR VOLUME: HYPERTONIC SODIUM INGESTION OR ADMINISTRATION

Rarely, there may be rapid oral intake of hypertonic sodium, as with ingestion of sea water during near-drowning, salt tablets, or very salty broths. The osmolality of sea water averages 1000 mosm/kg, and heavily salted soups have been reported to have osmolalities this high as well.

A hypertonic sodium solution is occasionally parenterally administered, as in inadvertent intravenous infusion of hypertonic sodium chloride during a therapeutic abortion. The 20% sodium chloride solution used for this procedure has an osmolality of almost 7000 mosm/kg and, if injected into the bloodstream, can acutely raise plasma sodium concentration by 20 meq/L. Likewise, intravenous hypertonic sodium bicarbonate

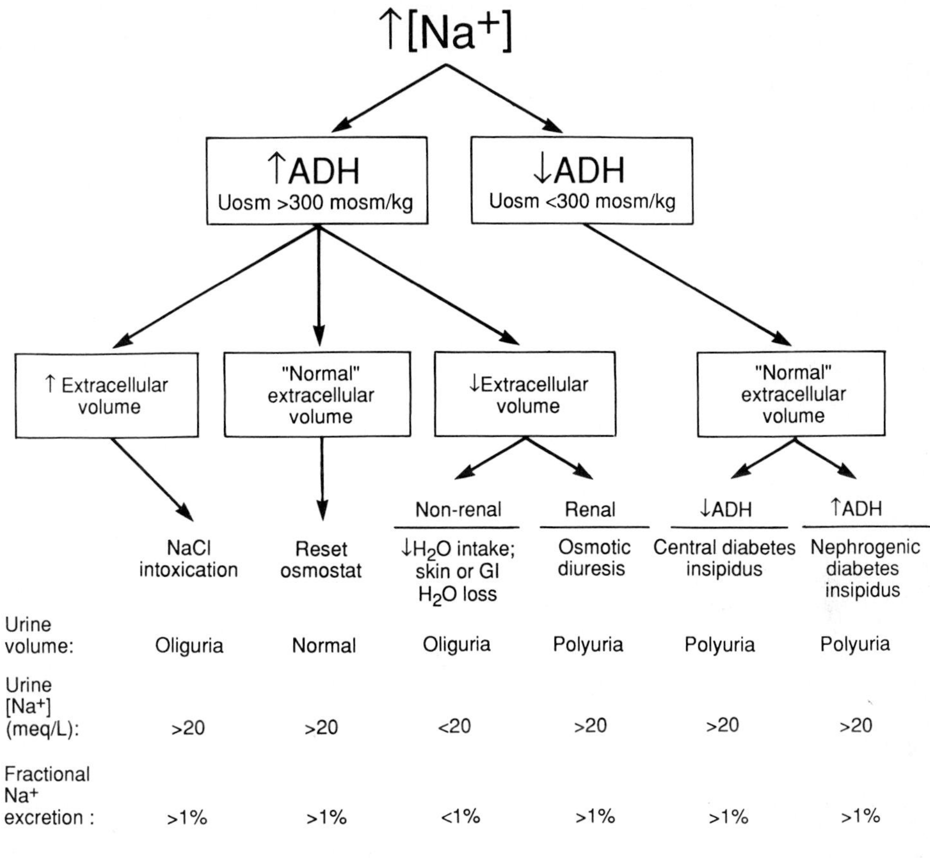

Figure 6–1. Diagnosis of hypernatremia.

administration during the management of acidemia or cardiopulmonary arrest can increase plasma sodium concentration, since the sodium concentration in each 50-mL ampule of 7.5% sodium bicarbonate is 892 meq/L.

During hemodialysis, sodium and water fluxes can be accidentally imbalanced if there is dialysate malproportioning. Modern dialysis monitoring systems prevent such a complication. During peritoneal dialysis, very rapid exchanges may cause water to diffuse faster than sodium from the blood into the peritoneal fluid and hence cause hypernatremia.

Finally, a slight, clinically insignificant rise in plasma sodium concentration occurs during the sodium retention of primary hyperaldosteronism or Cushing's syndrome, probably due to resetting of osmoregulation by chronic extracellular volume expansion (Fig 4–9).

In all these cases of sodium excess, the extracellular volume expansion predisposes toward hypertension and causes an appropriately high urinary sodium concentration (Fig 6–1). Hypernatremia causes strong stimulation of both ADH secretion and thirst. The high plasma ADH causes the urine to be concentrated and low in volume, unless the natriuresis is very brisk, in which case higher urine volume may be obligated (Fig 4–20).

NORMAL EXTRACELLULAR VOLUME: RESET OSMOSTAT

There are rare instances of hypothalamic injury in which thirst regulation and osmoregulatory control of ADH secretion are altered (Table 6–1). In some patients, the response to further increase in plasma osmolality is blunted but intact, but in others there is none. In general, nonosmotic control of ADH remains normal. Other pituitary disorders are common, affecting hormones released by the anterior as well as posterior pituitary. Sodium balance is usually well maintained in these disorders, so that normovolemia obtains and urinary sodium excretion is equivalent to intake.

Table 6-1. Causes of hypernatremia.

↑Sodium loads
- Hypertonic sodium chloride administration
- Salt water near-drowning
- Intravenous hypertonic sodium chloride during therapeutic abortion
- Salty broths
- Sodium chloride tablets
- Sodium bicarbonate infusion
- Hemodialysis with malproportioning of dialysate
- Rapid peritoneal dialysis exchanges
- Primary hyperaldosteronism
- Cushing's disease and syndrome

Reset osmostat (hypothalamic disorders)
- Ectopic pinealoma
- Dysgerminoma/germinoma
- Craniopharyngioma
- Teratoma
- Meningioma
- Pituitary adenoma
- Metastatic bronchial carcinoma
- Eosinophilic granuloma
- Schüller-Christian disease
- Sarcoidosis
- Granulomatous tumor
- Histiocytosis
- Hypothalamic neuronal degeneration
- Subarachnoid hemorrhage
- Posttraumatic carotid cavernous fistula
- Microcephaly
- Occult hydrocephalus
- Head trauma
- Aneurysmectomy (anterior communicating artery)

↑Insensible H_2O loss
- Heavy exercise
- Hot climate
- Diarrhea

Central diabetes insipidus
- Primary or idiopathic
- Head trauma or postsurgery
- Empty sella syndrome
- Tumors
  - Suprasellar cyst, craniopharyngioma, pinealoma
  - Metastatic
- Granulomatous and inflammatory diseases
  - Sarcoidosis
  - Wegener's granulomatosis
  - Tuberculosis
  - Syphilis
  - Eosinophilic granuloma
  - Schüller-Christian disease
- Infections
  - Encephalitis
  - Meningitis
  - Landry-Guillain-Barré syndrome
- Vascular
  - Aneurysms
  - Cerebral thrombosis or hemorrhage
  - Sickle cell disease
  - Postpartum (Sheehan's syndrome)
- Drugs
  - Ethanol
  - Opiate antagonists
  - Alpha-adrenergic agents (clonidine)
  - Phenytoin
  - Pregnancy (↑vasopressinase)

↓Concentrating ability and nephrogenic diabetes insipidus
- Congenital (familial)
- ↓GFR
- Osmotic diuresis
  - Glucose
    - Diabetes mellitus
    - Glucose infusion
  - Mannitol
  - Urea (postobstructive, post-acute tubular necrosis, posttransplant)
  - Radiocontrast dye
- ↓Thick ascending limb of Henle sodium reabsorption
- Loop diuretics

continued

Table 6-1. (cont'd). Causes of hypernatremia.

↑**Concentrating ability and nephrogenic diabetes insipidus (cont'd)**
Furosemide
Ethacrynic acid
Bumetanide
↓Urea
↓Protein intake
Urea washout following water diuresis
↓ADH function
Electrolyte disorders
Hypokalemia
Hypercalcemia
↑Prostaglandin E_2
Drugs
Lithium
Demeclocycline
Tubulointerstitial nephropathies chiefly affecting the medulla
Sickle cell disease
Sarcoidosis
Metabolic
Hyperuricemia
Hypercalciuria and nephrocalcinosis
Cystic
Medullary sponge kidney
Polycystic kidney
Obstructive uropathy
Acute bacterial infection
Amyloidosis
Sjögren's syndrome

LOW EXTRACELLULAR VOLUME: NONRENAL OR RENAL HYPOTONIC VOLUME LOSS

Impaired Water Intake

Insensible water loss from the respiratory tract, skin, and gastrointestinal tract amounts to about 0.5 L/d (Fig 4-3). In addition, there is renal water loss, usually about 1 L/d, that may or may not represent physiologic free water loss, depending on how much sodium and potassium are in the urine. Thus, obligate water loss from the body is at least 0.5 L/d and usually more. Hypernatremia may arise if water intake is insufficient to match this normal insensible water loss due to impaired water access, swallowing, or absorption. In these cases, the patient is appropriately very thirsty, and the urine is highly concentrated as a result of normal osmoregulatory stimulation of ADH (Fig 6-1).

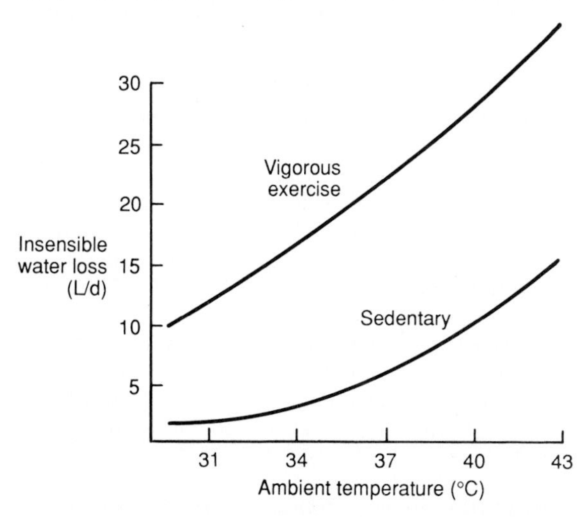

Figure 6-2. Insensible water loss as a function of environmental temperature.

Excessive Insensible Water Loss From the Skin or From the Respiratory or Gastrointestinal Tracts

The volume of sweat increases with exercise and environmental heat, as illustrated in Fig 6-2. Sweat is hypotonic to blood (sodium plus potassium is approximately 70 meq/L), so that net water loss occurs with heavy sweating, predisposing to hypernatremia if water ingestion cannot keep pace. Water loss from the respiratory tract is also exertionally and thermally regulated and makes a contribution, usually minor (~10%), to the increased insensible water loss, which can acutely approach 25 mL/min. Over time in a hot environment—at least 2-5 days—an acclimatization process occurs to reduce sweat volume, mediated by hypovolemia-induced hyperaldosteronism.

Stool water is slightly hypotonic (sodium plus potassium of 130 meq/L), though it tends to become closer to isotonicity with increasing flow in diarrheal states, especially in children. Again, there is a predisposition to hypernatremia if water ingestion fails to compensate for hypotonic volume loss in diarrhea.

In all these conditions, the renal response to a low effective circulating volume is to avidly conserve water and sodium (Fig 6–1). Antidiuresis, with oliguria and high urine osmolality, and antinatriuresis are therefore characteristically seen during these hypovolemic forms of hypernatremia.

Excessive Renal Water Loss During Osmotic Diuresis

As shown in Fig 4–20, the urine becomes essentially isotonic to blood as excretion increases of a poorly reabsorbed solute. The urinary solute may represent glucose, urea, mannitol, or radiocontrast dye.

As urine flow increases during osmotic diuresis, sodium excretion is also obligated. A steady state is approached in which the urinary sodium concentration averages 60 meq/L. Although the urinary potassium concentration varies depending on the extracellular volume status and aldosterone level (Chapter 7), the sum of sodium and potassium in urine is invariably less than the sum in blood. With osmotic diuresis, the kidney cannot maximally concentrate the urine despite a high level of ADH induced by both hypernatremia and decreased effective circulating volume (Fig 4–20). In fact, the maximal urine osmolality approaches that of blood (ie, is isosthenuric, ~290 mosm/kg). Thus, the urine is cryoscopically isotonic (due to the contributions of sodium, urea, potassium, and the osmotically active solute such as glucose, mannitol, or radiocontrast dye) but physiologically hypotonic (based on sodium and potassium concentrations only). Therefore, there is free water loss into the urine during osmotic diuresis, which predisposes to hypernatremia if water intake is insufficient to compensate.

Glycosuria may occur in the course of uncontrolled diabetes mellitus or following excessive glucose infusion. A urea diuresis may occur with excessive protein ingestion or parenteral feeding or when GFR is rapidly repaired in a previously azotemic individual, as in the diuretic recovery phase following acute tubular necrosis, following relief of urinary tract obstruction, or after a successful kidney transplant. Finally, a nonreabsorbable solute is sometimes intravenously administered, such as mannitol or a radiocontrast dye.

STATES WITH ABNORMALLY LOW ADH ACTIVITY: DIABETES INSIPIDUS

Excretion of a dilute urine in the face of hypernatremia is distinctly abnormal, reflecting either an abnormality of ADH release from the neurohypophysis, called central diabetes insipidus, or resistance to ADH at the level of the medullary collecting tubule, called nephrogenic diabetes insipidus.

In all of the states of diabetes insipidus, there is polyuria. Differentiation of central versus nephrogenic forms rests on measurement of basal ADH level, the renal response to deliberate water restriction with induction of dehydration, and subsequent response of urinary concentration to exogenously administered ADH.

CENTRAL DIABETES INSIPIDUS

Etiology

As shown in Table 6–1, traumatic, neoplastic, infiltrative, infectious, or vascular disorders that destroy the region of ADH production (the hypothalamus) or its transport track (the median eminence) can cause ADH deficiency. There are also familial and idiopathic forms.

A triphasic change in ADH sometimes occurs with traumatic or surgical lesions of the hypothalamus, in which there is a decrease in ADH followed by a transient increase (perhaps reflecting release from dying cells) followed by a permanent decrease. Removal of the storage area for ADH—the posterior pituitary—causes only transient reduction in ADH secretion so long as the hypothalamus remains uninjured.

Clinical Characteristics

With complete lack of ADH, the urine is highly dilute, with polyuria averaging about 5–12 L/d. Nocturia in adults and enuresis in children are very common. Thirst is strongly stimulated in these patients, with a preference for ice water, and is sufficient to compensate for these large urinary volumes. Plasma sodium concentration may therefore be maintained in the normal range and hypernatremia, if present, is of modest degree, with plasma sodium concentration no more than 150 meq/L. The patients remain in sodium balance and have a normal extracellular volume (Fig 6–1).

Renal Response to Dehydration

As shown in Fig 6–3, if further plasma hyperosmolality is induced in a patient with central diabetes insipidus by restricting water ingestion to induce dehydration equivalent to 5% of body weight, ADH does not rise appropriately. Without ADH, the urine osmolality continues to be quite dilute—50–250 mosm/kg, substantially below the plasma osmolality.

However, if exogenous ADH (Pitressin) is given following the period of dehydration, the kidney is normally responsive, so there is a marked rise in urine osmolality of greater than 50% and often as much as 500%. The increased urine osmolality after ADH administration (range, 250–600 mosm/kg) is still far less than the normal maximum, however, because of the preexisting water diuresis that causes washout of the medullary interstitium (Fig 4–21).

Partial Central Diabetes Insipidus

Some patients are not completely deficient in ADH and are able to mount a small ADH basal secretion rate. Basal urine volume tends to be less: 3–6 L/d. There is also an attenuated urinary concentration response to dehydration.

During water deprivation sufficient to cause a 5% reduction in weight, ADH level is measurable but substantially subnormal for the induced plasma hyperosmolality (Fig 6–3). The small increase in ADH permits urinary concentration to rise to about 300–600 mosm/kg—a heightened response from the basal value but nevertheless subnormal. Since the endogenous level of ADH is still low, exogenous ADH supplementation following the period of dehydration elicits a modest further (15–50%) increase in urinary osmolality.

NEPHROGENIC DIABETES INSIPIDUS

Etiology

A rare, X-linked congenital form of diabetes insipidus occurs in which ADH neurohypophyseal release is normal but the hydro-osmotic response of the medullary collecting tubule to ADH is impaired. The cellular mechanism of the ADH unresponsiveness is unknown, whether at the level of the V_2 receptor, G_s protein, adenylate cyclase, cAMP, or water channel (Fig 4–12). to here

Acquired forms of ADH resistance are common but not usually severe. As listed in Table 6–1 and as was shown in Fig 4–18, the principal causes of inability to concentrate the urine are diminished GFR, inhibition of transport in the thick ascending limb of Henle, disruption of medullary hypertonicity, and functional or anatomic disruption of the action of ADH on the medullary collecting tubule.

Clinical Characteristics

With complete inability of ADH to act on the medullary collecting tubule, there is a highly dilute urine with polyuria, usually 6–10 L/d (Fig 6–1). The resistance is usually incomplete, however, so some degree of urinary concentration is possible (though less than 300 mosm/kg) and urine volume does not exceed 3–4 L/d.

Central hypothalamic osmoregulatory mechanisms are normal, and the plasma ADH level is appropriate for the prevailing plasma osmolality and sodium concentration (Fig 6–3). Only slight hypernatremia usually occurs if any, since thirst mechanisms are also intact.

Renal Response to Dehydration

If water intake is curtailed, the subsequent dehydration causes a further increase in ADH, but no tubular response is possible, and the urine remains dilute (Fig 6–3). As expected, no substantial increment in urinary concentration is elicited if ADH is exogenously supplemented following the period of dehydration.

SYMPTOMS

NERVOUS SYSTEM

Central Nervous System

Initially, increased plasma osmolality causes water to leave brain cells, acutely shrinking the brain. As shown in Fig 6–4, brain shrinkage progressively causes mild symptoms of restlessness, lethargy, and headache, to moderate symptoms of disorientation and confusion. Severe brain shrinkage can lead to local hemorrhagic complications, seizures, stupor, and coma. The mortality rate from chronic hypernatremia greater than 160 meq/L has been reported to exceed 60%.

In response to prolonged increase in plasma osmolality, brain cells synthesize solutes, originally termed **idiogenic osmoles** and now known to be predominantly amino acids and to a lesser extent polyol sugars and methylamines. These solutes serve to increase intracellular osmolality and allow osmotic flow of water back into the cell, restoring cell volume. The synthesis of these intracellular solutes in response to extracellular hypertonicity begins within about 4–6 hours and takes several days to reach a steady-state level. Thus, brain water content is lowest following acute hypernatremia but is only slightly reduced during chronic hypernatremia as a result of this process of cell volume regulation.

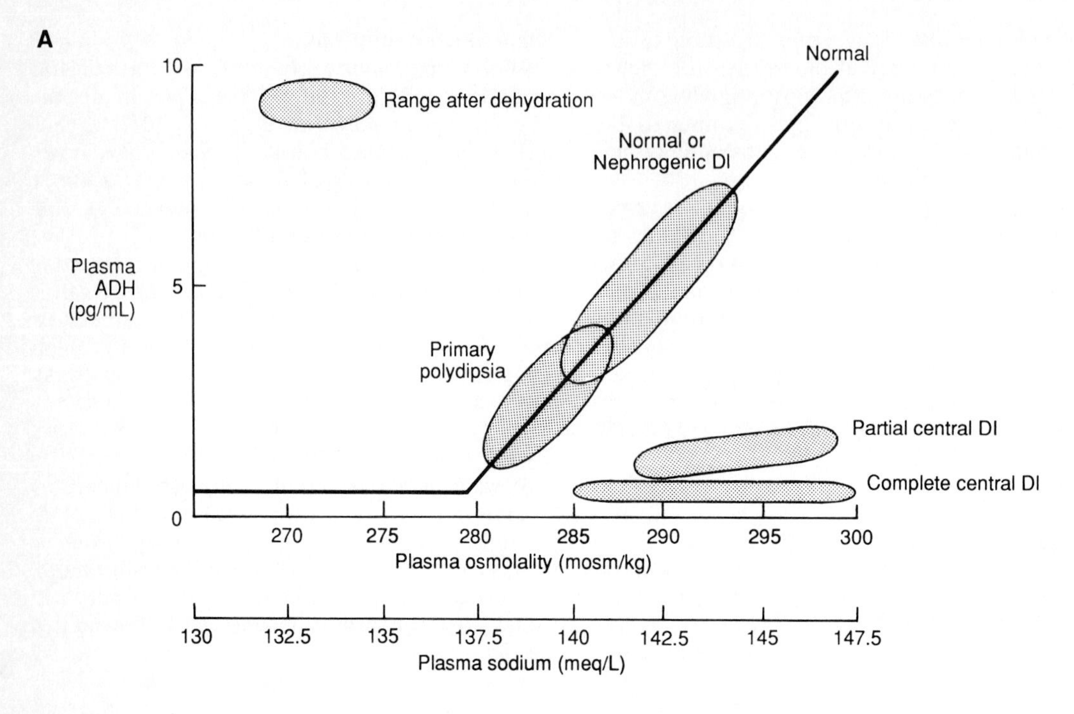

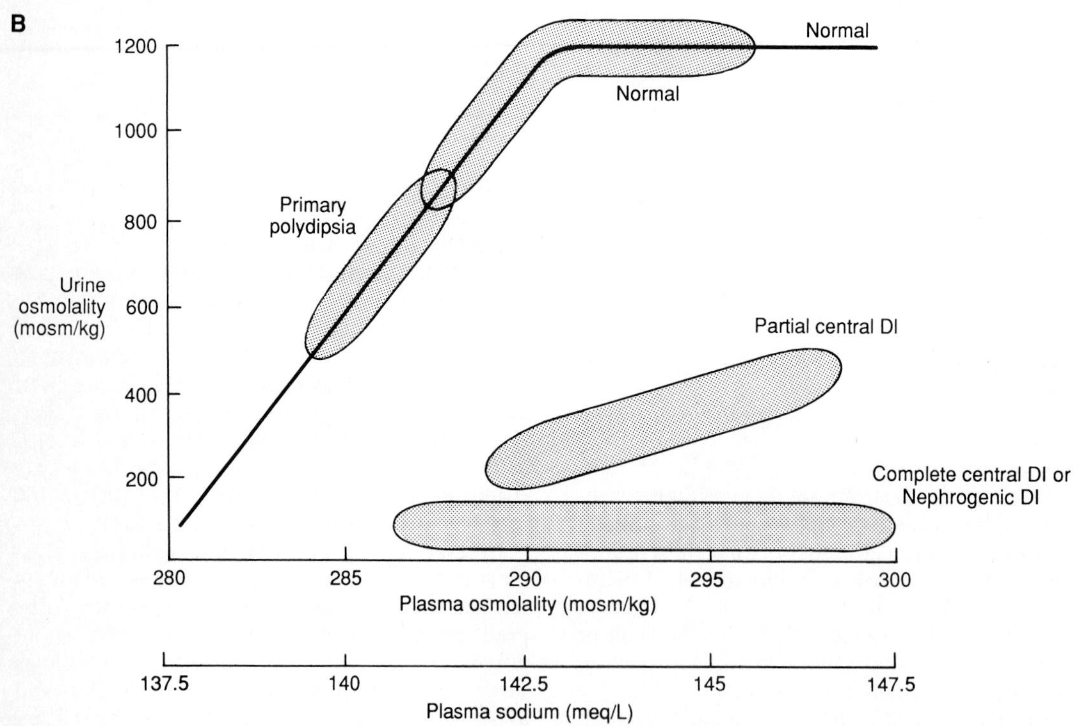

Figure 6-3. Diagnosis of diabetes insipidus (DI).

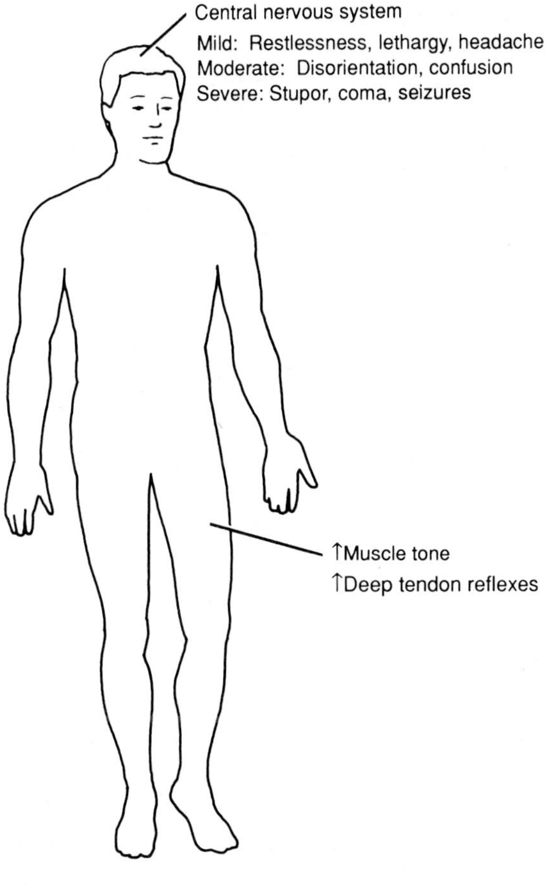

Central nervous system
Mild: Restlessness, lethargy, headache
Moderate: Disorientation, confusion
Severe: Stupor, coma, seizures

$\uparrow$Muscle tone
$\uparrow$Deep tendon reflexes

Figure 6-4. Signs and symptoms of hypernatremia.

Recognition that chronic hypernatremia induces production of these idiogenic osmoles in the brain is the rationale for slowly correcting hypernatremia. If hypernatremia is too rapidly corrected—before the idiogenic osmoles are allowed to dissipate—the plasma may become relatively hypotonic to the brain cells. This osmotic imbalance causes water to enter brain cells, causing brain swelling and potentially severe neurologic impairment.

Peripheral Nervous System

The consequences of hypernatremia on the peripheral neuromuscular system include increased muscle tone, including spasticity, and heightened deep tendon reflexes.

CARDIOVASCULAR

Hypernatremia does not cause direct hemodynamic manifestations, but changes can occur depending on the degree of associated alteration in extracellular volume. There are no systematic electrocardiographic changes.

DIAGNOSIS

APPROACH TO ETIOLOGIC DIAGNOSIS

History & Physical Examination

A careful assessment of water intake and urine volume is of obvious importance in the evaluation of hypernatremic states. Polyuria is defined as excretion of more than 3 L of urine per day. Medical conditions—especially those predisposing to diabetes mellitus, excessive sweating, or diarrhea and hence to hypovolemia—are assessed in the usual manner (Chapter 2). A dietary history is also of use in assessing electrolyte and protein intake.

Drug History

Drugs that can interfere directly or indirectly with the action of ADH, such as loop diuretics, lithium, and democlocycline, or cause tubulointerstitial nephritis, such as analgesics, should be carefully inquired about in the history.

Intracranial Abnormalities

As listed in Table 6–1, hypothalamic or pituitary abnormalities of ADH release are inherent in several causes of hypernatremia (reset osmostat, central diabetes insipidus). A meticulous neurologic examination and CT scan or MRI of the brain are often critical in the diagnostic workup.

Tubulointerstitial Nephropathies

Certain metabolic and toxic agents have a predilection for injury of the renal medulla and papilla, with relative sparing of cortical structures (Table 6–1). They predictably lead to impairment of the countercurrent system and hence renal concentrating mechanisms, even before changes in GFR or electrolyte concentrations appear.

The fact that the renal medulla is one of the most hypoxic, hypertonic, and acidic environments in the body often accounts for this clinical presentation, including propensity for sickling of red cells, precipitation of calcium, oxalate, and uric acid, deposition of certain drugs such as analgesics (analgesic nephropathy), and invasion of bacteria in acute pyelonephritis. The renal medulla is also the site in which functional effects of inherited cystic diseases are sometimes first manifested, including medullary sponge kidney and polycystic kidney disease, as well as other chronic inflammatory or obstructive disorders such as sarcoidosis, amyloidosis, and obstructive nephropathy.

RENAL RESPONSE

Basal Urine Osmolality & Volume

As shown in Fig 6–1, the presence or absence of functional ADH action during hypernatremia can be distinguished simply by assessing urine osmolality and volume. If ADH secretion is appropriately stimulated and the kidney is normally responsive, the urine is concentrated and the urine volume is low or normal. This renal response occurs in the settings of sodium intoxication, the reset osmostat syndrome, and hypotonic skin or gastrointestinal volume losses. The only exception is the renal response to osmotic diuresis in which ADH is high but the urine may be less than appropriately concentrated owing to the high rate of obligate solute (glucose, urea, mannitol, or radiocontrast dye) excretion.

In contrast, lack of ADH or its tubular action is signaled by an inappropriately dilute urine with polyuria in the setting of hypernatremia. The distinction between the two pathophysiologic alternatives of functional ADH deficiency, central and nephrogenic diabetes insipidus, is discussed below.

Plasma ADH Level

The differentiation of many of the hypernatremic states is greatly facilitated by direct radioimmunoassay of plasma ADH (Fig 6–3). Unfortunately, this measurement is difficult, and even specializing commercial laboratories are sometimes unreliable. Indirect tests of ADH action are therefore employed and are usually sufficient. If ADH is measured directly during pregnancy, phenanthrolene (0.1 mL of a 60 mg/mL solution per 10 mL sample) must be used during blood collection to prevent degradation by plasma vasopressinase, markedly increased during gestation.

Differentiation of Polyuric States

When glycosuria or another cause of osmotic diuresis is excluded, there are only four major causes of polyuria: primary polydipsia, complete or partial central diabetes insipidus, and nephrogenic diabetes insipidus. The first usually predisposes to hyponatremia (Chapter 5) and the others to mild hypernatremia. Nevertheless, the plasma sodium concentration does not reliably differentiate these conditions, due both to scatter in the range of normal plasma sodium concentration and to the impressive efficiency of coupling of osmoregulated thirst mechanisms with renal water excretion.

The test most often employed to make the diagnosis of a polyuric condition is the **dehydration test.** Other tests exist—including osmotic stress with hypertonic saline or nonosmotic stress with hypovolemic, emetic, or glycopenic stimuli—but are less well standardized. In the dehydration test, water is withheld for approximately 16–18 hours—less in severely polyuric patients—while weight is allowed to decline by about 3–5% and there is a commensurate rise in plasma osmolality. Urine osmolality eventually becomes constant, changing less than 30 mosm/kg per hour. The patient must be observed closely because hypovolemia may ensue if polyuria persists. At the end of the dehydration period, the plasma osmolality and the accompanying urine osmolality and volume are recorded along with the plasma ADH level, if an accurate assay is available. The plasma osmolality and the urine osmolality and volume response 60 minutes following administration of ADH (aqueous vasopressin, 5 units subcutaneously) are then determined.

Using this dehydration test, as shown in Fig 6–3 and Table 6–2, normal subjects usually concentrate their urine maximally—to 800–1200 mosm/kg—due to an appropriate rise in endogenous ADH, without further increment when exogenous ADH is subsequently administered. Hospitalized patients and patients with primary polydipsia have a normal plasma ADH response to dehydration but a submaximal urine osmolality response,

Table 6–2. Diagnosis of polyuric syndromes.

	Typical Urinary Osmolality After Dehydration (mosm/kg)	Typical Further Urinary Osmolality Change After ADH
Normal	800–1200	0%
Primary polydipsia	500–900	0%
Partial central diabetes insipidus	300–600	15–50%
Complete central diabetes insipidus	100–250	50–500%
Nephrogenic diabetes insipidus	100–400	0%

due to malnutrition and preexisting water diuresis (Fig 4-21), respectively. Patients with complete central diabetes insipidus have no plasma ADH or urine concentrating response to dehydration but have a marked rise in urine osmolality when exogenous ADH is subsequently administered. Those with incomplete central diabetes insipidus have an intermediate response, with a blunted rise in urine osmolality during dehydration and a further small increase in response to exogenous ADH. Finally, patients with nephrogenic diabetes insipidus have a very high level of circulating ADH but no urinary concentrating ability in response to dehydration, and the urine remains dilute following further supplementation of ADH.

THERAPY

SPECIFIC APPROACHES

Hypernatremia With Hypervolemia due to Hypertonic Sodium Intoxication

As summarized in Fig 6-5, when hypertonic sodium is ingested and renal function is normal, simple water administration is sufficient therapy for hypernatremia since excess sodium will be rapidly excreted in minimal urine volume. If GFR is low, diuretics may facilitate sodium excretion, though urinary concentrating ability will be impaired and a greater quantity of water will be needed. If GFR is severely reduced, hemodialysis may be necessary to remove sodium and to correct the hypernatremia.

Hypernatremia With Normovolemia: Reset Osmostat

Water is the acute treatment of hypernatremia due to hypodipsia. Then a regular schedule of water drinking is usually adequate therapy. Adjunctive simulation of thirst has been reported in this condition with chlorpropamide, though without uniform success.

Hypernatremia With Hypovolemia: Skin, Gastrointestinal, or Renal Hypotonic Loss

The primary goal of treatment of hypernatremic, hypovolemic conditions is to replenish the extracellular volume and restore circulatory fullness. Isotonic saline is therefore used first (Chapter 2). The high level of ADH will cause administered water to be retained. Only as euvolemia is approached clinically should the replacement solution be changed to half-normal saline and finally to 5% dextrose in water.

Hypernatremia Due to Central Diabetes Insipidus

When this diagnosis is made, renal water retention can be effected with synthetic compounds

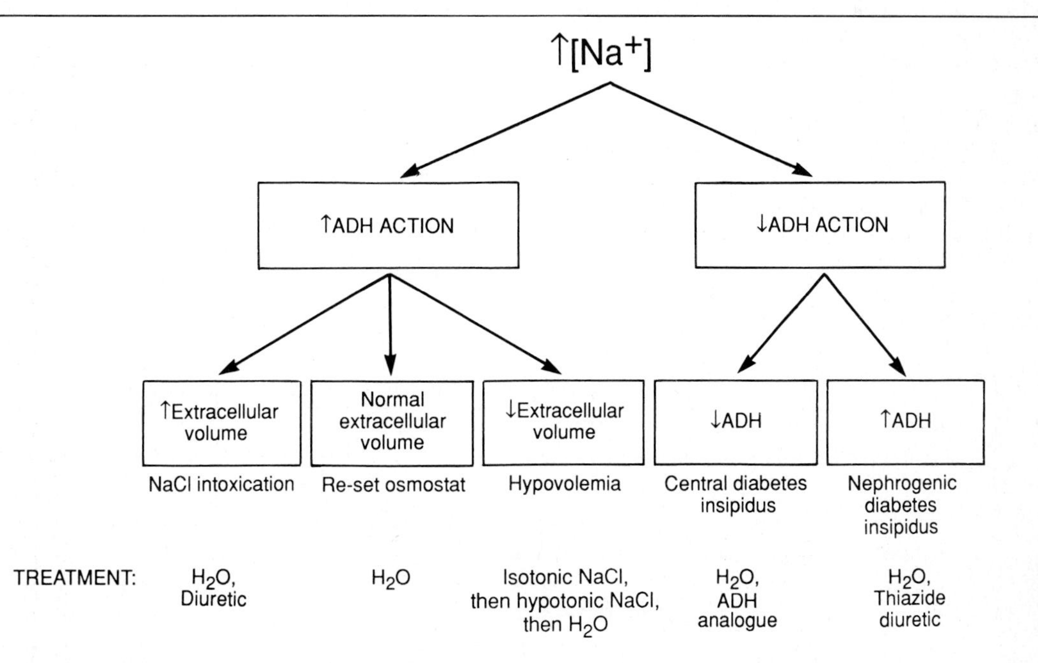

Figure 6-5. Treatment of hypernatremia.

that mimic the action of ADH, as shown in Table 6–3. For acute use, **aqueous vasopressin** (Pitressin) is used, and dosage is adjusted to maintain urine volume less than 3 L/d. Longer-lasting antidiuresis can be obtained with oil-based ADH analogues, but these retain the pressor effect of ADH.

For routine chronic use for treating a patient with central diabetes insipidus, the ADH analogue **desmopressin acetate (DDAVP)** is the recommended agent. The molecular substitutions of desmopressin make it antidiuretic but lacking in the pressor effect of ADH (ie, it binds to V_2 but not V_1 ADH receptors) and with a longer duration of action. Ten micrograms of desmopressin is equivalent to 40 units of vasopressin. Desmopressin is administered by nasal spray in one or two daily doses to maintain urine volume less than 3 L/d. Overadministration can cause hyponatremia (resembling SIADH). Although usually well-tolerated, there can occasionally be symptoms of mild nasal congestion and abdominal cramps.

For partial central diabetes insipidus, chlorpropamide has sometimes proved useful both by increasing ADH release and by enhancing its hydroosmotic action. Carbamazepine or clofibrate is sometimes effective in boosting residual ADH release, but neither drug is widely used because both have undesirable toxic effects.

Hypernatremia Due to Nephrogenic Diabetes Insipidus

When possible, culpable drugs (lithium, demeclocyline, or loop diuretics) should be stopped and electrolyte disturbances (hypokalemia, hypercalcemia) repaired. If lithium cannot be stopped because of psychiatric indications, amiloride, 5–20 mg/d, has been shown to be successful in blocking its uptake into medullary collecting tubule cells and diminishing the concentrating defect.

In severe nephrogenic diabetes insipidus, it is helpful to reduce extracellular volume and hence the volume of tubular fluid delivered to the cortical diluting segment with hydrochlorothiazide (Table 6–3).

ESTIMATION OF WATER DEFICIT

Acute Hypernatremia

With acute dehydration without solute loss, free water loss can be easily equated with weight loss. With hypotonic volume loss, the free water deficit may be less than the weight loss depending on the sodium and potassium concentrations of the fluid lost.

Chronic Hypernatremia

Estimation of water deficit is similar conceptually to the calculation of water excess in hyponatremia:

$$\text{Water deficit} = \text{Total body water} \times \text{Fractional change in plasma } [\text{Na}^+]$$

Since total body water can be estimated in a man as 0.6 × body weight and the fractional change in plasma sodium concentration (P_{Na^+}) can be approximated by (P_{Na^+} − 140)/140, the equation above can be reduced as follows:

$$\text{Water deficit} = (0.6 \times \text{Body weight}) \times \frac{(P_{Na^+} - 140)}{140}$$

When estimating total water replacement, both the initial deficit—calculated as above—plus ongoing water loss during the repletion period must be considered.

GENERAL APPROACH: RATE & ROUTE OF HYDRATION

Slow Correction of Chronic Hypernatremia

The danger of hypernatremia in causing significant, sometimes irreversible neurologic impair-

Table 6–3. Therapy of diabetes insipidus.

	Dose	Route	Onset	Duration
ADH analogues for central diabetes insipidus				
Aqueous vasopressin (Pitressin)	5–10 units	SC	0.5–1 hour	4–6 hours
Vasopressin tannate (Pitressin Tannate in Oil)	2–5 units	IM	2–4 hours	24–72 hours
Desmopressin acetate (DDAVP)	5–20 μg	Nasal spray	0.5–1 hour	12–24 hours
Adjunctive therapy for partial central diabetes insipidus				
Chlorpropamide (Diabenese)	250–750 mg/d	PO	1–2 hours	24–36 hours
Adjunctive therapy for nephrogenic diabetes insipidus				
Hydrochlorothiazide (Esidrix, Hydro-Diuril)	50–100 mg/d	PO	1–2 hours	12–24 hours

ment requires its correction. However, the adaptive increase in brain osmolality due to idiogenic osmole synthesis that occurs during hypernatremia which has persisted longer than 6–12 hours mandates cautious water replacement. Thus, rapid correction of extracellular osmolality is very dangerous, potentially causing water to rush into brain cells. Brain edema with severe neurologic symptoms—including seizures and death—may result. Thus, once the water deficit is calculated, the time for replenishment should be prolonged, usually over the course of more than 48 hours, depending on severity. The rate of correction of plasma sodium concentration should not exceed 1 meq/L per hour.

As mentioned previously, when hypovolemia is present, the first intravenous solution administered should be normal saline followed by half-normal saline. Even normal saline, with an osmolality of 308 mosm/kg is often less than that of the patient. Following repair of hypovolemia, the solution may be switched to a more dilute one.

Given the hazards of hypernatremia itself as well as the overly rapid correction of hypernatremia, frequent monitoring of the plasma sodium concentration is essential. Such frequent assessment is required because estimation of water deficit is imprecise and there are often unpredictable ongoing water and solute losses.

Intravenous Versus Enteral Water Replacement

With symptomatic hypernatremia, the route of water administration should be intravenous to ensure accurate and quantitative therapy. Free water is given in the form of **5% dextrose in water.** The 5% sugar is needed to avoid inducing hemolysis on contact with blood; it is rapidly metabolized once in the bloodstream.

When hypernatremia is not severe and there are no problems with gastrointestinal absorption or motility, water may be given orally. A nasogastric tube may also be used, but if the patient is comatose a cuffed endotracheal tube should be in place to avoid aspiration.

SECTION II REFERENCES: WATER HOMEOSTASIS & DISORDERS: HYPONATREMIA & HYPERNATREMIA

Anderson RJ: Hospital-associated hyponatremia. Kidney Int 1986;29:1237–1247.

Edelman IS, Leibman J: Anatomy of body water and electrolytes. Am J Med 1959;27:256–277.

Berl T: Treating hyponatremia. Kidney Int 1990;37:1006–1018.

Jamison RL, Oliver RE: Disorders of urinary concentration and dilution. Am J Med 1982;72:308–322.

Pollock AS, Arieff AI: Abnormalities of cell volume regulation and their functional consequences. Am J Physiol 1980;239:F195–F205.

Robertson GL: Physiology of ADH secretion. Kidney Int 1987;32(Suppl 21):S20–S26.

Rose BD: New approach to disturbances in plasma sodium concentration. Am J Med 1986;81:1033–1040.

Robertson GL, Aycinena P, Zerbe RL: Neurogenic disorders of osmoregulation. Am J Med 1982;72:339–353.

Schmale H, Fehr S, Richter D: Vasopressin biosynthesis: From gene to peptide hormone. Kidney Int 1987;32(Suppl 21):S8–S13.

Section III: Potassium Homeostasis & Disorders: Hypokalemia & Hyperkalemia

Normal Potassium Homeostasis 7

INTRODUCTION

TOTAL BODY POTASSIUM

Potassium is the second most abundant cation within the body, contributing about 50 meq/kg body weight in a young adult male. The body of an average-size man of 70 kg therefore contains about 3500 meq of potassium. Females have somewhat less potassium (about 40 meq/kg). Body potassium content declines as a function of age by about 2 meq/kg per decade, chiefly as a result of decrease in muscle mass.

As the most abundant intracellular cation, potassium is critical for many metabolic functions of cells, such as for optimal action of many enzymes, for growth and division, and for maintaining normal cell volume. The distribution of potassium between the internal and external environments of a cell is also important because it defines the electrical potential across the cell membrane and thus affects excitation and contraction in neuromuscular cells.

Ratio of Intracellular to Extracellular Potassium

Our ability to estimate total body potassium is hampered because 98% of all potassium is inside cells, as shown in Fig 7–1. The potassium concentration within cells is very high, about 120–150 meq/L, while that outside of cells is only about 4 meq/L. The clinical estimation of body potassium by sampling the blood serum is therefore inexact, since it is representative of only that portion of total body potassium—about 2%—resident in the extracellular fluid. Serum potassium measurement may not accurately reflect the status of total body potassium content for at least two reasons: the distribution of potassium between the intra- and extracellular compartments may change even when the total body content of potassium remains normal; or change in potassium concentration in each of the two compartments may not be proportionate when total body potassium content is abnormally high or low.

Serum Potassium

Potassium concentration, usually measured in the serum, is normally regulated close to 4.0 meq/L, with 95% confidence limits of **3.5–5.0 meq/L.** Maintaining this concentration within such narrow bounds is a compelling physiologic responsibility, since even a minute change in the 49:1 ratio of intracellular:extracellular potassium can greatly alter the serum potassium concentration. The extracellular fluid could be flooded with potassium if even a small fraction of the intracellular potassium stores were allowed to exit cells.

The problem of maintaining potassium homeostasis is made even more complex because potassium continually derived from the diet is added to the extracellular compartment more quickly than the kidneys can excrete it. The dietary intake of potassium in a single day may equal or exceed the total potassium normally present within the extracellular fluid. To illustrate, consider the consequence of adding 80 meq of dietary potassium to the 20-L extracellular space, which already contains 80 meq of potassium at a concentration of 4 meq/L. If this newly added potassium were not efficiently and rapidly translocated into cells, a doubling of serum potassium concentration to a life-threatening level of 8.0 meq/L would obviously ensue. Intracellular storage of potassium "buffers" the extracellular potassium concentration until relatively sluggish renal mechanisms can excrete the potassium load.

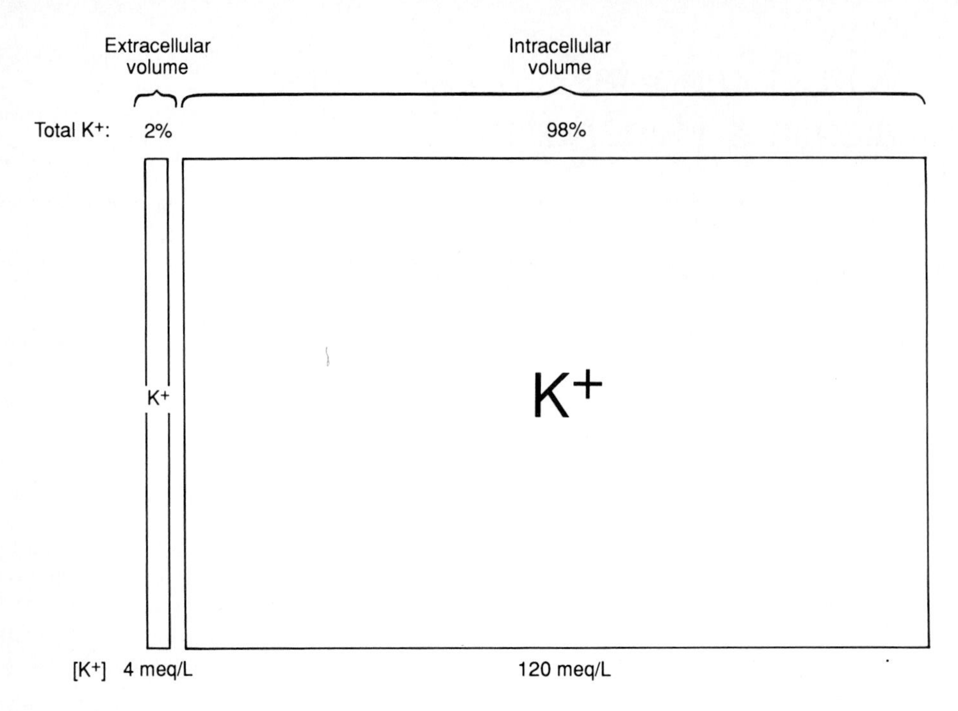

Figure 7-1. Distribution of body potassium.

POTASSIUM INTAKE & DISTRIBUTION

POTASSIUM INTAKE

The amount of potassium ingested in industrialized Western countries is about 50-100 meq/d, or roughly 0.7-1.3 meq/kg body weight daily (Fig 7-2). In certain cultures, however, intakes as low as 10 meq/d and as high as 400 meq/d have been recorded. There is no necessary relationship between the amounts of potassium and of sodium in the diet.

Whereas chloride intake is generally equivalent to sodium intake, the anion accompanying potassium in most natural foods is usually nonchloride, such as phosphate or an organic anion. Since nonchloride anions are poorly reabsorbed by the kidney, ingestion of these potassium salts helps the kidneys to subsequently eliminate potassium. Ingestion of potassium as the chloride salt usually occurs only in the form of salt substitutes—used to flavor low-sodium foods—and by prescription to combat diuretic-induced potassium wastage.

POTASSIUM DISTRIBUTION & TRANSLOCATION INTO CELLS

As shown in Fig 7-3, only about 50% of a potassium load administered orally or intravenously is excreted by the kidneys within 6-8 hours. To avoid a buildup of potassium within the extracellular fluid during the period required for renal elimination, potassium is temporarily stored in cells such as those of muscle, liver, red cells, and bone. As shown in Fig 7-2, these combined intracellular storage reservoirs are enormous (3500 meq) compared to the extracellular pool (60 meq) or to the usual amount of potassium ingested in the diet (80 meq/d).

Critical Role of Na^+/K^+-ATPase

As shown in Fig 7-4, the entry step for potassium across the cell membrane is via the Na^+/K^+-ATPase pump. Na^+/K^+-ATPase activity is controlled by several hormones, notably insulin, catecholamines, and aldosterone. A small rise in extracellular potassium—as a consequence of dietary intake, exercise, or acute hyperosmolality—stimulates secretion of these hormones, which activate the Na^+/K^+-ATPase to allow movement of potassium into cells. The transfer of potassium into cells thereby returns extracellular potassium concentration to a normal value.

Hormonal activation of the Na^+/K^+-ATPase pump increases the intracellular potassium con-

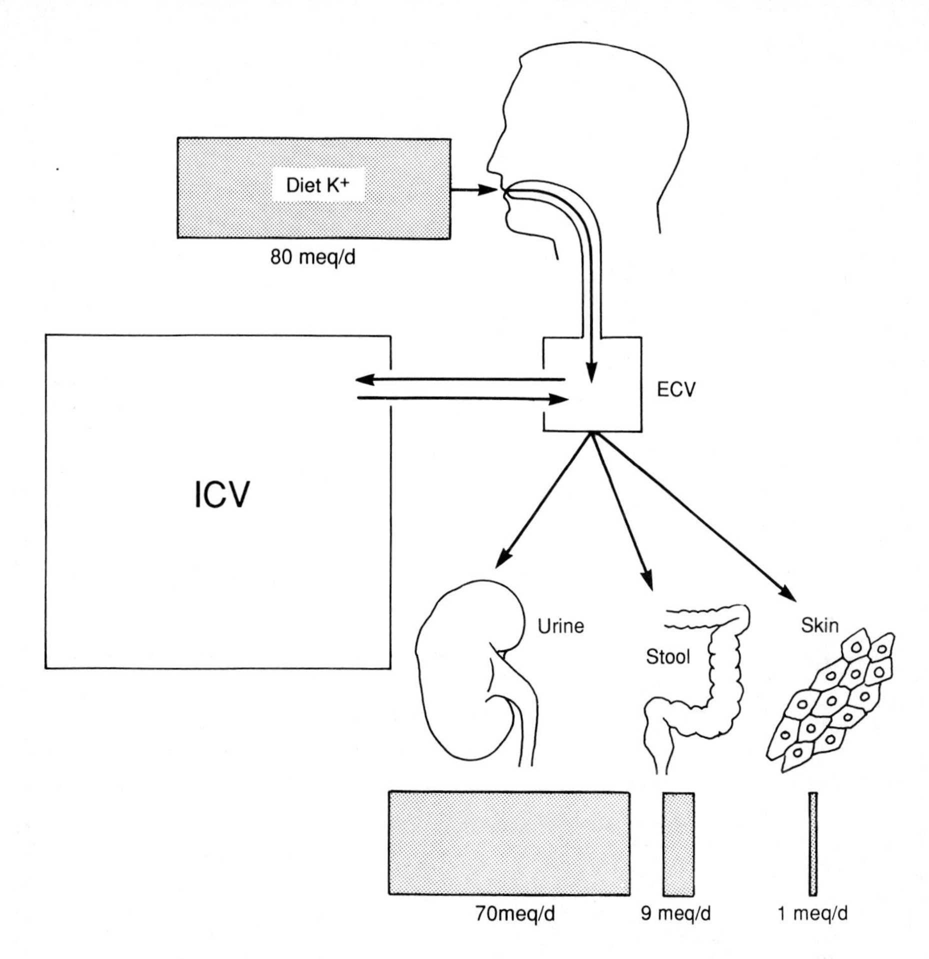

Figure 7-2. Normal potassium intake and routes of excretion.

centration by two mechanisms. First, stimulation of the pump directly transports more potassium into the cell. Second, the Na^+/K^+-ATPase is electrogenic, pumping three positive charges (sodium) out of the cell compared with two positive charges (potassium) in. Therefore, pump stimulation causes the cell interior to become more electrically negative (hyperpolarized), thus diminishing the outward leakage of potassium through membrane channels that conduct potassium.

As shown in Fig 7-5, if any one of these hormones were absent or inhibited, the Na^+/K^+-ATPase pump activity is reduced and potassium is then allowed to exit the cell, inducing a small (0.5-1.0 meq/L) rise in the steady-state serum potassium level. In this circumstance of hormone deficiency

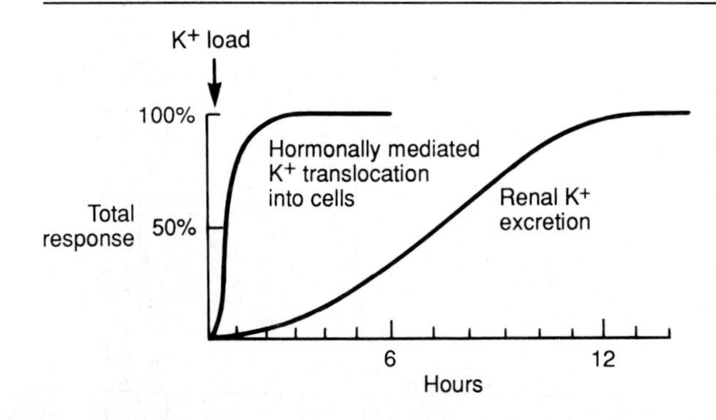

Figure 7-3. Time course of distribution and renal excretion following a potassium load.

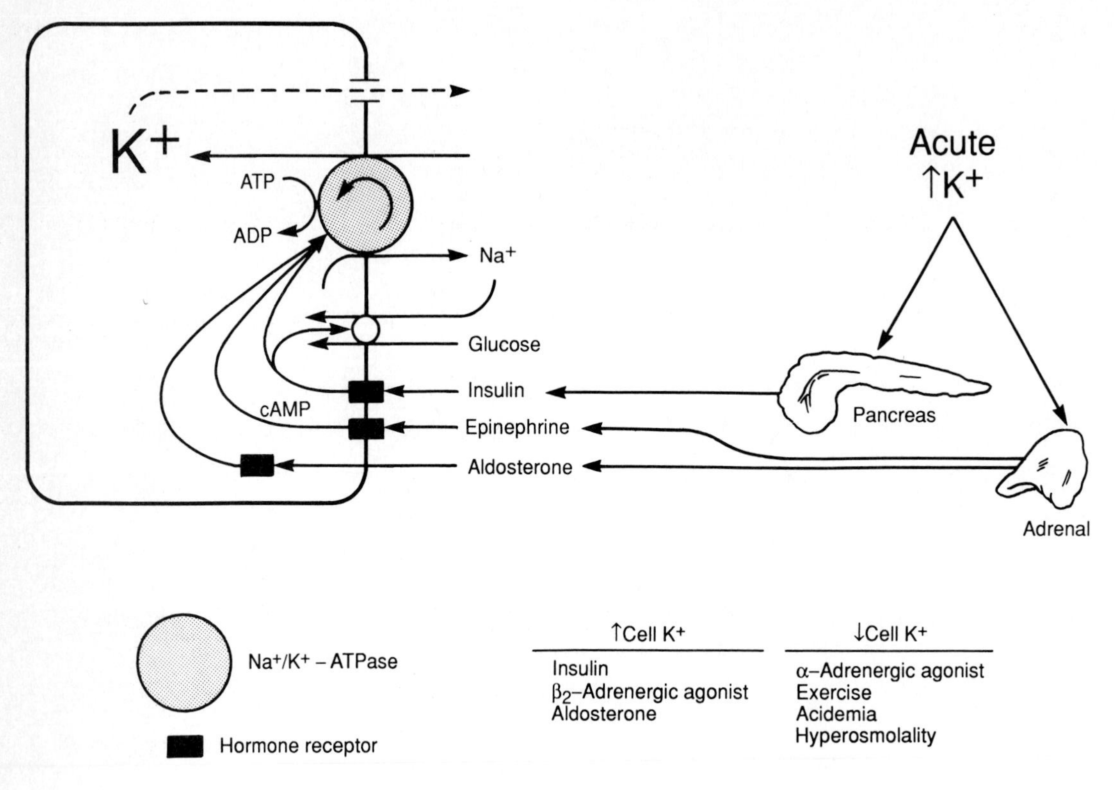

Figure 7-4. Factors regulating potassium uptake into cells.

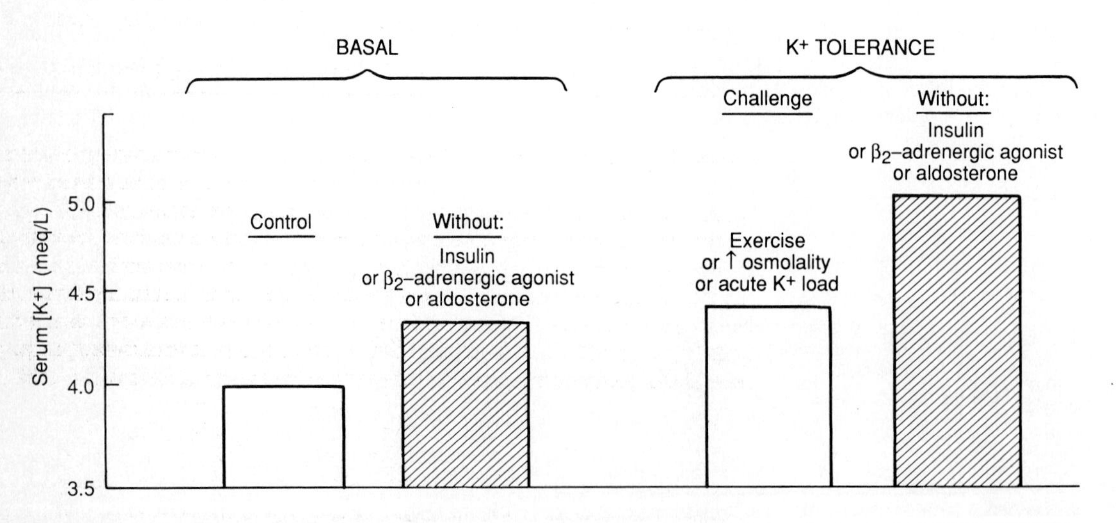

Figure 7-5. Role of insulin, beta-adrenergic agonists, and aldosterone in regulating serum potassium concentration.

or inactivity, an even larger increment in serum potassium concentration would occur if an acute load were encountered.

MAJOR PHYSIOLOGIC REGULATORS OF NA^+/K^+-ATPASE ACTIVITY

Insulin

Insulin plays a central role in maintenance of a normal extracellular-intracellular distribution of potassium. Although the peripheral blood insulin level does not rise unless the serum potassium concentration is acutely increased by more than 1 meq/L, it is probable that pancreatic insulin secretion and the portal insulin level increases as a feedback response to even a small rise in extracellular potassium concentration.

The targets for insulin's action are predominantly cells in liver, muscle, and adipose tissues. As illustrated in Fig 7–4, in each of these cells, binding of insulin to its cell surface receptor causes stimulation of the Na^+/K^+-ATPase, though the second messenger is unknown. A change in intracellular sodium or calcium or in enzymatic activity of the insulin receptor itself (which is a tyrosine kinase) may play a role in functionally coupling the receptor to the Na^+/K^+-ATPase. Insulin may also affect the channels by which potassium exits the cell.

The insulin-mediated enhancement of Na^+/K^+-ATPase activity and potassium influx into cells is mechanistically independent of the effect of insulin to increase glucose uptake into the cell (Fig 7–4). Recruitment of glucose transporters into the cell membrane and enhanced glucose influx by insulin is not required for—and can be dissociated from—change in potassium transport. Similarly, other effects of insulin, such as stimulation of Na^+/H^+ antiporter activity and cell growth, may contribute indirectly to change in intracellular potassium content but are not inherently linked to the primary change in potassium transport.

$Beta_2$-Adrenergic Agonists

The most important of the endogenous $beta_2$-adrenergic agonists, **epinephrine**, is at least permissive for normal potassium homeostasis. Other selective $beta_2$ agonists, including terbutaline and salbutamol, have similar effects. Conversely, inhibition of $beta_2$-adrenergic agonist activity by means of sympathectomy, adrenalectomy, or pharmacologic blockade not only increases the basal potassium level but also impairs tolerance to a potassium load (Fig 7–5). However, it is not yet certain that a feedback loop exists whereby a small rise in serum potassium concentration increases catecholamine release and allows potassium transfer into cells.

As shown in Fig 7–4, binding of epinephrine or another $beta_2$-adrenergic agonist to its receptor, located on cells of skeletal muscle and perhaps liver, initiates a rapid cascade, including (1) activation of membrane-associated adenylate cyclase (via a stimulatory G protein); (2) generation of the second messenger cAMP; (3) stimulation of the Na^+/K^+-ATPase; (4) and enhanced potassium uptake into the cell. Phosphodiesterase inhibitors block the breakdown of cAMP and potentiate the effect of a $beta_2$-adrenergic agonist. Stimulation of the Na^+/K^+-ATPase pump by epinephrine is additive to insulin, since they use different second messengers.

Aldosterone

There is increasing evidence that aldosterone has at least a permissive role in the extrarenal regulation of potassium distribution. Compared to insulin and catecholamines, aldosterone blood level is more sensitively regulated by the extracellular potassium concentration. However, the time necessary for aldosterone to exert its action is longer, measured in hours rather than minutes.

Aldosterone acts on the salivary gland, sweat glands, and colon, but these effects cannot account for the induced tolerance to potassium loading. More important sites of action are probably muscle, liver, and adipose cells. The mechanism of action of aldosterone in extrarenal tissues is similar to that in the kidney. The mineralocorticoid hormone first binds to a high-affinity receptor in the cytosol, and the hormone-receptor complex migrates to the nucleus, induces DNA translation, and enhances protein synthesis. The amount of Na^+/K^+-ATPase may be increased directly or its activity indirectly augmented by enhanced sodium entry into the cell.

OTHER FACTORS THAT REGULATE NA^+/K^+-ATPASE ACTIVITY

Potassium distribution can also be affected by other factors, but they may not exert much influence on minute-to-minute potassium homeostasis under physiologic conditions.

Alpha-Adrenergic Agonists

Potassium loss from the cell is enhanced by high alpha-adrenergic agonist tone. There is normally little tonic influence since an alpha-antagonist has little effect on serum potassium under basal conditions. Administration of a catecholamine with mixed alpha and $beta_2$-adrenergic activity, such as epinephrine, causes a brief (1- to 3-minute) rise in serum potassium followed by a prolonged fall. This latter phase signifies that $beta_2$ activity eventually outweighs alpha activity of such a mixed adrenergic agonist.

However, increased alpha-adrenergic tone causes deterioration of tolerance to potassium loads, accounting at least in part for the rise in serum potassium that usually occurs during exercise (Fig 7–5). Serum potassium rises only slightly (< 1 meq/L) with mild to moderate exercise but may increase as much as 2 meq/L with prolonged exercise that nears the point of exhaustion.

Arterial pH

A decrease in extracellular pH tends to cause potassium release from cells, although a unifying explanation for these phenomena is not presently available. Actually, only acidemia induced by hydrochloric acid infusion has any substantial impact on serum potassium concentration, as summarized in Table 7–1.

A reasonable hypothesis to explain the apparent ion coupling is that the high extracellular hydrogen ion concentration (low pH) slows the operation of the Na^+/H^+ antiporter present on all cell membranes, as shown in Fig 7–6 (middle). As a result of reduction in Na^+/H^+ exchange, there is diminished entry of sodium into the cell and reduction of intracellular sodium concentration, which in turn slows the normal rate of potassium uptake into the cell via the Na^+/K^+-ATPase. Passive potassium exit continues that depletes cell potassium content. The anion companion, chloride, cannot permeate the cell membrane and remains outside the cell. The overall result of the decrease in external pH is then net potassium extrusion from the cell. A decrease in pH of 0.10 unit by hydrochloric acid infusion increases the serum potassium concentration by 0.7 meq/L (Table 7–1).

In contrast, metabolic acidosis due to increase in organic acid production (ketoacidosis or lactic acidosis) does not substantially increase the serum potassium concentration. In these disorders, entry of hydrogen ion into the cell is accompanied by the organic anion that can easily permeate the cell membrane, as shown in Fig 7–6 (bottom). Intracellular release of the hydrogen ion coupled with ongoing acidification of the cell by metabolic processes prevents slowing of the Na^+H^+ antiporter and hence of the Na^+/K^+-ATPase and does not induce cellular potassium release.

Metabolic alkalosis tends to have a small opposite effect on serum potassium concentration. Serum potassium concentration falls by about 0.3 meq/L per increase in arterial pH by 0.10 unit. For instance, in severe metabolic alkalosis with arterial pH 7.60, serum potassium decreases by about 0.6 meq/L owing to the alkalemia itself, though a greater degree of hypokalemia is invariably observed owing to concomitant renal potassium wasting (Chapter 13). The fall in serum potassium during metabolic alkalosis is accentuated if there is preexisting hyperkalemia, reaching 1.0–1.5 meq/L per 0.1 unit increase in pH.

Both respiratory acidosis and respiratory alkalosis cause very little change in the serum potassium concentration—less than 0.1–0.2 meq/L per pH change of 0.1 unit (Table 7–1).

Table 7–1. Changes in serum potassium concentration attributable to arterial blood pH changes.

	ΔpH	Δ[K^+] (meq/L)
Metabolic acidosis		
Mineral (HCl)	↓0.1	↑0.7
Organic	↓0.1	0
Respiratory acidosis	↓0.1	↑0.1
Metabolic alkalosis	↑0.1	↓0.3
Respiratory alkalosis	↑0.1	↓0.2

Plasma Osmolality

Acute plasma hyperosmolality redistributes potassium out of cells. As shown in Fig 7–7, when a solute that cannot permeate the cell membrane such as glucose (without insulin) or mannitol is added to the blood, the increased extracellular osmolality causes water to flow out of cells. The shrinkage of the cell concentrates its potassium, which is the most abundant intracellular solute. The rise in intracellular potassium concentration causes potassium to diffuse out of the cell.

A 1.0- to 1.5-meq/L increment in serum potassium concentration occurs in response to an acute increase in plasma osmolality of only 10%, or 30 mosm/kg. The hyperosmolality-induced rise in extracellular potassium is normally ameliorated by the opposing actions of insulin, catecholamines, and aldosterone, so that a larger rise may obtain in diabetics lacking these hormones (Fig 7–5).

With addition to the blood of a solute that can permeate the cell membrane, such as urea or ethanol, plasma hyperosmolality occurs but there is no transcellular osmotic gradient. Consequently, no water or potassium shifts occur in these hyperosmolal states. The reverse circumstance, hypo-osmolality, usually occurs slowly and does not affect the serum potassium concentration.

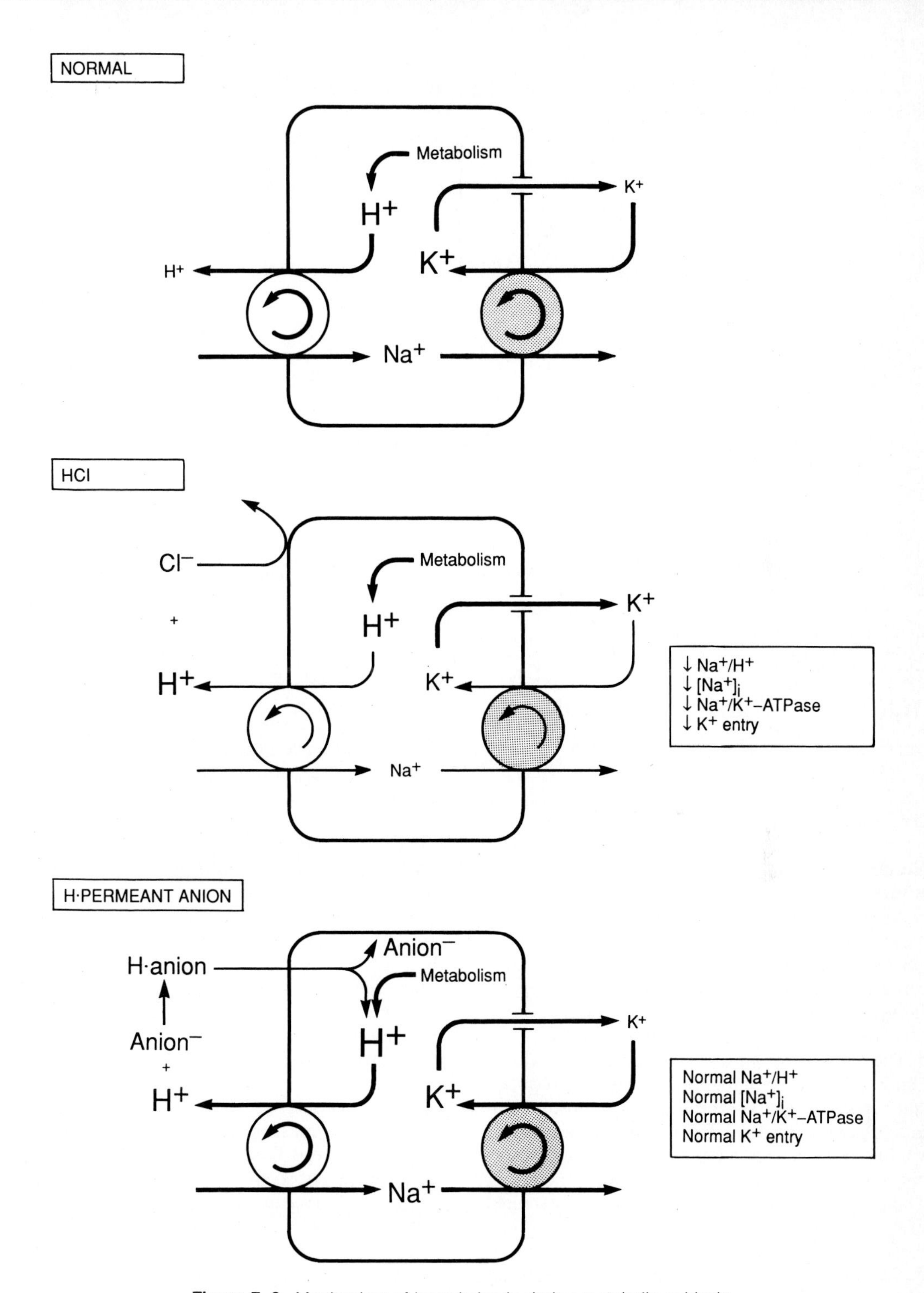

Figure 7-6. Mechanism of hyperkalemia during metabolic acidosis.

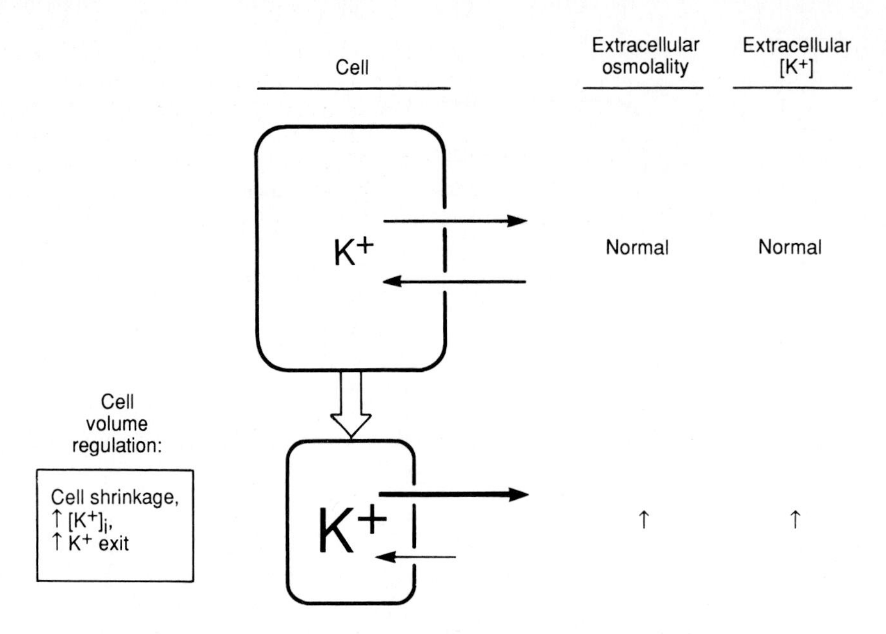

Figure 7-7. Mechanism of hyperkalemia during hyperosmolar states.

RENAL POTASSIUM EXCRETION

OVERVIEW

Nonrenal & Renal Excretory Routes

After being stored in cells, ingested potassium is excreted from the body (Fig 7-3). Before we undertake a detailed consideration of the major route of potassium excretion—ie, the kidney—it should be appreciated that there are small obligatory losses in sweat and in stool (Fig 7-2).

The concentration of potassium in sweat is about 9 meq/L, so that about 1-2 meq/d of potassium is always lost via the skin in the normal volume of sweat (0.1-0.2 L/d). An increase in the volume of sweat induced by hot climate or exercise causes an increase in potassium excretion. There is also about 9 meq/d of potassium excreted in stool (90 meq/L in 0.1 L of stool water per day; see Table 1-2). Increase in stool water content, as during diarrheal illness, can increase the amount of potassium lost by this route.

By far the most important route for eliminating potassium from the body is the kidney. For example, if dietary potassium intake were 80 meq/d, about 90% (or 70 meq/d) would be excreted by the kidneys (Fig 7-2), though several hours are required (Fig 7-3). Unlike the relatively fixed sweat and stool losses, the kidney adjusts its excretory rate to dietary potassium intake with precise accuracy. Dietary potassium loads of 10 meq/d to over 400 meq/d can be excreted by the kidney without significant alteration in serum potassium concentration.

Filtration, Near-Complete Reabsorption, & Differential Secretion

Unlike most electrolytes, urinary potassium excretion is governed by tubular secretion rather than reabsorption. Filtered potassium is efficiently reabsorbed in the proximal tubule and loop of Henle, and urinary potassium is derived from readdition of potassium to the luminal fluid predominantly in the cortical collecting tubule.

That regulation of renal potassium excretion is governed by secretion rather than reabsorption may at first seem surprising. Given that the kidney has an enormous GFR—about 150 L/d—the free filtration of potassium at 4.0 meq/L amounts to 600 meq/d. Since excretion of a dietary intake of only 100 meq/d or less is usually required, it would seem that incomplete (85%) reabsorption would easily do the trick. Regulation of potassium excretion by a secretory rather than reabsorptive process may be viewed as beneficial, however, in those unusual situations of renal dysfunction when the filtered potassium is less than potassium intake.

SITES OF POTASSIUM REABSORPTION & SECRETION

The sites of potassium transport are summarized in Fig 7–8. Potassium is freely filtered at the glomerulus; ie, the concentration of potassium in the glomerular ultrafiltrate is virtually identical to that in plasma water.

Proximal Tubule

About 40–50% of the filtered potassium is reabsorbed in the proximal convoluted tubule in proportion to sodium and water, since potassium concentration in the tubular fluid changes little. Potassium reabsorption is mediated probably by both active and passive mechanisms. Further reabsorption may also occur in the proximal straight tubule.

Descending Limb of Henle

As tubular fluid flows down the thin descending limb of Henle, potassium is added by passive diffusion from the medullary interstitium. This potassium is derived from reabsorption occurring in adjacent nephron segments—the thick ascending limb of Henle and the medullary collecting tubule. Potassium from the latter source in turn comes primarily from that secreted into the lumen by the cortical collecting tubule. Thus, the loop of Henle participates in recycling potassium (see below).

Thick Ascending Limb of Henle

Potassium is avidly reabsorbed in the thick ascending limb of Henle. The luminal membrane transport mechanism involves translocation of a potassium ion in conjunction with a sodium ion and two chloride ions (Fig 1–25). Virtually all of the potassium escaping proximal reabsorption as well as that which has leaked into the descending limb of Henle is reabsorbed by the thick ascending limb of Henle. Only about 10% (range, 5–15%) of the filtered potassium load remains by the time tubular fluid enters the distal convoluted tubule. A diuretic that acts on the thick ascending limb of Henle—eg, furosemide, bumetanide, or ethacrynic acid—blocks normal potassium reabsorption.

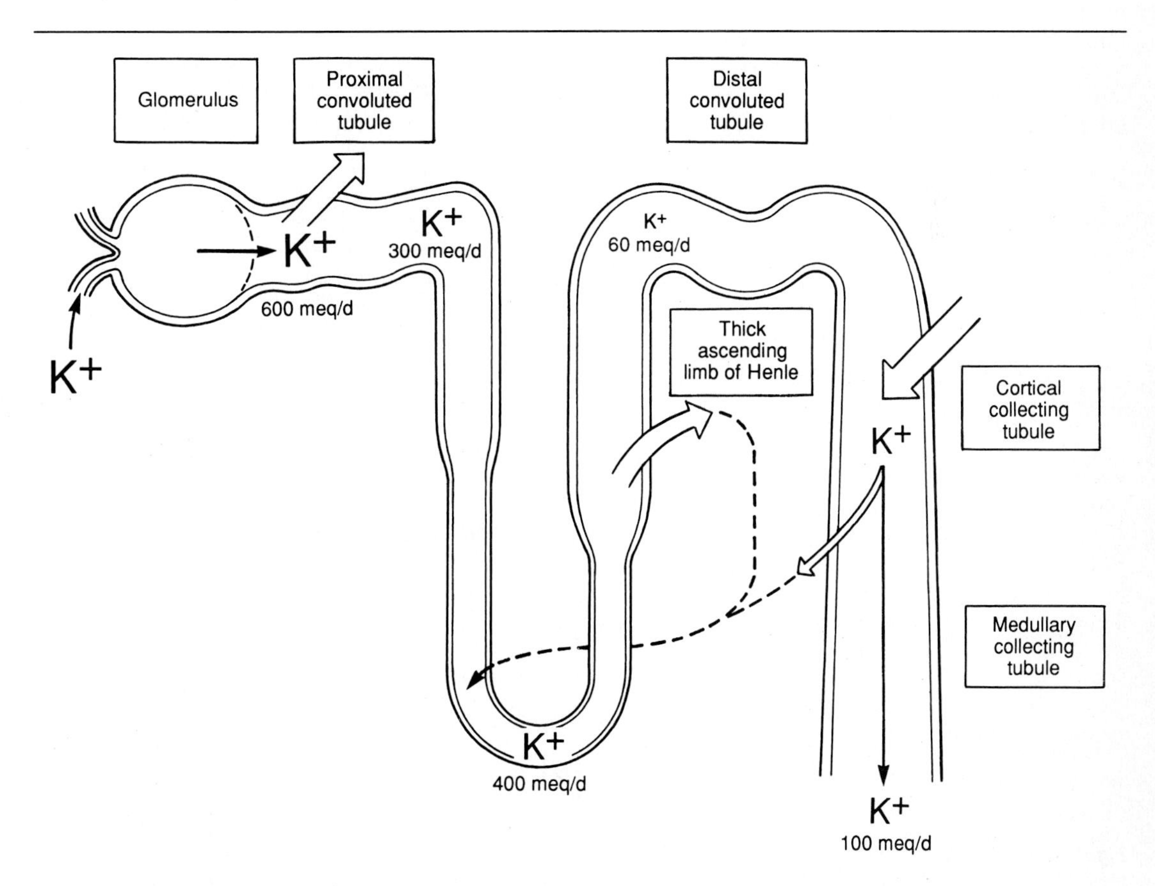

Figure 7-8. Normal amounts of potassium filtered, reabsorbed, secreted, and excreted.

Cortical Collecting Tubule

The major site of renal potassium secretion is the cortical collecting tubule. The regulation of transport in this segment will be discussed in greater detail below. Since the cell responsible for potassium secretion, the **principal cell,** also resides in the terminal portion of the distal convoluted tubule, the connecting tubule, and the outer stripe of the outer medulla, potassium secretion occurs in these segments as well.

Another cell type in the cortical collecting tubule—probably the **intercalated cell**—reabsorbs potassium. Usually, the principal cell dominates the overall transport process, with the result that net potassium secretion occurs. However, net potassium reabsorption can happen when transport by the principal cell is turned off.

Medullary Collecting Tubule

Finally, a relatively small amount of potassium is reabsorbed by cells of the medullary collecting tubule. The mechanism of potassium reabsorption is not fully understood but may be mediated at least in part by a potassium-activated H^+-ATPase similar to the one in the stomach.

Potassium reabsorbed in the medullary collecting tubule is derived predominantly from secretion upstream in the cortical collecting tubule. Some of the reabsorbed potassium transferred into the medullary interstitial fluid, together with potassium derived from reabsorption in the thick ascending limb of Henle (see dotted lines in Fig 7–8), is secreted into the descending limb of Henle. While much of this recycled potassium is reabsorbed again by the thick ascending limb, some escapes into the distal convoluted tubule and therefore amplifies the amount of potassium secreted directly by the cortical collecting tubule. When cortical collecting tubule potassium secretion is very low, potassium reabsorption in the medullary collecting tubule participates in potassium retention.

POTASSIUM TRANSPORT IN THE CORTICAL COLLECTING TUBULE

Overview

Fig 7–9 presents a schematic representation of potassium transport in the principal cell of the cortical collecting tubule. Most potassium transport in the principal cell occurs transcellularly; little potassium leaks around the cell. Potassium is pumped into the cell from the blood side by the Na^+/K^+-ATPase located on the peritubular or basolateral membrane. After entry into the cell, potassium is secreted across the luminal membrane through a channel that is selective for potassium. The driving force that governs how much potassium crosses this channel is the electrochemical gradient, defined as the sum of the cell-to-lumen potassium concentration gradient plus the cell-to-

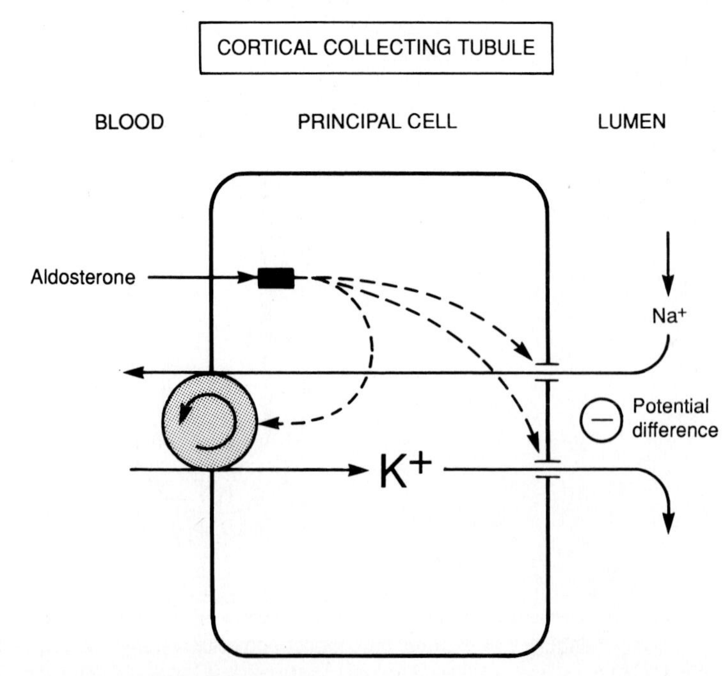

Figure 7-9. Cellular mechanism of potassium secretion in the cortical collecting tubule.

lumen electrical potential difference. The luminal potential difference is defined largely by the rate of sodium reabsorption by the principal cell. Aldosterone is a potent regulator of most of these steps in potassium secretion.

Electrochemical Driving Force

The difference in potassium concentrations that exist on either side of the luminal membrane of the principal cell (eg, 150 meq/L in the cell compared to 15 meq/L in the lumen) favors the diffusive migration of potassium from the cell into the lumen, as shown in Fig 7–10. This is the chemical component of the electrochemical driving force that "pushes" potassium out of the cell through its selective channel.

The difference in electrical potential across the luminal cell membrane tends to "pull back" potassium and keeps potassium within the cell, as illustrated in Fig 7–11 (top). This electrical component of the electrochemical gradient represents the difference in the cell and lumen electrical potentials, each relative to blood. The cell is relatively more electronegative than the lumen, so the positively charged potassium ion is restrained from crossing the luminal membrane.

Under physiologic circumstances, the potential within the cell probably varies relatively little (ie, is close to −90 mV compared to blood). However, the potential in the lumen can vary markedly, from 0 to over −40 mV compared to blood, as shown in Fig 7–11 (bottom), but it never exceeds the negative value within the cell. Therefore, in the above examples, the electrical force across the luminal membrane varies from −90 mV to −50 mV, but always in the antisecretory (absorptive) direction.

The lumen-negative potential is generated by the amount of concurrent sodium reabsorption (Fig 7–9). Absorption of the positively charged sodium across the luminal membrane occurs through a selective sodium channel driven by the lumen-to-cell sodium concentration gradient. The low intracellular sodium concentration is maintained by the Na^+/K^+-ATPase. Since sodium flows through its channel "by itself," without an accompanying anion, its transport is electrogenic and the lumen-negative electrical potential is thereby created.

Change in sodium reabsorption and in the magnitude of the lumen-negative electrical potential can greatly mitigate the overall electrical restraining force on potassium secretion. Consider, for example, the first case illustrated previously in which the cell is interiorly negative at −90 mV and sodium absorption is low, so that the lumen has a potential of 0 mV (both voltages referent to blood). The transluminal membrane potential difference—obtained by subtracting the luminal from the cell potentials—is −90 mV (−90 mV − 0 mV) (Fig 7–11, top). On the other hand, if sodium reabsorption were high as a result of a rise

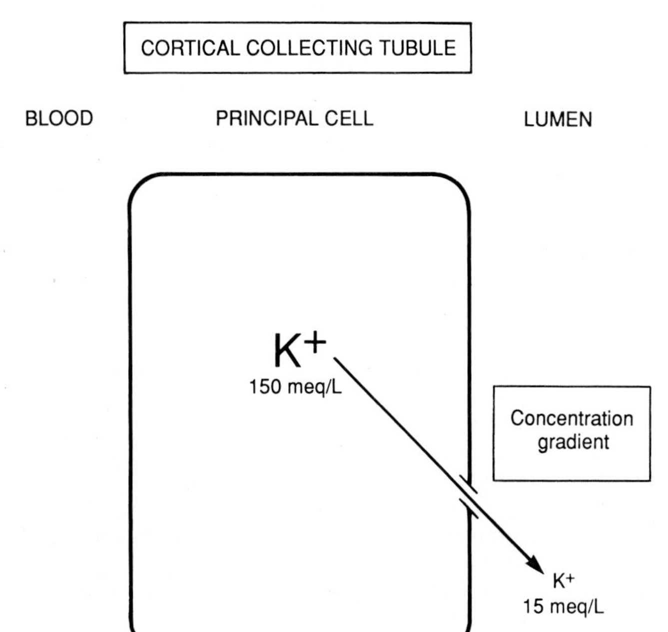

Figure 7–10. Chemical concentration driving force for K^+ secretion.

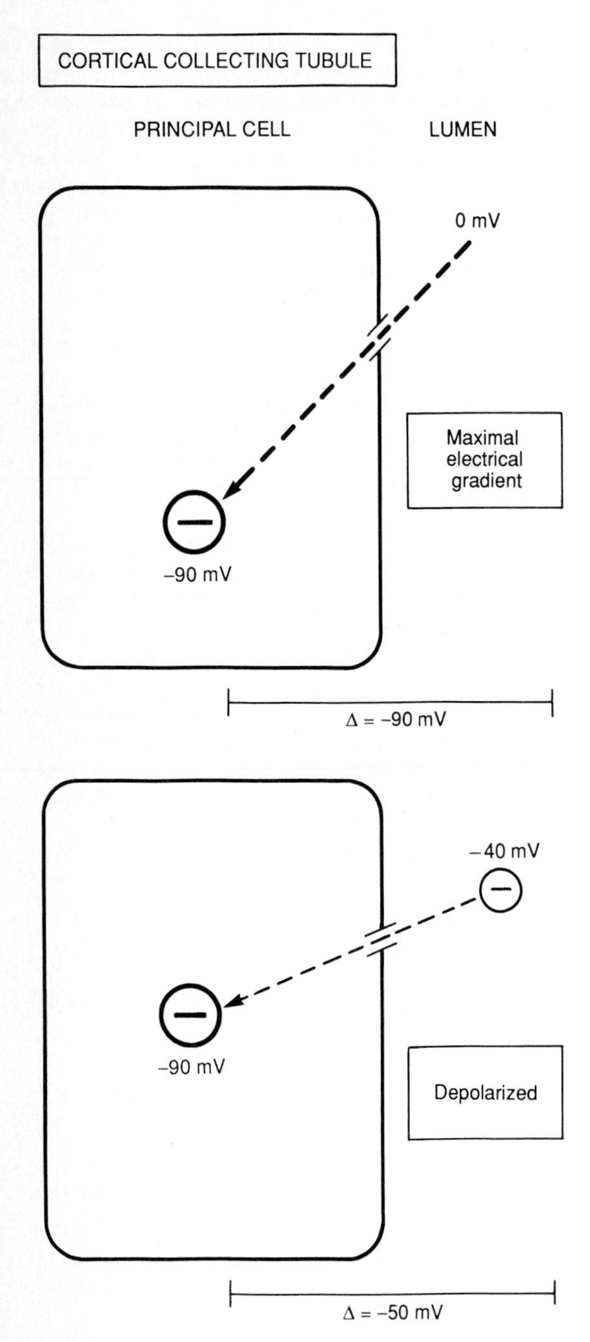

Figure 7-11. Electrical driving force for K^+ secretion.

in aldosterone level, so that the luminal potential difference is −40 mV, the potential difference across the luminal membrane would be depolarized to only −50 mV (−90mV −[−40mV]) (Fig 7-11, bottom). The potassium concentration gradient favoring secretion is assumed to be the same (Fig 7-10) and dominates the electrical driving force in both cases. Therefore, the net electrochemical driving force (the sum of the electrical and chemical components) is much greater in the latter than in the former case, since the electrical restraining force is attenuated. Thus, sodium reabsorption indirectly stimulates potassium secretion by modifying the lumen potential difference.

These chemical concentration and electrical forces can be quantitated and related to each other in a formal biophysical expression. The concentration gradient can be converted to an equivalent electrical potential by using the Nernst equation:

$$\text{Equivalent electrical potential (mV)} = -61 \log \frac{[K^+]_i}{[K^+]_o}$$

where $[K^+]_i$ and $[K^+]_o$ are the intracellular and luminal potassium concentrations, respectively. The total cell-to-lumen electrochemical gradient is then the sum of the electrical potential difference in mV and the chemical potential, also expressed in mV, using the Nernst equation. For instance, if the potassium concentrations in the cell and lumen were 150 meq/L and 15 meq/L, respectively (Fig 7-10), the equivalent Nernst potential difference would be −61 [−61 log(150/15)]. If the potential difference across the luminal membrane were −40 mV (Fig 7-1, bottom), the total electrochemical gradient (lumen-negative) favoring potassium secretion would be the sum of these potentials or −61 mV − −40 mV = −21 mV.

Luminal Membrane Potassium Permeability

The luminal membrane potassium channel is the conduit that allows potassium permeation. The absolute amount of potassium secreted for any electrochemical driving force is governed by the number as well as the activity of these channels, whether these ion gates are open or closed. Aldosterone regulates both the number and activity of these potassium channels.

Control of Electrochemical Gradient & Potassium Permeability by Aldosterone

As shown in Fig 7-9, aldosterone regulates many determinants of potassium secretion, including the luminal membrane sodium and potassium permeabilities and the Na^+/K^+-ATPase activity.

Sodium and potassium transport changes occur several hours after aldosterone enters the cells and binds to a high-affinity cytosolic receptor. The hormone-receptor complex is translocated into the nucleus, binds to specific sites on DNA, and initiates RNA transcription and protein translation.

An increase in luminal membrane potassium

permeability in response to aldosterone may be a direct effect or may secondarily result from altered sodium transport. Aldosterone increases sodium transport both by increasing luminal membrane sodium permeability and by enhancing Na^+/K^+-ATPase activity. A change in intracellular sodium concentration could cause a rise in cell calcium (Chapter 15) or intracellular electrical potential, known regulators of some kinds of potassium channels.

Whether primary or secondary, the combined functional effects of these aldosterone-initiated events is enhanced luminal membrane potassium permeability and increased electrochemical driving force for potassium secretion (due to stimulated sodium reabsorption and increased cellular potassium concentration). Aldosterone thus is a very potent promoter of potassium secretion.

Potassium Flux

In Fig 7–12, the propensity for potassium to cross the luminal membrane is analogous to a ball on a ramp. The steepness of the ramp is governed by both electrical and chemical driving forces, one of which tends to tilt the ramp more steeply (the "push" or chemical concentration component) while the other is tending to flatten the ramp (the "pull-back" or electrical component). The absolute amount of potassium secreted is defined by three factors: number of balls (ie, potassium ions); the number of ramps (ie, potassium channels in the membrane); and the steepness of the ramps (ie, the electrochemical gradient). These three parameters are in turn governed by aldosterone and other cellular and luminal determinants acting in concert.

DETERMINANTS OF POTASSIUM SECRETION IN THE CORTICAL COLLECTING TUBULE

Luminal Determinants

As listed in Table 7–2, there are several characteristics of the tubular fluid that act as determinants of potassium flux, including the potassium and sodium concentrations, anion composition, and flow rate.

First, **luminal potassium concentration** is obviously important in helping to define the chemical concentration gradient favoring potassium secretion. For any electrical gradient and cellular potassium concentration, more potassium is secreted if luminal potassium concentration is reduced (Fig 7–10). In general, however, the luminal potassium concentration at the beginning of the cortical collecting tubule does not vary markedly, being held at a relatively constant low level by virtue of the avid potassium reabsorptive processes upstream in the proximal tubule and thick ascending limb of Henle.

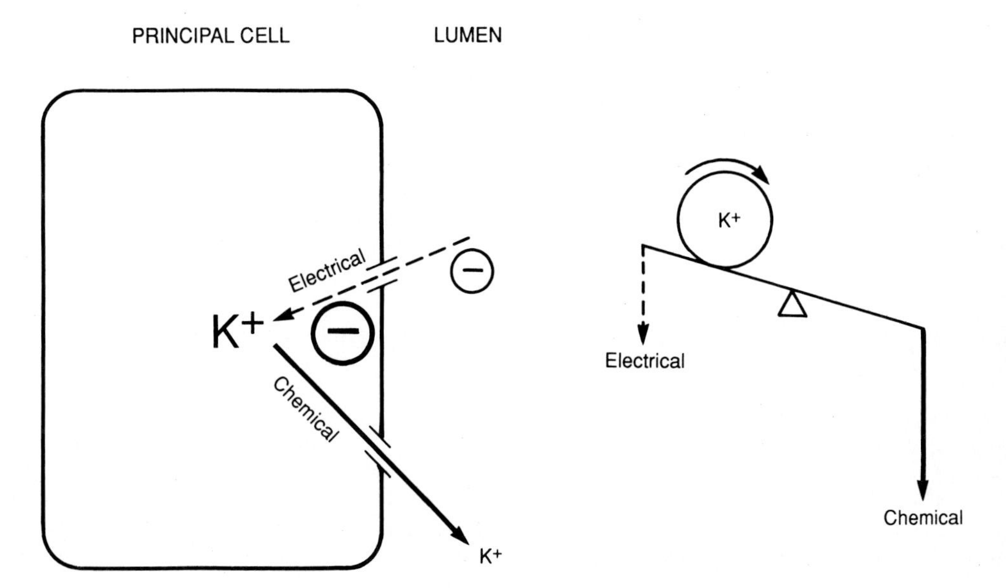

Figure 7–12. Electrochemical driving force for K^+ secretion.

Table 7-2. Determinants of K^+ secretion in the cortical collecting tubule.

Luminal	Cellular
Potassium concentration	Plasma potassium concentration
Sodium concentration	Membrane permeabilities for sodium and potassium
Reabsorbability of anion	Aldosterone activity
Flow rate	

Second, the importance of **sodium concentration and delivery** is shown in Fig 7-13 (top). Sodium delivery may be increased by diuretics, during osmotic diuresis, or during acid-base disorders. As discussed above, electrogenic sodium transport generates a lumen-negative electrical potential, which regulates the amount of potassium secreted (Fig 7-11). The fraction of sodium reabsorbed as a function of sodium delivery to the cortical collecting tubule is relatively constant. Therefore, the fraction that remains, which becomes the urinary sodium, is also an index of sodium delivery to the collecting tubule and hence of the stimulus to potassium secretion, as shown in Fig 7-13 (bottom).

The third luminal determinant, **reabsorbability**

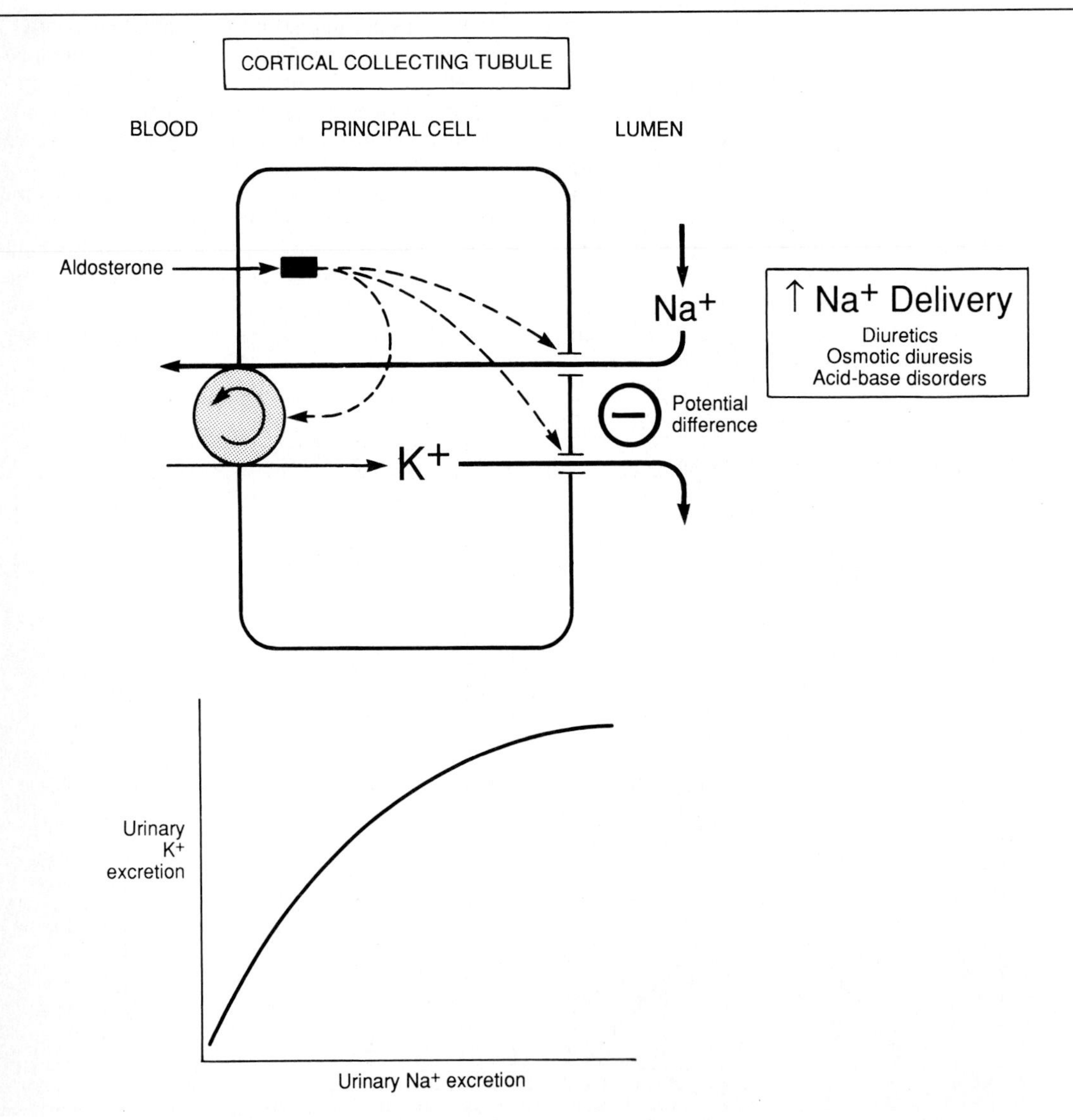

Figure 7-13. Regulation of potassium secretion by sodium delivery.

of the accompanying anion, modifies the impact of the lumen-negative potential generated by sodium reabsorption. For practical purposes, the only easily reabsorbable anion is chloride. As shown in Fig 7–14A, chloride reabsorption diminishes the lumen-negative potential that would otherwise drive potassium secretion. Stated another way, chloride reabsorption tends to "rob" from potassium the electrochemical gradient generated by sodium reabsorption. When accompanied by a nonreabsorbable anion (B^-), anion reabsorption no longer "competes" with potassium secretion for the potential, represented in Fig 7–14B. Delivery to the cortical collecting tubule of sodium accompanied by a poorly reabsorbed anion (such as *β*-hydroxybutyrate, bicar-

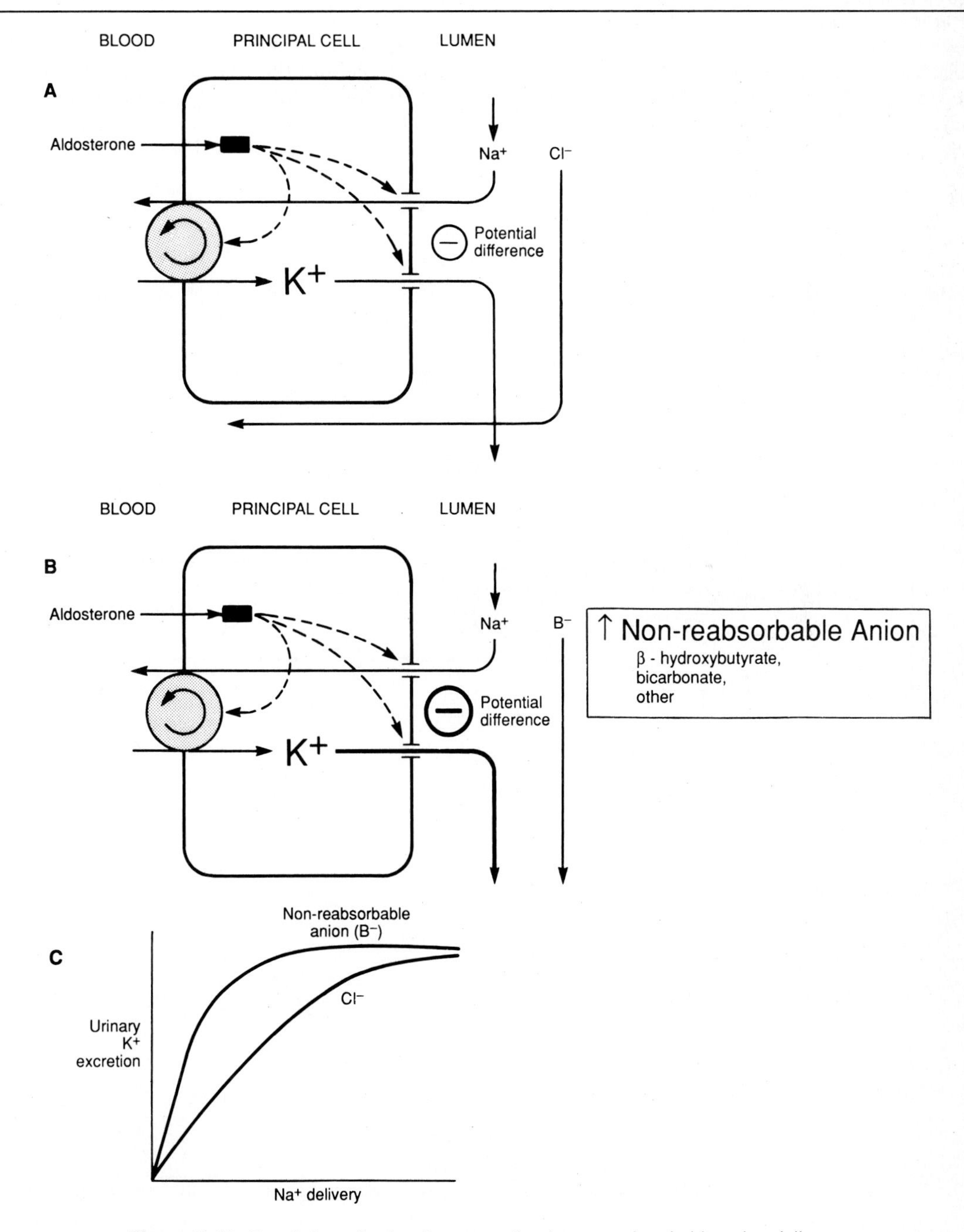

Figure 7–14. Regulation of potassium secretion by nonreabsorbable anion delivery.

bonate, or others) therefore elicits more potassium secretion than occurs with the same sodium delivery accompanied by the reabsorbable chloride anion, as shown in Fig 7–14C.

Finally, the fourth luminal determinant of potassium secretion is the **luminal flow rate,** which influences potassium secretion by two major means. First, flow rate and sodium delivery are generally tightly coupled. An increase in flow rate allows more sodium delivery and thus a more favorable electrical driving force for potassium secretion (Fig 7–13).

Second, the faster the luminal flow, the less the luminal potassium concentration buildup that occurs as potassium is secreted. By lessening dissipation of the cell-to-lumen potassium concentration gradient, higher flow therefore augments potassium secretion along the entire length of the collecting tubule.

Cellular Determinants

Potassium secretion is affected by the intracellular concentration of potassium, especially as defined by the activity of the Na^+/K^+-ATPase, the luminal membrane sodium and potassium conductances or channels, and aldosterone activity (Table 7–2).

First, the **intracellular potassium concentration** critically defines the cellular contribution to the electrochemical gradient for potassium secretion, including both the chemical and the electrical components. As shown in Fig 7–15 (top), factors

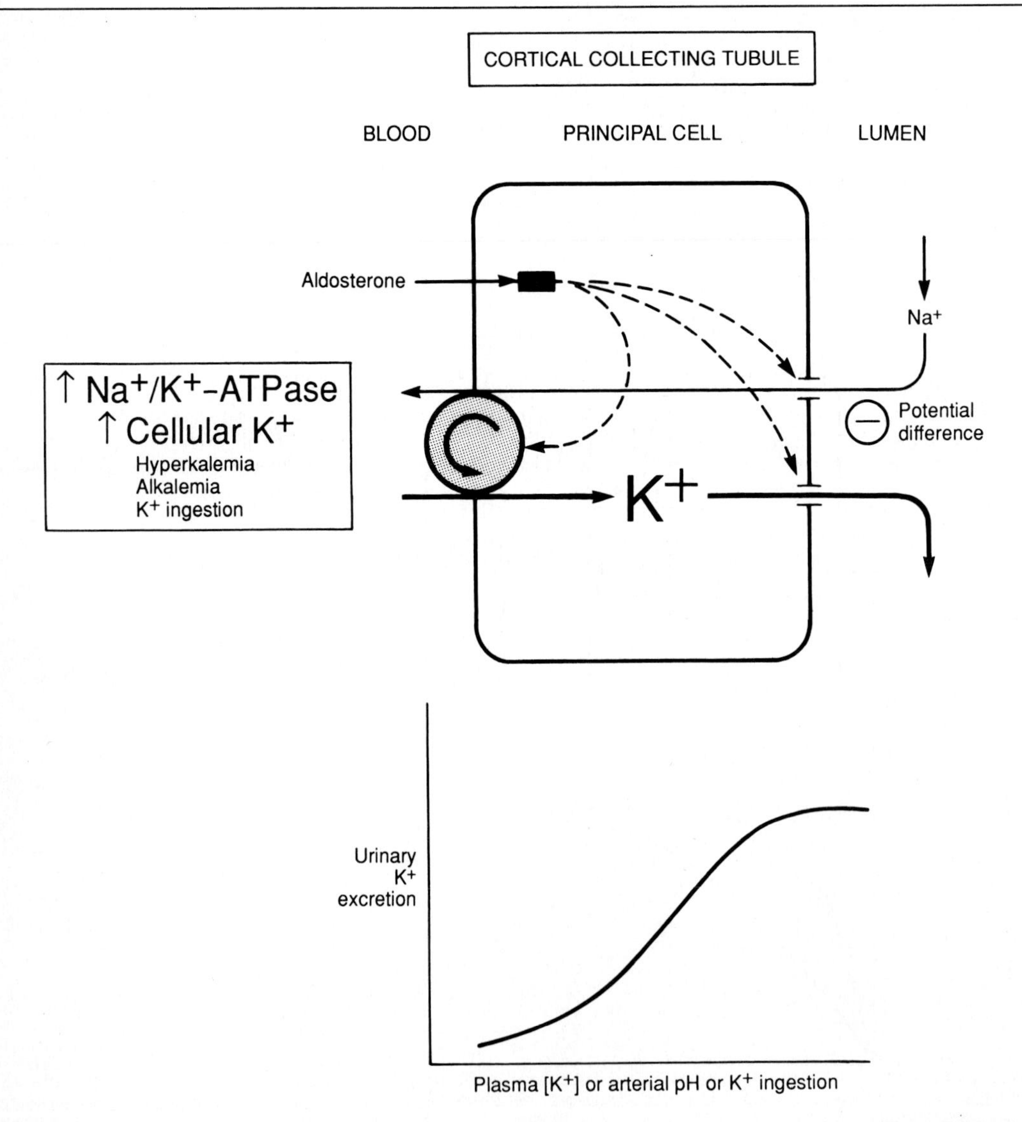

Figure 7–15. Regulation of potassium secretion by Na^+/K^+-ATPase activity and intracellular potassium concentration.

that promote potassium uptake into cells—including hyperkalemia, alkalemia, and chronic increase in potassium ingestion)—affect urinary potassium excretion (Fig 7–15, bottom). These changes are mediated by Na^+/K^+-ATPase-dependent and -independent mechanisms

Second, the **sodium and potassium conductances** in the luminal membrane are the selective ion channels ("gates") that allow sodium uptake into the cell and potassium secretion into the tubular fluid, respectively. An increase in the number of sodium or potassium channels allows more sodium entry or potassium exit for the respective electrochemical gradients. Regulation of the number of these channels is a function of aldosterone (see below). Certain diuretics—specifically, amiloride and triamterene—can block the sodium channel (Chapter 3).

Finally—and critically important—is the sensitive control by **aldosterone** of the sodium and potassium conductances in the luminal membrane and the Na^+/K^+-ATPase on the basolateral membrane of the principal cell. As shown in Fig 7–16 (top), aldosterone controls the "tone" of all cellular determinants responsible for potassium secretion mentioned above. Potassium excretion in the urine is potently regulated by aldosterone (Fig 7–16, bottom).

Other hormones also regulate potassium secretion by the principal cell, though their physiologic

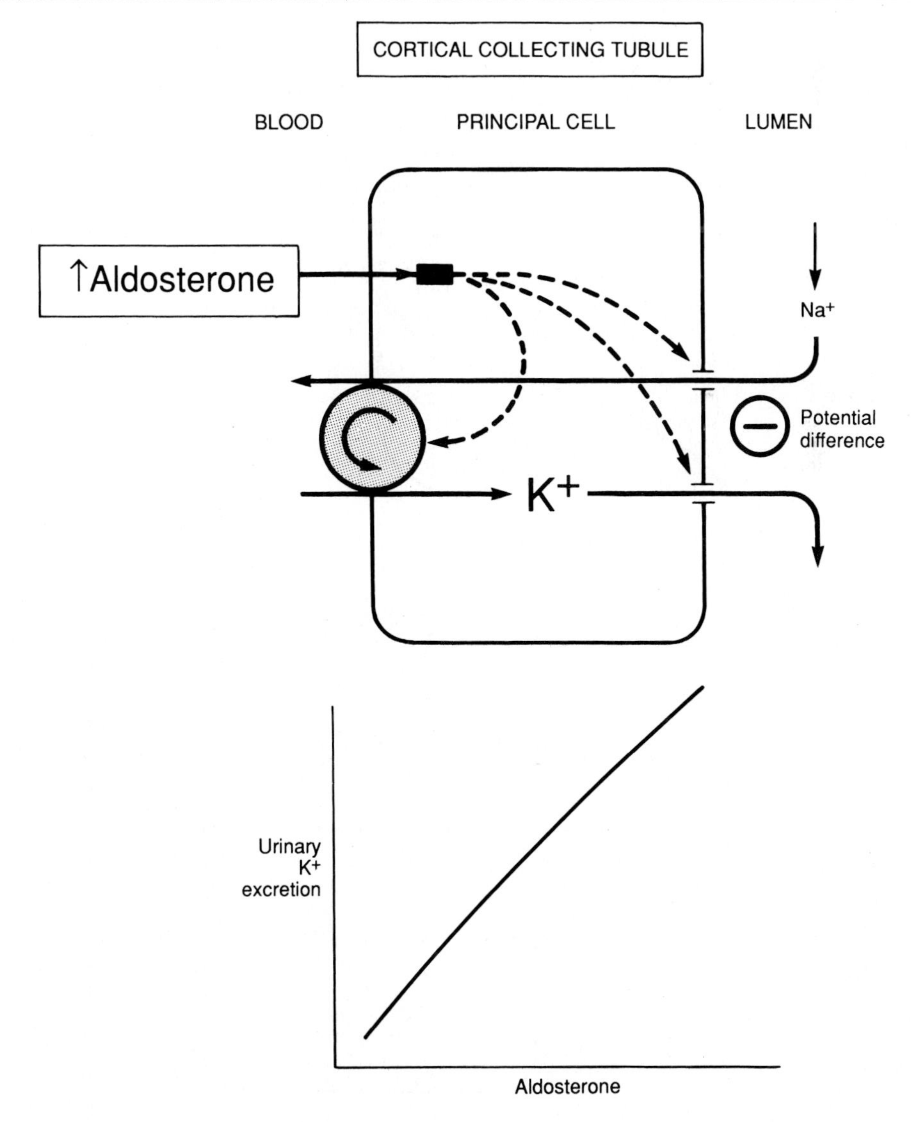

Figure 7–16. Regulation of potassium secretion by aldosterone.

importance is uncertain. Vasopressin increases while epinephrine decreases net potassium secretion.

REGULATION OF RENAL POTASSIUM EXCRETION

Under normal circumstances, renal potassium excretion is approximately equal to intake (minus small losses in the gastrointestinal tract and skin) (Fig 7-2). There is a circadian rhythm to renal potassium excretion—lowest in the early morning and highest in the late afternoon—which is not strictly related to potassium ingestion.

Coupling of Potassium Ingestion to Urinary Excretion

When potassium is ingested and absorbed from the gastrointestinal tract, a slight—often unmeasurable—rise in serum potassium concentration occurs, as shown in Fig 7-17. Neurohumoral mechanisms translocate potassium into cells, including the principal cell of the cortical collecting tubule. Simultaneously, aldosterone secretion increases because the cell in the portion of the adrenal gland responsible for producing mineralocorticoid, the zona glomerulosa, is under sensitive control by serum potassium. The increase in principal cell potassium concentration (Fig 7-15) and in plasma aldosterone level (Fig 7-16) serve to increase potassium secretion (Fig 7-17).

Another factor that operates to ensure that dietary potassium is quantitatively excreted is that

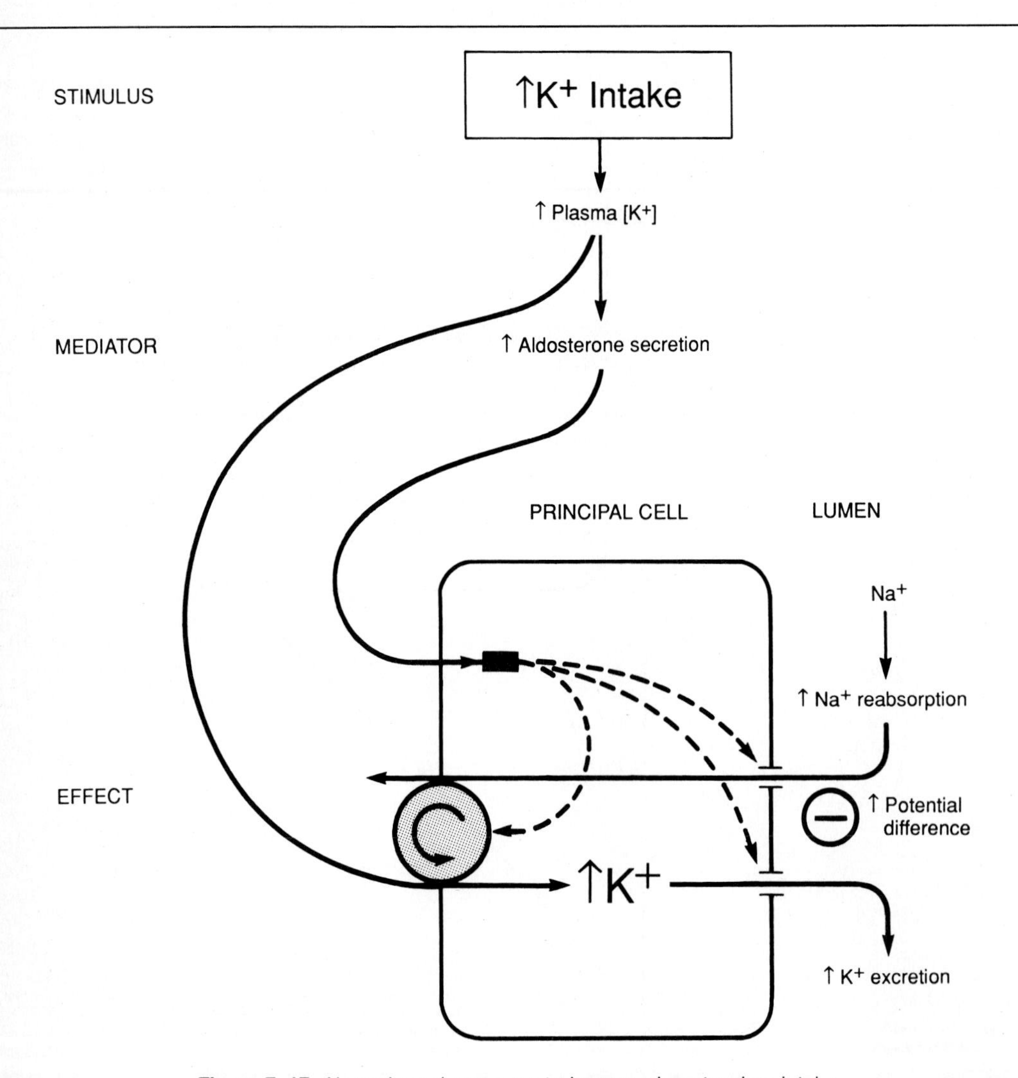

Figure 7-17. Normal renal response to increase in potassium intake.

anions accompanying potassium in most natural foodstuffs are organic or mineral anions which are nonchloride and therefore relatively nonreabsorbable in the cortical collecting tubule. Thus, the obligate delivery of nonreabsorbable anions equal in magnitude to the ingested potassium promotes an appropriate driving force for potassium excretion by the principal cell (Fig 7–14).

Independence of Potassium Excretion From Sodium Excretion

Urinary potassium excretion can be normally efficiently regulated independently of dietary sodium intake and urinary excretion. Dietary sodium and potassium intakes often vary widely and sometimes in opposite directions. Autonomy of urinary potassium excretion is possible because luminal sodium delivery and aldosterone both promote potassium secretion by the cortical collecting tubule but are affected oppositely by a change in sodium ingestion.

Fig 7–18 exemplifies the dual control of potassium secretion by sodium delivery and aldosterone. First consider excretion of dietary potassium (70 meq/d) when sodium intake is relatively low (70 meq/d), as shown in Fig 7–18A. Sodium reabsorption in the proximal tubule and loop of Henle is avid, so that sodium delivery to the cortical collecting tubule is low (Chapter 1). Considering the reciprocal relationship of sodium intake and renin-angiotensin-aldosterone levels (Chapter 1), such a diet is also associated with a relatively high aldosterone level, which serves to stimulate potassium secretion. As shown by point X in Fig 7–18C, potassium excretion is kept at a level of 70 meq/d—equal to intake—by a high aldosterone level that compensates for low sodium delivery in the tubular fluid.

Conversely, as shown in Fig 7–18B, when sodium intake is higher (175 meq/d), there is a rise in sodium delivery to the cortical collecting tubule, but increased dietary sodium simultaneously lowers the aldosterone level. Consequently, in this case, potassium secretion is maintained by a high sodium delivery and transport rate that offsets the impact on potassium transport of a low aldosterone level (Fig 7–18B). Thus, as shown by point Y in fig 7–18C, there is no change in potassium excretion despite a much higher sodium intake.

Urinary K^+:Na^+ Ratio

The examples above also serve to show how urinary potassium and sodium concentrations can be used as a crude index of prevailing aldosterone level. Since aldosterone enhances sodium reabsorption and potassium secretion, the urine concentration ratio of K^+ to Na^+ tends to reflect the aldosterone level. For instance, with a dietary intake of potassium and sodium of 70 and 175 meq/d, respectively, the urinary K^+:Na^+ ratio is 0.4 (point Y, Fig 7–18C). With the same potassium intake but a more restricted sodium chloride intake (eg, 70 meq/d), a higher aldosterone level obtains (point X, Fig 7–17C) with a urinary K^+:Na^+ ratio of 1.0. A urinary K^+:Na^+ ratio exceeding 1.0 reflects either an unusual diet or, more commonly, a state of effective extracellular volume contraction, unrelated to diet, which has caused hyperreninemia and hence a high aldosterone level. Thus, the urinary K^+:Na^+ ratio is a physiologic reflection of distal nephron potassium secretion compared to sodium reabsorption and is a "poor man's" estimate of aldosterone level.

LIMITS OF POTASSIUM CONSERVATION & EXCRETION

Response to Potassium Deprivation

When dietary potassium intake is curtailed or body potassium stores are depleted by other means, the kidney slowly reduces but cannot eliminate potassium from the urine. As shown in Fig 7–19, when potassium intake is abruptly changed from 100 to 5 meq/d (vertical interrupted line at day 0), absolute urinary potassium excretion declines exponentially with a half-time of about 2–3 days (Fig 7–19A). Potassium excretion never falls below a minimal threshold of 5–15 meq/d. Fractional potassium excretion, calculated as for sodium (see Chapter 1), $(U/P)_{K^+}/(U/P)_{cr}$, declines from a normal value of 0.17 to about 0.03–0.05 (Fig 7–19B).

Thus, a finite amount of potassium is excreted even when body homeostasis demands potassium conservation. Since urinary potassium cannot be eliminated in response to severe dietary potassium restriction, negative potassium balance occurs (shaded area between excretion and intake in Fig 7–19A). The obligate urinary potassium loss results in a gradual reduction in serum potassium concentration to frankly hypokalemic values (Fig 7–19C). Since potassium comes out of cells as a result of the hypokalemia, total body potassium also progressively declines. In the example of Fig 7–19, the cumulative potassium deficit (integrated shaded area, the sum of the daily differences between excretion and intake) is about 400 meq over 2 weeks, corresponding to a decrease in serum potassium concentration of approximately 0.1 meq/L.

By comparison, renal conservation of sodium is qualitatively more efficient, such that urinary sodium excretion can normally be reduced to less than 1 meq/d when demanded physiologically (Chapters 1 and 2). Even if renal potassium conservation were perfect, extrarenal potassium losses in the gastrointestinal tract and skin would

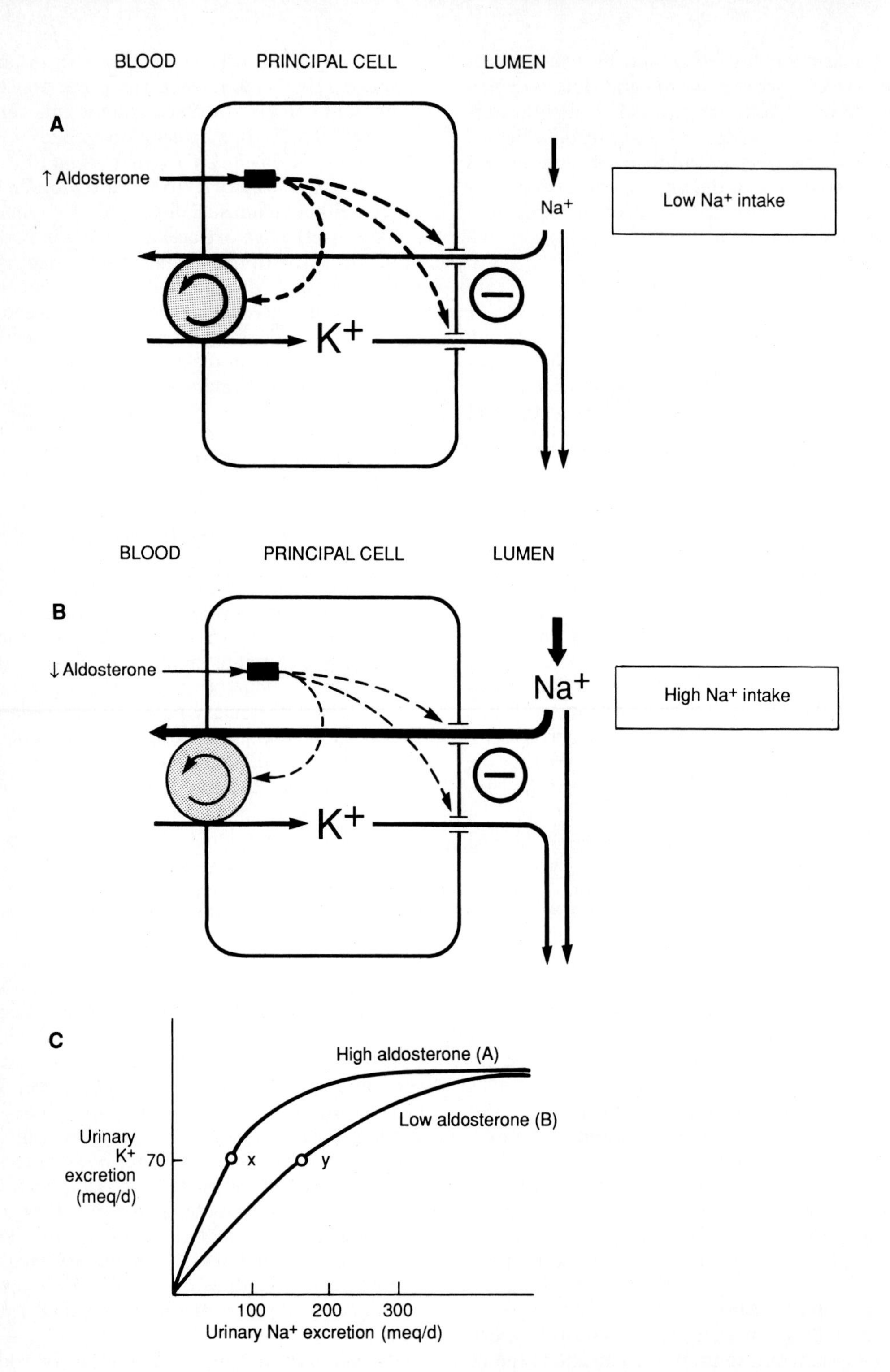

Figure 7-18. Independence of potassium homeostasis from change in sodium intake.

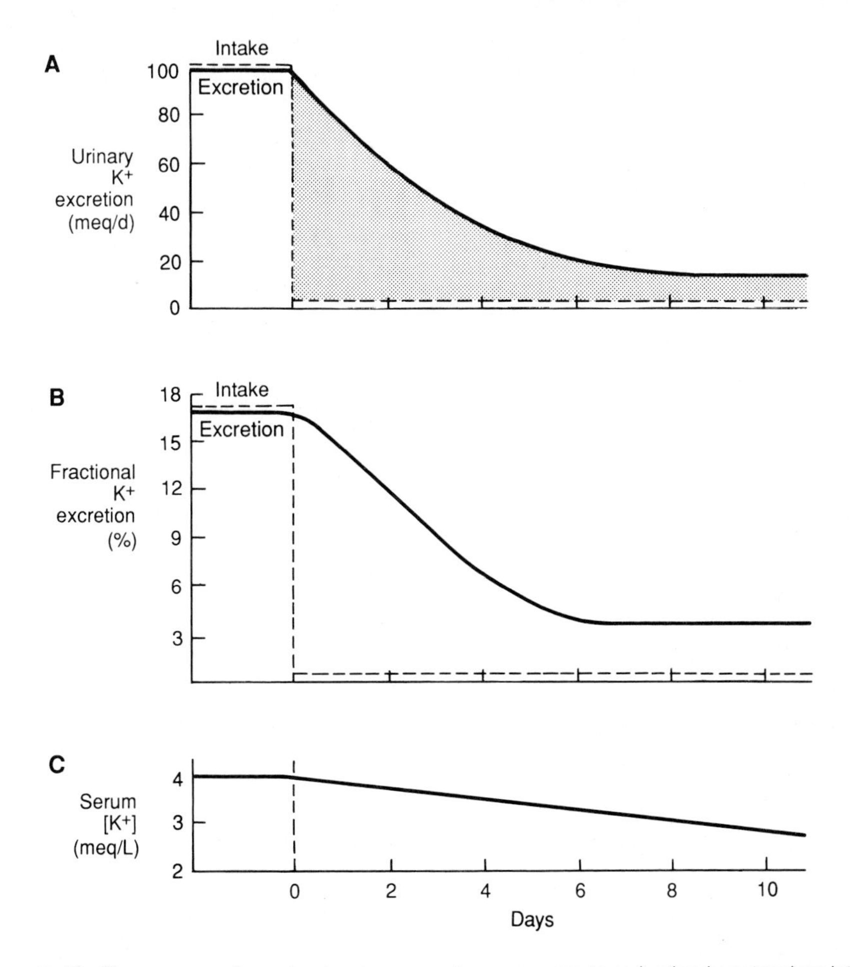

Figure 7–19. Time course of renal potassium excretory response to reduction in potassium intake.

decline (to about 10 meq/d) but not to zero in response to dietary potassium restriction. Thus, as a result of obligate renal and nonrenal potassium losses despite maximal signals for potassium conservation, a minimal potassium intake of about 10–25 meq/d is necessary to maintain potassium balance.

Response to a Potassium Load & Adaptation

Although temporary storage of a surfeit of potassium can be effected by intracellular translocation (Fig 7–4), the kidney is ultimately responsible for ridding the body of the excess.

As shown in Fig 7–20, an abrupt doubling of potassium intake from 100 to 200 meq/d causes renal potassium excretion to appropriately double, with a half-time of only 1 day or less, rapidly reaching steady-state balance within 2 days. Positive potassium balance (shaded area in Fig 7–20A) is quite small, about 75 meq. Serum potassium changes very little under these conditions, increasing by only 0.2 meq/L (Fig 7–20C).

When an increase in potassium input is sustained, adaptive mechanisms in the colon as well as in the kidney come into play to augment potassium excretion. The cellular mechanism of potassium secretory adaptation in the colon is generally the same as in principal cell of the collecting tubule. In both, enhanced potassium secretion follows expansion of basolateral membrane surface area concomitant with increased synthesis and membrane insertion of Na^+/K^+-ATPase units. Potassium loading induces this adaptive process by both aldosterone-dependent and aldosterone-independent means, though the mechanism of the latter process is unknown.

The capacity for renal adaptation of potassium excretion is enormous. If potassium intake rises gradually from 100 to 600 meq/d, urinary potassium excretion increases commensurately. Some of the increase occurs in the colon, and potassium

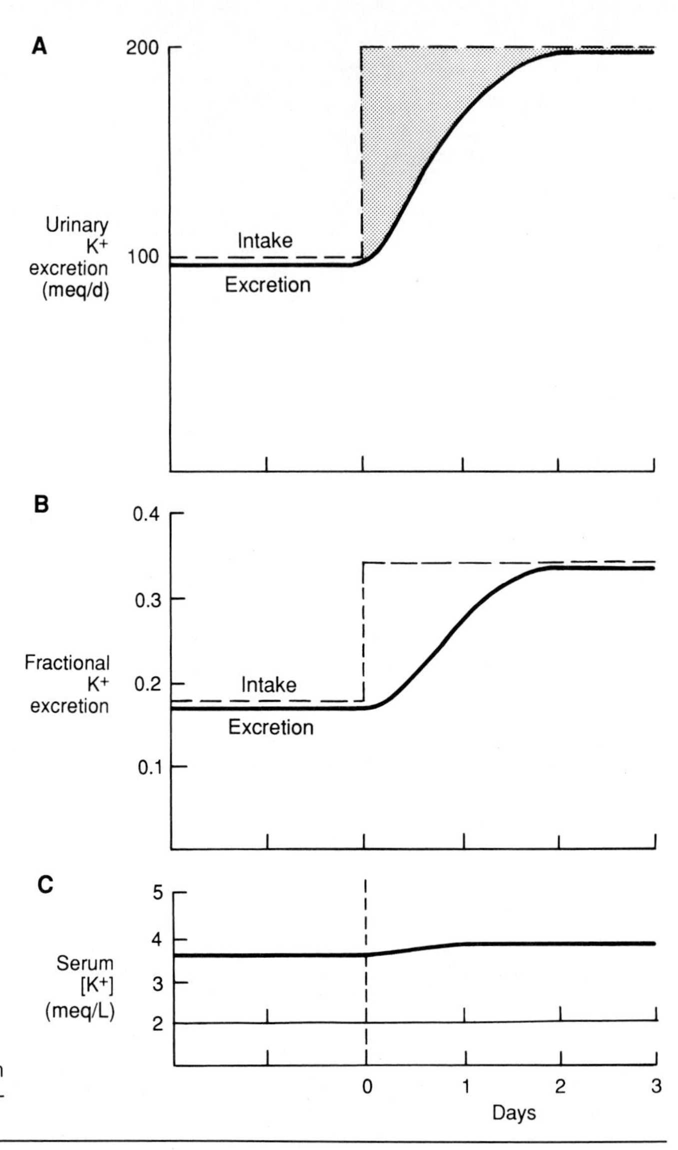

Figure 7-20. Time course of renal potassium excretory response to increase in potassium intake.

excretion rate more than 5 times normal can occur. But the quantitatively most important adaptive response resides with the kidney. If GFR remains constant, fractional potassium excretion can be calculated to rise from about 17% with an intake of 100 meq/d to 100% when intake is 600 meq/d, as shown in Fig 7-21A. Fractional potassium excretion exceeding 100% in humans probably can occur but is very unusual because potassium load from diet or cellular breakdown rarely exceeds 600 meq/d.

The process of potassium adaptation assumes great clinical importance when the number of nephrons is reduced. in the course of chronic renal disease. To maintain potassium homeostasis and prevent hyperkalemia, the diminished number of functioning nephrons strive to excrete a normal amount of ingested potassium using the same adaptation processes described above. Each residual surviving nephron unit excretes more potassium. For example, as shown in Fig 7-21B, to remain in balance with a fixed level of dietary potassium, a reduction in nephron population to one-sixth the original number demands that each residual nephron transport six times more potassium. Expressed in fractional excretion terms per nephron, potassium excretion would again have to increase about sixfold (eg, from 17% to 100%). Thus, in this latter case (Fig. 7-21B), one-sixth of the original number of nephrons excrete the same absolute amount of potassium, whereas in the former case (Fig 7-21A), the same number of nephrons excreted a sixfold greater amount of potassium.

The limit of the adaptive kaliuretic response is

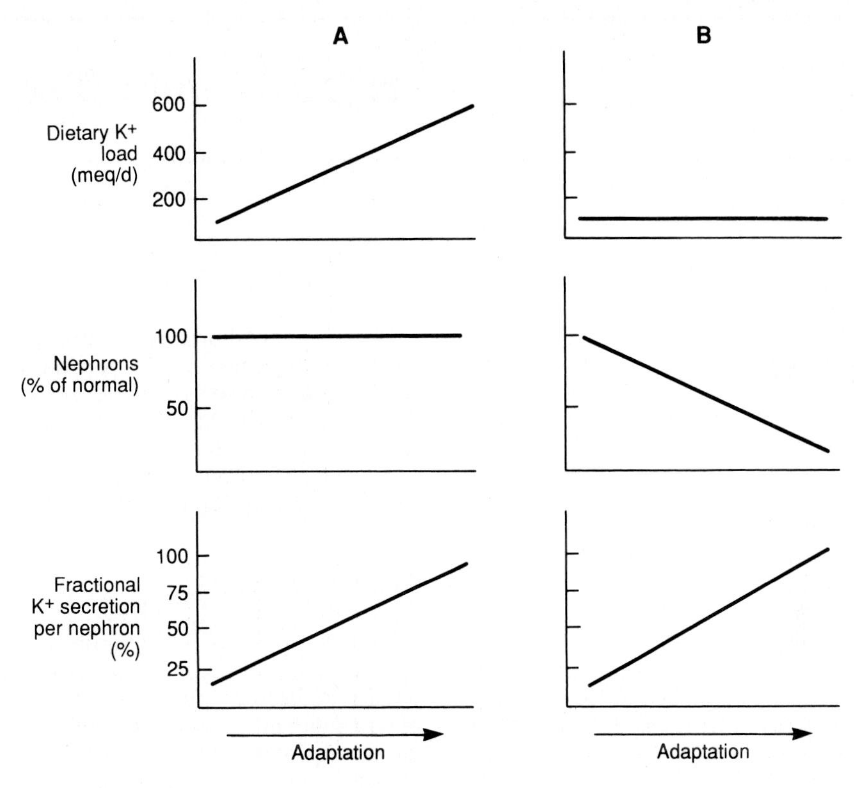

Figure 7-21. Adaptation of renal potassium secretion in response to increased dietary potassium or reduction in nephron number.

not exactly known but is probably in the range of a tenfold increase. When this maximal adaptation is exceeded, a further chronic increase in potassium intake or a superimposed acute potassium challenge cannot be tolerated. Thus, if potassium intake were to increase more than tenfold (if intake exceeded 1000 meq/d) or if the number of nephrons were to decrease more than tenfold (if GFR fell to < 10 mL/min), potassium excretion can fail to equal potassium input, resulting in hyperkalemia.

8 Hypokalemic Disorders

INTRODUCTION

DEFINITION

Hypokalemia is diagnosed when the potassium concentration, usually measured in the serum, is **less than 3.5 meq/L.**

Rule Out Pseudohypokalemia

The only cause of a spuriously low potassium concentration has been described in leukemic patients with white blood cell count higher than 100,000/μL. White cells in blood from such patients, if allowed to stand at room temperature, extract potassium. Promptly centrifuging blood specimens from such patients avoids this problem.

Relation to Total Body Potassium Stores

Hypokalemia does not necessarily imply total body potassium depletion. Extracellular potassium can be translocated into the intracellular compartment without changing total body potassium content. Conversely, total body potassium depletion may occur without hypokalemia. For instance, during diabetic ketoacidosis, hyperosmolality and hypoinsulinemia drive potassium out of cells and preserve normality of serum potassium concentration at the same time there is renal potassium wasting.

Relation to Acid-Base Disorders

Changes in potassium balance can cause acid-base derangements, and vice versa. Some of the relationships between potassium and acid-base disorders can be explained by the proximity in the cortical collecting tubule of the principal and intercalated cells responsible for secretion of potassium and hydrogen ions, respectively; their dual control by aldosterone; and their mutual dependence on sodium reabsorption. In other cases, the association is less direct. In any event, evaluation of a deranged plasma potassium concentration is usually intimately linked with elucidation of the cause of the concurrent acid-base disorder.

Classification of potassium disorders in this and the following chapter will be according to the pathophysiologic basis of potassium loss or retention, but it should be acknowledged that an efficient scheme for categorizing potassium problems can also be based on the accompanying acid-base abnormality (see Chapters 12 and 13).

PATHOPHYSIOLOGY

Table 8-1 outlines the pathophysiologic factors that can cause disorders of potassium homeostasis. These disorders can be traced to (1) changes in potassium intake or cellular release that exceed the normal ability of the kidney to compensate; (2) disturbances of the hormones, pH, and osmolality that regulate the normal distribution of potassium between extracellular and intracellular pools; and (3) inappropriate levels of the determinants of potassium secretion in the cortical collecting tubule.

INADEQUATE INTAKE OR EXCESSIVE NONRENAL LOSS OF POTASSIUM

The causes of hypokalemic disorders are summarized in Table 8-2 and Fig 8-1.

INADEQUATE POTASSIUM INTAKE

Dietary potassium must be at least 10 meq/d and usually closer to 20-30 meq/d to compensate for obligate nonrenal (5-15 meq/d) and renal (5-20 meq/d) losses (Fig 7-19). Dietary potassium intake of less than 20-30 meq/d may be due to complete starvation, sometimes for purposes of dieting or as a symptom of anorexia nervosa. Insufficient dietary input may also result from diets without potassium, as occurs in elderly patients who exist on a "tea and toast" diet, in alcoholics who consume only alcoholic beverages

Table 8-1. Causes of potassium disorders.

Primary Physiologic Disturbance	Hypokalemic Disorders	Hyperkalemic Disorders
Intake or endogenous source	↓	↑
Redistribution		
Insulin, beta-adrenergic agonists, aldosterone	↑	↓
ECV [H^+], osmolality	↓	↑
Renal excretion		
Luminal determinants		
Na^+ delivery and flow	↑	↓
Nonreabsorbable anion delivery	↑	↓
Cellular determinants		
Aldosterone	↑	↓

Table 8-2. Hypokalemic disorders.

Insufficient intake
- Starvation
- Low-potassium diet

Excessive nonrenal loss
- Sweat
  - Intensive exercise in hot climate
- Gastrointestinal
  - Diarrhea
  - Enteric fistula

Translocation into cells
- Hyperinsulinemia
  - Glucose load
  - Exogenous insulin administration
  - Insulinoma
- ↑Beta-adrenergic activity
  - Recovery from exercise
  - Post-head trauma or surgery
  - Delirium tremens
  - Pheochromocytoma
  - Sympathomimetic drugs
  - Phosphodiesterase inhibitors
- Hyperaldosteronism
- Other hormones or factors
  - Hypokalemic periodic paralysis
  - Physical conditioning
  - Hypothermia
  - Metabolic or respiratory alkalosis
  - ↑Na^+/K^+-ATPase units
    - Treatment of anemia
    - Leukemia
  - ↓Potassium channel activity
    - Barium

Excessive renal loss
- ↑Sodium delivery and flow
  - ↑GFR
    - Dietary protein
    - Atrial natriuretic factor
  - ↓Proximal sodium chloride reabsorption
    - Osmotic diuresis
      - Hyperglycemia
      - Mannitol
      - Radiocontrast dye
      - Urea
    - Metabolic or respiratory acidosis
  - ↓Loop of Henle sodium chloride reabsorption
    - Diuretics
      - Furosemide
      - Ethacrynic acid
      - Bumetanide
    - Bartter's syndrome
  - ↓Distal convoluted tubule sodium chloride reabsorption
    - Diuretics
      - Thiazides
- ↑Nonreabsorbable anion delivery
  - Bicarbonate
    - Generation of metabolic alkalosis
    - Proximal (type I) renal tubular acidosis
    - Respiratory alkalosis
  - Ketones
    - Diabetes
    - Starvation
    - Alcoholism
  - Other
    - Antibiotics
      - Penicillin
      - Carbenicillin
- ↑Mineralocorticoid activity
  - Primary ↑renin
    - Renal artery stenosis
    - Vasculitides
    - Renin-secreting tumors
    - Magnesium deficiency
    - Primary ↑aldosterone
  - Cushing's disease or syndrome
    - Adrenogenital syndromes
    - Primary hyperaldosteronism
    - Glycyrrhizic acid ingestion
      - Licorice
      - Chewing tobacco
      - Carbenoxolone
    - Other mineralocorticoids
      - Fludrocortisone
      - Liddle's syndrome
- Other
  - ↑Potassium channels
    - Amphotericin B
  - ↑Electrochemical gradient
    - Classical (type II) distal renal tubular acidosis

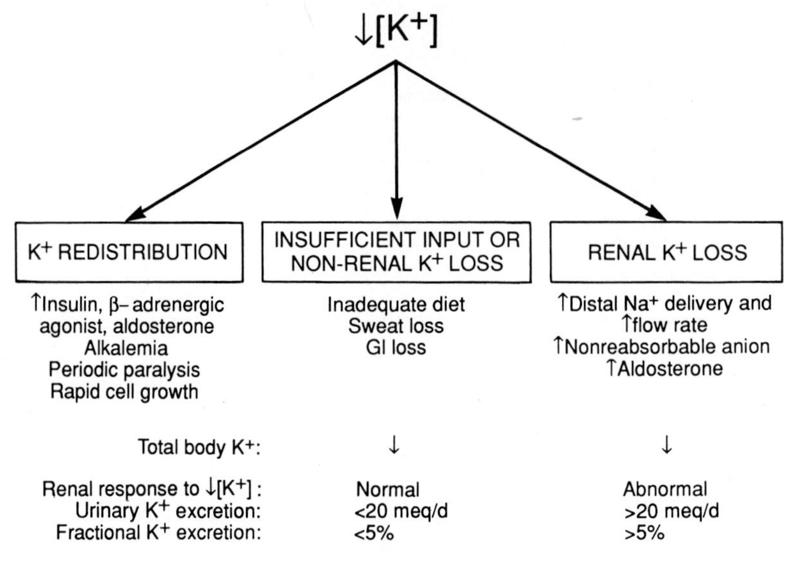

Figure 8-1. Diagnosis of hypokalemia.

(which contain little potassium) without solid food, or in patients who ingest certain forms of clay that bind dietary potassium in the gastrointestinal tract. In all these cases, urinary potassium excretion is appropriately low (< 20 meq/d or a fractional potassium excretion of < 5%), as shown in Fig 8-1. When both dietary potassium and protein are simultaneously restricted, a modest amount of potassium spillage in the urine occurs, which is derived from muscle cells undergoing atrophy.

Accompanying acid-base disorders of hypokalemia are discussed more fully in Chapters 12 and 13. Potassium deficiency due to caloric malnutrition may be accompanied by a mild starvation-induced ketoacidosis, indicated by acidemia, a small increased anion gap, and ketonuria. Alcoholics with potassium depletion also may have ketoacidosis induced by alcohol. Metabolic alkalosis due to vomiting occurs in patients with anorexia nervosa, so that renal potassium wastage exacerbates the dietary inadequacy.

In all of these conditions, total body potassium is depleted, so it should be replenished according to the general guidelines set forth below. If there is concurrent caloric malnutrition due to starvation, anorexia nervosa, or alcoholism, carbohydrate repletion should be initiated cautiously because carbohydrate-stimulated insulin secretion can exacerbate the hypokalemia.

SWEAT LOSS OF POTASSIUM

Normally, only about 1–2 meq/d of potassium is lost in the normal volume of sweat, about 0.1–0.2 L/d (Fig 7-2). Intensive physical conditioning in hot climates has been documented to increase sweat volume up to 100-fold, or as much as 12 L/d (Fig 6-2). Since the concentration of potassium remains relatively normal at about 9 meq/L, absolute potassium loss via sweat can approach 100 meq/d. When exercise is prolonged, net potassium loss of more than 500 meq over the course of 6 weeks has been reported.

The route of potassium loss in the setting of severe exercise in a hot climate is not solely sweat, however. Hyperaldosteronism occurs, which is unsuppressible by high sodium intake and prevents appropriate potassium retention by the kidney in response to the sweat potassium loss. The combination of potassium deficiency, extreme physical exertion, and hyperpyrexia predispose such individuals to heat stroke and rhabdomyolysis with myoglobinuria, sometimes leading to acute renal failure. Precautions to reduce exercise in heat, diminish sweat volume, and maintain adequate potassium intake are advised to prevent these complications.

GASTROINTESTINAL LOSS OF POTASSIUM

Diarrhea & Intestinal Fistulas

Potassium normally is secreted into the colon under the control of aldosterone until it reaches a

Table 8-3. Average K^+ concentration in gastrointestinal fluids.

	$[K^+]$ (meq/L)
Gastric	10
Duodenal	15
Biliary	5
Pancreatic	5
Jejunal	6
Ileal	8
Colonic	90
↑flow	35–60

concentration of about 90 meq/L. The volume of stool water is about 0.1 L, so that approximately 9 meq/d of potassium is normally excreted by this route (Fig 7–2). As stool volume increases in diarrheal states, the potassium concentration tends to decline, usually to a value of about 35–60 meq/L (Table 8–3). Nevertheless, potassium excretion in severe diarrheal states can be extreme, exceeding 100 meq/d in some instances. Hyperaldosteronism due to extracellular volume contraction exacerbates potassium loss in the stool and also prevents normal renal potassium reclamation. Negative potassium balance and hypokalemia ensue if dietary intake becomes less than the level of gastrointestinal and renal potassium excretion.

Potassium loss by the colon can potentially be induced by all causes of osmotic diarrheas (eg, due to laxatives, sugar malabsorption, generalized malabsorption) or of secretory diarrheas (due to toxins, bile acids, laxatives, gastrointestinal secretory hormones, villous adenoma, infections, etc).

Potassium loss can also occur via fistulas of the gastrointestinal tract that drain externally. Absolute potassium loss is generally less than with diarrhea, because potassium concentration is less in fluid from upstream segments of the gastrointestinal tract. For instance, using information in Tables 2–2 and 8–3, the normal potassium output from the jejunum can be estimated to be about 18 meq/d (3 L/d with a concentration of 6 meq/L) and from the ilium about 5 meq/d (0.6 L/d with a concentration of 8 meq/L). Drainage fluids from pancreatic and biliary fistulas have a low potassium concentration—about 5 meq/L.

The acid-base derangement associated with hypokalemia induced by diarrhea is usually hyperchloremic metabolic acidosis (Chapter 12), due to colonic bicarbonate secretion and chloride absorption, though exceptions exist.

In all cases of gastrointestinal potassium loss, renal potassium conservation should be normal—unless there is concurrent hyperaldosteronism induced by extracellular volume contraction—with urinary potassium excretion less than 20 meq/d (Fig 8–1).

POTASSIUM REDISTRIBUTION INTO CELLS

A fall in extracellular potassium concentration can occur as a result of increased activity or number of Na^+/K^+-ATPase units responsible for translocation of potassium into cells, as shown in Fig 8–2. Such shifts are usually rapid in onset—within minutes—but last only a short time.

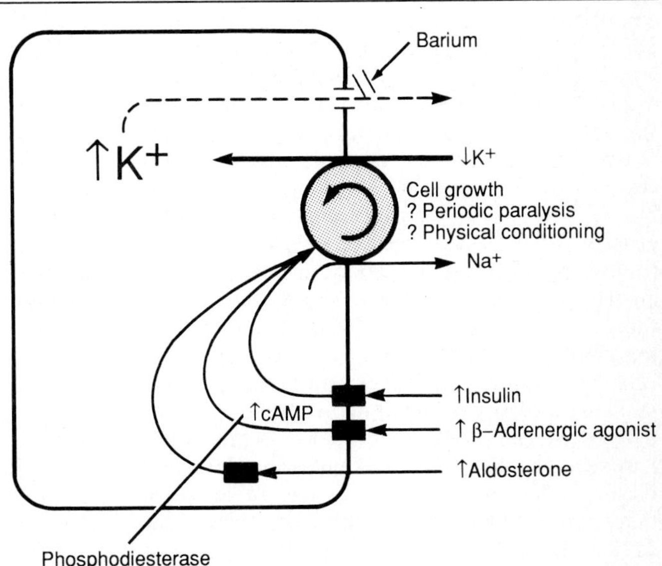

Figure 8–2. Causes of hypokalemia due to enhanced potassium translocation into cells.

INCREASED NA^+/K^+-ATPASE ACTIVITY

Insulin

As an important regulator of Na^+/K^+-ATPase activity, hyperinsulinemia can lead to rapid cellular uptake of potassium. Hypokalemia due to excess insulin may arise in a normal individual given an acute glucose load; in a diabetic given exogenous insulin; or in a patient with an insulinoma.

Beta-Adrenergic Agonists

Increased amount or activity of a beta-adrenergic agonist can also cause a rapid shift of potassium into cells. Clinical examples of hypokalemia probably due to increased endogenous release of epinephrine include the recovery period following exercise, post-head trauma, post-open heart surgery, the active phase of delirium tremens, or a secreting pheochromocytoma. Pharmacologic administration of a beta-adrenergic agonist in asthma or in the treatment of premature labor can also induce hypokalemia. Since the effect of epinephrine on the Na^+/K^+-ATPase is mediated by cAMP, administration of a phosphodiesterase inhibitor (such as aminophylline or theophylline) also can create hypokalemia.

Aldosterone

Although primary hyperaldosteronism creates hypokalemia predominantly by causing renal potassium wasting, the possibility should be entertained that extrarenal kaliopenic effects may also occur.

Other Disorders

A disease in which the cause of transfer of potassium from the extracellular to the intracellular compartments is unknown is hypokalemic periodic paralysis. Affected individuals suffer abrupt attacks of moderate to severe hypokalemia lasting 6–24 hours associated with flaccid muscular paralysis and sometimes arrhythmias. Abnormal sensitivity to insulin, beta-adrenergic agonists, or mineralocorticoid leading to heightened activity of Na^+/K^+-ATPase is one hypothesis for the pathogenesis of this disease, because attacks sometimes follow carbohydrate ingestion, meals, exercise, anxiety, or exposure to cold. The disease can be familial, with an autosomal dominant mode of inheritance, or associated with thyrotoxicosis, especially in Orientals. Effective therapy includes acetazolamide (250–750 mg/d) in the familial form or beta-adrenergic blockade in the thyrotoxic form.

A relative redistribution of potassium into cells occurs following exercise training. In long-distance runners and other conditioned athletes, despite a resting total muscle mass and intracellular potassium concentration higher than normal, there is a slight reduction in extracellular potassium concentration during rest (serum potassium still rises during exercise itself). These findings suggest increased Na^+/K^+-ATPase activity, but the signal is unknown. Hypokalemia can also occur during hypothermia by unknown mechanisms.

Finally, alkalemia causes potassium to enter cells by indirectly accelerating the Na^+/K^+-ATPase pump. Acutely, alkalemia may accelerate Na^+/H^+ antiporter exchange, raise cell sodium, stimulate the activity of the Na^+/K^+-ATPase, and thus transfer more potassium into cells (see converse of these pathogenetic events portrayed in Fig 7–6). The quantitative relationship is that the serum potassium concentration decreases about 0.3 meq/L per 0.1 unit increase in pH in metabolic alkalosis (Table 7–1). More important over the long run, however, is that metabolic alkalosis is virtually always associated with total body potassium depletion caused by renal potassium wastage, as discussed subsequently (see also Chapter 13).

INCREASED NUMBER OF NA^+/K^+-ATPASE UNITS

A large increase in the absolute number of Na^+/K^+-ATPase pumps due to rapid cell proliferation necessitates potassium transfer into new cells. Extracellular potassium may be depleted faster than can be replenished from diet—a form of relative dietary insufficiency. Such rapid cell growth may occur clinically during treatment of iron-deficiency anemia or megaloblastic anemia or in some leukemias. Potassium ingestion must be increased in these situations.

DECREASED POTASSIUM CHANNEL ACTIVITY

An unusual cause of hypokalemia is barium intoxication. Barium blocks the potassium channel by which potassium exits across the cell membrane. Therefore, a normal rate of Na^+/K^+-ATPase pump activity leads to excessive accumulation of potassium within the cell, because the normal potassium leak pathway has been blocked. Specific treatment consists of oral sodium sulfate administration (30 g in 200 mL of water) to bind and precipitate the barium, and forced saline diuresis.

RENAL POTASSIUM LOSS

The most common causes of hypokalemia are attributable to renal potassium wastage, when renal potassium excretion is uncoupled from potassium ingestion and the prevailing level of plasma potassium (Table 8–1). The potassium loss is often not of apparent benefit to the individual but occurs as a by-product of renal compensation for an underlying disorder (eg, resulting from secondary hyperaldosteronism signaled by extracellular volume contraction). Obviously, renal potassium loss must exceed intake to induce negative potassium balance and hypokalemia.

Confirmation that hypokalemia is due to renal potassium loss rests with establishing that the renal response to hypokalemia is inappropriate, ie, that the kidney is not optimally and physiologically retaining potassium in response to hypokalemia. Steady-state renal potassium loss may not be very high, but the diagnosis is confirmed if potassium excretion is not reduced appropriately, below 20 meq/d (Fig 8–1). Thus, renal potassium excretion of more than 20 meq/d or fractional potassium excretion greater than 5% establishes the kidney as the origin of the hypokalemia.

DISORDERED CORTICAL COLLECTING TUBULE FUNCTION

The pathogenesis of renal potassium loss is best understood by considering the disordered determinants of potassium secretion by the principal cell in the cortical collecting tubule, as summarized in Table 8–4. Renal potassium loss is usually due to increased sodium delivery to the cortical collecting tubule (especially when accompanied by a poorly reabsorbed anion such as bicarbonate or ketone), hyperaldosteronism, or both.

Hyperaldosteronism is an important factor in the genesis of all potassium secretory disorders, and clinically significant potassium wasting is difficult to produce in the absence of aldosterone (Fig 7–16). Aldosterone may be elevated either as a consequence of hyperreninemia due to extracellular volume contraction or as a primary condition.

INCREASED SODIUM DELIVERY WITH SECONDARY HYPERALDOSTERONISM

Any disorder that increases sodium chloride filtration or decreases the reabsorption of sodium chloride upstream to the potassium secretory site in the cortical collecting tubule may be a cause of hypokalemia, as shown in Fig 8–3. These conditions were previously discussed in detail in Chapter 2 and summarized in Fig 2–4.

Table 8–4. Pathophysiology of hypokalemic disorders due to renal potassium wasting.

	Distal Delivery				
	NaCl	$NaHCO_3$	Na·Ketone	Aldosterone	K^+ Secretion
Osmotic diuresis	↑	—	—	—	↑
Metabolic alkalosis					
Gastric (vomiting, nasogastric suction)	—	↑↑	—	↑	↑↑↑
Exogenous base	—	↑	—	—	↑
Renal					
Diuretics	↑↑	—	—	↑	↑↑↑
↑Aldosterone	—	—	—	↑↑↑	↑↑↑
Respiratory alkalosis	—	↑	—	—	↑
Metabolic acidosis					
Hyperchloremic (diarrhea, RTA)	↑	—	—	±↑	↑
Normochloremic (ketoacidosis)	↑	—	↑	↑	↑↑↑
Respiratory acidosis	↑	—	—	±↑	↑

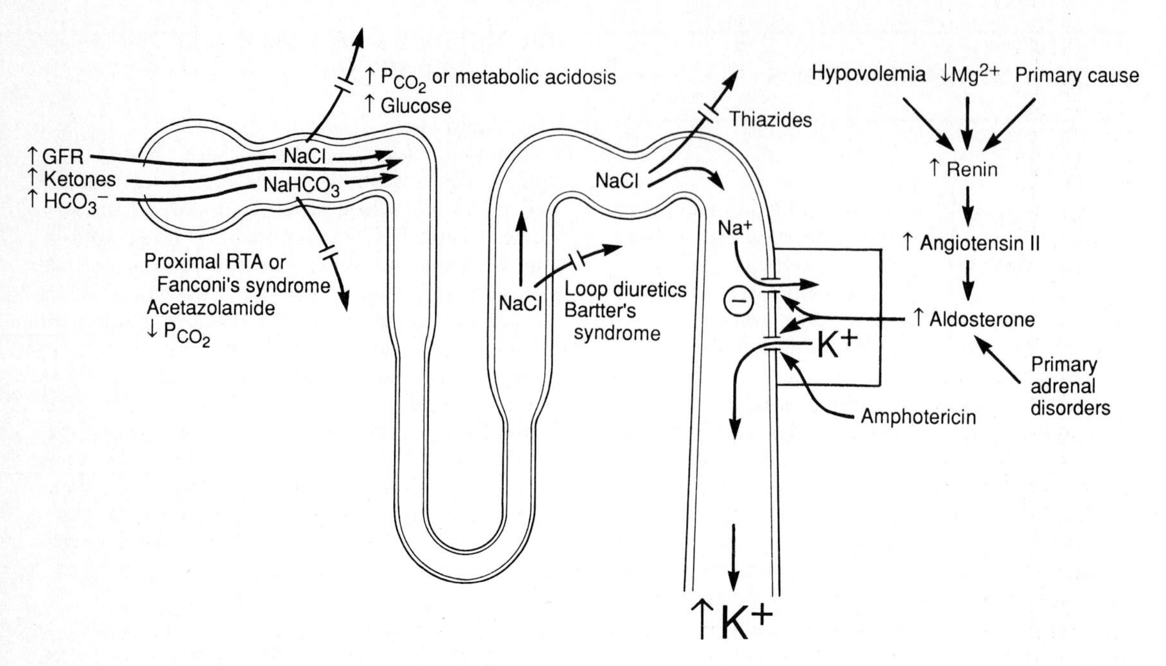

Figure 8-3. Causes of hypokalemia due to enhanced renal potassium loss.

Acute Glomerular Hyperfiltration

Increased GFR and filtration of sodium chloride sufficient to overwhelm sodium reabsorptive systems is unusual but can be induced acutely by a dietary protein load or by pharmacologic administration of atrial natriuretic factor. Sustained kaliuresis does not occur, however, so that an increase in GFR does not usually significantly alter the plasma potassium concentration.

Decreased NaCl Reabsorption in the Proximal Tubule

During osmotic diuresis, a poorly reabsorbed solute—such as glucose, mannitol, radiocontrast dye, or urea—achieves a sufficient concentration within the lumen of the proximal tubule to constrain sodium chloride and water reabsorption. Mannitol or radiocontrast dye is usually not given for long enough periods to cause significant potassium loss. In the setting of diabetic ketoacidosis, however, prolonged hyperglycemia and glycosuria causes renal potassium wasting additively or even synergistically with acidemia, increased delivery of poorly reabsorbable anion (ketones), and volume contraction-induced hyperaldosteronism (Table 8–4). An osmotic diuresis may also occur when there is swift recovery of GFR following a prolonged period of renal insufficiency, as occurs following relief of bilateral renal obstruction or in the resolution phase of acute tubular necrosis. Urea and other solutes that had been retained during the period of renal dysfunction may then be filtered in amounts which overwhelm recovering reabsorptive mechanisms.

Sodium chloride reabsorption in the proximal tubule is defective during any metabolic acidosis (Fig 1–23). The shift of potassium from the intracellular to the extracellular space in some forms of acidemia (Fig 7–6) serves to mitigate the impact on plasma potassium concentration that would otherwise occur as a result of enhanced urinary potassium excretion. Potassium wasting in chronic metabolic acidosis therefore tends to be short-lived, relatively mild, and usually insufficient to cause hypokalemia.

There is also a sodium chloride reabsorptive defect in the proximal tubule induced during the generation of respiratory acidosis. Unlike metabolic acidosis, hypercapnia appears to affect the active component of sodium chloride reabsorption, but the mechanism is unknown. Potassium wasting in respiratory acidosis is mild and transient, occurring only during the period of increasing P_{CO_2} but not in the steady-state condition. Overt hypokalemia is usually not observed.

Decreased Sodium Chloride Reabsorption in the Thick Ascending Limb of Henle

Inhibition of sodium chloride transport in the thick ascending limb of the loop of Henle occurs with administration of a loop diuretic, such as furosemide, bumetanide, or ethacrynic acid, as shown in Fig 8-3. These diuretics directly block the function of the luminal Na^+-K^+-$2Cl^-$ cotransporter, and cause potassium wasting for several reasons. First, enhanced potassium secretion occurs in the cortical collecting tubule because of increased sodium chloride delivery to this downstream potassium secretory site. Second, because transport in the thick ascending limb of Henle normally includes potassium, net potassium reabsorption is inhibited. In fact, net potassium secretion can even appear when the reabsorptive system is pharmacologically blocked by a loop diuretic, since there is a channel in the luminal membrane that allows potassium to leave these cells. Third, aldosterone tends to rise because of hyperreninemia induced by urinary sodium chloride loss and extracellular volume contraction.

The stimulation of urinary potassium excretion induced by a loop diuretic is relatively modest and occurs for about a week following institution of continuous daily therapy. With 40–80 mg/d of furosemide, the average potassium deficit is highly variable, ranging from 0 to 400 meq, with a mean of 200–300 meq, or about 6–9% of total body potassium stores. The average fall in serum potassium level is therefore about 0.5 meq/L, and a level below 3.0 meq/L occurs infrequently—in only about 1% of patients treated for hypertension. Counterbalancing factors come into play that mitigate kaliuresis, including the reduction in serum potassium concentration itself as well as reduction in GFR and enhancement of the fraction of sodium reabsorbed in the proximal tubule consequent to the extracellular volume contraction (Chapter 7). Usually, metabolic alkalosis is present as well.

A rare intrinsic defect of transport in the thick ascending limb of Henle also occurs in Bartter's syndrome, discussed in detail in Chapter 13. The pathophysiology of potassium loss is generally similar to that engendered with a loop diuretic.

Decreased Sodium Chloride Reabsorption in the Distal Convoluted Tubule

Finally, an increase in sodium chloride delivery and flow rate to the cortical collecting tubule can occur when sodium transport in its adjacent upstream neighbor, the distal convoluted tubule, is inhibited by a thiazide diuretic (Fig 8–3). The degrees of potassium depletion and of metabolic alkalosis with commonly used doses of thiazides is roughly similar to that of loop diuretics. Negative potassium balance of about 200–300 meq is usually achieved within the first week of therapy. Significant hypokalemia induced by thiazides is—for unknown reasons—somewhat more common than with loop diuretics, and a serum potassium level under 3.0 meq/L occurs in about 7% of patients.

INCREASED SODIUM DELIVERY WITH POORLY REABSORBED ANION

Bicarbonate

Increased bicarbonate delivery to the cortical collecting tubule may occur during the generation phase of metabolic alkalosis from gastric causes, such as vomiting or nasogastric suction (Table 8–4). As discussed in more detail in Chapter 13, the rise in blood bicarbonate concentration causes the filtered load of bicarbonate to transiently exceed the reabsorptive capacity of the proximal tubule. Excess sodium bicarbonate is then delivered to the collecting tubule. Only some of the surfeit of bicarbonate is reabsorbed by the cortical collecting tubule, so that potassium secretion is stimulated (Fig 7–14). Equally importantly, the concomitant extracellular volume contraction causes secondary hyperaldosteronism, potentiating potassium secretion.

The increased bicarbonate delivery out of the proximal tubule and resulting kaliuresis are short-lived, lasting only about a day following a single episode of vomiting or nasogastric suction. After that, compensatory mechanisms occur, either reduction in GFR or enhanced bicarbonate reabsorption (as discussed in Chapters 10 and 13), so that bicarbonate delivery out of the proximal tubule is normalized, bicarbonaturia disappears, and urinary potassium excretion falls back to the level of potassium intake. With repeated episodes of vomiting or chronic gastric drainage, recurrent periods of kaliuresis occur, and the total potassium deficit can be enormous, reaching several hundred or even 1000 meq.

Exogenous alkali administration can potentially cause the same pathophysiologic result as vomiting. Potassium wasting in this case is usually much less severe—if it occurs at all—because the aldosterone level is not increased and may actually be decreased if extracellular volume becomes expanded.

Increased bicarbonate delivery to the cortical collecting tubule can also occur as the result of an abnormality in proximal bicarbonate reabsorption (Fig 8–3), as an isolated problem in type I proximal renal tubular acidosis, or as part of the spectrum of defects in Fanconi's syndrome. As discussed in Chapter 12, until sufficient hypobicarbonatemia occurs to reduce the amount of bicarbonate escaping from the proximal tubule to a normal level, excessive sodium bicarbonate delivery to the cortical collecting tubule stimulates potassium secretion. When bicarbonate therapy is used to normalize the blood bicarbonate concentration, the amount of bicarbonate that escapes proximal reabsorption and is delivered to the cortical collecting tubule is tremendous. Massive potassium wasting occurs, and a large amount of potassium as bicarbonate or citrate (a better-tolerated base equivalent) is required for repletion.

Proximal bicarbonate reabsorption can also be transiently decreased physiologically during the generation of acute respiratory alkalosis. The resulting potassium wasting is mild and abates when the new steady-state P_{CO_2} is reached, so that overt hypokalemia is rarely precipitated. However, recurrent bouts of acute changes in P_{CO_2}—for instance, when P_{CO_2} is constantly fluctuating in patients requiring ventilatory support in the intensive care setting—cause repetitive kaliuretic episodes.

Ketones

Increased delivery of the sodium salts of poorly reabsorbable ketones to the cortical collecting tubule is one of the mechanisms by which potassium secretion is stimulated in ketoacidosis (Fig 8–3). Reabsorption of sodium without the accompanying ketone anion increases the lumen-negative potential difference and thereby augments potassium secretion (Fig 7–14). As mentioned above, increased sodium delivery induced by glycosuria during ketoacidosis, in conjunction with hyperaldosteronism, also contributes to the kaliuretic effect of ketones (Table 8–4).

Although marked potassium depletion is common—averaging 5 meq/kg of body weight (usually about 350 meq) in large series of patients with diabetic ketoacidosis—hypokalemia does not commonly result since concomitant potassium shift out of cells is caused predominantly by hyperosmolality and insulinopenia and, less importantly, by acidemia. However, when saline and insulin therapy of the hyperglycemia and ketoacidosis is instituted, profound hypokalemia may be induced for two reasons. First, normalization of the plasma glucose and insulin levels serves to redistribute potassium back into cells. Second, restoration of normal circulating plasma volume tends to increase GFR and thus enhance delivery of sodium and ketones to the cortical collecting tubule, still under stimulation by high aldosterone level, thereby increasing the kaliuresis. Potassium replacement during ketoacidosis should therefore begin as soon as the serum potassium level has fallen to a normal value if initially elevated and should be sufficient to compensate for both estimated preexisting loss and ongoing urinary excretion measured during therapy.

Antibiotics

High-dose sodium-penicillin therapy or administration of carbenicillin may cause potassium wasting due to the nonreabsorbable anion effect of the drugs themselves. There is insufficient substantiation, however, that these drugs usually achieve an intratubular concentration high enough to significantly augment potassium secretion. Other antibiotic-induced changes in renal function, such as magnesium depletion, may play a larger role in potassium transport than the direct effect of the antibiotic itself.

PRIMARY INCREASE IN MINERALOCORTICOID ACTIVITY

A large increase in mineralocorticoid activity can independently drive the principal cell of the cortical collecting tubule to secrete potassium in excess of physiologic demands (Tables 8–2 and 8–4 and Fig 8–3).

When the aldosterone level become disjoined from control of extracellular volume, the homeostatic level of potassium excretion—equivalent to potassium intake due to the reciprocal relationship of sodium delivery and aldosterone level described previously (Fig 7-18)—no longer obtains. By controlling both the principal and the intercalated cells of the collecting tubule, hypermineralocorticoidism increases rates of sodium reabsorption and of hydrogen ion and potassium secretion, causing sodium chloride retention and hypertension (without edema, however, due to escape), metabolic alkalosis, and hypokalemia. The detailed causes of hypermineralocorticoid syndromes are considered in Chapters 3 and 13, and only a broad categorization will be repeated below

Primary Renin Overproduction

Increased mineralocorticoid activity is the predictable cause of elevation by renin of aldosterone's major functional agonist, angiotensin II, as with renal artery stenosis, renal vasculitides, or renin-secreting tumors. In these cases, elevated concentrations of both renin and aldosterone occur.

Another very important cause of primary hyperreninism is magnesium deficiency. The mechanism by which hypomagnesemia increases renin production to cause hyperaldosteronism is unknown. Nevertheless, the resultant potassium wasting can be substantial and difficult to treat until the magnesium deficiency is repaired. The potassium deficiency ascribed to several drugs—eg, aminoglycosides, cisplatin, cyclosporine—is probably attributable in large measure to their propensity to cause magnesium wastage. Causes and treatment of magnesium deficiency are covered in Chapter 22.

Primary Adrenal Mineralocorticoid Overproduction

There are also situations in which mineralocorticoid is overproduced by the adrenal gland without stimulation by angiotensin II. Examples include Cushing's disease and syndrome, in which the primary mineralocorticoids are deoxycorticosterone and corticosterone instead of aldosterone; hereditary adrenal enzymatic defects usually associated with the adrenogenital syndromes; and primary hyperaldosteronism due to adenoma or hyperplasia. In these cases, aldosterone is increased whereas renin is suppressed by the extracellular volume expansion.

With sustained elevation in aldosterone level, after an initial antinatriuresis, sodium delivery to the collecting tubule returns to normal—the phenomenon called "mineralocorticoid escape" (Chapter 3). Should sodium delivery be further increased with diuretics, marked potassium wasting and profound hypokalemia can occur. Thus, a patient with a serum potassium level under 2.5 meq/L following diuretic therapy should be suspected of an autonomous mineralocorticoid excess syndrome.

Functional hypermineralocorticoidism may arise from glycyrrhizic acid-containing compounds that inhibit the enzyme—11β-hydroxysteroid dehydrogenase—which normally prevents endogenous glucocorticoids from occupying the intracellular mineralocorticoid receptor in the cortical collecting tubule. Examples include ingestion of licorice or chewing of tobacco or administration of the drug carbenoxolone. Inhalation of a nasal spray containing the mineralocorticoid fludrocortisone can also induce potassium wasting. An unknown mineralocorticoid is secreted in Liddle's syndrome, a rare disorder that symptomatically resembles primary aldosteronism but in which aldosterone is undetectable. In all these conditions, measurements of both renin and aldosterone are low.

ABNORMALITIES OF OTHER DETERMINANTS OF POTASSIUM SECRETION

There are unusual cases in which other determinants of potassium secretion besides those considered above are altered.

Luminal Potassium Channel

Potassium wasting occurs predictably with amphotericin B because the antibiotic is itself an ion channel. Amphotericin B molecules insert into the luminal cell membrane and thereby provide more channels for potassium permeation from cell to lumen. Aminoglycosides and other antibiotics may also increase potassium excretion by a similar mechanism, though this is poorly documented. Therapy consists of discontinuation of the antibiotic, if possible, followed by potassium repletion.

Electrochemical Gradient

Potassium wasting and hypokalemia occur also with type II classical distal renal tubular acidosis, though the mechanism is not completely clear. According to one hypothesis, the decrease in the lumen-positive component of the potential difference due to electrogenic hydrogen ion secretion by intercalated cells of the cortical collecting tubule allows a higher level of lumen negativity due to electrogenic sodium reabsorption. The greater lumen-negative potential then helps to drive more potassium secretion. Potassium wasting tends to be ameliorated when the acidosis is treated.

SYMPTOMS

Given the crucial role of potassium for cell metabolism and neuromuscular membrane potential, it is not surprising that manifestations of intracellular potassium deficiency are manifold, as summarized in Fig 8–4.

NEUROMUSCULAR

The interior-negative electrical potential difference in muscle is largely defined by the potassium diffusion potential, proportionate to the intracellular:extracellular potassium concentration ratio. With mild to moderate degrees of hypokalemia, resting muscle potential becomes more negative, or hyperpolarized. Intracellular potassium concentration may be decreased, but hyperpolarization still occurs because the extracellular potassium concentration is relatively more depressed. Hyperpolarization interferes with neuromuscular function by raising the threshold potential for rapid depolarization, which is necessary for neural conduction and muscular excitation. When rapid depolarization does occur, cell repolarization is also slowed. With severe potassium depletion, intracellular membrane voltage falls, apparently because of an increase in sodium permeability, and paralysis results.

With mild to moderate degrees of hypokalemia, symptoms of depressed neuromuscular function include muscular weakness, fatigue, and muscle cramps. Gastrointestinal tract motility is also affected, resulting in constipation or ileus.

Severe hypokalemia (< 2.5 meq/L) may cause more serious symptoms, such as tetany with paralysis of the legs or even of the trunk, leading to respiratory paralysis. Ileus may also occur. Rhabdomyolysis can occur, but this is predominantly

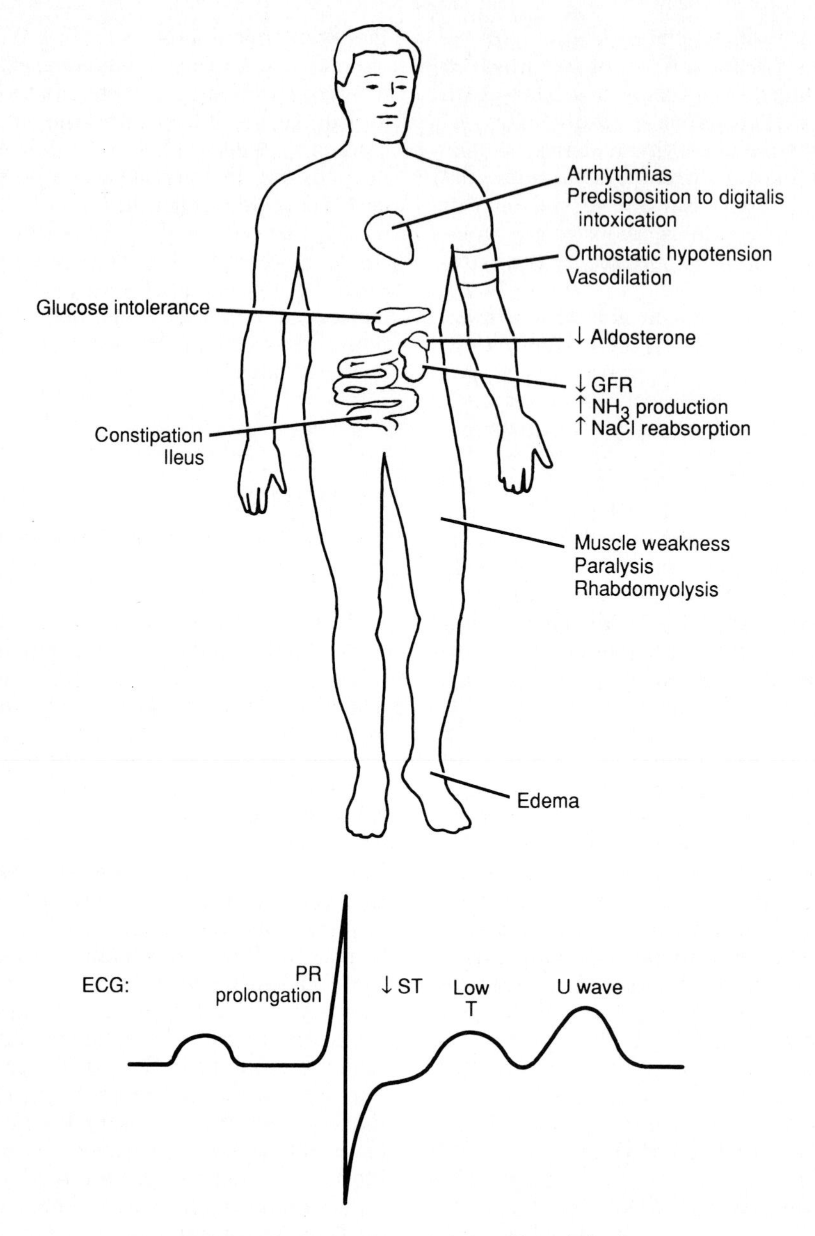

Figure 8-4. Symptoms and signs of hypokalemia.

due to relative vascular insufficiency and ischemia in most cases (see below).

CARDIOVASCULAR

Heart

Potassium deficiency-induced electrocardiographic changes include flattening of the T wave, depression of the S-T segment, and the appearance of a U wave (Fig 8-4). Increased predisposition to atrial and ventricular arrhythmias, atrioventricular block, and escape rhythms are feared consequences of hypokalemia—even without total body potassium depletion—especially if the patient is receiving digitalis.

Peripheral Vasculature

Potassium deficiency tends to relax basal vascular tone, decrease peripheral vascular resistance, and depress blood pressure. Vessels are simultaneously rendered unresponsive to vasoconstriction by pressors—in part due to prostaglandin (especially PGE_2) production—or further dilation by ischemia. The vascular unresponsiveness to ischemia can have devastating consequences during muscular exertion. Since potassium deficiency alters intracellular energy metabolism and prevents the normal increase in blood flow needed for oxygen supply and nutrient delivery during exercise, rhabdomyolysis can be precipitated, which can cause acute renal failure. Elevated creatine kinase levels reflecting muscle cell injury can be detected even with moderate potassium deficiency.

RENAL

GFR

Potassium deficiency causes a decrease in renal blood flow and GFR despite a tendency to cause renal hypertrophy. Although potassium deficiency acts to decrease systemic vascular resistance, it increases renal vascular resistance by inducing the intrarenal synthesis of the potent vasoconstrictors angiotensin II (via stimulation of renin) and thromboxane.

Water Metabolism

Potassium depletion inhibits the ability of ADH to increase the hydraulic water permeability in the medullary collecting tubule (Fig 4–12) and thereby impairs maximal urinary concentration capacity. Symptoms include polyuria and nocturia. Increased thirst induced by potassium depletion contributes to the vasopressin-resistant polyuric state (Chapter 6).

Sodium Chloride Transport

Potassium deficiency stimulates sodium chloride reabsorption by undefined mechanisms operating at unknown nephron sites. Edema can actually ensue (Chapter 3). Potassium deficiency-induced renal salt wasting and interstitial nephritis occur in animals, but these conditions have been difficult to confirm in humans.

Acid-Base Metabolism

Potassium depletion increases ammoniagenesis (Chapter 10), which can affect renal acid-base balance as well as predispose an individual with liver disease to coma. When potassium deficiency occurs during simultaneous restriction of dietary sodium chloride to prevent extracellular volume expansion, a moderate degree of metabolic alkalosis occurs. The role of potassium deficiency to maintain metabolic alkalosis is considered in greater detail in Chapters 10 and 13.

ENDOCRINE

Potassium depletion depresses insulin release and results in glucose intolerance. Hypokalemia tends to increase renin production, but the counterbalancing direct inhibition of aldosterone secretion usually causes mineralocorticoid activity to fall. Potassium deficiency also impairs utilization of nitrogen. If protein intake is even moderately reduced, negative nitrogen balance, muscle wasting, and growth retardation in children can occur.

DIAGNOSIS

APPROACH TO ETIOLOGIC DIAGNOSIS

History & Physical Examination

A dietary history to ascertain potassium intake is very important. Attention should be directed toward a history or signs of conditions associated with potassium redistribution into cells, such as hypoglycemia (eg, due to overproduction or overadministration of insulin), adrenergic hyperactivity and hypertension (eg, due to pheochromocytoma or a phosphodiesterase inhibitory drug such as aminophylline or theophylline), or thyrotoxicosis.

The possibility of extrarenal potassium loss should be considered if there is a history of diarrhea or other form of external gastrointestinal drainage. Surreptitious laxative abuse can be difficult to diagnose, but helpful clues can be obtained if a red color is obtained following alkalinization of fecal fluid, indicating the presence of phenolphthalein-containing laxative, or if the fecal fluid magnesium concentration is over 12 meq/L, indicating the presence of a magnesium-containing laxative.

All acid-base disorders are associated with renal potassium wasting, at least to a limited degree (Table 8–4). A history of diuretic ingestion, vomiting, nasogastric suction, or hypertension (due to primary hypermineralocorticoidism) should be sought when hypokalemia occurs in conjunction with metabolic alkalosis. Renal potassium loss almost always occurs in the setting of uncontrolled diabetes with ketoacidosis. Causes of the three forms of hypokalemic, hyperchloremic metabolic acidosis—diarrhea, type I proximal RTA, and type II classical distal RTA—are discussed in Chapter 12 and listed in Tables 12–4 and 12–5.

Urinary Potassium

Important in the approach to the diagnosis of total body potassium depletion is whether the kidney is responding in an appropriate physiologic manner to the hypokalemia by conserving potassium (inadequate net input and nonrenal loss categories in Fig 8–1) or whether the kidney is part of the problem by not retaining potassium appropriately (renal loss category in Fig 8–1). Appropriate renal conservation of potassium, as defined above, includes depression of urinary potassium loss to the minimal level of less than 20 meq/d, representing a fractional potassium excretion rate of less than 5%.

Finding an appropriately low urinary potassium excretory rate does not necessarily exonerate the kidney as the cause of the hypokalemia. The stimulus to renal potassium wasting may have subsided. For instance, during the active phase of vomiting or during the active use of a diuretic, hypokalemia may ensue as a consequence of renal potassium wastage. But if the vomiting should abate or the diuretic be discontinued, the hypokalemia will persist while the kidney resumes normal potassium conservation. An equally important consideration is that the urine potassium excretion need not be "high" to implicate the kidney as the major source of potassium loss; what is more significant is that urinary potassium is not appropriately low (< 20 meq/d), representing an unphysiologic response to the body's need to conserve potassium.

Arterial Blood Gases & Other Laboratory Tests

As mentioned above, disorders of potassium metabolism are frequently associated with acid-base disorders, and vice versa (Table 8–4). Other useful tests that can aid in the diagnosis of hypokalemic disorders include urinary sodium and chloride concentrations (diuretics), urinary and plasma glucose or ketones (diabetic ketoacidosis), plasma magnesium (hypomagnesemia), and hormone determinations (as indicated for investigation of disordered regulation of insulin, thyroid hormone, renin, aldosterone, or cortisol).

THERAPY

DECISION TO TREAT

Redistribution Versus Total Body Deficit

Assessment of the cause of hypokalemia is absolutely critical before a rational decision can be made about whether or how to treat the disorder. If potassium is merely shifted into cells and total body potassium is not depleted, potassium supplementation is not only not indicated but can be dangerous. If total body potassium depletion is the cause of the hypokalemia, then obviously the reason for the potassium wasting should be corrected if at all possible and the potassium deficit repaired.

Degree of Hypokalemia

Although some controversy exists, most physicians would correct a serum potassium concentration of less than 3.5 meq/L, especially if there is underlying cardiac disease, documented arrhythmias, digitalis therapy, or recent myocardial infarction. The data substantiating such a recommendation for mild hypokalemia of 3.0–3.5 meq/L are far from complete, but if untreated, an increased incidence of ventricular arrhythmias has been documented in several studies.

ROUTE, RATE, & AMOUNT OF POTASSIUM REPLACEMENT

Choice of Potassium Salt

When there is concurrent severe bicarbonate depletion (eg, in diarrheal states or type I proximal and type II classical distal RTA), potassium bicarbonate or its better-tolerated equivalent potassium citrate is appropriate therapy. Similarly, when there is severe phosphate depletion (eg, in diabetic ketoacidosis), potassium phosphate replacement is rational therapy.

In all other cases, however, potassium repletion should be in the form of the **chloride salt.** Potassium is better retained in the extracellular space if given with chloride. Renal retention of potassium is also better when given as the chloride salt. If potassium is administered with a nonreabsorbable anion such as gluconate, bicarbonate, or citrate, there is a likelihood of resecretion of potassium in the cortical collecting tubule and loss into the urine, with resulting attenuation of the therapeutic value. For the same reason, repletion of a large potassium deficit using food with high potassium content is difficult since the anions accompanying potassium in almost all natural foods are nonchloride.

Assessment of Potassium Deficit

The total amount of potassium depletion may be estimated by reference to Fig 8–5. A deficit of about 350 meq occurs for each meq/L decrement in serum potassium concentration below a level of 4.0 meq/L. According to this formulation, a patient may have a serum potassium concentration

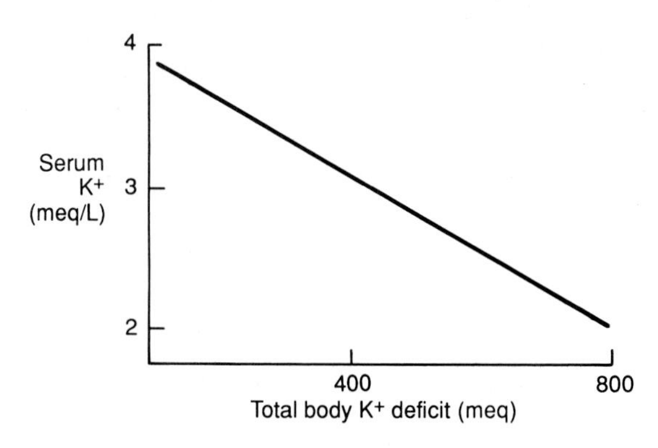

Figure 8-5. Magnitude of hypokalemia as a function of total body potassium deficit.

within the normal range (3.5 meq/L) and still have a 175 meq potassium deficit.

It is important to remember that concurrent changes in blood pH (Table 7-1), hormones (insulin, beta-adrenergic agonists, and aldosterone), and osmolality independently modify this relationship. Serum potassium concentration tends to decrease by 0.3 meq/L for each 0.1 unit rise in blood pH during metabolic alkalosis and to increase by about 0.3–0.5 meq/L for every 10 mosm/kg rise in serum osmolality. For example, a patient undergoing intractable vomiting with a serum potassium concentration of 2.0 meq/L and a blood pH of 7.5 might be assumed to have a serum potassium concentration of 2.3 meq/L corrected for blood pH and therefore a potassium deficit of about 600 meq (350 meq per 0.1 meq/L change × [4.0 meq/L − 2.3 meq/L]).

In addition, these parameters are dynamic and

Table 8-5. Oral potassium replacement.

Amount	meq of K^+	Anion	Names
Liquids			
15 mL	10	Cl^-	5% potassium chloride
15 mL	15	Cl^-	7.5% potassium chloride (Potassine)
15 mL	20	Cl^-	10% potassium chloride (Cena-K, EM-K-10%, Kaochlor 10%, Kaochlor S-F, Kay Ciel, Klor-10%, Klorvess, Potachlor 10%, Potasalan, Potassine 10%, SK-potassium)
15 mL	30	Cl^-	15% potassium chloride (Rum-K)
15 mL	40	Cl^-	20% potassium chloride (Kaon-Cl- 20%, Klor-Con, Potachlor 20%, SK-potassium chloride)
15 mL	20	Gluconate⁻	Potassium gluconate (Kaon, Kaylixir, K-G Elixir, My-K Elixir)
15 mL	20	Gluconate⁻/ Citrate⁻	(Bi-K, Twin K)
15 mL	20	Gluconate⁻/ Cl^-	(Duo-K, Kolyum)
15 mL	15	Gluconate⁻/ Citrate⁻/ Cl^-	(Twin-K-Cl)
15 mL	45	Acetate⁻, HCO_3^-, Citrate⁻	(Trikates, Tri-k)
POWDERS			
Packet	15	Cl^-	(K-Lor)
Packet	20	Cl^-	Potassium chloride (Kato, Kay Ciel, K-Lor, Klor-Con, Potage)
Packet	25	Cl^-	(Klor-Con/25, K-Lyte/C1)
Packet	20	Cl^-/HCO_3^-	(Klorvess Effervescent Granules)
Packet	20	Gluconate⁻/ Cl^-	(Kolyum)

Table 8-6. Potassium content of foods.

	Very High (12-20 meq)	High (5-12 meq)	Moderate (3-5 meq)	Low (< 3 meq)
Milk (1/2 cup or as stated)			Buttermilk Ice cream Milk Yogurt	Cheese (1 oz) Cream cheese (1 oz) Sherbet Sour cream (1 tbsp)
Beans and nuts (1/2 cup or as stated)	Garbanzos Soy beans	Kidney, lima, navy, pinto beans	Almonds (20) Peanut butter (2 tbsp)	Bean sprouts Pecans (20 halves) Walnuts (15) Peanuts (1 tbsp)
Meat (2 oz or as stated)			Fish Ground beef Lamb chop Pork chop Sirloin steak Veal cutlet	Chicken (3 1/2 oz) Egg (1)
Fruit (1/2 cup or as stated)	Papaya (1 medium)	Apricots (3 halves) Banana (6 inch) Cantaloupe (1/4-inch slice) Honeydew melon (1/4-inch slice) Orange (3-inch diameter) Orange juice Pear (1 large) Prunes (4) Prune juice Rhubarb	Apple (1) Apricot (1 large) Cherries Dates (3) Figs (3) Fruit cocktail Grapefruit(1) Grapefruit juice Lemon juice Peach (1 medium) Pineapple juice Raisins (2 tbsp) Strawberries (12)	Apple juice Applesauce Blueberries Cranberries Grapes Grape juice Lemon (1) Lime (1) Pear Pineapple Plum (1) Raspberries Tangerine (1) Watermelon
Vegetables (1/2 cup or as stated)		Artichoke (1) Avocado (1/4) Brussels sprouts Carrot (7 1/2 in) Chard Potato (1 baked, 1 broiled, 10 fries, 1/2 cup mashed) Pumpkin Spinach Sweet potato (1) Tomato (1) Turnip Yam	Beets Broccoli Cauliflower Celery (8-inch stalk) Corn (1 ear) Eggplant Green pepper (1) Lettuce (4 leaves) Mushrooms Mustard greens Squash	Asparagus (3) Cabbage Corn kernels Cucumber (6 slices) Green beans Green peas Onions (2 tbsp) Radishes (4)
Grains			Bran muffin (1) Bran flakes (1 cup)	Bread (1 slice) Cereals (1 1/2 cup) Noodles (1 cup) Rice (1 cup) Crackers (4)
Other	Bouillon, unsalted (1 cube) Salt substitute (1 tsp)	Potato chips (10) Ketchup (1 Tbsp) Tomato or vegetable juice (1/2 cup)	Chocolate candy (1 oz) Coffee (1 cup) Tea (1 cup)	Beer (12 oz) Butter (1 pat) Soft drinks (1) Honey, jam, jelly Mayonnaise (1 tbsp) Mocha mix (1/2 cup) Whiskey (1 1/2 oz) Wine (4 oz)

often change during the course of therapy. Therefore, frequent monitoring of potassium replacement therapy is mandatory. During rapid replenishment, the serum potassium concentration should be checked frequently—every 2 or 3 hours. With less rapid potassium administration, monitoring every few hours or days may be sufficient. It is often wise to measure ongoing potassium losses in the stool and urine, since they need to be replaced in addition to preexisting deficit. Potassium depletion in the setting of counteracting influences of hyperosmolality, acidemia, and insulin deficiency can actually be associated with hyperkalemia (eg, diabetic ketoacidosis). In such circumstances, potassium repletion must be withheld until the serum potassium is normalized.

Therapy in Acute Situation

With life-threatening cardiovascular or neuromuscular complications of hypokalemia—malignant ventricular arrhythmias, severe digitalis intoxication, or muscular paralysis—potassium repletion should be swift, administered as the chloride salt, and given by the intravenous route. Rates up to 40 meq/h of potassium in intravenous solutions containing up to 60 meq/L of potassium chloride can be given, but continuous electrocardiographic monitoring, preferably in the intensive care setting, should be maintained along with frequent (every 3–6 hours) rechecking of the serum potassium level.

Therapy in Nonacute Situation

With less urgent indications in the mildly symptomatic or asymptomatic hypokalemic patient, potassium losses may be replaced more slowly by the oral route, usually achieved by using liquid, powder, tablet, or slow-release tablets or capsules, as set forth in Table 8–5. Unless there are reasons for not doing so as mentioned above, chloride-containing forms of therapy should be given.

Liquid potassium chloride has an unpleasant taste, and enteric-coated tablets of potassium chloride may cause gastric and duodenal mucosal abnormalities in affected patients or induce ulceration. Liquids are better tolerated when added to other beverages such as fruit juice. Slow-release tablets and capsules also have the rare potential for causing ulceration, especially when intestinal motility is impaired. Salt substitutes, which contain about 50– 65 meq of potassium chloride per level teaspoon, are well tolerated form of therapy.

Twenty to 160 meq/d of potassium can be given orally in divided doses, depending on the estimate of present or anticipated loss. As mentioned above, dietary sources of potassium may be used in some circumstances, although the nonchloride nature of the natural forms of potassium diminish their usefulness. Nevertheless, modest potassium loss may be prevented by dietary therapy, as guided by Table 8–6.

In patients receiving a diuretic, a useful adjunct to oral potassium replacement is dietary sodium restriction. Reduction of sodium chloride intake to less than 100 meq/d may actually obviate the need for potassium supplementation.

Intravenous administration of potassium can be used when oral medications cannot be given, but potassium by this route tends to be sclerosing, and a concentration greater than 40 meq/L and rate exceeding 10 meq/h in peripheral veins should be avoided if possible. In a large central vein, the concentration of infused potassium can be increased, but intracardiac administration should be avoided because of the risk of precipitating arrhythmias. Concurrent glucose administration should also be avoided if possible, because insulin release will further depress the serum potassium concentration.

Agents are also available that block the action of aldosterone and thus prevent potassium secretion in the cortical collecting tubule. Such drugs are particularly useful in states in which the primary problem is hyperaldosteronism (eg, primary hyperaldosteronism), until the underlying disorder can be corrected. Their use in other situations, such as the amelioration of potassium wasting due to diuretics, is less optimal, since these drugs themselves have side effects, prevent normal homeostatic adjustments to changes in potassium intake, and predispose to hyperkalemia in patients with impaired renal function.

These potassium-sparing diuretics are considered in greater detail in Chapter 3. They include spironolactone (25–400 mg/d), which inhibits the binding of the aldosterone to its receptor; and triamterene (100–400 mg/d) and amiloride (5–40 mg/d), which block the sodium channel in the luminal membrane of the cortical collecting tubule. Monitoring of the serum potassium is essential while the drugs are titrated to the optimal level in each individual patient. Potassium supplementation should never be given concurrently with these potassium-sparing diuretics, since life-threatening hyperkalemia may be inadvertently induced. These diuretics should also never be used in patients with renal insufficiency, who are very dependent on mineralocorticoid function to maintain potassium homeostasis, especially diabetics with marginal aldosterone reserve.

9 Hyperkalemic Disorders

INTRODUCTION

DEFINITION & CLASSIFICATION

Hyperkalemia is diagnosed when the potassium concentration, usually measured in the serum, is **5.0 meq/L or more.** A value between 4.5 and 5.0 meq/L falls into a gray zone but is considered excessive by most clinicians.

Rule Out Pseudohyperkalemia

Spurious causes of hyperkalemia must be excluded before the diagnosis is accepted, as shown in Fig 9–1. These causes of pseudohyperkalemia are, in general, due to leakage of potassium out of cells into the serum of the blood sample while clotting. A small amount of leakage occurs normally, but the rise in serum potassium following coagulation is small (< 0.5 meq/L). Abnormally large leakage is usually attributable to too many cells or to defective membrane permeability. The problem can often be averted if potassium concentration is measured in plasma rather than in serum to obviate clotting.

Spurious hyperkalemia occurs commonly when there is hemolysis. The released hemoglobin is usually visible, though leaky red cell membranes (familial or in infectious mononucleosis) can sometimes allow exit of potassium from red cells without hemoglobin. Severe thrombocytosis, with platelet count over 1,000,000/μL, or severe leukocytosis, with white blood cell count over 500,000/μL, can also elevate serum potassium concentration. Prompt analysis of potassium concentration in a plasma rather than a serum sample—before potassium egress from the formed blood elements can occur (< 2 hours at room temperature)—usually avoids these problems.

Since muscle cells release potassium in response to exercise and ischemia, prolonged ischemic blood drawing using a tourniquet may cause more potassium to appear in the blood sample than is representative of the systemic circulation. Care should be taken to collect the sample after only brief tourniquet application with minimal fist clenching or, preferably, from free-flowing blood after removal of the tourniquet.

Association With Metabolic Acidosis

As a generality, hyperkalemic disorders are associated with metabolic acidosis. The hyperkalemic metabolic acidosis may be due to coordinately increased hydrogen ion and potassium production or release (eg, states of hypercatabolism, tissue necrosis, or diabetic ketoacidosis). Alternatively, both hyperkalemia and metabolic acidosis may be caused by common factors that affect both renal hydrogen ion and potassium excretion, as occurs with reduced sodium delivery to the cortical collecting tubule (states of diminished GFR) or functional hypoaldosteronism (deficient aldosterone release, production, or action). Thus, metabolic acidosis may be of the increased anion gap type (eg, ketoacidosis, lactic acidosis, or uremia) or the hyperchloremic type (eg, hypoaldosteronism).

Whatever the cause of the hyperkalemia, a sustained high blood potassium level diminishes ammonia production and thus depresses urinary net acid excretion and generates or contributes to chronic metabolic acidosis, as described more fully in Chapter 12. Finally, acidemia tends to further elevate the serum potassium by redistributing potassium out of cells (Fig 7–6). Thus, both hyperkalemia and metabolic acidosis may be caused in tandem by the same underlying disorder, or metabolic acidosis and hyperkalemia may be simply a consequence of each other.

PATHOPHYSIOLOGY

As shown in Fig 9–1, similar to the situation for hypokalemic states (Table 8–1), hyperkalemia may be attributable to (1) extreme alterations in potassium intake, (2) changes in intracellular-extracellular distribution, and (3) abnormal renal potassium excretion due to decreased GFR or diminished production, secretion, or action of aldosterone.

As part of the expected response to hyperka-

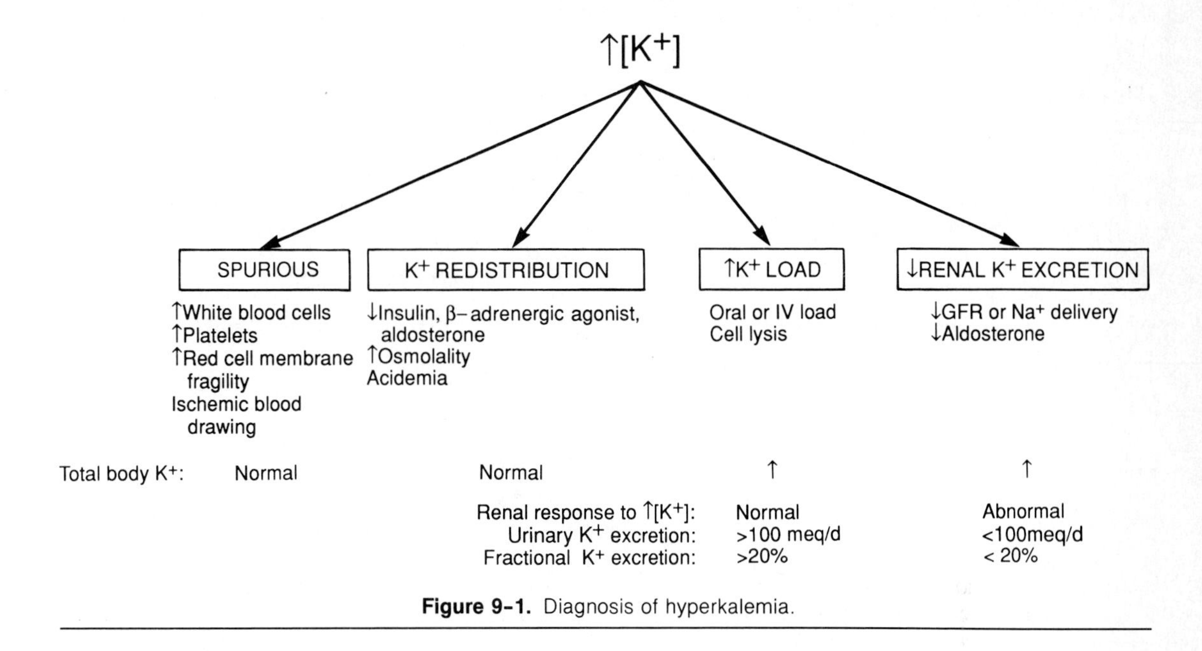

Figure 9-1. Diagnosis of hyperkalemia.

lemia if the kidney is normal and not responsible for generating the condition, urine potassium concentration should exceed 100 meq/L. Fractional potassium excretion should be well over 20% and usually over 100% if sufficient time has elapsed to promote the adaptation process (Fig 9-1). The urine potassium:sodium concentration ratio should be extremely high, certainly greater than 1 and sometimes greater than 50, reflecting maximal potassium secretion and sodium reabsorption due to high aldosterone activity in the cortical collecting tubule.

EXCESSIVE POTASSIUM LOAD

The causes of hyperkalemia are summarized in Table 9-1.

With normal renal function and time for the potassium adaptation phenomenon to occur, an ingested potassium load of at least 400 meq/d, and probably higher, can be excreted without marked alteration in the serum potassium concentration (see Figs 7-20 and 7-21). With more rapid loading, renal potassium excretory capacity can be overwhelmed, but such cases are unusual.

A massive potassium load may occur as a result of exogenous administration by either the oral (single dose of ≥ 300 meq) or intravenous (> 40 meq/h) routes; or from mobilization of endoge-

Table 9-1. Hyperkalemic disorders.

Potassium load
- Exogenous
  - Oral
  - Intravenous
- Endogenous
  - Cell lysis
  - Trauma
  - Rhabdomyolysis
  - Resolution of hematoma
  - Hemolysis
  - Hypercatabolism

Redistribution of potassium
- ↓Na^+/K^+-ATPase activity
  - ↓Insulin
  - ↓Beta-adrenergic agonist
  - ↓Aldosterone
  - Digitalis therapy
- Metabolic acidosis
- Hyperosmolality
  - Hyperglycemia
  - Mannitol
  - Radiocontrast media
- Other
  - Exercise
  - Depolarizing muscle relaxants
    - Succinylcholine
  - Fluoride intoxication
  - Hypothermia
  - Hyperkalemic periodic paralysis

Abnormal renal potassium excretion
- ↓Sodium delivery
  - Chronic renal failure
  - Acute renal failure
- ↑Sodium reabsorption
  - Hypovolemic states
  - Hypervolemic states
- ↓Aldosterone activity
  - ↓Aldosterone synthesis, release, or action (see Table 12-6)

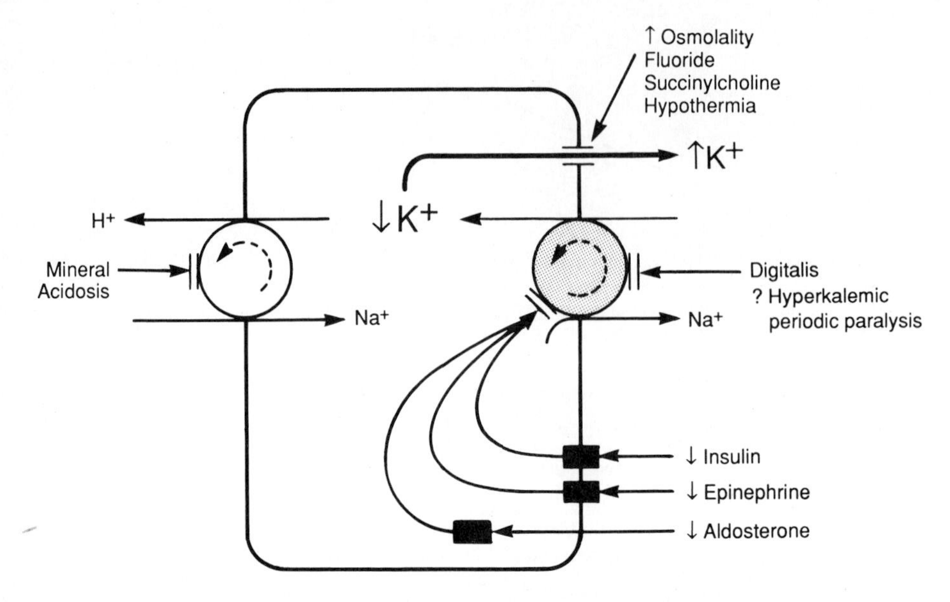

Figure 9–2. Causes of hyperkalemia due to enhanced potassium translocation out of cells.

nous stores, eg, following rapid cell lysis during treatment of tumors with massive cell mass, such as Burkitt's lymphoma or acute leukemia. Release of large endogenous potassium stores can also occur during extensive trauma, rhabdomyolysis, breakdown of a large hematoma or of a collection of blood from gastrointestinal hemorrhage, acute hemolysis, or a markedly hypercatabolic state.

In all of these conditions, urinary potassium concentration is appropriately elevated. Nevertheless, renal potassium excretion is insufficient to cope with the potassium load, especially because GFR is usually concomitantly depressed in these conditions owing to hypovolemia or acute tubular necrosis.

POTASSIUM REDISTRIBUTION OUT OF CELLS

DECREASED Na^+/K^+-ATPASE ACTIVITY

Potassium release from cells can be due to diminished Na^+/K^+-ATPase activity, as shown in Fig 9–2. These are the counterparts of the conditions that cause hypokalemia (Fig 8–2). Given the large intracellular pool of potassium compared to the extracellular pool—3500 meq versus 60 meq—it is obvious that even a small shift in potassium distribution can have a profound effect on serum potassium. For instance, the rapid movement from cells of 60 meq of potassium—less than 2% of the intracellular pool—would double the serum potassium content.

Deficient Insulin, Beta-Adrenergic Agonists, Or Aldosterone

Subnormal potassium uptake into cells can occur when there is deficiency of the three hormones that control Na^+/K^+-ATPase activity: insulin (in diabetes), catecholamines (in autonomic insufficiency or during administration of beta-blocking drugs), or aldosterone (in hyporeneninemic or hypoaldosteronemic states). Actually, a deficiency of any one of these hormones is usually without clinical consequence with respect to potassium homeostasis. But their combined absence, as frequently occurs in diabetes, may lead to an elevation in serum potassium concentration, and predictably so when there is a superimposed stimulus for potassium egress from cells (Fig 7–5). For instance, during hyperglycemia, the high extracellular concentration of glucose induces cell shrinkage and potassium extrusion into the extracellular volume (Fig 7–7), but the resulting hyperkalemia cannot be combated in those diabetics who coordinately lack insulin, catecholamine, and aldosterone activities (Fig 7–5). For the same reason, vigorous exercise—which causes potassium to leave cells—in an individual who is nondiabetic but is being treated with a beta-adrenergic blocking drug may cause marked hyperkalemia and even sudden death.

Pharmacologic inhibition of Na^+/K^+-ATPase activity can also occur during digitalis intoxication.

METABOLIC ACIDOSIS

Hyperchloremic metabolic acidosis due to hydrochloric acid or an equivalent mineral acid increases the serum potassium concentration (Table 7–1). Other metabolic or respiratory acidoses have little impact on the serum potassium concentration. Serum potassium concentration rises about 0.7 meq/L for each 0.1 pH unit decrease during a mineral acidosis. The high extracellular hydrogen ion concentration slows membrane Na^+/H^+ antiporter activity, reduces cell sodium concentration, and depresses potassium uptake via the Na^+/K^+-ATPase (Fig 7–6). Continuing potassium egress from the cell then results in a net increase in extracellular potassium in response to the acidemia.

HYPEROSMOLALITY

An acute increase in the extracellular concentration of a solute that is incapable of permeating cell membranes causes cell shrinkage, diffusion of potassium out of the cell, and hyperkalemia (Fig 7–7). Clinical examples of such impermeant solutes include glucose (eg, diabetic ketoacidosis), mannitol, and radiocontrast media. Serum potassium concentration rises about 0.3–0.5 meq/L for each 10 mosm/kg increase in plasma osmolality.

OTHER

Hyperkalemia may be induced acutely during exercise, following an increase in cell potassium permeability during therapy with depolarizing muscle relaxants such as succinylcholine, following fluoride intoxication, and following hypothermia during cardiac bypass.

In the rare autosomal dominant disease hyperkalemic periodic paralysis, hyperkalemia often accompanies attacks of paralysis, lasting from minutes to hours. These episodes are sometimes provoked during the relaxation period following exercise; by cold exposure, infection, fasting, excitement, or general anesthesia; or by an increase in cortisol level. There is presumably a functional impairment of Na^+/K^+-ATPase activity in this disorder. Salbutamol, a $beta_2$-adrenergic agonist, has recently been found useful in preventing attacks and acutely aborting attacks.

INADEQUATE RENAL POTASSIUM EXCRETION

The pathogenesis of hyperkalemic disorders due to insufficient renal potassium excretion represents the inverse of the pathogenesis of hypokalemic states (Table 8–1). Since depression of nonreabsorbable anion excretion is not a clinically recognized cause of subnormal potassium excretion, the renal causes of hyperkalemia can be condensed into only two categories: those due to decrease in sodium delivery to the cortical collecting tubule and those due to diminished aldosterone action, as summarized in Fig 9–3. The first category can be further simplified, since a depression in GFR is usually the cause of reduction in sodium delivery and flow rate severe enough to impair potassium secretion.

In either category, the rate of urinary potassium excretion is inappropriate for the high blood potassium level (Fig 9–1). The normal physiologic response to chronic hyperkalemia, with urinary potassium excretion exceeding 600 meq/d, does not occur in these disorders. Fractional potassium excretion is therefore also not physiologically elevated—ie, it is less than the expected 100%. Patients may be in potassium balance, excreting dietary potassium of 50–100 meq/d in the steady state, but the urinary potassium is not high enough to excrete the excess potassium.

DECREASED SODIUM DELIVERY TO THE CORTICAL COLLECTING TUBULE

Chronic Renal Failure

In the course of chronic renal disease, surviving nephrons have a remarkable adaptive ability to maintain potassium homeostasis (Fig 7–21). In the face of a six- to tenfold reduction in nephron number, potassium balance can be preserved by six- to tenfold higher potassium secretion by remaining nephrons. It is apparent, however, that a further decrease in GFR or a relative increase in potassium load—exogenously or endogenously—can overwhelm the ability of the remaining nephrons to secrete potassium. Thus, hyperkalemia can be observed both with a normal potassium load and severe reduction in GFR to less than 10 mL/min or with an increased potassium load in the setting of moderate chronic renal insufficiency. Therapy should consist of reduction in the potassium burden (if possible), ion exchange resin, diuretics, or dialysis, as discussed below.

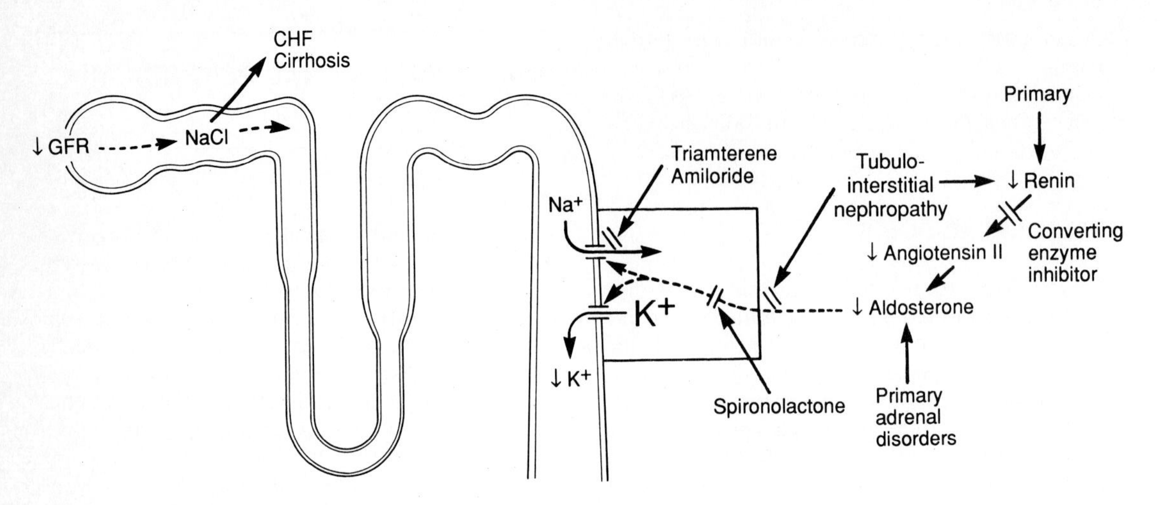

Figure 9-3. Causes of hyperkalemia due to inadequate renal potassium excretion.

Acute Renal Insufficiency

With abrupt decrease in GFR, there is both depression in filtered sodium and insufficient time for the potassium adaptation process (Fig 7–21) to occur. It is not uncommon for there to be a concomitant increase in endogenous potassium load with the condition that caused acute renal failure (eg, acute tubular necrosis due to bleeding and hypotension, rhabdomyolysis, or hemolysis). Therapy should be directed at restoring renal function—if possible, by extracellular volume resuscitation. Normalizing GFR allows faster excretion of excess potassium than any other adjunctive therapy. If depressed GFR is intrinsic and unrelated to ongoing extracellular volume deficit (eg, acute tubular necrosis), therapy with a potassium exchange resin or dialysis may eventually be required.

Intense Renal Sodium Chloride Reabsorption

Less commonly, intense reabsorption of sodium in the proximal tubule and loop of Henle results in insufficient delivery of sodium to the cortical collecting tubule to allow normal urinary potassium excretion. As illustrated in Fig 9–3, such avid sodium chloride reabsorption and hyperkalemia occurs occasionally in the course of congestive heart failure and cirrhosis and rarely in other hypovolemic or hypervolemic conditions associated with functional depression of effective circulating volume. A decrease in GFR is very common in these disorders and contributes to the propensity to hyperkalemia. In most cases, however, secondary hyperaldosteronism suffices to counteract this depression in sodium delivery and flow rate and thus to prevent subnormal potassium secretion and hyperkalemia (Fig 7–18). When hyperkalemia occurs in hypervolemic disorders, diuretic therapy (a thiazide or loop diuretic) combats the extreme sodium reabsorption and restores adequate sodium delivery to the cortical collecting tubule to normalize potassium secretion.

INSUFFICIENT ALDOSTERONE ACTIVITY

As shown in Fig 9–3, functional depression of mineralocorticoid activity may result from any of three causes: (1) depressed aldosterone release due to decrease in renin or angiotensin II; (2) abnormal biosynthesis of aldosterone by the adrenal gland; and (3) inability of aldosterone to act on its target organ, the cortical collecting tubule, due to intrinsic cellular damage or pharmacologic inhibition of mineralocorticoid action.

All these disorders exhibit the combined fluid and electrolyte abnormalities that result from lack of sufficient mineralocorticoid action on the principal cell (responsible for sodium reabsorption and potassium secretion) and intercalated cell (responsible for hydrogen ion secretion) of the cortical collecting tubule: hyperkalemia, renal salt wasting (Chapter 2), and hyperchloremic metabolic acidosis. This constellation of clinical findings is known as type IV generalized distal renal tubular acidosis. The causes are discussed in more detail in Chapter 12 (see Table 12–6) and will be only briefly summarized here.

Primary Decrease in Renin & Angiotensin II

Disorders associated with hyporeninemia predictably cause hypoaldosteronism. Subnormal renin output from the juxtaglomerular apparatus can be found in some patents with diabetes or other tubulointerstitial renal diseases. Mildly decreased GFR (20–50 mL/min) is almost always present. Since renin release is in part governed by renal nerve activity and prostacyclin production, hyporeninemia may occur in the course of autonomic insufficiency or administration of drugs, such as beta-blockers or prostaglandin synthesis inhibitors (such as nonsteroidal anti-inflammatory drugs). Because of the low plasma renin level, the plasma aldosterone level is initially low, but the latter sometimes returns to a "normal" level when hyperkalemia develops since serum potassium concentration is an independent stimulus for aldosterone secretion.

Administration of a converting enzyme inhibitor similarly decreases formation from angiotensin I of the major agonist for aldosterone release, angiotensin II. Functional hypoaldosteronism occurs as described above, although there is compensatory hyperreninemia. In patients who develop hyperkalemia when given a converting enzyme inhibitor, a low GFR is almost always a contributing pathogenetic factor.

Decrease in Aldosterone Production or Response

Hypoaldosteronism can result from primary adrenal diseases due to destruction or due to inherited or acquired enzymatic defects of the mineralocorticoid synthetic pathway. In these disorders, plasma renin is elevated while plasma aldosterone is low (Table 12–6).

Target organ resistance, or "pseudohypoaldosteronism," can result from congenital abnormalities, from destruction by certain tubulointerstitial diseases that have a predilection for injuring the cortical collecting tubule, or from drugs that antagonize mineralocorticoid activity. Spironolactone inhibits binding of aldosterone to the mineralocorticoid receptor, while triamterene and amiloride interfere with a principal action of aldosterone, increased luminal sodium channel activity. In all these disorders due to functional defects in aldosterone's target site and action, both renin and aldosterone levels are elevated.

Specific therapeutic approaches to hypoaldosteronism are discussed below and more fully in Chapter 12.

SYMPTOMS

As a corollary to what happens during hypokalemia, a rise in serum potassium concentration tends to depolarize cells. The resting cell potential is closer to the threshold for excitation. Symptoms are frequent when the serum potassium concentration exceeds 8.0 meq/L but may occur at a level between 6.0 and 8.0 meq/L. With a serum potassium concentration of less than 6.0 meq/L, symptoms tend to be relatively minor or absent.

HEMODYNAMIC

Cardiac

An important clinical manifestation of hyperkalemia is on cardiac conduction and provocation of arrhythmias. As shown in Fig 9–4, when serum potassium increases, electrocardiographic manifestations evolve from peaking of T waves → prolongation of the PR interval with bradycardia → disappearance of P waves (sinoventricular rhythm) → widened QRS complex → a full-blown sine wave pattern that can lead to ventricular fibrillation and asystole. Arrhythmias secondary to hyperkalemia are exacerbated by concurrent hyponatremia, hypocalcemia, and acidosis.

Peripheral Vasculature

A high-potassium diet even without overt hyperkalemia tends to decrease peripheral vascular resistance and modestly reduce blood pressure. It is debated whether a rise in serum potassium concentration causes vasodilatation directly or acts by diminishing angiotensin II receptor density.

NEUROMUSCULAR

Hyperkalemia decreases the intracellular:extracellular potassium concentration ratio that defines the resting membrane potential in nerve and muscle cells. This reduction in electrical potential brings it closer to the depolarization threshold value that initiates the action potential, so that nerve conduction and muscle contraction are more easily initiated. The clinical sequelae include paresthesias and weakness and eventually paralysis. When paralysis of the respiratory muscles occurs, hyperkalemia can be life-threatening.

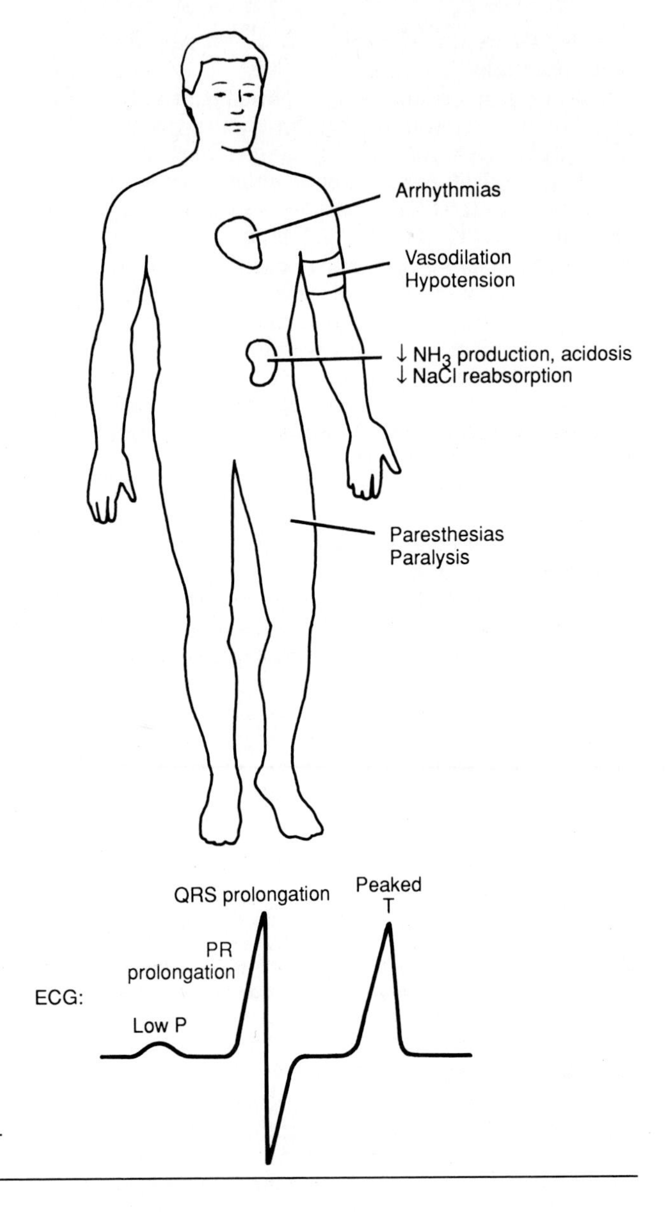

Figure 9–4. Symptoms and signs of hyperkalemia.

RENAL

Hyperkalemia elicits some effects opposite to those induced by hypokalemia. Reduced ammoniagenesis from hyperkalemia predictably reduces urinary ammonium excretion. Urinary net acid excretion is then insufficient to cope with dietary acid ingestion, and hyperchloremic metabolic acidosis ensues. A high potassium intake and hyperkalemia causes natriuresis, though the nephron site of diminished sodium reabsorption is unknown. In contrast to hypokalemia, however, hyperkalemia does not alter GFR or water metabolism.

DIAGNOSIS

APPROACH TO ETIOLOGIC DIAGNOSIS

History & Physical Examination

The approach to diagnosis of hyperkalemic disorders includes, of course, a careful history, physical examination, nutritional summary, and medication history (eg, for digitalis, beta-blockers, aldosterone antagonists, heparin, nonsteroidal anti-inflammatory drugs, nephrotoxins).

When assessing potassium intake, it is especially important to note dietary sources of potassium (Table 8-6), potassium consumed in the chloride form (such as salt substitutes, which contain about 10-13 meq/g or 50-65 meq/tsp), red clay ingestion (a potassium-rich substance), penicillin therapy (1.7 meq of potassium per million units), transfusions of stored blood (up to 30 meq/L), and potassium replacement preparations (Table 8-5). Potassium chloride preparations are more likely to cause hyperkalemia due to the reabsorbability of the accompanying chloride anion than potassium in food, which is usually accompanied by a nonreabsorbable anion.

Signs and symptoms should be sought of entities associated with potassium redistribution out of cells (such as diabetes, especially with hyperglycemia, acidemia, or adrenergic hypoactivity or blockade), cell lysis disorders (such as gastrointestinal and other bleeding, hemolysis, tumor lysis, conditions associated with hypercatabolism or tissue necrosis), or inadequate renal potassium excretion (acute or chronic renal failure, diminished effective circulating volume, or primary or secondary adrenal disease).

LABORATORY TESTS

The urinary potassium concentration should be checked to ascertain whether the kaliuretic response is normal (urinary potassium concentration >100 meq/L) or abnormal (Fig 9-1). Other helpful laboratory tests include urinary sodium concentration (renal salt wasting, diuretics), plasma osmolality and urinary glucose and ketones (hyperglycemia or acidosis), arterial blood gases (metabolic acidosis), and appropriate plasma hormone concentrations if indicated (insulin, catecholamines, renin, aldosterone, or cortisol).

THERAPY

TREATMENT FOR SEVERE, SYMPTOMATIC HYPERKALEMIA

Indications

Emergency reduction in serum potassium concentration is indicated when cardiac toxicity is apparent or imminent or when muscular paralysis has occurred. Emergency treatment aimed at driving potassium into cells is indicated when the electrocardiogram shows a peaked T wave or more serious disturbances. Severe elevation of the serum potassium—over 6.5-7.0 meq/L—even without electrocardiographic changes still requires rapid treatment.

Table 9-2 summarizes modalities available for the prompt treatment of hyperkalemia. Continuous electrocardiographic monitoring should be maintained throughout treatment since arrhythmias may occur or change at any time. Treatment of adrenal crisis is not listed (see Chapter 2), but if crisis is present treatment consists of prompt institution of hydrocortisone and saline.

Calcium

Infusion of calcium does not reduce the serum potassium concentration but immediately antagonizes the adverse cardiac conduction abnormalities. Calcium should be given over a 2-minute period. The calcium may be repeated after 5 minutes if no benefit is apparent. Further doses, unless there is hypocalcemia, are unlikely to be of benefit. Calcium should be given with extreme caution to patients receiving digitalis, since digitalis toxicity may be precipitated.

Bicarbonate

Bicarbonate rapidly drives potassium into cells. It is especially useful when systemic acidemia is present but is efficacious even when arterial pH is normal. Complications include hypernatremia (Chapter 6) and metabolic alkalosis (Chapter 13), so that intravenous use of over 132 meq (three ampules) should be done with caution and only after checking the serum sodium and bicarbonate concentrations. Bicarbonate should not be given in the same intravenous solution as calcium, since an insoluble precipitate forms.

Insulin

Insulin also drives potassium into cells, requiring about 15-60 minutes to do so, by mechanisms described previously (Fig 7-4). Glucose is usually used with insulin to prevent hypoglycemia. Glucose should not be used as a sole agent, however, since it may cause hyperglycemia and hyperosmolality and thus exacerbate the hyperkalemia.

TREATMENT OF ASYMPTOMATIC HYPERKALEMIA

The modalities described above are temporizing, preventing cardiac toxicity and acutely translocating potassium from the extracellular to the intracellular pools. The ultimate goal in treating hyperkalemia associated with total body potassium surfeit is, of course, removal of excess potassium. With asymptomatic elevation of the serum

Table 9-2. Treatment of hyperkalemia.

EMERGENCY

Modality	Mechanism of Action	Onset	Duration	Prescription	K^+ Removed From Body
Calcium	Antagonizes cardiac conduction abnormalities	0-5 min	1 hour	Calcium gluconate 10%, 5-30 mL IV; or calcium chloride 5% 5-30 mL IV	0
Bicarbonate	Distributes K^+ into cells	15-30 min	1-2 hours	$NaHCO_3$, 44-132 meq (1-3 ampules) IV	0
Insulin	Distributes K^+ into cells	15-60 min	4-6 hours	Regular insulin, 5-10 units IV, plus glucose 50%, 25 g (1 ampule) IV	0

NONEMERGENCY

Modality	Mechanism of Action	Duration of Treatment	Prescription	K^+ Removed From Body
Loop diuretic	↑Renal K^+ excretion	0.5-2 hours	Furosemide, 40-160 mg IV or PO with or without $NaHCO_3$, 0.5-3 meq/kg daily	Variable
Sodium polystyrene sulfonate (Kayexalate)	Ion exchange resin: Binds K^+	1-3 hours	Oral: 15-30 g in 20% sorbitol (50-100 mL) Rectal: 50 g in 20% sorbitol	0.5-1 meq/g
Hemodialysis	Extracorporeal K^+ removal	48 hours	Blood flow ≥ 200-300 mL/min Dialysate $[K^+] = 0$	200-300 meq
Peritoneal dialysis	Peritoneal K^+ removal	48 hours	Fast exchange, 3-4 L/h	200-300 meq

potassium concentration, potassium removal can be slower. The specific treatment of many forms of hyperkalemia have already been discussed. The available general methods for treating hyperkalemia are summarized in Table 9-2.

Estimation of Potassium Excess

The amount of potassium necessary for removal depends on the degree of hyperkalemia. The ability to "buffer" a change in the serum potassium concentration is less impressive in the upward than in the downward direction (Fig 8-5). The exact relationship between a surfeit of total body potassium and serum potassium concentration is not known. In general, serum potassium concentration rises about 1.0 meq/L for a potassium surplus of 50-200 meq. But the starting level influences the increment in serum potassium concentration to a given load. As basal serum potassium concentration rises, a progressively smaller amount of potassium causes the same absolute increment in serum concentration. A corollary of this general rule is that less potassium removal is required to promote a given decrement in serum potassium concentration when the initial concentration is very high. For example, if serum potassium concentration were 8.0 meq/L, reduction to a tolerable level of 6.0 meq/L might require removal of only about 100 meq of potassium. A reduction from 6.0 meq/L to 4.0 meq/L might require a more substantial amount of potassium removal—perhaps 200 meq.

As with hypokalemia, estimation of potassium excess must take into consideration concurrent alterations in serum pH, osmolality, or hormone levels. Serum potassium concentration rises about 0.7 meq/L for every decrease of 0.1 pH unit during mineral acidosis (but not organic metabolic acidosis or respiratory acidosis) and about 0.3-0.5 meq/L for each 10 mosm/kg increase in osmolality.

Mineralocorticoid Replacement

Physiologic mineralocorticoid replacement with fludrocortisone, 0.1 mg/d, is recommended therapy for hyperkalemic hyperchloremic metabolic acidosis and renal salt wasting due to deficient circulating aldosterone in primary adrenal diseases or hyporeninemic disorders (see Chapter 12). If there is concomitant tubulointerstitial disease, higher replacement doses may be required (up to 0.5 mg/d) to overcome the mineralocorticoid resistance associated with collecting tubule damage. Overtreatment, however, can lead to hypertension and edema.

When hyporeninemic hypoaldosteronism is already accompanied by hypertension, furosemide plus sodium bicarbonate is usually preferable to fludrocortisone as therapy (see below).

Dietary Potassium Restriction

Restricting potassium in the diet is useful as a sole or adjunctive therapeutic approach to many mild hyperkalemic disorders. Since protein intake is usually tightly coupled to potassium intake, caution should be exercised so that potassium intake is not reduced to an extent that causes protein malnutrition. Potassium intake can usually be reduced to 40–60 meq/d without adverse consequence. Table 8–6 may be used to counsel a hyperkalemic patient regarding the potassium content of common foods. Patients should also be instructed to avoid salt substitutes, which contain potassium chloride (50–60 mg/tsp).

Diuretics

If GFR is normal or only moderately reduced, rapid and efficient potassium removal can be afforded by a loop diuretic such as furosemide (40–160 mg every 4–6 hours). As an adjunct, saline should be administered to expand the extracellular volume and enhance sodium delivery to the cortical collecting tubule. Alternatively, an isotonic solution containing bicarbonate—sufficient to increase urine pH to over 7.5—can be used when metabolic acidosis is concurrently present, but treatment must be monitored carefully to avoid overshooting and inducing alkalemia.

To maintain a normal serum potassium concentration, the use of furosemide (40–160 mg orally every 4–6 hours) plus sodium bicarbonate (0.5–3.0 meq/kg orally per day) is frequently efficacious. This regimen is especially useful in patients with type IV generalized distal RTA who are hypertensive and therefore cannot be treated with mineralocorticoid alone.

Caution should be exercised when using diuretics in order to prevent extracellular volume contraction and a further lowering of GFR. A further reduction in GFR can compromise sodium delivery to the cortical collecting tubule and hence restrict potassium excretion. Obviously, any drug that interferes with normal operation of the renin-angiotensin-aldosterone axis should be avoided (such as beta-blockers, nonsteroidal anti-inflammatory drugs angiotensin-converting enzyme inhibitors, spironolactone, triamterene, and amiloride).

Exchange Resin

As an adjunctive measure to diuretics or as a sole agent when renal function is markedly impaired, a potassium exchange resin can be given orally or as a retention enema. About 0.5–1 meq of potassium is removed from the body for each gram of resin administered. A resin should not be used when ileus is present, since intestinal perforation can result. Sodium loading occurs with administration of an exchange resin, since sodium is exchanged for potassium in a ratio of about 1.5:1. Small amounts of calcium and magnesium are also removed by the resin. Resin administration may be repeated every 3–4 hours (oral) or 4–6 hours (rectal) up to four times per day. Rectal administration is facilitated by inserting a retention (eg, Foley) catheter with balloon inflation after resin instillation to ensure retention for at least 30–60 minutes and preferably several hours.

Dialysis

When potassium removal cannot be sufficiently effected by renal excretion or exchange resin, dialysis is indicated. Hemodialysis is quite efficient because potassium removal is a sensitive function of blood delivery to the dialyzer. Blood flow from the patient therefore should be high—at least 200 mL/min and optimally over 300 mL/min. The newer high-flux polysulfone membrane with higher potassium permeability than the conventional dialysis membrane should be used if available. Obviously, the dialysate should be potassium-free. Hemofiltration is a much less efficient form of therapy.

Peritoneal dialysis is about one-tenth as efficient as hemodialysis, so that it must be continued approximately ten times longer to remove the same amount of potassium. Rapid exchanges with large volumes—preferably 3–4 L/h—should be used. The choice of which dialytic method to use depends on the amount of potassium to be removed, the availability and the time required to institute the procedure, and other clinical considerations such as hemodynamic status and degree of catabolism.

SECTION III REFERENCES: POTASSIUM HOMEOSTASIS & DISORDERS: HYPOKALEMIA & HYPERKALEMIA

Adrogue HJ, Madias NE: Changes in plasma potassium concentration during acute acid-base disturbances. Am J Med 1981;71:456–467.

Brown RS: Extrarenal potassium homeostasis. Kidney Int 1986;30:116–127.

DeFronzo RA, Bia M, Smith D: Clinical disorders of hyperkalemia. Annu Rev Med 1982;33:521–554.

Field MJ, Giebisch G: Hormonal control of renal potassium excretion. Kidney Int 1985;27:379–387.

Hayslett JP, Binder HJ: Mechanism of potassium adaptation. Am J Physiol 1982;243:F103–F112.

Knochel JP: Neuromuscular manifestations of electrolyte disorders. Am J Med 1982;72:521–535.

Ponce SP et al: Drug-induced hyperkalemia. Medicine 1985;64:357–370.

Schambelan M, Sebastian A: Hyporeninemic hypoaldosteronism. Annu Rev Med 1979;24:385–405.

Schambelan M, Sebastian A, Biglieri EG: Prevalence, pathogenesis, and functional significance of aldosterone deficiency in hyperkalemic patients with chronic renal insufficiency. Kidney Int 1980;17:89–101.

Sterns RH et al: Internal potassium balance and the control of the plasma potassium concentration. Medicine 1981;60:339–354.

Tannen RL: Diuretic-induced hypokalemia. Kidney Int 1985;28:988–1000.

Section IV: Acid-Base Homeostasis & Disorders: Acidosis & Alkalosis

Normal Acid-Base Homeostasis

10

INTRODUCTION

BLOOD ACIDITY

The normal blood hydrogen ion concentration $[H^+]$) is 40 nmol/L, usually expressed in terms of its negative logarithm, pH of 7.40. The hydrogen ion concentration is maintained under normal physiologic conditions within narrower limits—± 4–5 nmol/L (corresponding to a ± 0.04–0.05 pH unit range)—than any other electrolyte in the blood.

The control of blood pH is critically important since modest swings—eg, 0.10–0.20 pH unit in either direction—can cause symptoms referable to impaired cardiopulmonary performance and neurologic function. More extreme pH changes can be fatal.

Relation of Blood pH to Intracellular pH & Cell Function

The primary focus of this chapter will be on the physiologic control of arterial blood pH, but it is actually the intracellular pH that is critical for cellular viability, normal enzyme function, and other metabolic processes. Cells have mechanisms by which they carefully defend their internal pH in response to changes in pH of the extracellular milieu. However, extreme alterations in extracellular pH disrupt the integrity of intracellular pH and can then interfere with cell metabolism.

Regulation of Blood pH by the Bicarbonate-P_{CO_2} Buffer System

The arterial pH is defined by the poise in the equilibrium relationship of the major buffering system in the blood, bicarbonate (HCO_3^-) and carbon dioxide (CO_2). The kidneys carefully regulate bicarbonate concentration, while the lungs sensitively control CO_2 tension. A decrement of plasma bicarbonate concentration or surfeit of CO_2 decrease the blood pH (acidemia). Conversely, an increase in plasma bicarbonate concentration or decrease in CO_2 cause blood pH to rise (alkalemia).

DEFENSE OF NORMAL ACID-BASE HOMEOSTASIS

Metabolic Acid Production & Renal Acid Excretion

The usual problem faced by the body every day is that the amount of acid accrued from a usual diet and endogenous metabolic processes titrates and eliminates a large proportion of the existing bicarbonate buffer reservoir. For instance, consider the situation when the average daily acid production, 60 meq/d, is added to the extracellular buffer pool consisting principally of 375 meq bicarbonate (25 meq/L × 15 L extracellular volume). It is obvious that buffer depletion with fatal acidosis would occur well within a week if bicarbonate were not repleted. Pulmonary adjustments serve to ameliorate changes in pH that would otherwise occur with changes in bicarbonate buffer. However, it is ultimately the kidney that must conserve existing bicarbonate and—importantly—regenerate the bicarbonate titrated by the daily onslaught of acid.

Alkali Addition & Renal Bicarbonate Excretion

Under unusual dietary circumstances but more commonly in pathologic conditions (eg, vomiting or diuretic overusage), there may be net addition of bicarbonate rather than acid to the body. The kidney can normally defend quite efficiently against such a surfeit. The kidney quantitatively excretes the excess bicarbonate so that it does not accumulate within the body.

CO_2 Production & Pulmonary CO_2 Excretion

In addition to metabolic, nonvolatile acids, cellular production of large quantities of CO_2 occurs normally, which increases the existing content of CO_2 in the blood, causing acidemia. Since dissolved CO_2 in the blood is in equilibrium with gaseous CO_2, it is considered a volatile acid. The lungs are responsible for eliminating excess CO_2. Renal compensation helps ameliorate disturbances in arterial blood pH when there is a primary disorder of pulmonary function.

BLOOD ACID-BASE COMPOSITION

DEFINITION OF BLOOD pH BY THE BICARBONATE-P_{CO_2} SYSTEM

Relationship of $[H^+]$ to Bicarbonate, Carbonic Acid, & Carbon Dioxide

The exquisite control of blood acidity is accomplished because the CO_2 tension (P_{CO_2}) and bicarbonate concentration ($[HCO_3^-]$) are very closely regulated in the blood. Actually, the hydrogen ion concentration ($[H^+]$) is defined by the ratio of the dominant acid-base pair in the blood, bicarbonate:carbonic acid (H_2CO_3):

$$H^+ + HCO_3^- \rightleftharpoons H_2CO_3$$

By mass balance relationship, $[H^+]$ is proportionate to $[H_2CO_3]/[HCO_3^-]$. The H_2CO_3 is, in turn, in equilibrium with CO_2, catalyzed by the enzyme carbonic anhydrase. Therefore, the complete relationship between these acid-base species is as follows:

$$H^+ + HCO_3^- \rightleftharpoons H_2CO_3 \rightleftharpoons CO_2 + H_2O$$

The $[H^+]$ at any given moment is then proportionate to the ratio of $[CO_2]/[HCO_3^-]$ and the carbonic acid dissociation constant. Finally, CO_2 dissolved in the blood is in equilibrium with CO_2 as a gas, the CO_2 tension (P_{CO_2}), related by the solubility coefficient of $CO_2(\alpha)$. Therefore, the blood acidity is proportionate to the ratio $(\alpha \cdot P_{CO_2})/[HCO_3^-]$ and the carbonic acid dissociation constant (K), to yield the **Henderson equation:**

$$[H^+] = K \frac{(\alpha \cdot P_{CO_2})}{[HCO_3^-]}$$

The product of K and α is 24 (omitting units), so that the Henderson equation can be restated:

$$[H^+] = 24 \frac{P_{CO_2}}{[HCO_3^-]}$$

Though we do not usually think of acidity in terms of $[H^+]$, this term bears a simple inverse relationship to the more familiar term pH. The normal $[H^+]$ is 40 nmol/L, remembered as the last two digits of the normal pH (7.40). $[H^+]$ falls by 10 units as pH rises by 0.10 units (so $[H^+]$ is 30 nmol/L when pH is 7.50) and vice versa ($[H^+]$ is 50 nmol/L when pH is 7.30). The normal relationship of the acid-base parameters is therefore 40 = 24 (40)/(24). The calculation of bicarbonate concentration if pH were 7.30 ($[H^+]$ = 50 nmol/L) and P_{CO_2} were 25 mm Hg would be 50 = 24 (25)/(X), or 12 meq/L.

Henderson-Hasselbalch Equation

More familiar is the negative logarithmic transformation of the Henderson equation to the Henderson-Hasselbalch equation with the appropriate numeric equivalents of K and α:

$$pH = 6.1 + \log \frac{[HCO_3^-]}{0.03\ P_{CO_2}}$$

The Henderson-Hasselbalch equation not only states the physicochemical definition of blood pH but also gives insight into the physiologic control of blood acid-base composition. The bicarbonate concentration is regulated chiefly by the kidney, while the P_{CO_2} is controlled by the lung. From a physiologic perspective, the Henderson-Hasselbalch equation is then equivalently stated as follows:

$$pH \approx \frac{\text{Kidney}}{\text{Lung}}$$

It is readily apparent that normal physiologic acid-base homeostasis is a coordinate effort by both the kidney and the lung. Acid-base disorders result when one or both of these two organs are overwhelmed or intrinsically disordered.

Normal Blood Acid-Base Values

As mentioned above, blood pH is normally carefully maintained within very strict limits. The 95% confidence limits for blood pH are **7.35–7.43** (hydrogen ion concentration 37–45 nmol/L). The corresponding normal limits for P_{CO_2} are **37–45 mm Hg** and for bicarbonate concentration **22–26 meq/L** under normal dietary and extracellular volume conditions.

Spontaneous variation in P_{CO_2} slightly affects the normal resting level of bicarbonate concentration. Unusual diets also affect the normal setpoint at which the plasma bicarbonate concentration is regulated. For instance, a very high protein diet that generates a large acid load decreases the steady-state plasma bicarbonate concentration, whereas large alkali ingestion in the setting of a sodium chloride-restricted diet increases steady-state plasma bicarbonate concentration.

SOURCES, BUFFERING, & RESPIRATORY COMPENSATION OF HYDROGEN ION ADDITION TO THE BODY

SOURCES OF ACID INPUT

There are three roughly equivalent sources for the normal daily addition of hydrogen ion to the body: diet, metabolism, and stool base loss, as shown in Fig 10–1. The sum of the hydrogen ion

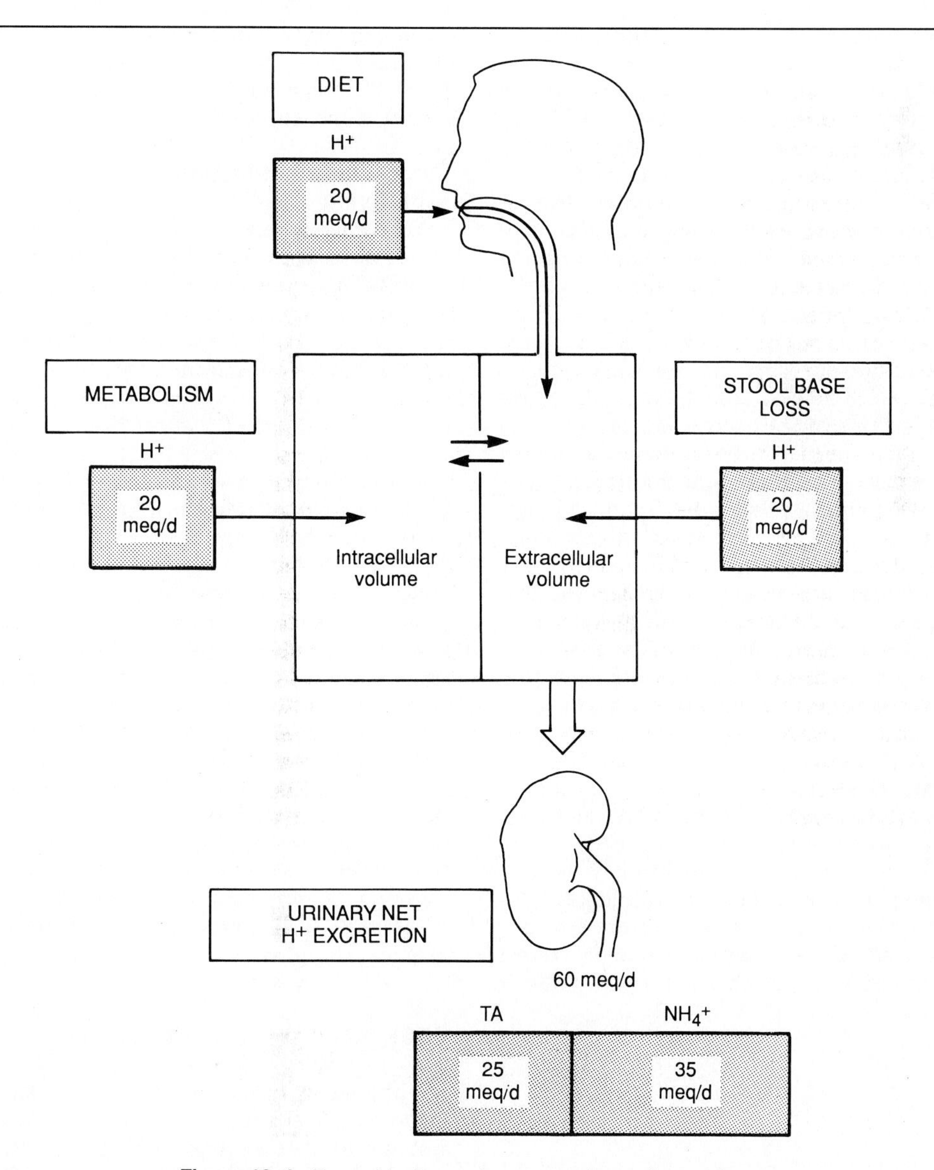

Figure 10–1. Normal hydrogen ion generation and excretion.

input on an average American diet is about 0.8–1.0 meq/kg per day, or about 55–70 meq/d. In a steady state, this acid load is quantitatively excreted by the kidney.

Dietary & Metabolic Acids

Several amino acids derived from dietary protein are metabolically converted to organic or mineral acids. In general, the quantity of acid is proportionate to the amount of animal protein ingested. The normal quantity of diet-derived acid is about 20–30 meq/d. In addition, normal metabolic processes, both dependent on and independent of diet, account for endogenous acid production of another 20–30 meq/d. Together, diet plus metabolism account for about 40–60 meq/d of acid load.

The composition of the dietary and metabolic acid loads is highly varied. The oxidation of organic sulfur in methionine and cysteine residues of protein forms sulfuric acid. Organic acids are formed from the incomplete oxidation of dietary carbohydrates, fats, and proteins as well as nucleic acids. Even during starvation, metabolism of endogenous protein results in organic acid production. Hydrolysis of phosphate esters in some proteins as well as nucleic acids yields phosphoric acid. Hydrochloride acid is generated when chloride salts of cationic amino acids, such as lysine and arginine, are metabolized to neutral products.

The normally modest amount of endogenous acid production can be increased under certain pathologic circumstances. For instance, keto acid formation increases in hypoinsulinemic states (diabetes), alcoholism, or starvation. Toxin and drug ingestion can result in accelerated formation of organic acids, such as formic acid from methanol, oxalic acid from ethylene glycol, and salicylic acid from aspirin. If the normal hepatic conversion to glucose of the enormous quantities (750–1500 meq/d) of lactic acid derived from muscle metabolism (Cori cycle) is interrupted, lactic acid accumulation rapidly ensues.

Stool Base Loss

About 20–30 meq/d of bicarbonate and base equivalents (organic anions) are normally lost in stool. For every molecule of base lost in stool, one hydrogen ion is retained in the extracellular fluid. Gastrointestinal base loss and therefore systemic acid load can be markedly increased in certain pathologic conditions, such as diarrheal states.

EXTRACELLULAR & INTRACELLULAR BUFFERING

Extracellular Distribution & Buffering

When hydrogen ions are added to the blood from endogenous or exogenous sources, there is rapid distribution within the extracellular volume, virtually complete in 20–30 minutes, as shown in Fig 10–2. As mentioned above, the principal extracellular buffer is bicarbonate:

$$H^+ + HCO_3^- \rightleftharpoons H_2CO_3 \underset{CA}{\rightleftharpoons} CO_2\uparrow + H_2O$$

The CO_2 that is formed following titration of bicarbonate and carbonic anhydrase (CA)-catalyzed dehydration of carbonic acid is exhaled by the lungs. There is also buffering of lesser importance by other buffers (eg, proteins and phosphate) in the extracellular space.

Intracellular Distribution & Buffering

Following extracellular distribution and buffering of an acid load, a second intracellular phase of buffering ensues, complete within several hours (Fig 10–2). The principal intracellular buffers (B^-) include hemoglobin, proteins, dibasic phosphate (HPO_4^{2-}), and bone carbonate:

$$H^+ + B^- \rightleftharpoons HB$$

The relative ratio of extracellular:intracellular buffering is about 1:1. Thus, the apparent volume of distribution of an acid load with respect to the total buffering process includes both the extracellular and intracellular volumes and is approximately equal to total body water. If the acid load is large or persistent, the extracellular:intracellular buffering contribution may increase to 1:2 or even higher. Both the bicarbonate and the nonbicarbonate buffering processes are reversible.

As discussed in Chapter 7 (Fig 7–6), the entry of hydrogen ion into cells causes potassium to simultaneously exit. With a mineral acid (in which the conjugate anion cannot enter the cell), potassium concentration rises in the extracellular space by about 0.6 meq/L for each 0.10 arterial pH unit of decrease, sometimes resulting in overt hyperkalemia.

RESPIRATORY COMPENSATION

Although extracellular and intracellular buffering mechanisms blunt the change in blood pH that would otherwise be produced from an acid load, a decrease in plasma bicarbonate concentration

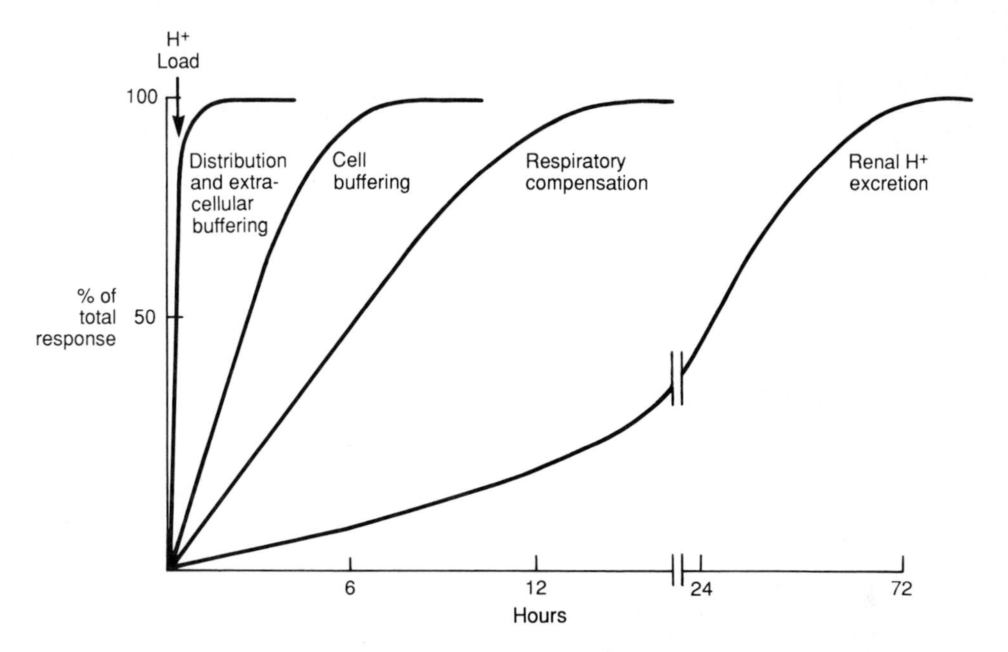

Figure 10-2. Time course of distribution, buffering, respiratory compensation, and renal excretion of an acid load.

and a small degree of acidemia nevertheless is inevitable. However, the Henderson-Hasselbalch equation dictates that the degree of acidemia produced by reduction in plasma bicarbonate concentration is less should there be a simultaneous decrease in P_{CO_2}. Thus, respiratory compensation, by increasing minute ventilation to diminish the P_{CO_2}, blunts the change in arterial pH that would otherwise occur as a result of the reduction in plasma bicarbonate concentration alone. Stated another way, pH tends to change less when there are parallel changes in bicarbonate concentration and P_{CO_2}.

Acidemia stimulates central nervous system centers responsible for control of ventilation. The signal is probably acidification of the interstitial fluid of the brain stem or the cerebrospinal fluid. Approximately 6–12 hours is required for completion of this process (Fig 10–2). The resulting increase in respiratory drive enhances pulmonary CO_2 excretion and lowers the arterial P_{CO_2}. As shown in Table 10–1, arterial P_{CO_2} decreases by about 1.25 mm Hg (range, 0.9–1.5 mm Hg) for each meq/L drop in plasma bicarbonate concentration.

This physiologic ventilatory response to a primary decrease in plasma bicarbonate concentration is shown graphically in Fig 10–3. This acid-base nomogram, to be discussed more fully in Chapter 11, superimposes normal physiologic responses to primary acid-base disturbances on the arterial pH, P_{CO_2}, and $[HCO_3^-]$ relationships defined by the Henderson-Hasselbalch equation.

For instance, if bicarbonate concentration decreased by 8 meq/L (from 24 to 16 meq/L), the normal ventilatory response would decrease the P_{CO_2} by 10 mm Hg (1.25 × 8), from 40 to 30 mm Hg, and only a mild acidemia (pH 7.35) would occur. This hypocapnic compensation moderates the degree of acidemia; if P_{CO_2} had not changed, the resulting pH would have been lower (7.25).

The ventilatory response is ameliorative but is never sufficient to completely normalize arterial pH. The normal 95% confidence limits of the respiratory response to a primary reduction in plasma bicarbonate concentration (metabolic acidosis) is shown in Fig 10–3.

RENAL ACIDIFICATION MECHANISMS & RESPONSE TO HYDROGEN ION ADDITION TO THE BODY

DUAL ROLES OF H^+ SECRETION BY THE KIDNEY IN ACID-BASE HOMEOSTASIS

Bicarbonate Reclamation

The kidney is normally confronted with two important tasks that help to preserve the normal acid-base composition of the body. First, the

Table 10-1. Physiologic response of pH, P_{CO_2}, and HCO_3^- in acid-base disorders.[1]

Respiratory alterations
- Acute:
  - Acidosis
    - ↑1 mm Hg P_{CO_2}
      - ↑$[HCO_3^-]$ = 0.1 meq/L
      - ↑$[H^+]$ = 0.8 nmol/L; ↓pH = 0.008 pH unit
  - Alkalosis
    - ↓1 mm Hg P_{CO_2}
      - ↓$[HCO_3^-]$ = 0.25 meq/L
      - ↓$[H^+]$ = 0.7 nmol/L; ↑pH = 0.007 pH unit
- Chronic:
  - Acidosis
    - ↑1 mm Hg P_{CO_2}
      - ↑$[HCO_3^-]$ = 0.5 meq/L
      - ↑$[H^+]$ = 0.25 nmol/L; ↓pH = 0.0025 pH unit
  - Alkalosis
    - ↓1 mm Hg P_{CO_2}
      - ↓$[HCO_3^-]$ = 0.5 meq/L
      - ↓$[H^+]$ = 0.3 nmol/L; ↑pH = 0.003 pH unit

Metabolic alterations
- Acidosis
  - ↓1 meq/L $[HCO_3^-]$
  - ↓P_{CO_2} = 1.25 mm Hg
  - ↑$[H^+]$ = 1.2 nmol/L; ↓pH = 0.012 pH unit
- Alkalosis
  - ↑1 meq/L $[HCO_3^-]$
  - ↑P_{CO_2} = 0.5 (0.2-0.9) mm Hg
  - ↓$[H^+]$ = 0.3-0.8 nmol/L; ↑pH = 0.003-0.008 pH unit

[1]Adapted from Cogan MG, Rector FC Jr: Acid-base disorders. Pages 457–517 in: *The Kidney,* 3rd ed. Brenner BM, Rector FC (editors). Saunders, 1986.

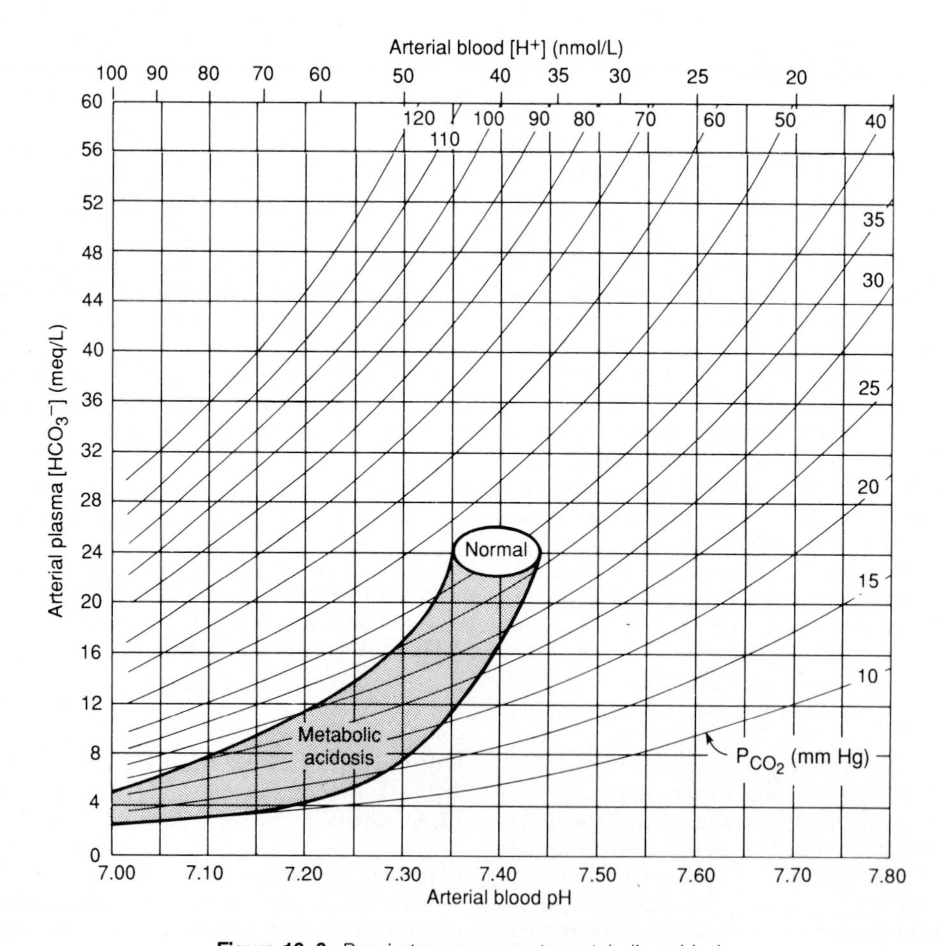

Figure 10-3. Respiratory response to metabolic acidosis.

enormous filtered bicarbonate load must be quantitatively reclaimed. Because GFR is quite high, about 4500 meq/d of bicarbonate (25 meq/L $\times$ 180 L/d) is filtered and must be reabsorbed to prevent loss into the urine. This process of bicarbonate reclamation is accomplished by hydrogen ion secretion predominantly in the proximal tubule.

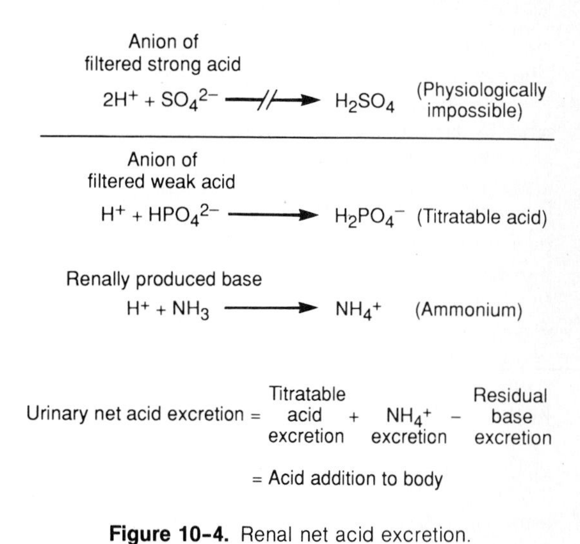

Figure 10–4. Renal net acid excretion.

New Bicarbonate Generation by Urinary Titratable Acid & Ammonium Excretion

The second task of the kidney is to replenish the bicarbonate titrated by normal dietary and metabolic processes, about 60 meq/d (Fig 10–1). To accomplish this task, hydration of CO_2 occurs in renal tubular cells, which yields a hydrogen ion that is secreted into the tubular fluid lumen and is eventually excreted into the urine plus an equivalent amount of new bicarbonate that enters the blood.

$$CO_2 + H_2O \longrightarrow H^+ \text{ (Tubule lumen)} + HCO_3^- \text{ (Blood)}$$

Ideally, the kidney would simply reprotonate all anions of dietary and metabolic acids and excrete these acid conjugates into the urine. For instance, when sulfuric acid is generated by metabolism, it titrates the available blood buffer bicarbonate with formation of the sodium salt of sulfate:

$$H_2SO_4 + 2NaHCO_3 \longrightarrow 2CO_2\uparrow + 2H_2O + Na_2SO_4$$

The simplest process by which bicarbonate could be reformed would be if the kidney simply reversed this sequence. That is, if Na_2SO_4 underwent glomerular filtration and subsequent protonation, the H_2SO_4 could be excreted into the urine with simultaneous regeneration of an equivalent amount of bicarbonate added to the blood.

Unfortunately, as illustrated in Fig 10–4, simple reversal of sulfuric acid accrual cannot occur physiologically. The lowest pH attainable in the collecting tubule luminal fluid and in the urine is about 4.5. Since the pK_a of sulfuric acid is about 1.0, the left-to-right reaction by which sulfuric acid is formed in Fig 10–4 cannot occur—the steady-state equilibrium of this reaction is actually shifted far to the left. Thus, if the pK_a of an acid—in this case sulfuric acid—is substantially below the lowest attainable urinary pH (4.5), its conjugate anion cannot be reprotonated. The anion will instead be excreted unchanged into the urine as its sodium (or potassium) salt (eg, sodium sulfate).

There are, however, conjugate anions of some weak acids with a higher pK_a that can be titrated at the pH attainable in the luminal fluid of the collecting tubule and in the urine. The most important example is the anion formed when phosphoric acid is generated by metabolism and has reacted with blood bicarbonate. The residual anion, dibasic phosphate (HPO_4^{2-}), can be reprotonated to form monobasic phosphate ($H_2PO_4^-$) under physiologic conditions, since the pK_a of this acid is 6.8. Thus, the reaction for hydrogen ion added to HPO_4^{2-} shown in Fig 10–4 can occur and is driven to the right at the usual urinary pH.

$$H^+ + HPO_4^{2-} \longrightarrow H_2PO_4^- \text{ (Titratable acid)}$$

In this case, a hydrogen ion can be added to a filtered conjugate anion of a weak acid with pK_a (6.8) substantially greater than the lowest attainable urinary pH (4.5). This process forms **titratable acid.** The term "titratable" implies that the conjugate anion can be titrated within the pH range which occurs physiologically in the urine.

To summarize, the kidney can secrete hydrogen ion into the urine and thereby generate and deliver new bicarbonate into the blood by directly reforming some, but not all, dietary and metabolic acids. Strong acids such as H_2SO_4 and HCl, with pK_a's substantially less than 4.5, cannot be excreted as such. Therefore, the kidney must secrete hydrogen ions along with another buffer in order to regenerate the requisite amount of bicarbonate. The way in which this is done is by the renal production of ammonia (NH_3) from glutamine. Hydrogen ion secretion by the kidney then allows ammonia to be protonated to form **ammonium** (NH_4^+).

$$H^+ + NH_3 \longrightarrow NH_4^+$$

Ammonium excretion thus compensates for the inability to excrete very strong acids in the urine, as shown in Fig 10–4.

Net Acid Excretion Equals Acid Input From Diet & Metabolism

Under normal conditions, the sum of the urinary excretion of acid in the form of titratable acid plus ammonium minus any residual base (bicarbonate or equivalent alkali) excretion, is called **net acid excretion.** Urinary net acid is equal to new bicarbonate generated and added to the blood and is physiologically controlled to be equal to the amount of bicarbonate that was titrated by dietary, metabolic acid (Fig 10–4). For instance, if dietary and metabolic, and gastrointestinal-derived acid production plus stool base loss is 60 meq/d, urinary net acid excretion is normally also 60 meq/d, of which about 25 meq/d is usually in the form of urinary titratable acid ($U_{TA}V$) and 35 meq/d in the form of urinary ammonium ($U_{NH4^+}V$) and there is no bicarbonaturia ($U_{HCO3^-}V$).

$$U_{TA}V + U_{NH_4^+}V - U_{HCO_3^-}V = \text{Urinary net acid excretion} = \text{Acid input}$$

OVERVIEW OF RENAL ACIDIFICATION

As shown in Fig 10–5, the two functional goals of the kidney with respect to acid-base homeostasis—reclamation of the filtered bicarbonate and urinary net acid excretion (protonation of filtered and endogenously produced nonbicarbonate buffers)—are geographically separated in the proximal and distal portions of the nephron, respectively.

Bicarbonate Reabsorption

Eighty to 85 percent of the filtered bicarbonate is reabsorbed in the proximal convoluted tubule. There is further reabsorption in the loop of Henle (~5%) and the distal convoluted tubule (~5%). The remaining bicarbonate (~5%) is reabsorbed in the collecting tubule, so that little bicarbonate normally escapes into the urine.

Titratable Acid Formation

Much of the titratable acid that is ultimately excreted is formed in the proximal tubule (Fig 10–5). The pH in the proximal tubule normally falls to about 6.8, which is the pK_a of the $H_2PO_4^-$/HPO_4^{2-} acid-base pair; therefore, about half of the phosphate leaving the tubule is in the acid form. Phosphate is partially reabsorbed in the loop of Henle, reducing titratable acid delivery to the collecting tubule. The ability of the collecting tubule to markedly lower pH allows further titration of the residual phosphate and final formation of urinary titratable acidity.

Ammonium Excretion

Most of the ammonium destined for urinary excretion is actually formed in the proximal tubule (Fig 10–5). Ammonia is derived from glutamine, which is transported into the proximal tubule cell. the deamination of glutamine is catalyzed by glutaminases within the proximal tubule cell, specifically the cell in the early (S_1) proximal tubule segment. Ammonia diffuses into the proximal tubule lumen, and simultaneous hydrogen ion secretion by the cell allows protonation of ammonia to form ammonium. Ammonium is also formed within the cell and is transported into the lumen via the Na^+/H^+ antiporter (substituting for H^+).

Because of water reabsorption during transit in the descending limb of the loop of Henle, the luminal concentration of bicarbonate that has escaped proximal reabsorption rises. This alkalinization of the luminal fluid (the pH is back to pH 7.4 at the tip of the loop of Henle) favors dissociation of ammonium into ammonia and a hydrogen ion. The hydrogen ion is buffered by bicarbonate and other luminal buffers while ammonia, which is a gas, diffuses freely into all medullary structures. In the medullary collecting tubule, the low pH in the luminal fluid created by active hydrogen ion secretion allows titration of ammonia back to ammonium. Ammonium, which is nondiffusible, is then excreted into the urine.

Response to Acidemia

The kidney under normal physiologic conditions is always responding to modest acid challenge in the form of dietary and metabolic acid. Circumstances in which the acid load increases are accompanied by augmentation of the usual renal acidification processes in both the proximal tubule and collecting tubule.

Acidemia increases ammoniagenesis. Ammonium excretion can increase by over tenfold, from the usual value of 35 meq/d to over 300 meq/d. If the acidemia occurs in the setting of titratable buffer production, as in ketoacidosis (β-hydroxybutyrate), titratable acid excretion can also increase by more than tenfold (to values $>$ 250 meq/d). Potassium depletion, if present, also increases ammoniagenesis.

There is also an adaptive increase in hydrogen ion secretion in the proximal tubule. The value of this compensation in the proximal tubule is limited, since the filtered bicarbonate load and thus the quantity of bicarbonate requiring proximal reabsorption in metabolic acidosis is low. The adaptation of hydrogen ion secretion in the collecting tubule is useful, however, in increasing the titration of filtered buffers (eg, phosphate or ketones during ketoacidosis) or endogenously produced buffer (eg, ammonia) to increase urinary net acid

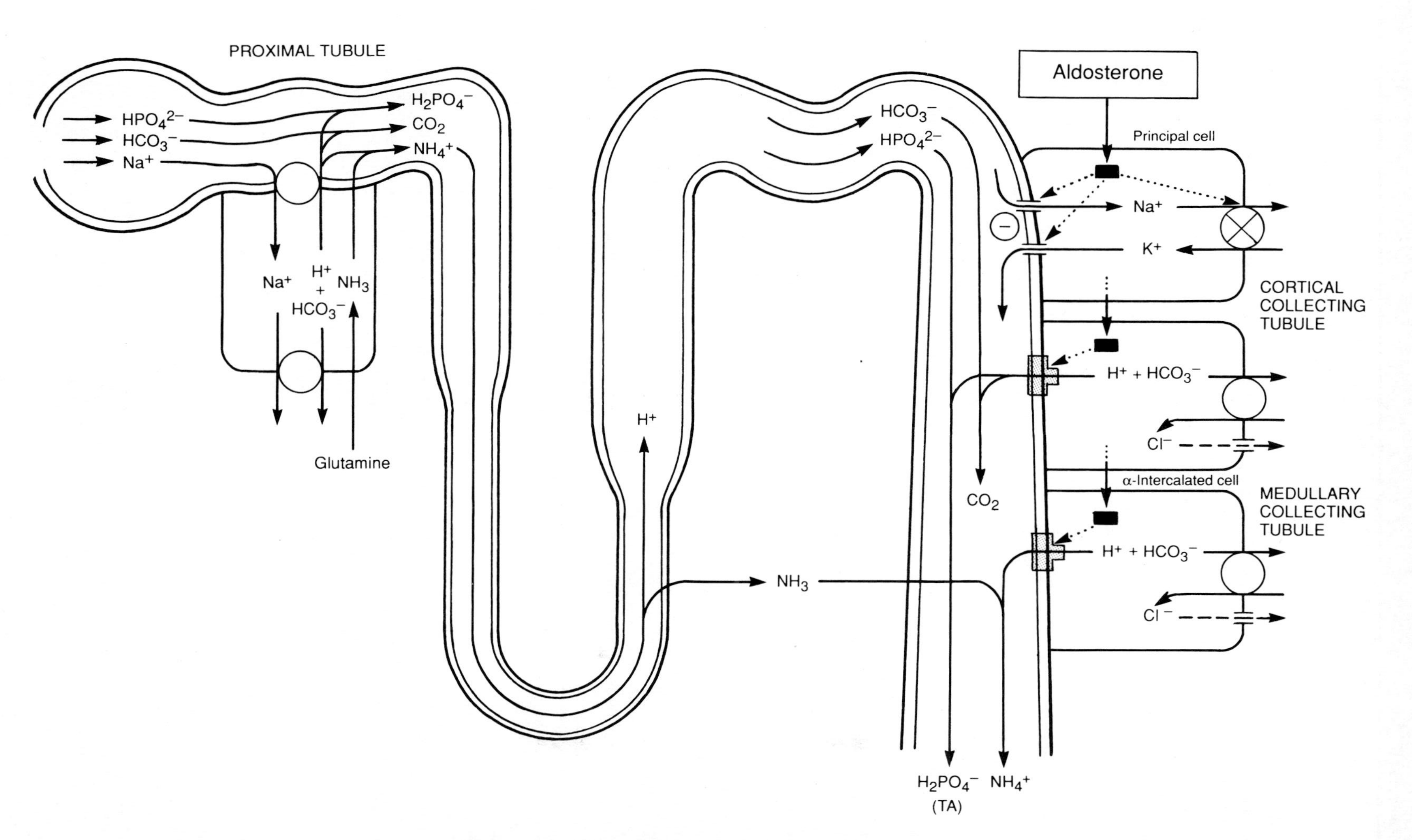

Figure 10–5. Sites and mechanisms of renal acidification.

excretion and hence generate more bicarbonate for addition to the blood.

ACIDIFICATION MECHANISMS & REGULATION BY NEPHRON SEGMENTS

A more detailed mechanistic appreciation of bicarbonate reabsorption and urinary net acid excretion can be gained by consideration of acid-base transport in each segment of the nephron.

PROXIMAL TUBULE

Cellular Mechanisms

As shown in Fig 10–6, hydrogen ion secretion in the proximal tubule (specifically, the proximal convoluted tubule) is effected by the luminal Na^+/H^+ antiporter. This antiporter exchanges a sodium ion for a hydrogen ion with a 1:1 stoichiometry—ie, the process is electroneutral. Hydrogen ion secretion is driven by the electrochemical gradient for sodium. Sodium moves down its concentration gradient from lumen to cell owing to the low intracellular sodium concentration (about tenfold lower than in the lumen) maintained by the basolateral Na^+/K^+-ATPase (discussed in Chapter 1). There may also be a H^+ translocating ATPase on the luminal membrane of the proximal tubule, but this is of minor importance.

Secretion of hydrogen ion into the luminal fluid of the proximal tubule performs three major functions (Figs 10–5 and 10–6). First, the quantitatively most important role is for titration of bicarbonate to carbonic acid. Carbonic acid is subsequently rapidly dehydrated to CO_2 by the isoenzyme (type IV) of carbonic anhydrase on the luminal membrane of the proximal tubule cell. The CO_2 then diffuses into the cell, to be hydroxylated to bicarbonate (see below). Second, hydrogen ion titrates ammonia to form ammonium. The ammonia derives from the deamination of glutamine by the action of intracellular (intramitochondrial) glutaminase. Some ammonia is also protonated to ammonium within the cell and transported into the lumen via the Na^+/H^+ antiporter (ammonium substitutes for hydrogen ion on the antiporter). Third, hydrogen ion secretion titrates filtered dibasic phosphate (HPO_4^{2+}) to form monobasic phosphate ($H_2PO_4^-$), the principal titratable acid. The cell of the early (S_1 subsegment) proximal tubule is responsible for more bicarbonate reabsorption and ammonium and titratable acid formation than the cell of the late (S_2 subsegment) proximal tubule.

The hydrogen ion destined for secretion into the lumen is derived from the splitting of water within the proximal tubule cell. A simultaneously liberated hydroxyl ion (OH^-) combines with CO_2 to form bicarbonate, catalyzed by the intracellular isoenzyme (type II) of carbonic anhydrase. The bicarbonate then exits the cell on a transporter, coupled with sodium with a stoichiometry of 3:1; ie, the process is electrogenic.

Thus, the net process of sodium bicarbonate reabsorption involves sodium reabsorption and bicarbonate loss from the lumen due to titration by secreted hydrogen ion, with the same amount of sodium bicarbonate exiting the cell to enter the peritubular blood. However, note that bicarbonate is not transported intact directly across the cell.

Regulation

Ninety-eight percent of hydrogen ion secretion in the proximal tubule is expended in the reabsorption of bicarbonate. This acidification process is responsible for reabsorbing fully 85% of the bicarbonate filtered. Thus, for practical purposes, hydrogen ion secretion by the proximal tubule and renal bicarbonate reabsorption can be thought of interchangeably. This process is affected by several luminal and peritubular factors, listed in Table 10–2.

The rate of bicarbonate reabsorption in the proximal tubule is markedly sensitive to luminal bicarbonate concentration and pH. A higher bicarbonate concentration and alkalinity in the luminal fluid establishes a more favorable gradient

Table 10–2. Factors influencing acidification and buffer availability in the proximal tubule.

Factor
Control of proximal H^+ secretion
Major determinants
Filtered HCO_3^- load
Luminal $[HCO_3^-]$ and pH
GFR and luminal flow rate
Pertibular $[HCO_3^-]$ and pH
Arterial P_{CO_2}
Angiotensin II
Minor determinants
Peritubular [protein]
Renal nerve activity
Potassium
PTH
Control of proximal NH_3 production
Arterial pH
Potassium
Control of proximal phosphate delivery
Plasma [phosphate]
GFR
PTH
Arterial pH

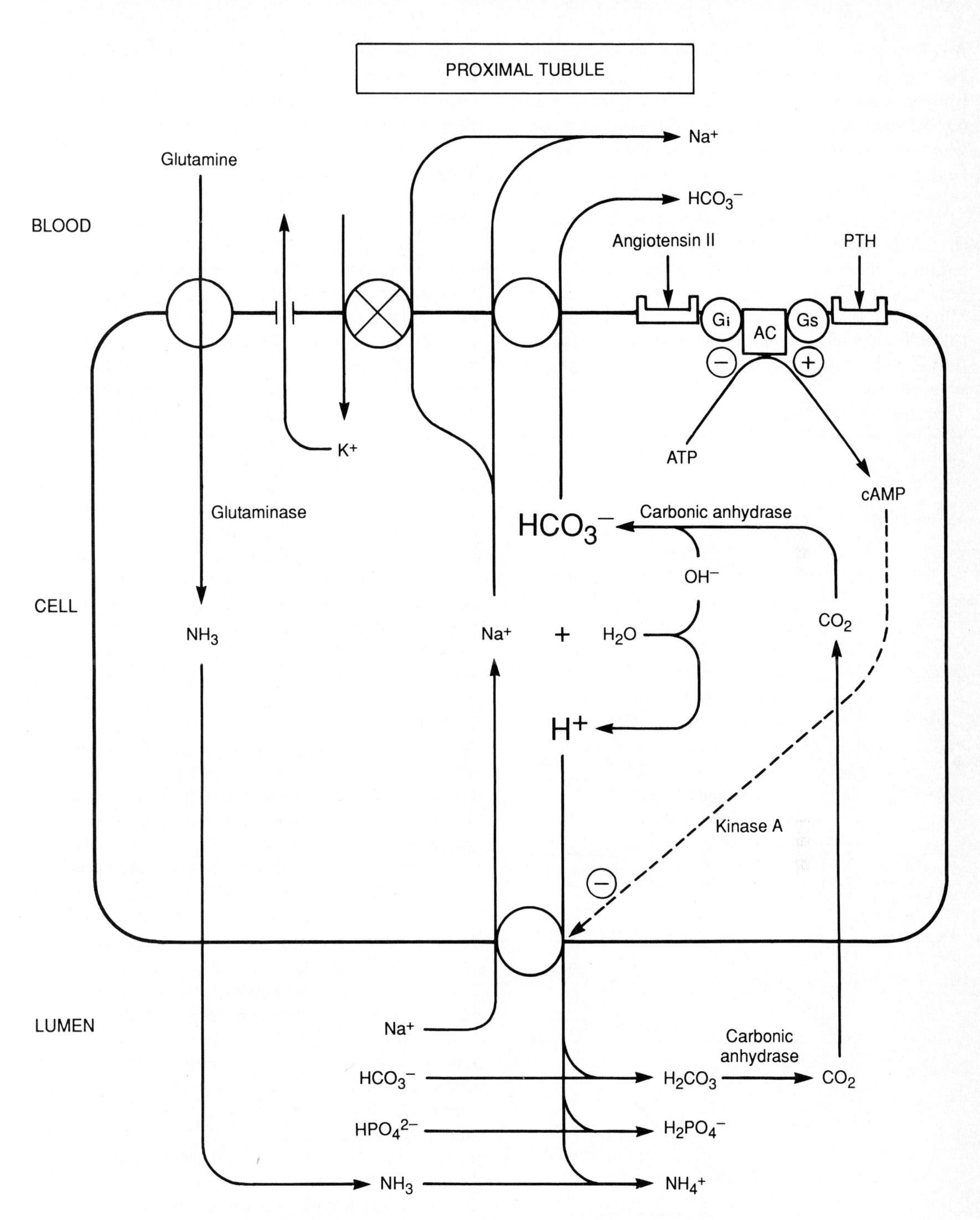

Figure 10-6. Cellular mechanisms of acid-base transport in the proximal tubule.

for hydrogen ion secretion. As shown in Fig 10–7 (top), selectively increasing luminal bicarbonate concentration at a constant flow rate (GFR) stimulates proximal bicarbonate reabsorption until saturation is reached. A similar effect is found with luminal fluid flow rate and GFR, shown in Fig 10–7 (middle). Increasing flow rate of the luminal fluid apparently serves to separate the luminal microvilli of the proximal tubule cell, enabling bicarbonate to diffuse more easily and be within ready access of the Na^+/H^+ antiporter. Since GFR is sensitive to extracellular volume status (Chapter 1), proximal bicarbonate reabsorption is also secondarily influenced by extracellular volume.

Because bicarbonate reabsorption is so readily affected by luminal bicarbonate concentration and flow rate, it is sensitively influenced by the product of the two, the filtered bicarbonate load, as shown in Fig 10–7 (bottom). The reabsorptive system is normally poised well below the saturation level, so that small changes in filtered bicarbonate load evoke proportionate change in reabsorption. This excellent load dependence of proximal bicarbonate reabsorption is also called glomerulotubular balance for bicarbonate.

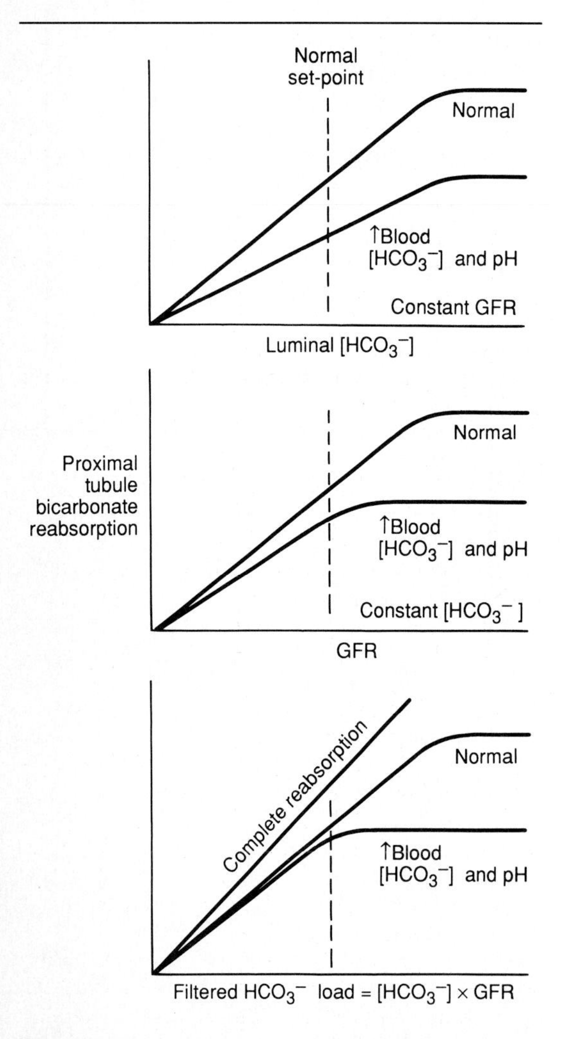

Figure 10–7. Impact of luminal and blood bicarbonate concentration and GFR on proximal bicarbonate reabsorption.

The blood bicarbonate concentration and pH have the opposite effect on bicarbonate reabsorption in the proximal tubule. A high bicarbonate concentration and pH on the basolateral side of the cell renders bicarbonate egress from the cell more difficult. The cell therefore alkalinizes, which diminishes the hydrogen ion supply needed for luminal secretion and slows bicarbonate reabsorption. This braking effect on proximal bicarbonate reabsorption by blood alkalinity is shown in Fig 10–7. The separate responses to luminal bicarbonate concentration (top panel) and GFR (middle panel) or to the product of the two—the filtered bicarbonate load (bottom panel)—are all markedly blunted in the setting of a high blood bicarbonate concentration and pH.

Since CO_2 is an important substrate for the intracellular formation of bicarbonate and influences intracellular pH (Fig 10–6), the blood P_{CO_2} markedly affects the rate of proximal bicarbonate reabsorption. As shown in Fig 10–8, reduction in P_{CO_2} reduces proximal bicarbonate reabsorption, although an acute increase in CO_2 has little effect. However, a chronic elevation in P_{CO_2} causes an adaptive stimulation of proximal bicarbonate reabsorption.

Finally, hydrogen ion secretion, particularly in the early (S_1) proximal convoluted tubule, is under the powerful regulation of angiotensin II. Angiotensin II receptor occupancy serves to activate an inhibitory G protein (G_i) that decreases adenylate cyclase activity. The subsequent reduction in intracellular cAMP increases bicarbonate reabsorption, since cAMP is an inhibitor of the Na^+/H^+ antiporter (Fig 10–6). Thus, angiotensin II acts to derepress the Na^+/H^+ antiporter. Parathyroid hormone (PTH) has the opposite effect, to enhance adenylate cyclase activity via a stimulatory G protein (G_s) and raise the intracellular cAMP level, thereby reducing Na^+/H^+ antiporter activity and bicarbonate reabsorption. Although PTH can influence the Na^+/H^+ antiporter activity, it probably has little influence within its physiologic concentration range.

There are minor effects attributable to peritubular protein concentration and renal nerve activity (Table 10–2). The peritubular capillary protein concentration (a component of the Starling forces), which is determined by the glomerular filtration fraction (see Chapter 1) in particular can markedly affect sodium chloride reabsorption.

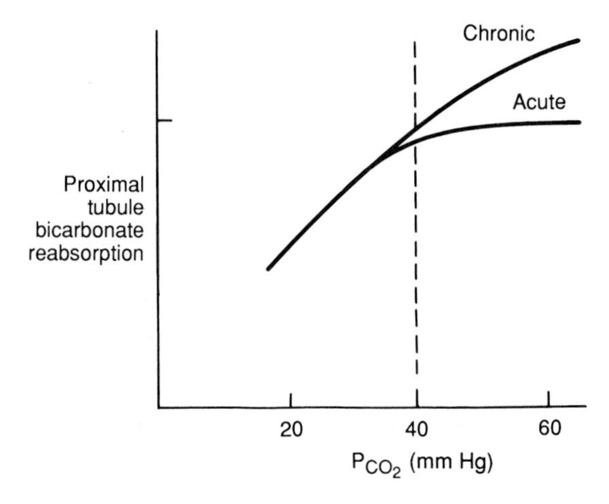

Figure 10-8. Regulation of proximal bicarbonate reabsorption by P_{CO_2}.

However, the influence of peritubular protein concentration on sodium bicarbonate transport is minimal even though there is an effect on the rate at which bicarbonate back-leaks through the paracellular pathway (ie, after reabsorption, bicarbonate returns from the blood to the lumen between cells). This back-leak component is quite small under basal conditions and has little impact on overall bicarbonate reabsorption. Likewise, renal nerve activity appears to be much more important for regulating sodium chloride than sodium bicarbonate reabsorption in the proximal tubule. The body potassium stores may also have a small influence on proximal bicarbonate reabsorption.

Rates of ammonium and titratable acid formation are affected by the rates of buffer generation and delivery. Ammoniagenesis is under the control of two major factors: blood pH and potassium (Table 10–2). Acidemia and potassium deficiency (hypokalemia) stimulate ammonia production and ammonium formation; alkalemia and hyperkalemia have the opposite effects. Titratable acid formation in the proximal tubule is regulated by the amount of phosphate filtered, determined by the plasma phosphate concentration and the GFR, and the quantity reabsorbed, sensitively controlled by PTH and arterial pH (Chapter 18).

LOOP OF HENLE & DISTAL CONVOLUTED TUBULE

About 15% of the filtered bicarbonate load escapes reabsorption in the proximal tubule. The bicarbonate concentration and pH at the end of the proximal tubule are about 8 meq/L and 6.8, respectively. As a result of water reabsorption in the descending limb of the loop of Henle, this residual bicarbonate is concentrated. By the tip of the loop of Henle, the luminal bicarbonate concentration and pH have risen to values similar to those of plasma, 24 meq/L and 7.4, respectively. Such alkalinization favors dissociation of ammonium to hydrogen ion plus ammonia in the luminal fluid. As shown in Fig 10–5, the hydrogen ion is captured by other luminal buffers while the ammonia, which is a gas, freely diffuses throughout all tissue in the medulla. Blood flow in the medulla is quite low, so that little ammonia is washed away by this route. Rather, ammonia is trapped as the nondiffusible ammonium in the most acidic medullary fluid compartment, the medullary collecting tubule (see below).

There are other acid-base transport processes in the loop of Henle not depicted in Fig 10–5. First, both the proximal straight tubule and the thick ascending limb of Henle reabsorb some bicarbonate—together, about 5% of the filtered load. Second, the thick ascending limb of Henle reabsorbs ammonium, and ammonium appears to be secreted in the adjacent descending limb of Henle, creating a countercurrent multiplication system for enhancing the medullary ammonia tension. Finally, some phosphate is reabsorbed in the loop of Henle, reducing titratable acid delivery.

Hydrogen ion secretion also occurs in the distal convoluted tubule. A small amount of bicarbonate is normally reabsorbed by this segment—approximately 5% of the filtered load—by mechanisms probably similar to those in the cortical collecting tubule.

CORTICAL COLLECTING TUBULE

Cellular Mechanisms

The three major cell types in the cortical collecting tubule are shown in Fig 10–9. The first is the **principal cell,** described in detail in Chapters 1 and 7. This cell reabsorbs sodium electrogenically, creating a transepithelial potential difference that is lumen-negative. Chloride is the only anion which for practical purposes can be reabsorbed in this segment. Chloride moves paracellularly, and this chloride reabsorptive process is driven by the potential difference. The principal cell is also responsible for secreting potassium. Function of the principal cell is exquisitely regulated by aldosterone. The principal cell does not participate directly in acid-base regulation.

Interspersed with the principal cell is the **α- intercalated cell.** This aldosterone-regulated cell has a proton-translocating ATPase on its luminal membrane that actively transports hydrogen ion into the lumen. Bicarbonate exits these cells into the blood by a 1:1 exchange for chloride, which

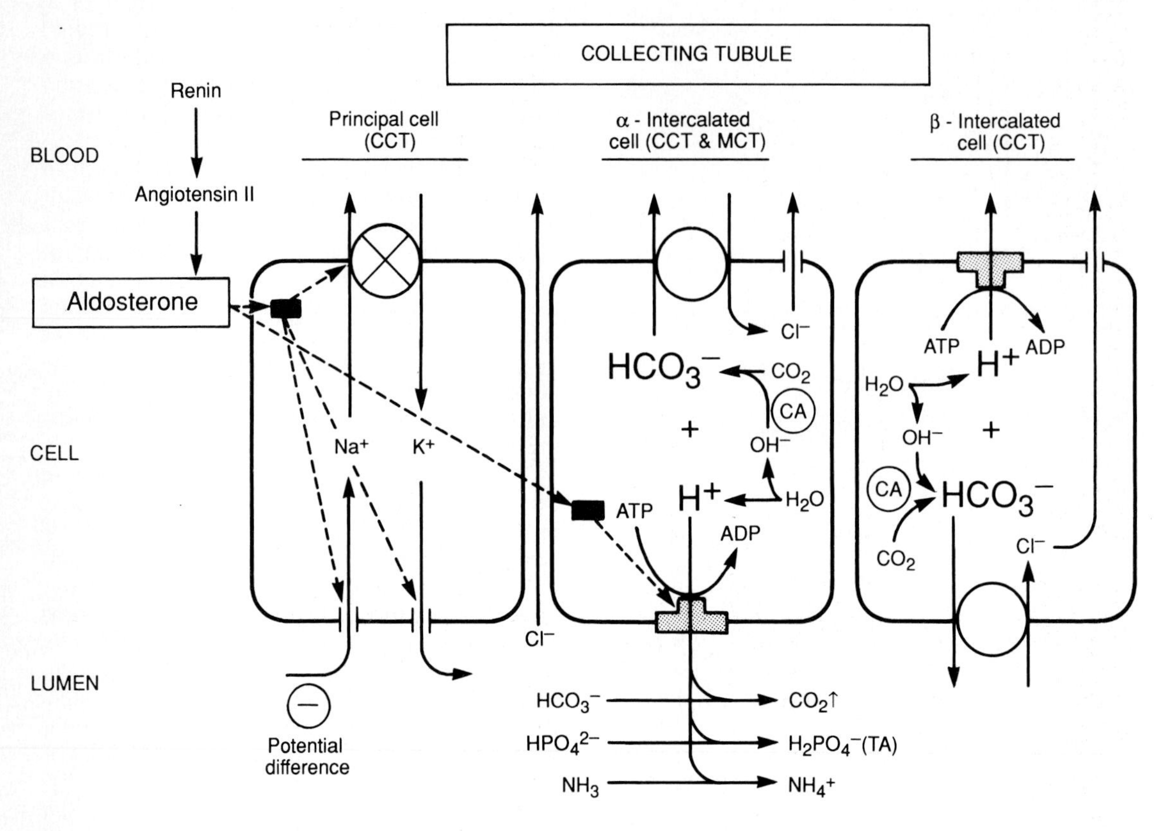

Figure 10-9. Cellular mechanisms of acid-base transport in the collecting tubule.

simply recycles through a basolateral conductive pathway. The luminal fluid can be acidified to about 3.0 pH units below blood—ie, pH ~4.5. By secreting the hydrogen ion electrogenically, this pump in itself creates a lumen-positive potential difference. Creation of both luminal fluid acidity and electrical positivity serves to self-inhibit the activity of the hydrogen ion pump. Hydrogen ion secretory activity of the α-intercalated cells is therefore enhanced by buffer delivery, which mitigates the fall of luminal fluid pH that would otherwise occur by hydrogen ion secretion. Hydrogen ion pumping is also enhanced by the electrical effects consequent to sodium reabsorption in the neighboring principal cell. The lumen-negativity created by sodium reabsorption mitigates the degree of lumen-positivity that might otherwise be produced by hydrogen ion secretion alone.

The purposes of hydrogen ion secretion in the cortical collecting tubule are threefold (Fig 10–9). First, hydrogen ion secretion permits reabsorption of the final amount of bicarbonate that has eluded the upstream reabsorptive system—usually about 5% of the filtered bicarbonate load. Second, lowering of luminal pH allows further titration of filtered buffers, notably HPO_4^{2-} to $H_2PO_4^-$, the formation of titratable acid. Third, some ammonia is trapped as ammonium in the cortical collecting tubule, though this function is performed largely in the medullary collecting tubule (see below).

A cell with functional polarity essentially opposite to the α-intercalated cell is called the **β-intercalated cell.** This cell secretes bicarbonate into the lumen in exchange for chloride (Fig 10–9). A proton-translocating ATPase exists on the basolateral membrane, and hydrogen ion is extruded into the blood. This β-intercalated cell is relatively quiescent under normal conditions but is functionally activated to secrete bicarbonate when the blood becomes alkaline. Operation of this cell tends to antagonize normal acidification processes in the cortical collecting tubule but may help to defend acid-base homeostasis during metabolic alkalosis.

Regulation

Factors that influence acid-base function in the cortical collecting tubule are listed in Table 10–3. Because hydrogen ion pumping of the α-interca-

Table 10-3. Major determinants of control of H^+ secretion in the collecting tubule.

Aldosterone
Sodium delivery
Reabsorbability of anion
Potassium stores
Luminal buffer delivery and pH
Arterial $[HCO_3^-$ and pH and P_{CO_2}

lated cell is electrogenic and sensitive to the luminal potential, it is stimulated when its neighboring principal cell reabsorbs more sodium, as shown in Fig 10–10 (top). The effect of sodium delivery and reabsorption is further enhanced if the sodium delivery occurs in the presence of a nonreabsorbable anion, a nonchloride anion, which further potentiates the lumen negativity associated with a given amount of sodium transport (Fig 7–14).

Aldosterone also importantly controls hydrogen ion secretion in the cortical collecting tubule directly, by regulating the proton-translocating ATPase of the α-intercalated cell. Potassium depletion potentiates the influence of aldosterone on hydrogen ion secretion, as shown in Fig 10–10 (middle).

Hydrogen ion secretion is markedly sensitive to luminal buffer delivery and pH, as shown in Fig 10–10 (bottom). If buffer delivery is increased—in the form of phosphate during phosphate loading or β-hydroxybutyrate during ketoacidosis—there is a commensurate increase in hydrogen ion secretion and titratable acid formation. As in the proximal tubule also, the blood bicarbonate concentration, P_{CO_2}, and pH affect the acidification process in the cortical collecting tubule, resulting in stimulation during acidemia and inhibition during alkalemia.

Medullary Collecting Tubule

The cell type in the outer medullary collecting tubule is the α-intercalated cell (Fig 10–9). Another cell, called the inner medullary collecting tubule cell, exists in the inner medulla, which appears to have functional characteristics similar to those of the α-intercalated cell. There are no principal cells or β-intercalated cells in the medullary collecting tubule. Sodium and anion delivery are therefore not important determinants of acidification function in this segment, and hydrogen ion secretion creates an unopposed, lumen-positive potential difference.

Under normal circumstances, most of the filtered bicarbonate has been reabsorbed and titration of nonbicarbonate buffers has occurred upstream of the medullary collecting tubule. However, since the prevailing ammonia tension is highest and luminal fluid pH lowest, substantial ammonium formation and trapping occurs along the length of the medullary collecting tubule (Fig 10–5). When medullary ammonia tension is enhanced, as during metabolic acidosis, hydrogen ion secretion and ammonium formation by the medullary collecting tubule increases proportionately.

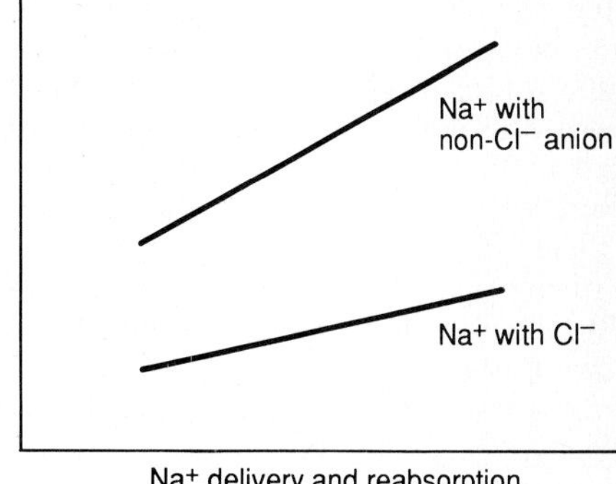

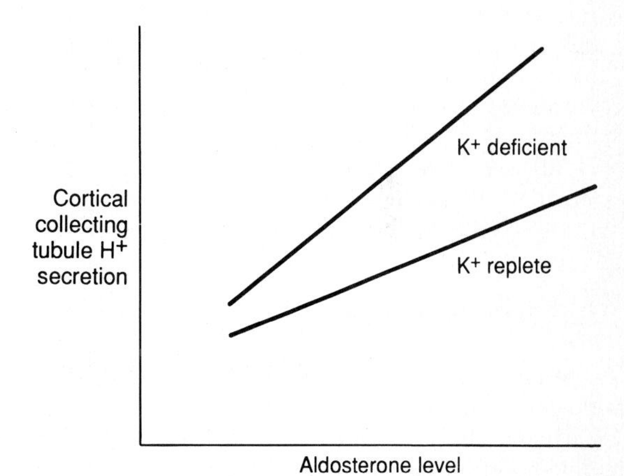

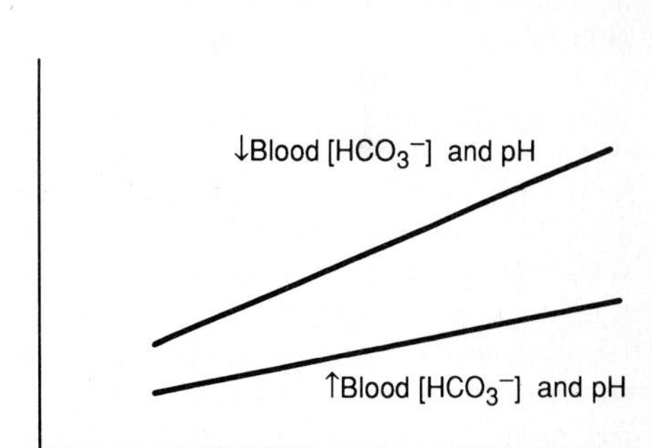

Figure 10–10. Determinants of hydrogen ion secretion in the cortical collecting tubule.

Summary

The hydrogen ion secretory process in the proximal tubule is a high-capacity system that generates only a small luminal pH gradient (< 1.0 pH unit below blood). The proximal tubule is responsible for reabsorbing most of the filtered bicarbonate and generating most of the titratable acid and ammonium eventually destined for urinary excretion. Hydrogen ion secretion in the collecting tubule is of low capacity but capable of generating a steep pH gradient (3.0 pH units less than blood). This distal nephron system is responsible for reabsorbing the final amount of bicarbonate, for titrating residual filtered buffer (titratable acid formation), and for reforming ammonium from ammonia lost in transit through the loop of Henle. Hydrogen ion secretion in both the proximal and collecting tubules is sensitive to luminal buffer concentration and P_{CO_2} and inversely to blood pH. Thus, both buffer generation (ammoniagenesis) and hydrogen ion secretion increase in the setting of acidemia and decrease in alkalemia. Hydrogen ion secretion is controlled by angiotensin II in the proximal tubule and by aldosterone in the collecting tubule.

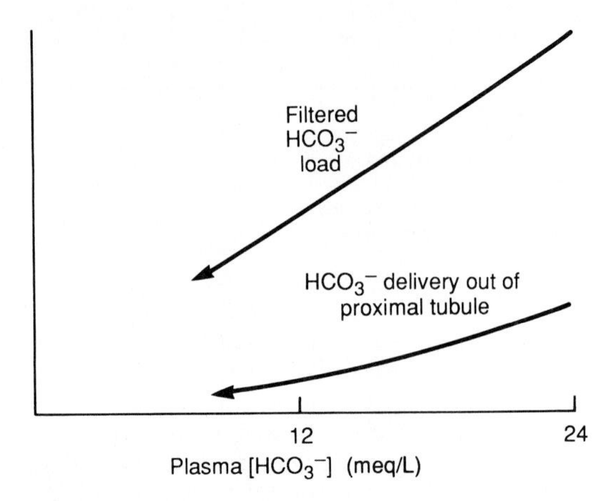

Figure 10-11. Proximal tubule bicarbonate transport during metabolic acidosis.

RESPONSE TO ACIDEMIA

A decrease in plasma bicarbonate concentration during metabolic acidosis lowers the filtered bicarbonate burden the kidney must reabsorb. Acidemia also increases intrinsic acidification capacity and ammoniagenesis to augment urinary net acid excretion.

Proximal Bicarbonate Reabsorption

The fraction of filtered bicarbonate reabsorbed in the proximal tubule is relatively constant and even increases slightly (from 80-85% to about 90%) as the filtered bicarbonate load declines. Stated another way, absolute bicarbonate reabsorption in the proximal tubule is proportionate to filtered bicarbonate load, as shown in Fig 10-7 (ie, glomerulotubular balance is preserved). Assuming GFR is constant, the absolute amount of bicarbonate leaving the proximal tubule diminishes in proportion to the decrease in filtered bicarbonate concentration during metabolic acidosis, as illustrated in Fig 10-11. With severe metabolic acidosis, the bicarbonate concentration in the tubular fluid leaving the proximal tubule can be very low, < 1 meq/L.

The decrease in delivery of bicarbonate out of the proximal tubule allows the cortical collecting tubule to expend hydrogen ion secretion on net acid formation and thus on generation of new bicarbonate rather than on reabsorption of residual bicarbonate escaping proximal reabsorption. Since most hydrogen ion secretion by the kidney as a whole is utilized in bicarbonate reabsorption, it can be calculated that absolute hydrogen ion secretion is reduced during metabolic acidosis. There is a shift, however, in the proportion used in reabsorbing bicarbonate to that used in titrating ammonium and phosphate.

Urinary Net Acid Excretion

To compensate for an acid load, the kidney endeavors to excrete more urinary net acid and regenerate plasma bicarbonate. As shown in Fig 10-12 (upper panel), acid loading acutely causes the plasma bicarbonate concentration to fall, despite intracellular buffering. Owing to a decrease in bicarbonate delivery out of the proximal tubule and probable stimulation by acidemia of hydrogen ion secretion of the α-intercalated cell in the collecting tubule, urinary pH acutely falls to less than 5.5 (Fig 10-12, middle panel). This aciduric response occurs quickly—within 2-4 hours of a systemic acid pulse.

With time, the most important effect of acidemia is stimulation of glutaminase activity to augment ammoniagenesis in the proximal tubule. The increase in ammonia delivery and enhanced hydrogen ion secretory capacity in the cortical and medullary collecting tubules allows for an increase in ammonium excretion by as much as tenfold (Fig 10-12, lower panel).

The full increase in urine ammonium excretion requires 3-5 days. During this period, plasma bicarbonate concentration begins to return toward normal despite continued acid loading owing to the increase in bicarbonate generated from enhanced urinary net acid formation.

Note that the increase in ammonia delivery causes urinary pH to rise despite increase in ammonium excretion (Fig 10-12, middle panel). Thus, urinary pH does not necessarily reflect absolute hydrogen ion excretion; the number of hydrogen ions in a buffered (ammonia-rich) urine

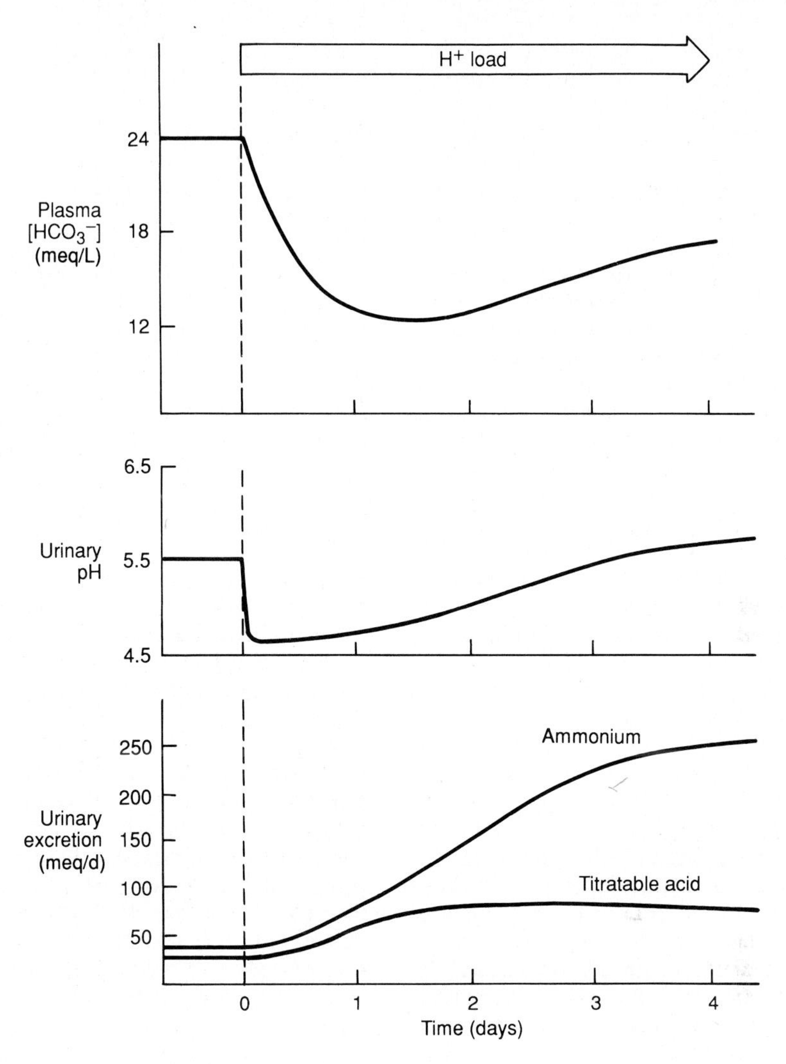

Figure 10–12. Renal acidification response to metabolic acidosis.

with a high pH can be far greater than in an unbuffered urine with a lower pH. This exemplifies the consideration of urine pH as a qualitative test of acidification, the hydrogen ion concentration gradient that can be developed. Urinary net acid excretion, however, is a quantitative assessment of the absolute number of hydrogen ions in the urine in conjunction with a given buffer delivery.

There is a mild rise in titratable acid excretion (Fig 10–12, lower panel) resulting from the inhibition of phosphate in the proximal tubule, due to acidemia itself as well as to a decrease in PTH activity (acidemia increases ionized calcium, as will be discussed in Chapter 15). In some forms of metabolic acidosis, acidemia is generated concomitantly with enhanced filtration of titratable buffer. The best example is the increased quantity of ketones filtered during ketoacidosis. In this case, the increase in urinary net acid excretion, which again can be over tenfold, is chiefly in the form of titratable acid.

SOURCES, BUFFERING, RESPIRATORY COMPENSATION, & RENAL RESPONSE TO HCO_3^- ADDITION TO THE BODY

While the usual physiologic task the kidney faces is urinary excretion of net acid, there are some circumstances, usually pathologic, when an alkali load must be excreted. The response to excess blood bicarbonate is again triphasic, with distribution and buffering, respiratory compensation, and renal excretion.

SOURCES OF ALKALI INPUT

Diet

Though not well documented, it is possible that the metabolism of some vegetarian diets may be primarily base-forming rather than acid-forming. For instance, in some societies the population subsists on foods rich in potassium, which is often in the form of salts in which the anion is a potential base (eg, potassium citrate, in which the citrate is subsequently converted in the body to bicarbonate).

Exogenous Administration

Bicarbonate, or an equivalent base which is converted by the body to bicarbonate, is sometimes administered therapeutically. Examples include antacid therapy (carbonate), parenteral fluid therapy for hyperalimentation (acetate), anticoagulation (citrate), hemodialysis (acetate or bicarbonate), and peritoneal dialysis (acetate or lactate).

H^+ Loss From the Body

There are two potentially acidic fluids in the body: the gastric juice and the urine. Excessive loss of hydrogen ion in either of these fluids is equivalent to a net gain of bicarbonate to the body. Examples include vomiting or nasogastric aspiration and conditions of hyperstimulation of collecting tubule acidification, as with diuretics or hypermineralocorticoidism.

EXTRACELLULAR & INTRACELLULAR BUFFERING

As with an acid load (Fig 10–2), an alkali load rapidly enters both the extracellular and the intracellular compartments. The proportioning is slightly different with alkali loading, since about two-thirds of the base is retained in the extracellular fluid and one-third is buffered in cells. A modest potassium shift into cells occurs, so that extracellular potassium concentration falls by about 0.4–0.5 meq/L per 0.1 unit pH increase (Table 7–1)—again opposite to the situation with acid buffering.

RESPIRATORY COMPENSATION

The ventilatory response to an acute rise in bicarbonate concentration is biphasic, with respiratory stimulation initially and then respiratory suppression.

Acute Response

Neutralization of an acute sodium bicarbonate load by nonbicarbonate buffers (H^+B^-) causes CO_2 to be liberated:

$$NaHCO_3 + H^+B^- \rightleftharpoons Na^+B^- + H_2CO_3$$
$$\rightleftharpoons H_2O + CO_2\uparrow$$

The acute increase in P_{CO_2}, which lasts for about an hour following a bicarbonate pulse, initially stimulates ventilatory drive. If pulmonary processes are disordered for any reason and cannot respond appropriately to this increased CO_2 production, hypercapnia can ensue.

Chronic Response

When the acute CO_2 generation subsides following a bicarbonate load, respiration becomes suppressed. Mirroring the effects of acidemia (Fig 10–3), alkalemia inhibits respiratory drive, which results in a secondary hypercapnic response that returns blood pH back toward (but not completely to) normal. As shown in Table 10–1 and Fig 10–13, the P_{CO_2} rises by about 0.5 mm Hg for each meq/L increase in plasma bicarbonate concentration. For instance, if bicarbonate concentration were to increase by 20 meq/L (from 24 to 44 meq/L), the physiologic rise in P_{CO_2} would be 10 mm Hg (from 40 to 50 mm Hg). This hypoventilatory response predictably leads to a commensurate fall in arterial P_{O_2}. In patients with chronic lung disease, frank respiratory failure can be induced by the respiratory suppressive effect of metabolic alkalosis.

RENAL RESPONSE

Though buffering and respiratory responses blunt the change in arterial pH that would otherwise ensue from an alkali load, it is the kidney that is ultimately responsible for excreting the excess bicarbonate.

In the Setting of Normal Extracellular Volume & Potassium Stores

Under normal physiologic conditions, the kidney excretes excess bicarbonate with great efficiency. Indeed, the bicarbonaturic response to an even slight increase in blood bicarbonate is so efficient that it is difficult to sustain even a small rise in blood bicarbonate concentration except by massive or very rapid bicarbonate administration.

When blood bicarbonate concentration rises,

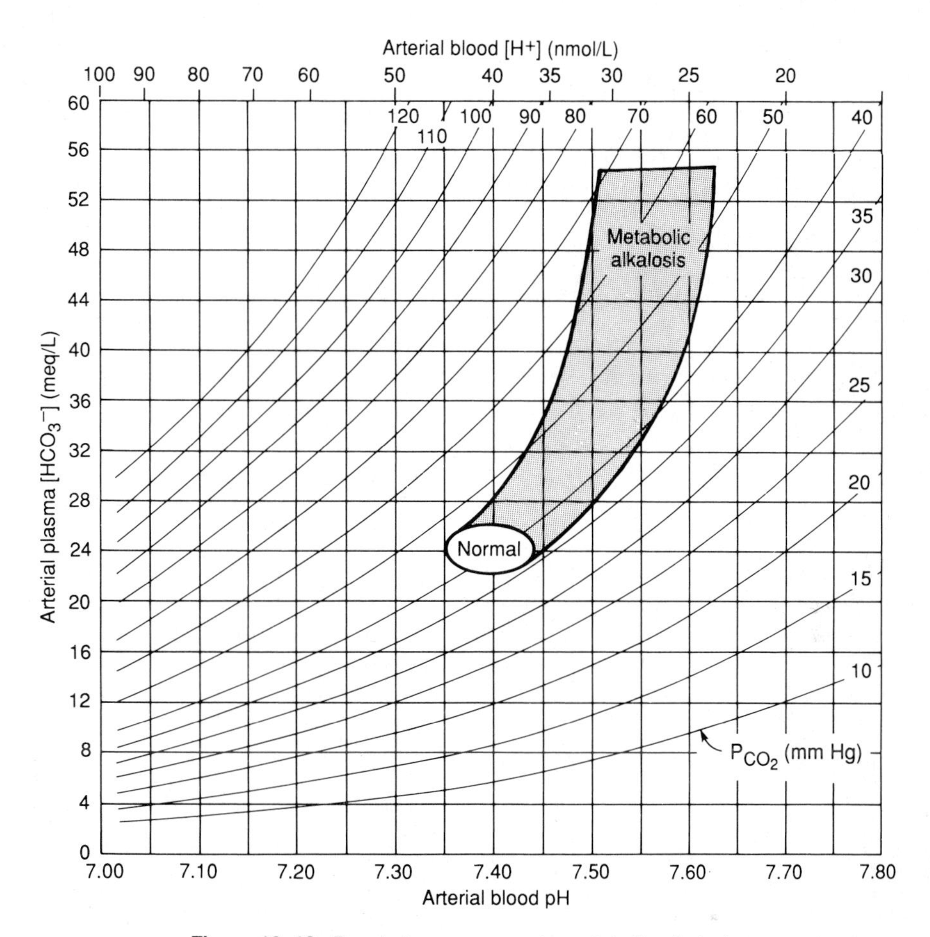

Figure 10–13. Respiratory response to metabolic alkalosis.

the concentration of bicarbonate increases both in the glomerular ultrafiltrate (thus increasing the filtered bicarbonate load if GFR is unchanged) as well as in the blood adjacent to the proximal tubule. As previously discussed (Fig 10–7), the rise in blood bicarbonate concentration and pH inhibits bicarbonate reabsorption in the proximal tubule and renders reabsorption static, unable to respond to the increase in filtered bicarbonate load (Fig 10–14, top panel). Thus, the increment in bicarbonate filtered is delivered out of the proximal tubule (Fig 10–14, middle panel).

The collecting tubule is a low-capacity reabsorptive system, and though hydrogen ion secretion in this segment can be load-responsive, it cannot cope with the flood of bicarbonate emerging from the proximal tubule. As a result, most of the increment in bicarbonate leaving the proximal tubule is unreabsorbed and excreted in the urine (Fig 10–14, lower panel). Titratable acid and ammonium excretion are quickly eliminated, and urinary net acid excretion (urinary titratable acid + $NH_4^+ - HCO_3^-$) is then negative (ie, there is net base excretion). The increase in urinary bicarbonate excretion serves to eliminate excess bicarbonate from the blood and to rapidly restore the plasma bicarbonate concentration to normal.

This brisk and highly efficient bicarbonaturic response of normal individuals in response to bicarbonate ingestion is shown in Fig 10–15. The kidney's bicarbonate excretory capacity is enormous and a sustained rise in plasma bicarbonate concentration is not tolerated. Indeed, this renal bicarbonaturic response to alkali administration is so efficient that individuals eating as much as 20 meq/kg per day of bicarbonate for several weeks sustain only a trivial rise in plasma bicarbonate concentration.

In the Setting of Reduced Extracellular Volume or Potassium Stores

The extraordinary ability of the kidney to defend against a rise in blood bicarbonate concentration and pH can be abrogated in the setting of diminished extracellular volume or potassium de-

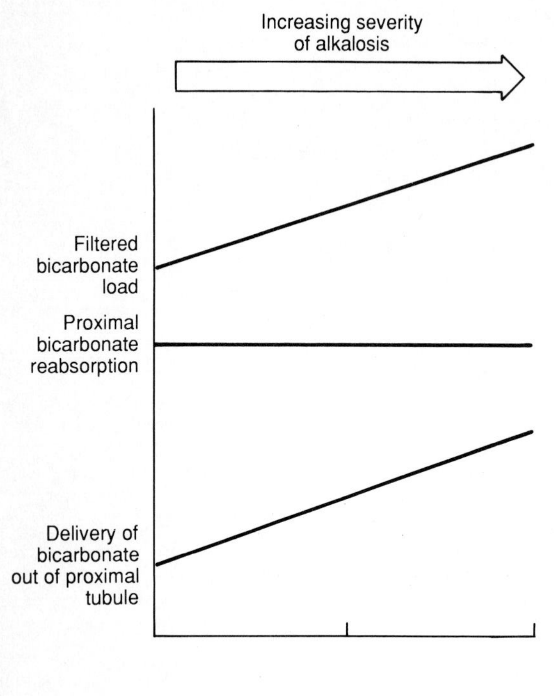

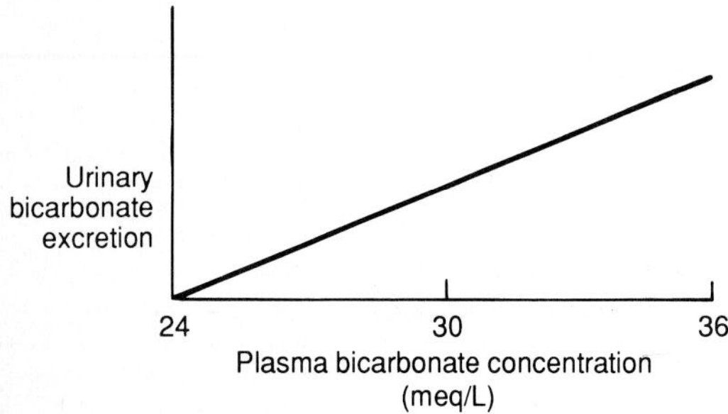

Figure 10–14. Renal bicarbonate handling in response to hyperbicarbonatemia.

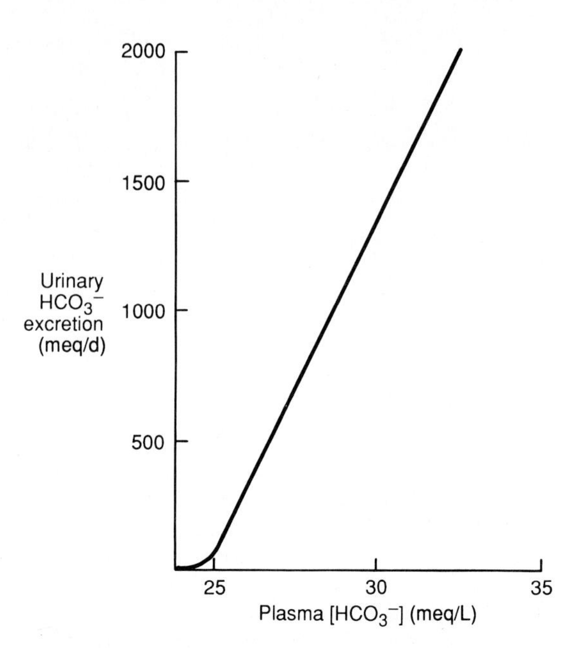

Figure 10–15. Normal bicarbonaturic response to hyperbicarbonatemia.

ficiency. As will be seen subsequently (Chapter 13), a rise in plasma bicarbonate concentration frequently occurs simultaneously with hypovolemia and reduced potassium stores. In this setting, the kidney is "reset," such that hyperbicarbonatemia and alkalemia no longer engender the normal corrective bicarbonaturic response. Instead, the high plasma bicarbonate concentration is sustained and not repaired. The kidney thus allows a new "set-point" at which the plasma bicarbonate concentration is regulated.

As shown in Table 10–4, three mechanisms may account for this change in renal response to elevated plasma bicarbonate concentration that involve alteration in renal hemodynamics (GFR) or tubular transport function. In each of these three cases, the hyperbicarbonatemia is sustained because the normal bicarbonaturic response does not occur.

In the first mechanism, a reciprocal reduction in GFR accompanies the rise in plasma bicarbonate concentration. The resulting filtered bicarbonate load (higher bicarbonate concentration × lower GFR) is therefore unchanged and remains at a normal level. A normal rate of bicarbonate reabsorption in both the proximal tubule and collecting tubule therefore suffice to reabsorb the filtered bicarbonate and prevent bicarbonaturia. The neurohumoral mechanisms by which extracellular volume depletion and hypovolemia reduces GFR were discussed at length in Chapter 1. Potassium deficiency has also been shown to decrease GFR in animals—though confirmation in humans has not been achieved—by increasing the potent vasoconstictors angiotensin II and thromboxane. According to this first mechanism, the increase in plasma bicarbonate concentration is then inversely proportionate to the decrease in GFR, as shown in Fig 10–16. For instance, a doubling of plasma bicarbonate concentration (from 24 to 48 mmol/L) could be maintained by a halving of GFR (from 100 to 50 mL/min). This mechanism has been confirmed in many animal and human models of metabolic alkalosis.

In the second mechanism (Table 10–4), stimulation of bicarbonate reabsorption in the proximal tubule might occur as a result of extracellular volume depletion or potassium depletion. According to this scheme, a decrease in GFR does not occur. The high plasma bicarbonate concentration is accompanied by a proportionate increase in filtered bicarbonate load which is reabsorbed in the proxi-

Table 10-4. Pathophysiology of metabolic alkalosis.

	Plasma [HCO_3^-]	GFR	Filtered HCO_3^- Load	Proximal Tubule	Collecting Tubule	Urinary HCO_3^- Excretion
Normal	↑	Normal	↑	Normal	Normal	↑
Sustained alkalosis						
1	↑	(↓)[1]	Normal	Normal	Normal	0
2	↑	Normal	↑	(↑)[2]	Normal	0
3	↑	Normal	↑	Normal	(↑)[2]	0

[1]Due to ↓ECV or ↓K^+.
[2]Due to ↑ aldosterone.

mal tubule. As mentioned above, however, it has been difficult to demonstrate that extracellular volume state or potassium stores have anything but a minor effect on proximal bicarbonate reabsorption (Table 10–2). Most experimental evidence in animals and humans does not support this mechanism.

In the third mechanism, stimulation of bicarbonate reabsorption in the collecting tubule occurs, attributable mainly to hyperaldosteronism acting synergistically with potassium deficiency (Fig 10–10, middle panel) despite alkalemia. In this mechanism, excessive mineralocorticoid activity acts both to generate as well as sustain a high plasma bicarbonate concentration. The collecting tubule is a relatively low-capacity acidification system even under maximal drive by aldosterone. Thus, increase in total renal bicarbonate reabsorption and hence in plasma bicarbonate concentration attributable to this mechanism is relatively small.

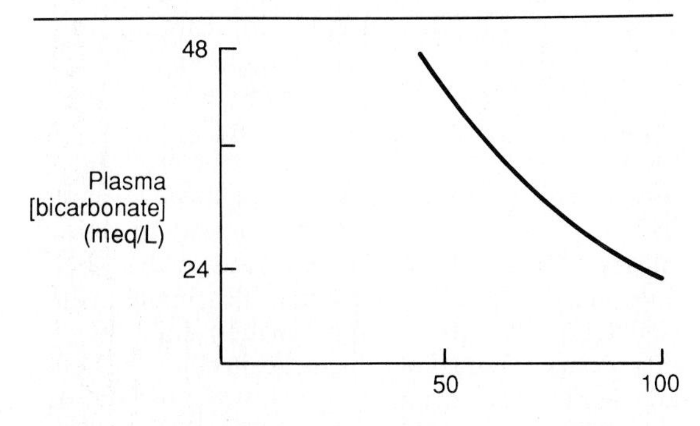

Figure 10–16. Glomerular hypofiltration during metabolic alkalosis.

BUFFERING & RENAL RESPONSES TO CHANGES IN P_{CO_2}

CAUSES OF PRIMARY ALTERATIONS IN P_{CO_2}

Normally, sensitive chemoreceptor control of ventilation maintains P_{CO_2} within very strict limits, 37–45 mm Hg. Intrinsic disturbances in ventilation either in the central control of ventilatory drive or in pulmonary gas exchange can cause P_{CO_2} to become disordered. Because of the large quantity of CO_2 produced and excreted (6–8 mmol/min), small ventilatory changes in the production-excretion balance produce rapid changes in arterial P_{CO_2}.

ACUTE BUFFERING PROCESSES

An acute increase in or loss of P_{CO_2} changes the blood pH rapidly—within 10 minutes. Nonbicarbonate buffering of the subsequent changes in carbonic acid concentration and pH occurs. This physicochemical change in blood acid-base composition causes a small change in bicarbonate concentration. In addition, acute hypocapnia increases organic acid production (eg, lactic and citric acids), which further reduces the blood bicarbonate concentration. As shown in Table 10–1 and Fig 10–17, bicarbonate concentration falls by about 0.25 meq/L for each millimeter of mercury fall in P_{CO_2} during acute respiratory alkalosis and rises 0.1 meq/L for each millimeter of mercury increase in P_{CO_2} during acute respiratory acidosis (Table 10–1). These small parallel changes in bicarbonate concentration provide only a minor defense of blood pH, which is more affected by the acute changes in P_{CO_2}.

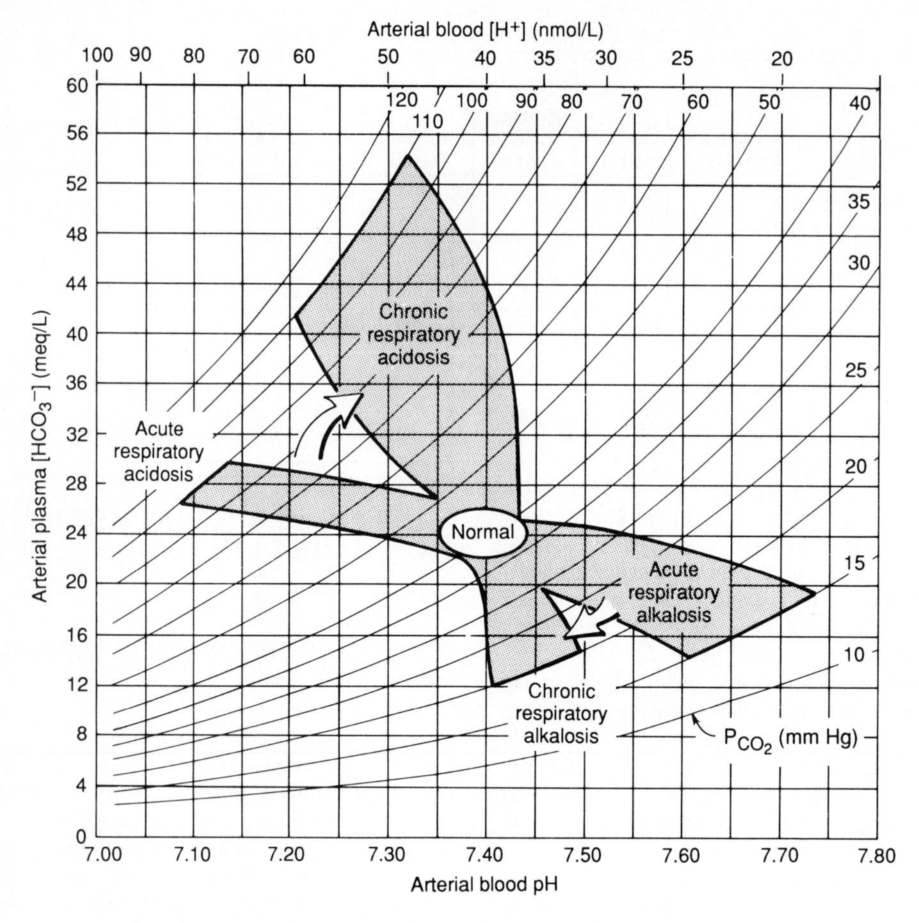

Figure 10–17. Renal response to respiratory disorders.

RENAL COMPENSATION

Renal Response to Chronic Hypocapnia

Over time, the kidneys adapt to a change in arterial P_{CO_2} to ameliorate the alteration in pH. As shown in Fig 10–8, hypocapnia reduces bicarbonate reabsorption in the proximal tubule. Hydrogen ion secretion in the collecting tubule is also diminished. As a result, hypocapnia increases urinary bicarbonate excretion and transiently reduces urinary net acid excretion. Plasma bicarbonate concentration therefore falls over a period of several hours to days. The increased delivery of sodium bicarbonate out of the proximal tubule results in kaliuresis (Chapter 7), with variable natriuresis and chloruresis (depending on dietary sodium chloride intake). In a steady state—as shown in Fig 10–17 and Table 10–1—plasma bicarbonate concentration falls by about 0.5 meq/L for each millimeter of mercury decrease in P_{CO_2} during chronic respiratory alkalosis. The arterial blood pH is brought back toward (but not to) normal, and the alkalemia is greatly ameliorated.

Renal Response to Chronic Hypercapnia

Chronic hypercapnia stimulates ammonia production and both proximal and collecting tubule hydrogen ion secretion. The enhanced ammoniagenesis and collecting tubule acidification causes urinary pH to fall and urinary ammonium excretion to be augmented, as shown in Fig 10–18 (middle and lower panels). There is a mild increase in titratable acid formation and an associated chloruresis. This process takes several days to complete. The increase in urinary net acid excretion causes new bicarbonate to be added to the blood and thus to increase the plasma bicarbonate concentration (Fig 10–18, upper panel) and filtered bicarbonate load. There is a rise in the bicarbonate reabsorptive capacity in the proximal tubule (Fig 10–8), so that the elevated filtered bicarbonate load is completely reabsorbed. There is no bi-

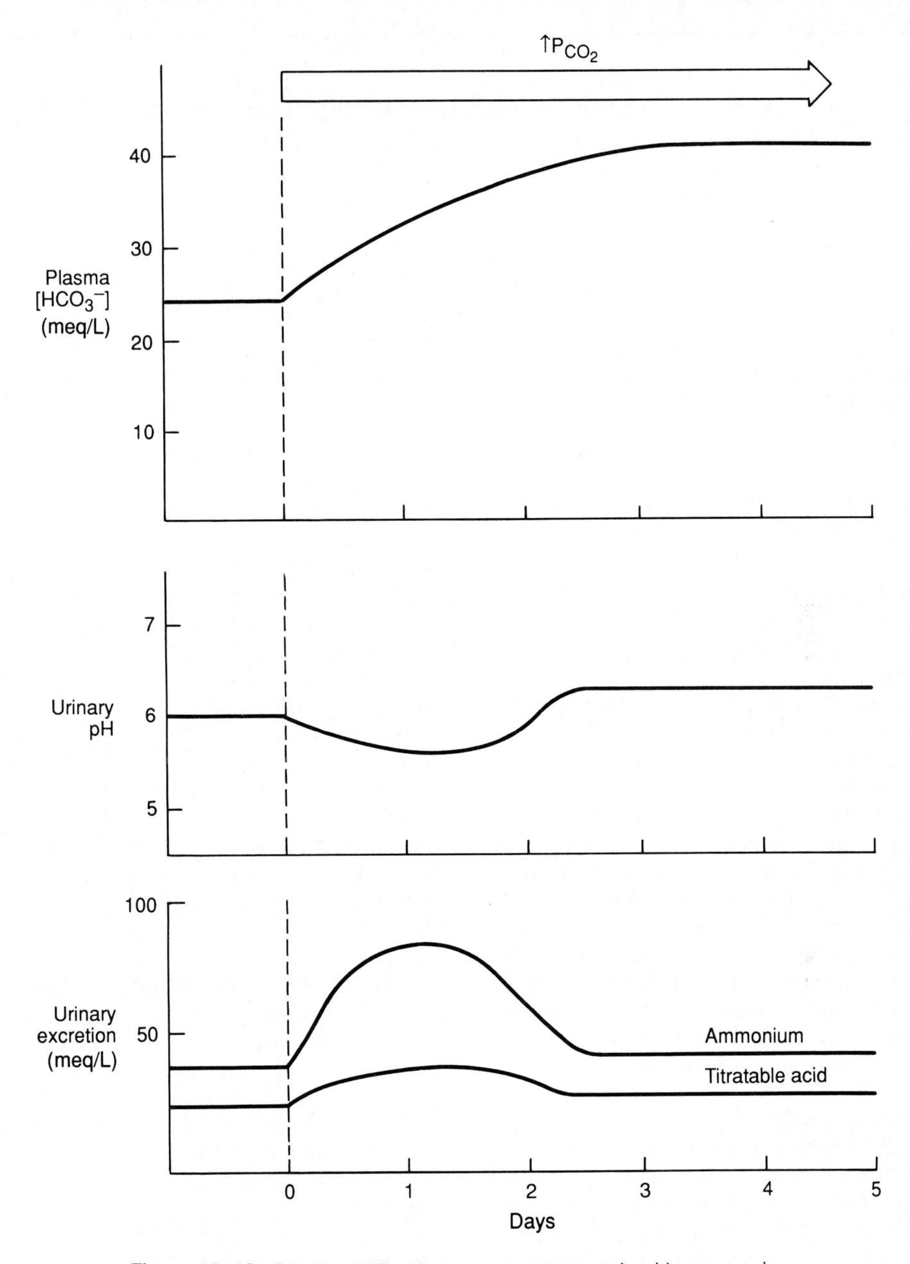

Figure 10-18. Renal acidification response to sustained hypercapnia.

carbonaturia, and the hyperbicarbonatemia is sustained.

As shown in Fig 10-17 and Table 10-1, plasma bicarbonate concentration is increased in the steady state by an average of 0.5 meq/L for each millimeter of mercury increase in P_{CO_2} during chronic respiratory acidosis. The increase in plasma bicarbonate concentration ameliorates the acidemia of chronic hypercapnia. A slight overshoot in arterial pH has been described in which so much bicarbonate is generated and retained during chronic hypercapnia that the blood becomes overtly alkaline at least when measured in the morning—following exacerbation of the normal nocturnal CO_2 retention and bicarbonate generation.

11 Diagnosis of Acid-Base Disorders

CLASSIFICATION OF ACID-BASE DISORDERS

DEFINITION OF SIMPLE ACID-BASE DISORDERS

Metabolic & Respiratory Disorders

A derangement in arterial blood pH may arise when there is an abnormality in the plasma bicarbonate concentration or in P_{CO_2}, or both. When the primary disturbance is that of the bicarbonate concentration, due to addition or loss of nonvolatile acid or alkali to or from the extracellular fluid, the resulting disorder is termed metabolic. A low arterial pH due to reduced bicarbonate concentration is termed **metabolic acidosis,** whereas a high arterial pH with increased bicarbonate concentration is termed **metabolic alkalosis.**

Similarly, when the primary disturbance is that of P_{CO_2}, reflecting a primary increase or decrease in alveolar respiration, it is termed respiratory. A low arterial pH due to increased P_{CO_2} is **respiratory acidosis,** and a high arterial pH due to reduced P_{CO_2} is **respiratory alkalosis.**

"-emia" Versus "-osis"

Changes in arterial pH in the downward or upward direction are known as acidemia or alkalemia, respectively. The metabolic or respiratory disturbance responsible for causing such change in arterial pH is known as acidosis or alkalosis, respectively. For instance, metabolic acidosis is termed "metabolic" because it involves a primary change in bicarbonate concentration and "acidosis" because it tends to cause acidemia. Likewise, respiratory alkalosis is termed "respiratory" because it involves a primary change in P_{CO_2} and "alkalosis" because it tends to cause alkalemia.

Primary Versus Compensatory Alterations

As discussed in detail in Chapter 10, a primary change in bicarbonate concentration evokes a compensatory change in respiration, and vice versa. The compensatory responses in each case are *not* considered separate acid-base disturbances. For instance, the acidemia associated with metabolic acidosis induces hyperventilation as part of its physiologic response and reduction in P_{CO_2} (Fig 10–3). This secondary hypocapnia is not a separate acid-base disorder—ie, is not a respiratory alkalosis—but rather is part of the normal physiologic compensatory response to the primary disorder. Metabolic acidosis with appropriate hypocapnia is thus a simple, physiologically compensated acid-base disorder.

Acute Versus Chronic Respiratory Disorders

A primary change in P_{CO_2} evokes a renal compensatory response that takes hours to several days to complete (Fig 10–18). This renal response—to depress the plasma bicarbonate concentration in chronic hypocapnia and elevate the plasma bicarbonate concentration in chronic hypercapnia—is again considered part of the normal physiologic response to a primary change in P_{CO_2}.

Primary respiratory disorders can therefore be further categorized depending on whether the appropriate time has elapsed for the renal response. Respiratory disorders are termed acute if the renal response has not yet occurred. The bicarbonate concentration is only mildly altered by nonbicarbonate buffering of changes in carbonic acid formed by CO_2 (Fig 10–17).

If the respiratory disorder has persisted long enough for the full renal response to occur with physiologic alteration in bicarbonate concentration, the term chronic is used. For instance, chronic hypercapnia is associated with a rise in bicarbonate concentration (Fig 10–17, which is a normal physiologic compensation by the kidney, not a separate metabolic alkalosis.

Acid-Base Nomogram

A detailed quantitative knowledge of the normal physiologic respiratory responses to primary metabolic disturbances (and vice versa) is needed to judge the appropriateness of bicarbonate concentration and P_{CO_2} changes in a given individual. The equations of Table 10–1 serve this purpose, or, equivalently, the acid-base nomogram shown in Fig 11–1 can be used.

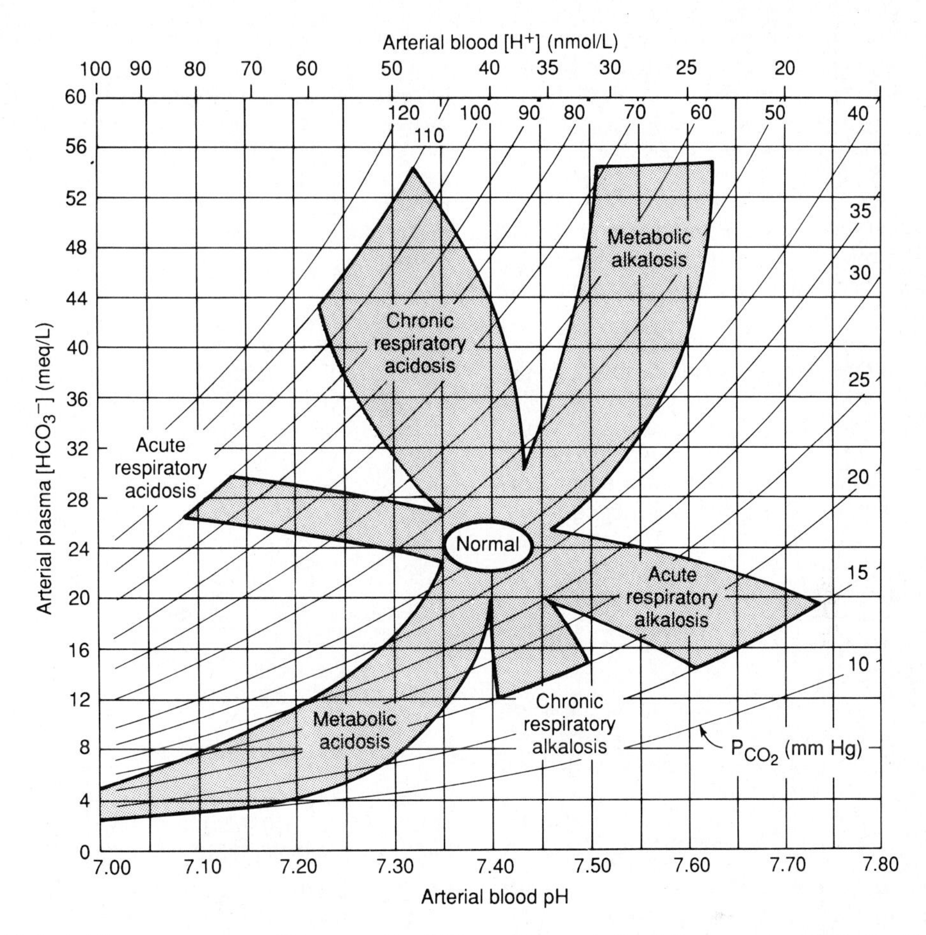

Figure 11-1. Acid-base nomogram. Shown are the 95% confidence limits of the normal respiratory and metabolic compensations for primary acid-base disturbances. (Reproduced, with permission, from Cogan MG, Rector FC Jr: Acid-base disorders. Pages 457–517 in: *The Kidney,* 3rd ed. Brenner BM, Rector FC Jr [editors]. Saunders, 1986.)

In this acid-base map, the bicarbonate concentration, hydrogen ion concentration and pH, and P_{CO_2} values intersect according to the Henderson or Henderson-Hasselbalch equations. The central open circle shows the 95% confidence limits for acid-base status in normal individuals. The shaded areas in this nomogram show the 95% confidence limits for the normal compensations to simple metabolic and respiratory disorders.

Several cautions are necessary when using this map, as discussed in greater detail subsequently in this chapter. First, it is assumed that sufficient time has elapsed for the full compensatory response (ie, 6–12 hours for the ventilatory compensation in primary metabolic disorders and as much as 3–5 days for the metabolic response to primary respiratory disorders). Second, finding acid-base values within a shaded area does not necessarily mean there is a simple acid-base disorder. Imposition of one disorder over another can sometimes result in values that lie within a third.

PLASMA ANION GAP

Definition

Especially useful in diagnosing metabolic acidoses is the anion gap. There is, of course, no real anion gap in the plasma (blood doesn't spark!). But only certain cations and anions are routinely measured by the clinical laboratory, such as sodium, bicarbonate, and chloride. Most other electrolytes and organic anions can be measured using special techniques, but these are not part of the rapid autoanalysis of blood used routinely by clinical laboratories.

For this diagnostic technique, the composition of plasma is considered to be simply as shown below:

$$[Na^+] + \text{Unmeasured cations} = [HCO_3^-] + [Cl^-] + \text{Unmeasured anions}$$

Rearranging, the plasma anion gap, which is the excess of unmeasured anions over unmeasured cations, is defined as follows:

$$\text{Anion gap} = \text{Unmeaured anions} - \text{Unmeasured cations}$$
$$= [Na^+] - ([HCO_3^-] + [Cl^-])$$

The normal value for the anion gap is **12 ± 4 meq/L.** The anion gap may increase if the unmeasured anions rise or the unmeasured cations fall, and vice versa.

The major unmeasured cations include calcium (5 meq/L), magnesium (2 meq/L), gamma globulins, and, for convenience, potassium (4 meq/L). The unmeasured anions are principally represented by albumin, a highly negatively charged molecule with a charge equivalency at normal pH of about 2 meq/L for each g/dL. Other contributors to the unmeasured anions include phosphate (2 meq/L), sulfate (1 meq/L), lactate (1-2 meq/L), and other organic anions (3-4 meq/L).

Etiology of Reduction in the Anion Gap

As shown in Table 11-1, a decrease in the anion gap occurs when unmeasured cations increase, as with combined hyperkalemia, hypercalcemia, and hypermagnesemia; with lithium intoxication; or with increased gamma globulin (if the isoelectric point of IgG is < 7.4) in multiple myeloma. A decreased anion gap also occurs with a reduction in unmeasured anions, as with hypoalbuminemia or reduction in the charge equivalency of albumin during acidemia (see below). Finally, artifactual decrease in the anion gap can arise from laboratory errors induced by hyperviscosity (underestimation of the true sodium concentration) or bromism (overestimation of the true chloride concentration).

Etiology of an Increase in the Anion Gap

A modest rise in the anion gap occurs with combined reduction in unmeasured cations, as occurs in hypomagnesemia (with reduced Mg^{2+}, Ca^{2+}, and K^+) and, hyperalbuminemia. Alkalemia also increases the anion gap because it increases the negative charge density on albumin by shifting the reaction below to the left:

$$\underset{\text{(Contributes to anion gap)}}{H^+ + \text{Albumin}^-} \underset{\text{Alkalemia}}{\overset{\text{Acidemia}}{\rightleftharpoons}} \underset{\text{(Does not contribute to anion gap)}}{H \cdot \text{albumin}}$$

Since only the charged albuminate (albumin⁻) but not the uncharged protonated form (H•albumin) contributes to the anion gap, alkalemia in-

Table 11-1. Disorders with alterations in anion gap.[1]

Anion gap = $Na^+ - (HCO_3^- + Cl^-)$

Decreased Anion Gap	Increased Anion Gap
Increased cations (not Na^+)	Decreased cations (not Na^+)
↑K^+, Ca^{2+}, Mg^{2+}	↓K^+, Ca^{2+}, Mg^{2+}
↑Li^+	
↑IgG	
Decreased anions (not Cl^- or HCO_3^-)	Increased anions (not Cl^- or HCO_3^-)
↓Albumin concentration	↑Albumin concentration
Acidosis	Alkalosis
	↑Inorganic anions
	Phosphate
	Sulfate
	↑Organic anions
	Lactate
	Ketones
	Uremic
	↑Exogenously supplied anions
	Toxins
	Salicylate
	Methanol
	Ethylene glycol
	Paraldehyde
	↑Unidentified anions
	Toxins
Laboratory error	Uremic
Hyperviscosity	Hyperosmolar, nonketotic states
Bromism	Myoglobinuric acute renal failure

[1]Adapted from Cogan MC, Rector FC Jr: Acid-base disorders. Pages 457-517 in: *The Kidney,* 3rd ed. Brenner BM, Rector FC Jr (editors). Saunders, 1986.

creases this charged component for any total amount of albumin.

More importantly, the anion gap is elevated when an acid other than hydrochloric acid is added to the plasma, so that the bicarbonate concentration is titrated and lost as CO_2 while the sodium salt of the conjugate unmeasured (nonchloride) anion of the acid remains in the blood.

For instance, consider the addition of 10 meq/L of an organic acid (HA)—any acid other than HCl—to blood with a simplified composition of 24 meq/L $NaHCO_3$, 104 meq/L NaCl, and 12 meq/L NaX (X = all unmeasured anions). The initial anion gap = total $Na^+ - (HCO_3^- + Cl^-)$ $= 140 - (24 + 104) = X = 12$ meq/L. The HA would cause an equivalent amount of bicarbonate to be titrated and lost as CO_2 with formation of the sodium salt of A^- (NaA):

$$\underset{10}{HA} + \underset{24}{NaHCO_3} + \underset{104}{NaCl} + \underset{12}{NaX} \longrightarrow$$

$$\underset{14}{NaHCO_3} + \underset{104}{NaCl} + \underset{10}{NaA} + \underset{12}{NaX}$$

A^- is not routinely measured by the clinical laboratory (ie, it is not HCO_3^- or Cl^-). Therefore, the new anion gap = total $Na^+ - (HCO_3^- + Cl^-)$ $= 140 - (14 + 104) = A + X = 22$ meq/L. The increase in the anion gap (from 12 to 22 meq/L) signals the addition to the blood of 10 meq/L of nonchloride (and nonbicarbonate) anion. Thus, the nonchloride acid has simultaneously caused the fall in bicarbonate concentration (from 24 to 14 meq/L) and rise in the conjugate anion A^- (from 0 to 10 meq/L). Note that there has been no change in the measured chloride concentration (104 meq/L). Such a metabolic acidosis is termed **high anion gap metabolic acidosis** or, synonymously, **normochloremic metabolic acidosis.** Examples of such acidoses are listed in Table 11–1 and will be discussed in greater detail in Chapter 12.

Alternatively, consider the addition of 10 meq/L of hydrochloric acid (HCl) to the same initial simplified blood. Again, HCl would cause an equivalent amount of bicarbonate to be lost with retention of the sodium salt of Cl^-:

$$\underset{10}{HCl} + \underset{24}{NaHCO_3} + \underset{104}{NaCl} + \underset{12}{NaX} \longrightarrow$$

$$\underset{14}{NaHCO_3} + \underset{114}{NaCl} + \underset{12}{NaX}$$

The new anion gap = total $Na^+ - (HCO_3^- + Cl^-) = 140 - (14 + 114) = X = 12$ meq/L. In this case, the conjugate anion (Cl^-) of the acid is routinely measured and thus does not contribute to the anion gap. The rise in chloride concentration (from 104 to 114 meq/L) and reciprocal fall in bicarbonate concentration (from 24 to 14 meq/L) maintains normality of the anion gap (12 meq/L) and thus signals that a chloride-containing acid or its equivalent has been added to the blood. Such a metabolic acidosis is termed **normal anion gap metabolic acidosis** or, synonymously, **hyperchloremic metabolic acidosis.** Such metabolic acidoses include those in which there is excess hydrogen ion that has occurred concomitantly with chloride retention, seen with gastrointestinal base loss (eg, diarrhea) and defects in renal acidification (renal tubular acidoses), as discussed in greater detail in Chapter 12.

DEFINITION OF COMPLEX ACID-BASE DISORDERS

Simple acid-base disorders are defined by the primary change in bicarbonate concentration or P_{CO_2} with the appropriate secondary change in the other parameter (Fig 11–1). Sometimes, the compensation may be too small or too large. In other cases, two or more disturbances may combine. When acid-base values do not follow the simple rules of Table 10–1 or lie outside the shaded areas of Fig 11–1, there is suspicion of a complex, or mixed, acid-base disorder.

It is necessary to first make certain that sufficient time has elapsed so that the appropriate compensatory changes have had time to occur. Given this temporal consideration, failure of the appropriate secondary compensation to a primary disorder is an example of a complex acid-base disorder. For instance, during metabolic acidosis with plasma bicarbonate concentration of 14 meq/L, lack of hyperventilation with maintained P_{CO_2} at the preexisting normal value (40 mm Hg) is inappropriate. Since the P_{CO_2} is higher than should be expected (25–30 mm Hg), the disorder would be considered respiratory acidosis superimposed on metabolic acidosis. The sustained "normal" P_{CO_2} causes acidemia to be worse (arterial pH 7.15) than expected from a simple metabolic acidosis alone had the P_{CO_2} fallen appropriately (arterial pH 7.30).

It sometimes happens that two disorders can combine with acid-base values that lie within the boundaries usually found in a third. For instance, consider a patient with sepsis that has caused both a lactic acid form of metabolic acidosis, reduced the plasma bicarbonate concentration from 24 to 14 meq/L (and simultaneously increased the anion gap from 12 to 22 meq/L) and acute respiratory alkalosis, which has induced primary hyperventilation that reduced arterial P_{CO_2} from 40 to 20 mm Hg. The resulting arterial pH is 7.45

and, according to the nomogram in Fig 11–1, the acid-base disturbance at first glance looks like chronic respiratory alkalosis. Both the clinical situation and the presence of an elevated anion gap, which does not occur in chronic respiratory alkalosis, lead to the correct acid-base diagnosis. Even the presence of a normal pH does not exclude the possibility that there has been superimposition of one acid-base disorder on another (see below).

Table 11–2. Diagnosis of acid-base disorders.

1. Verify that arterial blood pH, P_{CO_2}, and $[HCO_3^-]$ are mutually compatible.
   a. Henderson equation
   b. Henderson-Hasselbalch equation
   c. Acid-base nomogram
2. Identify primary disturbance
   a. Arterial blood pH, P_{CO_2}, $[HCO_3^-]$
3. Distinguish simple (appropriately compensated) disorder from complex disorder
   a. Acid-base nomogram
   b. Anion gap

DIAGNOSTIC STRATEGY IN ACID-BASE DISORDERS

VERIFICATION OF ACID-BASE VALUES

A given triad of arterial pH, P_{CO_2} and $[HCO_3^-]$ values must be verified as compatible. As listed in Table 11–2, one may use either the Henderson equation, the Henderson-Hasselbalch equation, or the nomogram of Fig 11–1.

SIMPLE VERSUS COMPLEX ACID-BASE DISORDERS

Identify Primary Disturbance

Simple acid-base disorders are identified by the direction of the pH change, since compensatory responses to primary derangements are incomplete. Thus, acidemia indicates acidosis and alkalemia indicates alkalosis. The change in bicarbonate concentration and P_{CO_2}—which directionally parallel each other in simple disorders—then gives the necessary subsequent information. For instance, acidemia with a low bicarbonate concentration and P_{CO_2} is metabolic acidosis; acidemia with a high bicarbonate concentration and P_{CO_2} is chronic respiratory acidosis.

Note that the bicarbonate concentration or P_{CO_2} alone is incomplete information for identification of an acid-base disturbance. A low bicarbonate concentration may represent metabolic acidosis *or* chronic respiratory alkalosis and a high bicarbonate concentration metabolic alkalosis *or* chronic respiratory acidosis.

Use of Acid-Base Nomogram & Calculation of Plasma Anion Gap

Plotting the acid-base values on the nomogram of Fig 11–1 quickly identifies whether an acid-base disturbance is appropriately compensated and thus simple. The plasma anion gap is also helpful in that it is elevated in some metabolic acidoses (Table 11–1) and is expected to be modestly raised in metabolic alkalosis but should be normal in all other disturbances.

The presence of an elevated anion gap signifies that a metabolic acidosis is present even if the arterial pH is normal or high. For instance, consider an alcoholic patient who developed metabolic alkalosis due to vomiting, with arterial pH 7.55, bicarbonate concentration 44 mmol/L, and P_{CO_2} 55 mm Hg. If alcoholic ketoacidosis with β-hydroxybutyric acid addition of 20 mmol/L subsequently developed and was superimposed, the bicarbonate concentration would fall to normal (24 meq/L). The P_{CO_2} would also be normal (40 mm Hg)—no longer elevated to compensate during the metabolic alkalemia—so the pH would then be normal (7.40). The clue to the complex acid-base disturbance, the combined metabolic alkalosis plus metabolic acidosis, would be the abnormally high anion gap of 32 meq/L (initial 12 meq/L plus 20 meq/L β-hydroxybutyrate).

Metabolic Acidosis

12

INTRODUCTION

DIAGNOSTIC STRATEGY

As with all acid-base disorders, arterial blood gases and serum electrolytes are essential starting points in the diagnostic approach to metabolic acidosis.

Hypocarbonatemia, Acidemia, & Appropriate Hypocapnia

Diagnosis of metabolic acidosis depends first on ascertaining that a disorder with a low plasma bicarbonate concentration is associated with acidemia. As shown in Fig 12–1, arterial pH easily distinguishes between chronic respiratory alkalosis and metabolic acidosis.

The arterial blood gases are also needed to ensure in a simple disorder that the P_{CO_2} is appropriately reduced (Table 10–1 and Fig 11-1). If enough time has elapsed for the full ventilatory response and P_{CO_2} is found to be lower or higher than expected for the bicarbonate concentration, there is a superimposed complex or mixed acid-base disorder—respiratory alkalosis or acidosis, respectively.

Anion Gap Normal or Elevated

Once a metabolic acidosis is diagnosed, calculation of the anion gap neatly separates two groups of metabolic acidoses, the relatively acute normochloremic organic acidoses from the relatively chronic hyperchloremic acidoses associated with diarrhea or renal acidification defects (Chapter 11).

RENAL ADAPTATION

As summarized in Fig 12–2 and previously shown in Figs 10–11 and 10–12, the kidney responds to acidemia and low filtered bicarbonate load by increasing hydrogen ion secretory capacity and augmenting ammoniagenesis. Avid, complete reabsorption of all filtered bicarbonate and enhanced urinary net acid excretion, usually predominantly in the form of ammonium but sometimes as titratable acid, generates new bicarbonate to replace that lost from extracellular stores.

HYPERCHLOREMIC METABOLIC ACIDOSES

OVERVIEW

A hyperchloremic metabolic acidosis may derive from (1) rapid dilution of the extracellular buffer stores with solutions devoid of bicarbonate; (2) rapid addition to the body of hydrochloric acid or its metabolic equivalent, which overwhelms the renal capacity to eliminate acid; or (3) an inherent abnormality in renal acidification mechanisms.

As shown in Table 12–1, in dilutional acidosis, acidosis due to an excessive hydrochloric acid or organic acid load, or acidosis due to gastrointestinal bicarbonate loss, renal acidification mechanisms are intact and the acidemia elicits a normal increase in urinary net acid excretion. In contrast, when one or more of the acidification mechanisms by the renal tubules are primarily disordered and cannot even cope with normal (not increased) dietary and metabolic acid addition to the body, the disorder is called a **renal tubular acidosis (RTA).**

In practice, the gastrointestinal loss of bicarbonate and the RTAs are the most common causes of hyperchloremic metabolic acidosis. A complete list is presented in Table 12–2.

DILUTIONAL ACIDOSIS

In the resuscitation of effective circulating volume during some hemodynamic emergencies (eg, sepsis, burns, or trauma), it is sometimes necessary to administer a great quantity of volume

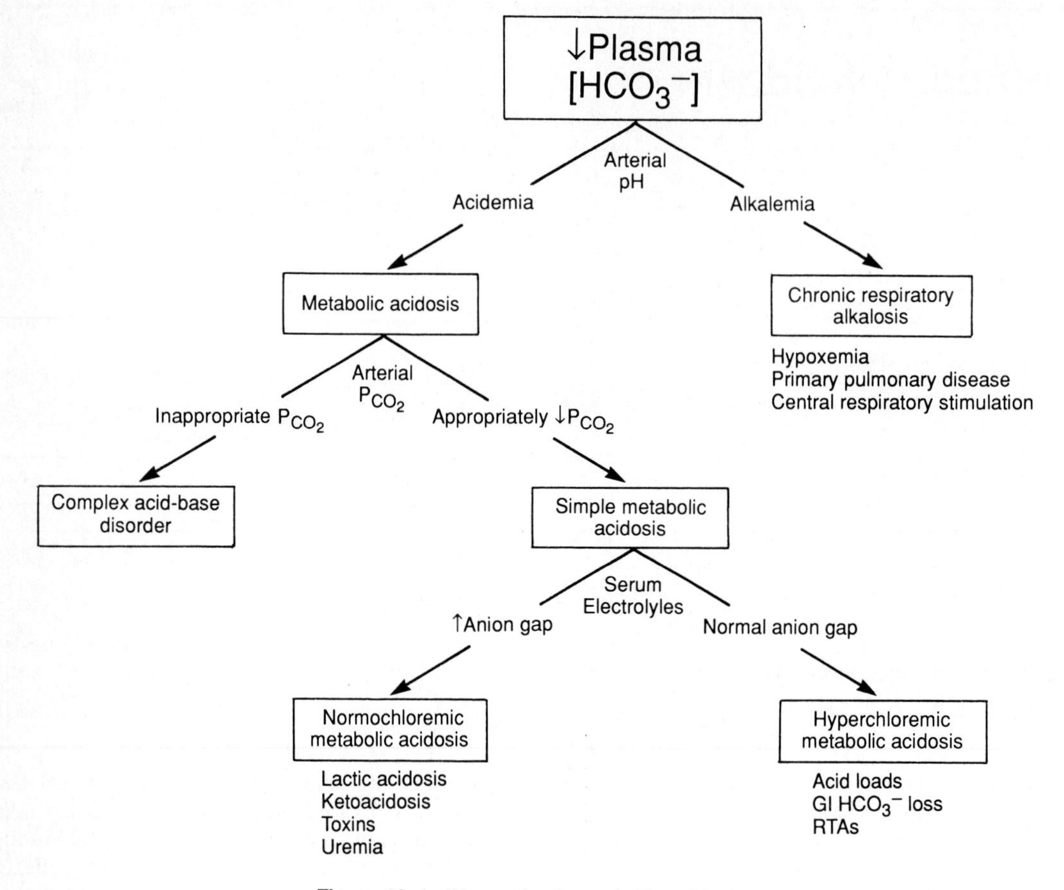

Figure 12-1. Diagnosis of metabolic acidosis.

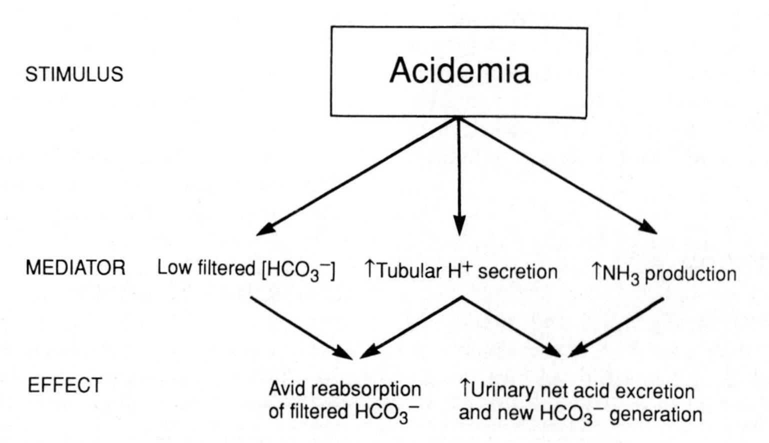

Figure 12-2. Pathophysiology of metabolic acidosis.

Table 12-1. Renal response in hyperchloremic metabolic acidoses.

	Dilution Acidosis	Acid Load: HCl	Acid Load: Organic Acidosis	Gastro-intestinal HCO_3^- Loss	Renal Tubular Acidosis (RTA)
Urinary net acid excretion	↑	↑	↑	↑	Normal or ↓

intravenously. When the administered solution is devoid of buffer, such as normal saline, existing bicarbonate in the extracellular volume is diluted. As a result of shift of potassium out of cells and hydrogen ion into cells, the degree of bicarbonate dilution is less than would be calculated from the extracellular bicarbonate pool and the amount of administered volume. Nevertheless, massive saline administration (> 10 L) predictably results in metabolic acidosis. Dilutional acidosis is mild, and plasma bicarbonate concentration usually does not fall below 15–20 meq/L. Correction of the acidosis occurs by renal mechanisms, with normal augmentation of urinary net acid excretion over the course of days, or by resolution of the primary disorder with natriuresis and diuresis to renormalize the extracellular volume and plasma bicarbonate concentration.

Table 12-2. Causes of hyperchloremic, normal anion gap metabolic acidosis.

Dilution
Acid load
- Ammonium chloride
- Arginine hydrochloride
- Lysine hydrochloride
- Methionine sulfate
- Hyperalimentation
- Sulfur
- Ketoacidosis with urinary ketone loss and renal chloride retention

Bircarbonate loss
- Gastrointestinal
  - Diarrhea
  - Pancreatic
  - Ureterosigmoidostomy
  - Drugs
    - Calcium chloride
    - Magnesium sulfate
    - Cholestyramine
- Renal
  - Posthypocapnia
  - Post-carbonic anhydrase inhibition

Defects in renal acidification (RTA)
- Type I, proximal RTA: Decreased bicarbonate reclamation
  - Selective (unassociated with Fanconi syndrome)
  - Generalized (associated with Fanconi syndrome)
- Type II, classical distal RTA: Selective distal acidifying defect
- Type III, reduced buffer delivery distal RTA
  - Glomerular insufficiency
  - Severe phosphate restriction
- Type IV, generalized distal RTA: Combined Na^+, K^+, and H^+ transport defect
  - Primary hypoaldosteronism
  - Hyporeninemic hypoaldosteronism
  - Hypoangiotensinemia
    - Converting enzyme inhibition
  - Aldosterone-resistant states
    - Congenital
    - Tubulointerstitial diseases

ACID LOAD

Administration of HCl or Metabolic Equivalent

Rapid administration of a large quantity of hydrochloric acid or its metabolic equivalent, such as ammonium chloride or calcium chloride, causes acidosis by overwhelming the renal mechanisms responsible for acid elimination. In practice, this form of acidosis may occur during administration of hyperalimentation solution, since some cationic and other amino acids are acid equivalents following their metabolism in the body.

Organic Acidosis

During the induction of an organic acidosis such as ketoacidosis, the hydrogen ion titrates the bicarbonate while the conjugate anion (eg, B^- representing β-hydroxybutyrate) remains as the sodium salt.

$$H^+B^- + NaHCO_3 \rightleftharpoons CO_2 + NaB$$

At this stage, there is metabolic acidosis (due to the loss of bicarbonate) of the high anion gap variety (due to retained B^- in the plasma). However, if the organic acid production is not extreme and GFR is reasonably normal, filtration and excretion of NaB may be brisk and unaccompanied by reabsorption. Although the kidney is poor at reabsorbing most organic anions, including ketones, it can efficiently compensate by reabsorbing chloride, especially since some degree of extracellular volume contraction usually exists concurrently. Sodium chloride reabsorption is therefore enhanced. Thus, B^- is lost into the urine as NaB—or as KB or NH_4B depending on collecting tubule transport processes—and NaCl is retained. The acidosis is thereby converted to a hyperchloremic metabolic acidosis (ie, bicarbonate is still low and the plasma B^- has been replaced by Cl^-).

This process of renal excretion of conjugate base with chloride retention also occurs frequently in the recovery phase of diabetic ketoacidosis. As the patient—who has ketoacidosis and hypovolemia—is treated with saline, GFR rises toward normal, resulting in ketonuria while there is still avid renal retention of chloride. While bicarbonate remains low, there is a resulting exchange of chloride for keto anion in the composition of the plasma. Insulin is of course simultaneously administered, but the renal excretion of ketones can be faster than their insulin-mediated metabolic conversion to bicarbonate. Thus, by normalization of GFR, a high anion gap, normochloremic metabolic acidosis is converted to a normal anion gap, hyperchloremic metabolic acidosis.

GASTROINTESTINAL BICARBONATE LOSS

As discussed in Chapter 1 and reviewed in Fig 12–3, the small intestine—especially the ileum—alkalinizes the initially acidic luminal contents

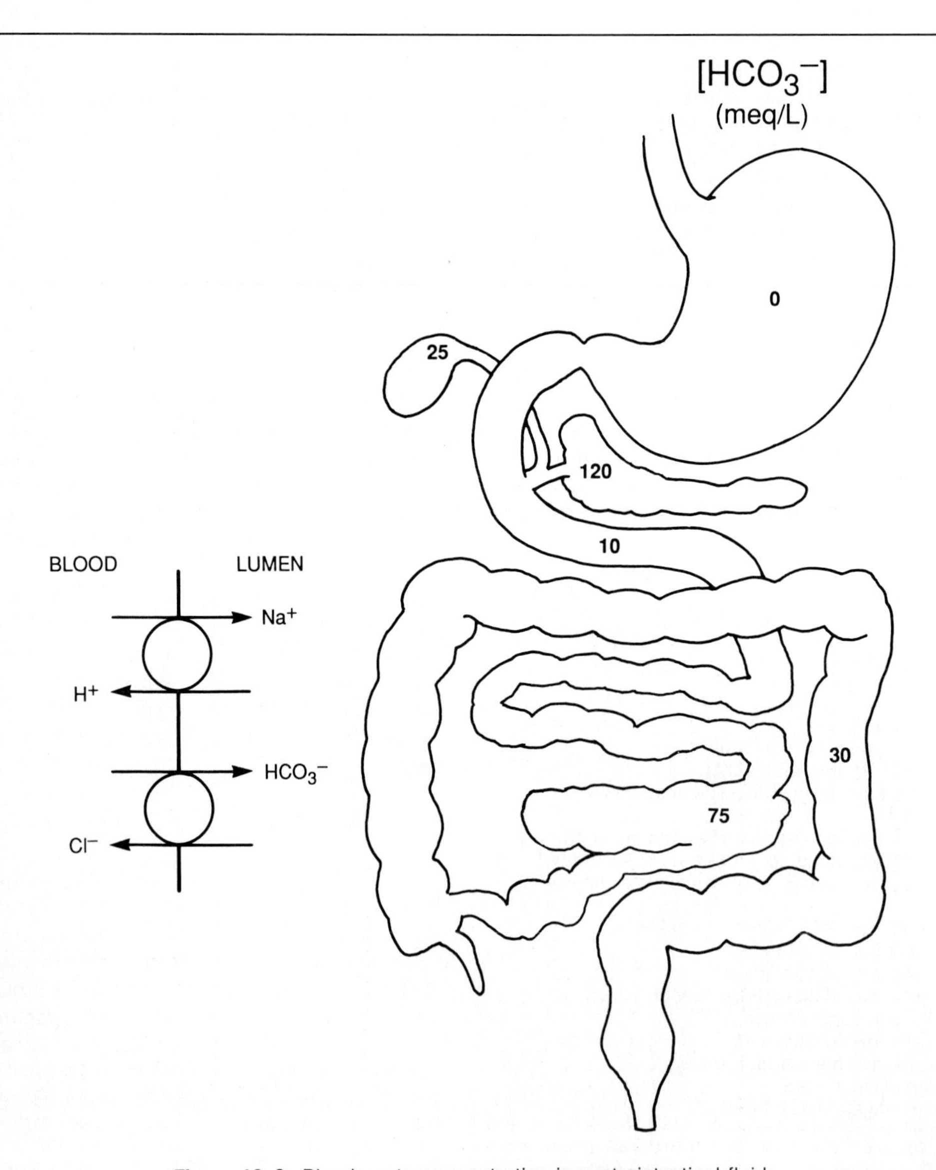

Figure 12-3. Bicarbonate concentration in gastrointestinal fluids.

emerging from the stomach. Sodium bicarbonate is added to the lumen, and hydrochloric acid is simultaneously added to the blood by means of parallel Na^+/H^+ and Cl^-/HCO_3^- exchangers, shown on the left in Fig 12–3. During secretory diarrhea, there is excessive operation of these transporters, with the result that a large amount of sodium bicarbonate is added to the diarrheal fluid and an equal amount of hydrochloric acid to the extracellular volume, thus causing a hyperchloremic metabolic acidosis.

As also shown in Fig 12–3, another overtly alkaline intestinal fluid is that of the pancreas. External pancreatic drainage may also result in hyperchloremic metabolic acidosis.

In cases of gastrointestinal sodium bicarbonate loss, electrolyte and volume losses lead to extracellular volume depletion and stimulation of renin and aldosterone. Potassium losses occur both in diarrheal fluid and in the urine, owing to hyperaldosteronism (see Chapter 7).

The renal response to the acidemia is to increase hydrogen ion secretion and ammonia production (due to acidemia as well as to hypokalemia). Urinary net acid excretion increases markedly (Fig 12–2), with a tendency for urinary pH to rise, sometimes as high as 6.0 consequent to the large increase in buffer (ammonia) delivery (Fig 10–12). Another index of the high rate of urinary ammonium excretion is the highly negative value of the urinary anion gap, as discussed below (Fig 12–11).

RENAL BICARBONATE LOSS

Posthypocapnia

Hypocapania inhibits bicarbonate reabsorption in the proximal tubule (Fig 10—8) and enhances urinary bicarbonate excretion so that plasma bicarbonate concentration falls (Fig 10–17). Should P_{CO_2} subsequently return to normal with resolution of the ventilatory stimulus, plasma bicarbonate concentration initially remains subnormal while the arterial pH converts from alkalemic to acidemic. At this point, the individual has a hyperchloremic metabolic acidosis. The situation will be corrected over the course of 3–5 days (Fig 10–12) as the kidneys generate bicarbonate to replace that lost during the hypocapnic stage.

Post-Carbonic Anhydrase Inhibition

Proximal bicarbonate reabsorption can also be inhibited by a carbonic anhydrase inhibitor, with induction of type I proximal renal tubular acidosis (see below). When such an agent is discontinued, a hyperchloremic metabolic acidosis is present, now with normal proximal tubule function. Normalization of acid-base status by renal mechanisms eventually occurs as in the posthypocapnic situation.

RENAL TUBULAR ACIDOSIS

In each renal tubular acidosis (RTA), the acidification process (hydrogen ion secretion or buffer delivery) is inherently defective (Table 12–1). There is abnormal reclamation of filtered bicarbonate or insufficient new bicarbonate generation due to reduced urinary net acid excretion. Metabolic acidosis ensues, since the kidney is unable to cope with a normal rate of acid production from diet, metabolism, and stool base loss.

Overview

There are four major RTAs, depending on the site and kind of acidification defect: one proximal type and three distal (collecting tubule) types.

As shown in Fig 12–4, a disorder of acidification in the proximal tubule is termed type I (defective bicarbonate reabsorption) proximal RTA; a disorder in which there is a selective defect in hydrogen ion pumping in the distal nephron (cortical or medullary collecting tubule) is known as type II (selective H^+ secretory abnormality) classical distal RTA; a disorder in which intrinsic collecting tubule H^+ pumping is intact but ammonia delivery is disordered in the course of renal disease is known as type III (buffer deficiency) distal RTA; and a disorder in which there is global transport dysfunction in the collecting tubule, including sodium reabsorption, potassium secretion, and hydrogen ion secretion, is known as type IV (combined Na^+, K^+, and H^+ transport defect) generalized distal RTA.

The four kinds of RTA can be distinguished by several accompanying characteristics, summarized in Table 12–3. These features include GFR, the serum potassium concentration, proximal H^+ secretion and the response to bicarbonate loading, distal H^+ secretion and the urinary pH and net acid excretion during acidosis (urinary ammonium is estimated by the urine anion gap, as discussed below), and the response to therapy.

Type I (Defective Bicarbonate Reabsorption): Proximal RTA

A defect in acidification in the proximal tubule necessarily impairs the ability of the kidney to reabsorb the normal quantity of filtered bicarbonate. Usually, the hydrogen ion secretory defect occurs in association with inability to reabsorb other sodium-cotransported solutes in the proximal tubule, including glucose, amino acids, phosphate, urate, and low-molecular-weight proteins, a constellation of defects known as **Fanconi's syn-**

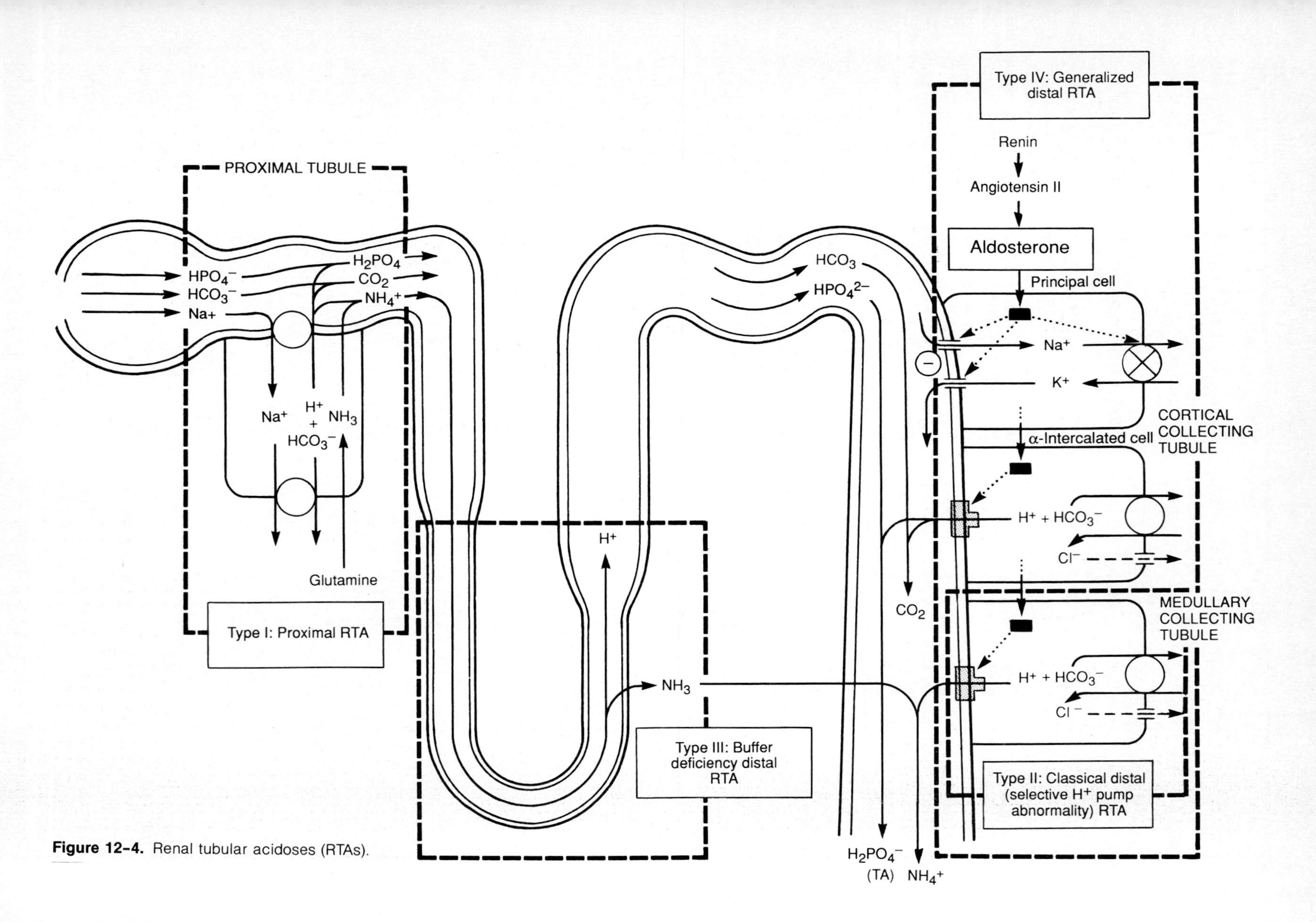

Figure 12-4. Renal tubular acidoses (RTAs).

Table 12-3. Hyperchloremic, normal anion gap metabolic acidoses.

	Renal Defect	GFR	Serum [K^+]	Proximal H+ Secretion[1]	Distal H^+ Secretion: Minimal Urine pH	Distal H^+ Secretion: Urinary NH_4^+ + Titratable Acid	Urinary Anion Gap	Treatment
Gastrointestinal HCO_3^- loss	None	↓	↓	Normal	±< 5.5	↑↑	Negative	Na^+, K^+ and HCO_3^- as required
Renal tubular acidosis								
I. Proximal	Proximal H^+ secretion	Normal	↓	↓[2]	< 5.5	Normal	Zero or positive	$NaHCO_3$ or $KHCO_3$ (10–15 meq/kg/d), thiazide
II. Classical distal:	Distal H^+ secretion	Normal	↓	Normal	> 5.5	↓	Zero or positive	$NaHCO_3$ (1-3 meq/kg/d)
III. Buffer deficiency distal	Distal NH_3 delivery	↓	Normal	Normal	< 5.5	↓	Zero or positive	$NaHCO_3$ (1-3 meq/kg/d)
IV. Generalized distal	Distal Na^+ reabsorption, K^+ secretion, and H+ secretion	↓	↑	Normal	< 5.5	↓	Zero or positive	Fludrocortisone (0.1–0.5 mg/d), dietary K^+ restriction, furosemide (40–160 mg/d), $NaHCO_3$ (1–3 meq/kg/d)

[1]HCO_3^- reabsorption during HCO_3^- loading.
[2]Fractional excretion of bicarbonate > 15% during bicarbonate loading; usually associated with Fanconi's syndrome.

drome. Thus, proximal RTA may exist in isolation but more usually occurs with glycosuria, aminoaciduria, phosphaturia, uricosuria, and tubular (low-molecular-weight and nonalbumin) proteinuria. Excess delivery of sodium bicarbonate out of the proximal tubule enhances potassium secretion by the principal cell in the cortical collecting tubule and results in hypokalemia (Chapter 8).

A normal plasma bicarbonate concentration and filtered bicarbonate load cannot be sustained owing to the defect in bicarbonate reabsorption in the proximal tubule. However, as shown in Fig 12–5, during metabolic acidosis with reduced plasma bicarbonate concentration (eg, 15 meq/L) and filtered bicarbonate load, the residual level of proximal bicarbonate reabsorptive capacity (eg, 60% instead of the normal 85%) allows enough bicarbonate reabsorption to prevent excess bicarbonate delivery out of the proximal tubule. The collecting tubule "sees" a normal bicarbonate load emerging from the proximal tubule and can lower urine pH normally. Urinary net acid excretion is then "normal"—equivalent to acid production, so that acid-base steady-state persists—but is actually physiologically inappropriate because there is sustained, uncorrected acidemia.

As shown in Fig 12–5, if the plasma bicarbonate concentration is progressively normalized by exogenous bicarbonate infusion or ingestion, the limited proximal bicarbonate reabsorptive capacity becomes apparent with massive bicarbonaturia. Fractional excretion (excretion divided by filtered load) of bicarbonate during this period of normal filtered bicarbonate load is greater than 15%, by definition signifying a reabsorptive defect in the proximal tubule (Table 12-3).

The clinical presentation of this hypokalemic, hyperchloremic metabolic acidosis often includes the full Fanconi syndrome with bone disease, especially osteomalacia and rickets in children and osteopenia in adults. GFR at least in the early stages may be normal. Nephrolithiasis is unusual, perhaps because urinary citrate excretion remains normal.

This proximal form of RTA is relatively rare. As shown in Fig 12–6, it is potentially caused by

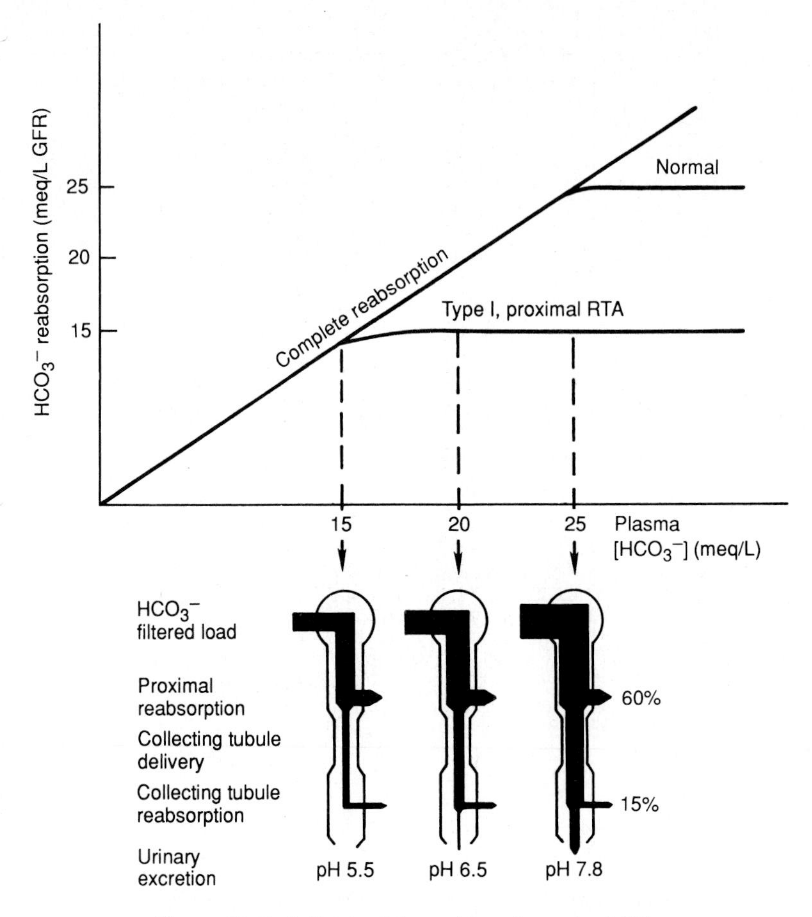

Figure 12-5. Pathophysiology of type I, proximal RTA.

(1) a defect in the luminal Na^+-H^+antiporter; (2) a defect in the basolateral $Na(HCO_3)_3$ cotransporter; (3) inhibition or deficiency of either luminal or intracellular carbonic anhydrase; (4) a defect in the cell sodium concentration due to increased cell sodium permeability or impaired energy supply to or functional activity of the Na^+/K^+-ATPase; or (5) an imbalance in the hormonal regulation of intracellular cAMP concentration.

Clinical causes of proximal RTA are shown in Table 12–4 and include primary, drug-induced (carbonic anhydrase inhibitors), genetic (eg, cystinosis), dysproteinemic (multiple myeloma), toxic (heavy metals), and some secondary hyperparathyroid/vitamin D-deficient states.

Treatment requires a large amount of bicarbonate, often more than 10–15 meq/kg per day to compensate for the massive bicarbonaturia that occurs when plasma bicarbonate concentration is normalized (Fig 12-5). The alkali is often better tolerated as citrate, which is converted by the body to bicarbonate, and as a mixed sodium and potassium salt of citrate. Potassium is used in therapy because potassium wastage increases as plasma bicarbonate concentration returns toward normal (Chapter 8). Thiazide administration is often also helpful in contracting the extracellular volume, to reduce GFR and filtered bicarbonate load and thereby lessening the bicarbonate reabsorptive burden (12–5).

Type II (Selective H^+ Secretory Defect): Classical Distal RTA

A selective defect in hydrogen ion secretion by the α-intercalated cell of the cortical or medullary collecting tubules impairs urinary net acid excretion. As summarized in Table 12–3, owing to the defect in hydrogen ion pumping capacity, urine pH cannot be lowered properly. Urine pH is always greater than 5.5, even in the face of acidemia. The inability to form a lumen-to-blood pH gradient impairs the collecting tubule's ability to reabsorb the residual bicarbonate escaping proximal reabsorption, to fully titrate filtered phosphate and form titratable acid, and to capture ammonia for the urinary excretion of ammonium.

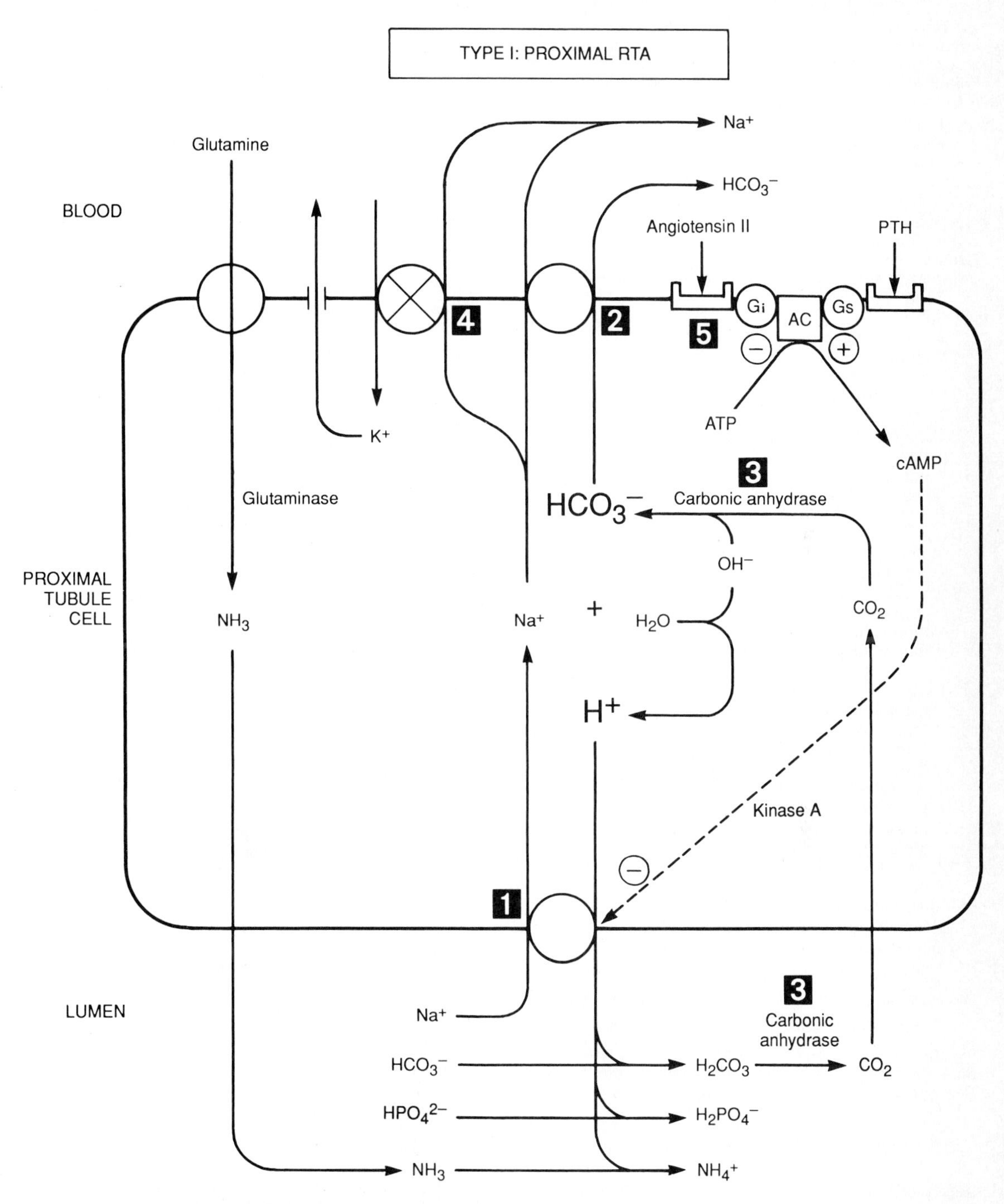

Figure 12-6. Possible cellular mechanisms responsible for type I, proximal RTA.

The urine anion gap, as a reflection of urine ammonium, is therefore zero or positive (see below).

The proximal tubule is normal, so that if plasma bicarbonate concentration and filtered bicarbonate load are normalized by bicarbonate administration, no significant bicarbonaturia results, and Fanconi's syndrome is absent. Potassium secretion is actually enhanced, perhaps because the electrochemical gradient generated by sodium transport in the cortical collecting tubule is no longer "used" by hydrogen ion secretion and is then preferentially available to stimulate potassium secretion. Abnormal calcium metabolism with hypercalciuria, nephrocalcinosis, and

Table 12-4. Disorders with defective bicarbonate reclamation—Type I proximal renal tubular acidosis.[1]

- **Selective (unassociated with Fanconi's syndrome)**
  - Primary
    - Transient (infants)
    - Idiopathic or genetic
  - Carbonic anhydrase II deficiency
  - Drugs
    - Acetazolamide
    - Sulfanilamide
    - Mafenide acetate
- **Generalized (associated with Fanconi's syndrome)**
  - Primary (without associated systemic disease)
    - Genetic
    - Sporadic
  - Genetically transmitted systemic diseases
    - Cystinosis
    - Lowe's syndrome
    - Wilson's disease
    - Tyrosinemia
    - Galactosemia
    - Hereditary fructose intolerance (during fructose ingestion)
    - Metachromatic leukodystrophy
    - Pyruvate carboxylase deficiency
  - Dysproteinemic states
    - Multiple myeloma
    - Monoclonal gammopathy
  - Secondary hyperparathyroidism with chronic hypocalcemia
    - Vitamin D deficiency or resistance
    - Vitamin D dependency
  - Drug or toxic nephropathy
    - Drugs
      - Outdated tetracycline
      - Tricromyl
      - Streptozocin
      - Glue
      - Gentamicin
    - Heavy metals
      - Lead
      - Mercury
  - Tubulointerstitial diseases
    - Sjögren's syndrome
    - Medullary cystic disease
    - Renal transplantation
  - Other renal and miscellaneous diseases
    - Nephrotic syndrome
    - Amyloidosis
    - Osteopetrosis
    - Paroxysmal nocturnal hemoglobinuria

[1]Adapted from Cogan MG, Rector FC Jr: Acid-base disorders. Pages 457–517 in: *The Kidney,* 3rd ed. Brenner BM, Rector FC Jr (editors). Saunders, 1986.

nephrolithiasis (due in part to diminished citrate excretion) are prominent features—either as a cause or a result—of this form of distal RTA.

The clinical presentation of type II classical distal RTA is therefore that of a hypokalemic, hyperchloremic metabolic acidosis with inappropriately high urinary pH (> 5.5) and low urinary net acid excretion in the setting of acidemia.

The cellular defect in collecting tubule acidification may be, as shown in Fig 12–7, (1) a deficiency or abnormality in the luminal proton-translocating ATPase; (2) an abnormality in the basolateral HCO_3^-/Cl^- exchanger; or (3) an increase in the luminal membrane hydrogen ion permeability. The last mechanism disallows development of a pH gradient, since the secreted hydrogen ion can back-leak—for example, as induced by the antibiotic amphotericin B (which intercalates in the luminal membrane and acts as a protonophore).

Causes of this selective form of distal RTA are listed in Table 12–5 and include genetic forms, au-

Table 12-5. Disorders with selective defect in net acid excretion—Type II classical distal renal tubular acidosis.[1]

- **Primary (without associated systemic disease)**
  - Genetic
  - Idiopathic
- **Genetically transmitted systemic diseases**
  - Ehlers-Danlos syndrome
  - Hematologic disorders
    - Hereditary elliptocytosis
    - Sickle cell anemia
    - Carbonic anhydrase I deficiency or alteration
  - Medullary cystic disease
  - With nerve deafness
  - Glycogenosis type III
- **Autoimmune diseases**
  - Hypergammaglobulinemia
    - Hyperglobulinemic purpura
    - Cryoglobulinemia
    - Familial
  - Sjögren's syndrome
  - Thyroiditis
  - Pulmonary fibrosis
  - Chronic active hepatitis
  - Primary biliary cirrhosis
  - Systemic lupus erythematosus
- **Diseases associated with nephrocalcinosis**
  - Primary hyperparathyroidism
  - Vitamin D intoxication
  - Hyperthyroidism
  - Idiopathic hypercalciuria
    - Hereditary
    - Idiopathic
  - Hereditary fructose intolerance (after chronic fructose ingestion)
  - Medullary sponge kidney
  - Fabry's disease
  - Wilson's disease
- **Drug or toxic nephropathies**
  - Amphotericin B
  - Toluene
  - Glue
  - Analgesics
  - Cyclamate
  - Balkan nephropathy
- **Tubulointerstitial diseases**
  - Chronic pyelonephritis secondary to urolithiasis
  - Obstructive uropathy
  - Renal transplantation
  - Leprosy
  - Hyperoxaluria
- **Miscellaneous**
  - Hepatic cirrhosis
  - Empty sella syndrome

[1]Adapted from Cogan MG, Rector FC Jr: Acid-base disorders. Pages 457–517 in: *The Kidney,* 3rd ed. Brenner BM, Rector FC Jr (editors). Saunders, 1986.

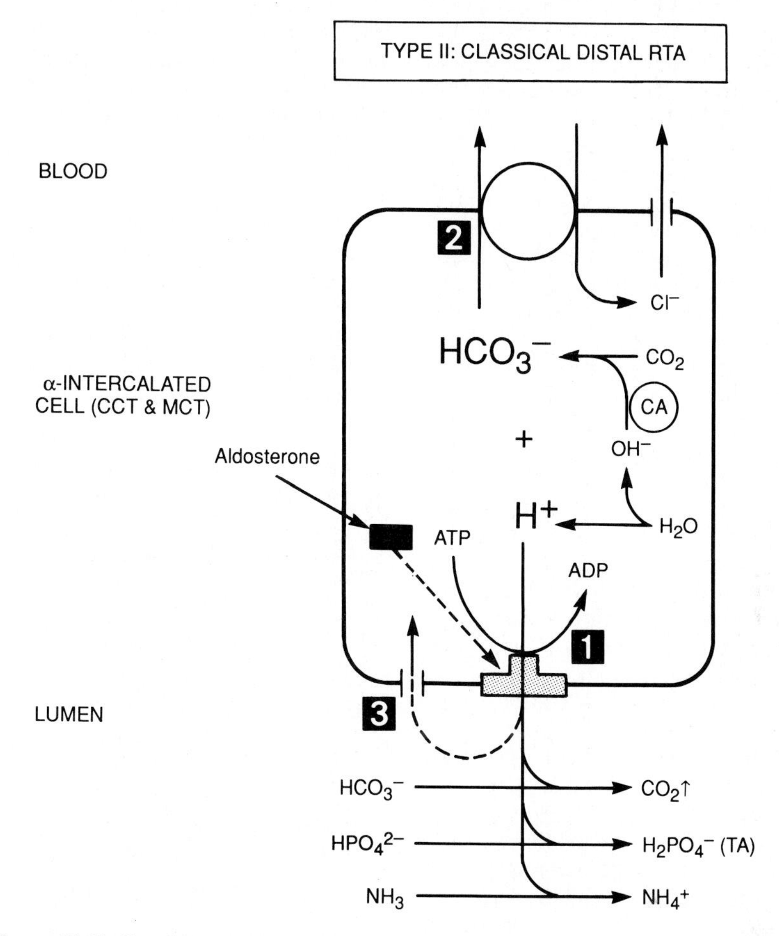

Figure 12-7. Possible cellular mechanisms responsible for type II, classical distal RTA.

toimmune hypergammaglobulinemic diseases, disorders of calcium metabolism, drugs (especially amphotericin B), and various tubulointerstitial diseases.

Treatment consists of remedying the underlying cause of the RTA if possible. In lieu of this, sodium bicarbonate is used to compensate for the contribution the collecting tubule normally makes to urinary net acid excretion and bicarbonate regeneration—about 1-3 meq/kg per day (Table 12-3). Bicarbonate therapy reduces potassium wastage and allows for normal (even catch-up) growth in children and for prevention of further nephrolithiasis, nephrocalcinosis, and bone disease in adults.

Type III: Deficient Buffer Delivery Distal RTA

As shown in Fig 12-4, the transfer of ammonia into the medullary interstitium is critical for subsequent trapping as ammonium in the luminal fluid of the collecting tubule. The liberation of ammonia into the medullary interstitium is in turn dependent on alkalinization of the fluid in the descending limb of Henle as a result of operation of the countercurrent system and water abstraction. As nephrons are lost in the course of any chronic renal disease, the countercurrent system and ability to concentrate the urine become defective (Fig 4-19). Ammonia may be produced normally by the residual surviving nephrons but—because of lack of alkalinization of loop fluid—is not properly delivered to and concentrated in the renal medullary interstitium. Instead, ammonium is delivered into the cortical portion of the ascending limb of Henle and is there reabsorbed and washed away by the high renal cortical blood flow.

This disorder—in which renal medullary ammonia tension is reduced as a result of defective countercurrent function of the loop of Henle—is found in states of **glomerular insufficiency,** when GFR has fallen to 20-30 mL/min. When GFR decreases further, ammonia transport remains abnormal but the normochloremic, high anion gap form of uremic acidosis supervenes (see below).

When GFR is 20-30 mL/min, the ability of the

hydrogen ion secretory system in the cortical collecting tubule to lower luminal fluid and urine pH is normal, so that urinary pH is less than 5.5 during acidemia (Table 12-3). Proximal tubule transport is also reasonably normal. The ability to reabsorb sodium in the cortical collecting tubule is normal or only mildly impaired, and potassium homeostasis is preserved (see Chapter 7), in contrast to the patient with type IV generalized distal RTA (see below). The patient therefore has a normokalemic, hyperchloremic (non-anion gap) metabolic acidosis. Renal osteodystrophy and abnormal vitamin D and PTH metabolism are frequent concomitants of chronically diminished renal function, in part due to acidemia.

A similar buffer deficiency state involving phosphate rather than ammonia delivery can occur with severe hypophosphatemia. With severe phosphate restriction, titratable acid excretion rather than ammonium excretion in the urine is subnormal. Similar to the case in which GFR is reduced, urinary net acid excretion is less than acid input from diet and metabolism, so that metabolic acidosis ensues.

Treatment of type III distal RTA consists simply of sodium bicarbonate therapy sufficient to replace the usual daily urinary net acid excretion, about 1–3 meq/kg per day (Table 12-3). If too much sodium bicarbonate is inadvertently given, the kidney is able to excrete the excess (in contrast to a sodium chloride load, which tends to be retained). Correction of acidemia slows the progress of the bone disease associated with chronic renal insufficiency.

Type IV (Combined Na^+, K^+, & H^+ Transport Defects): Generalized Distal RTA

Defective aldosterone function has both direct and indirect effects on hydrogen ion secretion in the cortical and medullary collecting tubules (Chapter 10). First, as shown in Fig 12-4, aldosterone directly regulates the hydrogen ion secretory rate of the proton-translocating ATPase pump of the α-intercalated cell. Second, it affects sodium transport in the adjacent principal cell of the cortical collecting tubule and hence the magnitude of the lumen-negative transepithelial potential difference that secondarily enhances hydrogen ion secretion. Third, aldosterone both directly and indirectly controls potassium secretion by the principal cell (see Chapter 7), which regulates plasma potassium concentration and body potassium stores and thus affects ammoniagenesis.

A defect in aldosterone concentration or action leads to a predictable triad of transport consequences in the collecting tubule with diminished sodium reabsorption and potassium and hydrogen ion secretion. There is also secondarily depressed ammonia production (due to hyperkalemia) in the proximal tubule, which contributes to the insufficiency of urinary net acid excretion. Thus, this type IV generalized distal RTA is attributable both to diminished hydrogen ion secretion and buffer delivery and therefore has characteristics of types II and III distal RTA.

As shown in Fig 12-8, insufficient aldosterone activity can occur as a result of disorders that affect adrenal production or release of aldosterone or the ability of aldosterone to act on its target organ, the cortical collecting tubule. Functional hypoaldosteronism may arise from (1) hyporeninism (diabetes mellitus, tubulointerstitial nephrop-

Table 12-6. Disorders with generalized dysfunction of distal nephron—Type IV generalized distal renal tubular acidosis.[1]

Primary mineralocorticoid deficiency
- Combined deficiency of aldosterone, deoxycorticosterone, and cortisol
  - Addison's disease
  - Bilateral adrenalectomy
  - Bilateral adrenal destruction
    - Hemorrhage or carcinoma
- Congenital enzymatic defects
  - 21-Hydroxylase deficiency
  - 3β-ol-Dehydrogenase deficiency
  - Desmolase deficiency
- Isolated aldosterone deficiency
  - Familial hypoaldosteronism
    - Corticosterone methyloxidase deficiency, types 1 and 2
  - Primary zona glomerulosa defect
  - Transient hypoaldosteronism of infancy
  - Chronic idiopathic hypoaldosteronism
  - Persistent hypotension
  - AIDS
  - Drugs
    - Converting enzyme inhibitors
      - Captopril
      - Enalapril
      - Lisinopril
    - Chronic heparin

Hyporeninemic hypoaldosteronism
- Diabetic nephropathy
- Tubulointerstitial nephropathies
- Nephrosclerosis
- Drugs
  - Nonsteroidal anti-inflammatory agents

Mineralocorticoid-resistant hyperkalemia
- Without salt wasting
- With salt wasting
  - Childhood forms
  - Adult forms with tubulointerstitial nephritis and renal insufficiency
    - Methicillin
    - Obstructive nephropathy
    - Transplantation
    - Sickle cell disease
  - Drugs
    - Spironolactone
    - Amiloride
    - Triamterene

[1]Adapted from Cogan MG, Rector FC Jr: Acid-base disorders. Pages 457–517 in: *The Kidney,* 3rd ed. Brenner BM, Rector FC Jr (editors). Saunders, 1986.

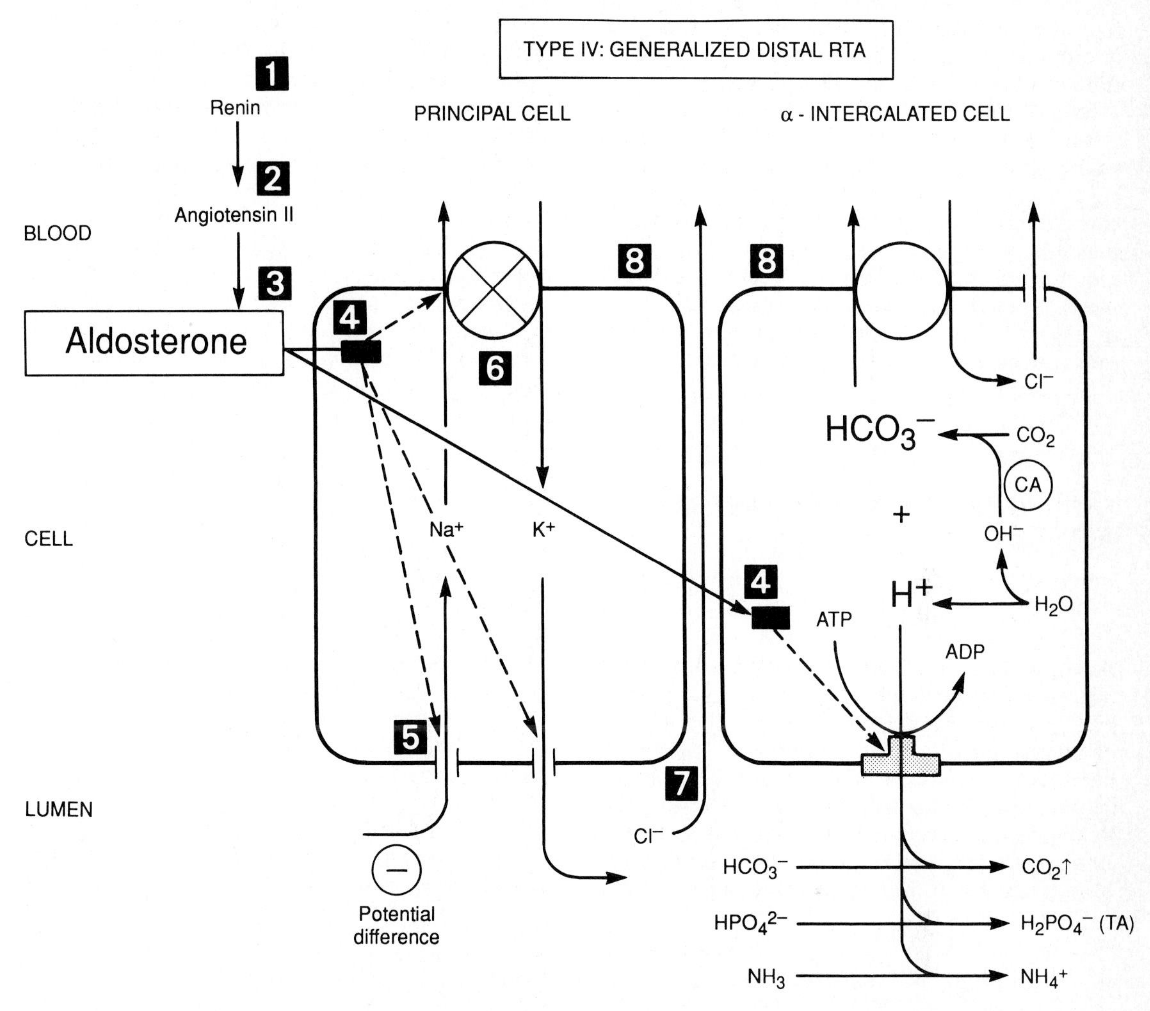

Figure 12-8. Possible cellular mechanisms responsible for type IV, generalized distal RTA.

athies, nephrosclerosis); (2) abnormal conversion of angiotensin I to angiotensin II (converting enzyme inhibition); (3) abnormal aldosterone and other mineralocorticoid synthesis (parenchymal or enzymatic adrenal diseases); (4) blockade of the mineralocorticoid receptor (spironolactone); (5) blockade of the sodium channel in the cortical collecting tubule (amiloride or triamterene); (6) Na^+/K^+-ATPase inhibition; (7) enhanced chloride permeability in the collecting tubule with shunting of the transepithelial potential difference; or (8) destruction of the target cell (tubulointerstitial nephropathies). A full list of entities associated with type IV generalized distal RTA is presented in Table 12-6.

In all of these cases of aldosterone deficiency, the coordinate transport abnormalities cause hyperkalemic hyperchloremic metabolic acidosis, usually with the syndrome of renal salt wasting. The clinical expression of the renal salt wasting depends on the severity of the defect and whether dietary sodium chloride intake is sufficient to compensate for the renal leak (see Chapter 2).

As summarized in Table 12-3, residual hydrogen ion secretory capacity is often sufficient in the setting of poor buffer (ammonia) delivery to lower the urine pH to an appropriate value (< 5.5) during acidemia. Individuals with hyporeninism or a tubulointerstitial nephropathy usually have a modest reduction in GFR due to their primary disease. Proximal bicarbonate reabsorption is normal.

Exclusion of primary adrenal disease is the first priority in diagnostically approaching this hyperkalemic metabolic acidosis, since prompt treatment with cortisol is essential if adrenal insufficiency is present. In other states of hypoaldosteronism—especially those due to hypereninism if hypertension is absent—mineralocorticoid supplementation daily with a large, supraphysiologic dose of fludrocortisone (0.1–0.5 mg/d) is appropriate because of the frequent presence of tubular unresponsiveness (Table 12–3).

Simple dietary potassium restriction, gastrointestinal potassium removal (potassium exchange resin), or facilitated renal excretion (furosemide, 40–160 mg/d) to selectively lower plasma potassium ameliorates the metabolic acidosis by enhancing ammoniagenesis and hence urinary ammonium excretion. Sodium bicarbonate (1–3 meq/kg per day) is sometimes necessary as a sole or adjunctive agent with furosemide to compensate for the depressed net acid excretion.

NORMOCHLOREMIC METABOLIC ACIDOSIS

As shown in Table 12–7, there are four major forms of normochloremic (high anion gap) metabolic acidoses. Except for the metabolic acidosis of uremia, they share some common features in that they generally represent relatively acute disorders associated with organic acid overproduction. Renal mechanisms for bicarbonate reabsorption and urinary net acid (especially ammonium) excretion in these disorders are intact (Fig 12–2) but insufficient to compensate for the excess acid load.

The diagnostic hallmark of each of these entities is that increased organic acid generation results in titration of plasma bicarbonate and simultaneous retention in the plasma of the sodium salt of the conjugate anion, causing elevation of the plasma anion gap (Chapter 11). Though not routinely measured by the serum electrolyte autoanalyzer, tests for each of these conjugate anions can be specially performed to provide the exact diagnosis when doubt exists about the cause of an abnormal plasma anion gap. The full list of normochloremic, high anion gap metabolic acidoses is given in Table 12–8.

LACTIC ACIDOSIS

Lactic acid is normally produced by muscle in great quantities—as much as 12–30 meq/kg per day, or 750–1500 meq/d. As shown in Fig 12–9, the lactic acid is quantitatively converted by the liver to glucose. This gluconeogenic-lactic acid generation coupling is known as the Cori cycle. Given this enormous daily output and metabolism of lactic acid, it can be readily seen that increased plasma lactic acid concentration could arise in either of two ways: overproduction or underutilization of lactic acid.

The normal metabolism of lactate is to pyruvate with the conversion of NAD^+ to NADH and liber-

Table 12–7. Normochloremic, high anion gap metabolic acidoses.

Type	Acid	Unmeasured Plasma Anion	Treatment[1]
Lactic acidosis	Lactic acid	Lactate	Treat underlying disorder.
Ketoacidosis Diabetic	β-Hydroxybutyric acid, acetoacetic acid	β-Hydroxybutyrate, acetoacetate	Insulin saline (K^-, phosphate, Mg^{2+} repletion)
Alcoholic			Glucose and saline (K^+, phosphate, Mg^{2+} repletion)
Starvation			Carbohydrate
Toxins[2] Salicylate	Salicylic acid	Salicylate	Alkaline diuresis, hemodialysis
Ethylene glycol	Oxalic acid	Oxalate	Ethanol, thiamine and pyridoxine, ECV expansion, hemodialysis
Methanol	Formic acid	Formate	Ethanol, ECV expansion, hemodialysis
Uremia	Organic acids	Organic anions	$NaHCO_3$ (1–3 mg/kg/d)

[1]$NaHCO_3$ for severe symptomatic acidemia.
[2]Also generate lactic acidosis and ketoacidosis.

Table 12-8. Causes of normochloremic, high anion gap metabolic acidosis.[1]

- **Lactic acidosis**
  - Overproduction
    - Severe exercise
    - Seizures
    - Sepsis
    - Leukemias
  - Underutilization
    - Type A: Poor tissue perfusion or oxygenation
      - Shock
        - Cardiogenic
        - Hemorrhagic
        - Septic
    - Type B: Various common disorders
        - Diabetes mellitus
        - Renal failure
        - Liver disease
        - Infection
      - Drugs or other toxic substances
        - Phenformin
        - Metformin
        - Ethanol
        - Methanol
        - Salicylates
        - Sorbitol
        - Xylitol
        - Dithiazinine iodide
        - Streptozocin
        - Isoniazid
        - Cyanide
        - Nitroprusside
      - Hereditary forms
        - Glucose-6-phosphate deficiency (type I glycogenosis)
        - Fructose-1,6-diphosphatase deficiency
        - Pyruvate carboxylase deficiency
        - Pyruvate dehydrogenase deficiency
        - Oxidative phosphorylation deficiencies
        - Methylmalonic aciduria
      - Miscellaneous
        - Ingestion of "lactic acid milk"
        - D-Lactic acidosis in short bowel syndrome
- **Ketoacidosis**
  - Diabetic
  - Ethanol intoxication
  - Starvation
  - Inborn errors of metabolism
- **Intoxications**
  - Salicylates
  - Ethylene glycol
  - Methanol
  - Paraldehyde
- **Uremia (late)**

[1]Adapted from Cogan MG, Rector FC Jr: Acid-base disorders. Pages 457–517 in: *The Kidney,* 3rd ed. Brenner BM, Rector FC Jr (editors). Saunders, 1986.

ation of a hydrogen ion. The normal lactate:pyruvate concentration in the cell is therefore proportionate to the redox ratio NADH:NAD$^+$. Conditions that primarily affect the redox state, the NADH:NAD$^+$ ratio, also affect lactate metabolism, and vice versa.

The causes of lactic acidosis are shown in Table 12–8 to be associated with lactic acid **overproduction** (severe exercise, seizures, leukemia) or **hepatic underutilization,** due to tissue hypoperfusion and hypoxia (type A) or hereditary diseases, drugs, toxins, or other disorders that interfere with normal hepatic performance (type B).

The basic approach to therapy of lactic acidosis is that the underlying condition should be corrected if at all possible. If disordered, restoration of hepatic perfusion and oxygenation is critical. Dichloroacetate has been recently promoted for the treatment of lactic acidosis to enhance lactic acid utilization. Though short-term improvement occurs, there is little change in the overall poor prognosis. This therapeutic experience emphasizes that lactic acidosis is a symptom of a serious underlying disorder and not a primary entity in itself.

Use of sodium bicarbonate in therapy of acidosis will be discussed below. In general, if the acidemia is severe (eg, arterial pH < 7.10) and there is symptomatic compromise of hemodynamic function, most physicians would advocate use of sodium bicarbonate. It should be recognized, however, that experimental evidence in animals suggests that exacerbation of intracellular acidosis and of lactic acid production with worsening of the acidemia may paradoxically result from bicarbonate administration in this setting.

KETOACIDOSIS

There are three common forms of ketoacidosis: diabetic, alcoholic, and starvation (Tables 12–7 and 12–8). Some very rare inborn errors of metabolism are also associated with ketoacidosis.

Diabetic

Absolute or relative insulin deficiency leads to hyperglycemia and increased production of the ketoacids β-hydroxybutyric and acetoacetic acids. The ratio of β-hydroxybutyrate to acetoacetate in the plasma is proportionate to the redox potential, the NADH:NAD$^+$ ratio. The nitroprusside reaction for ketone detection measures only the acetoacetate in the plasma and may be falsely negative if the redox state is markedly altered (ie, the β-hydroxybutyrate:acetoacetate ratio is very high).

Diabetic ketoacidosis is accompanied by glycosuria and consequent osmotic diuresis, extracellular volume contraction, hyperreninemic hyperaldosteronism, kaliuresis, magnesuria, and phosphaturia. Treatment therefore consists of insulin therapy, saline administration, and potassium, magnesium, and phosphate repletion.

Alcoholic

Discontinuation by a nondiabetic alcoholic individual of food intake with maintenance of alco-

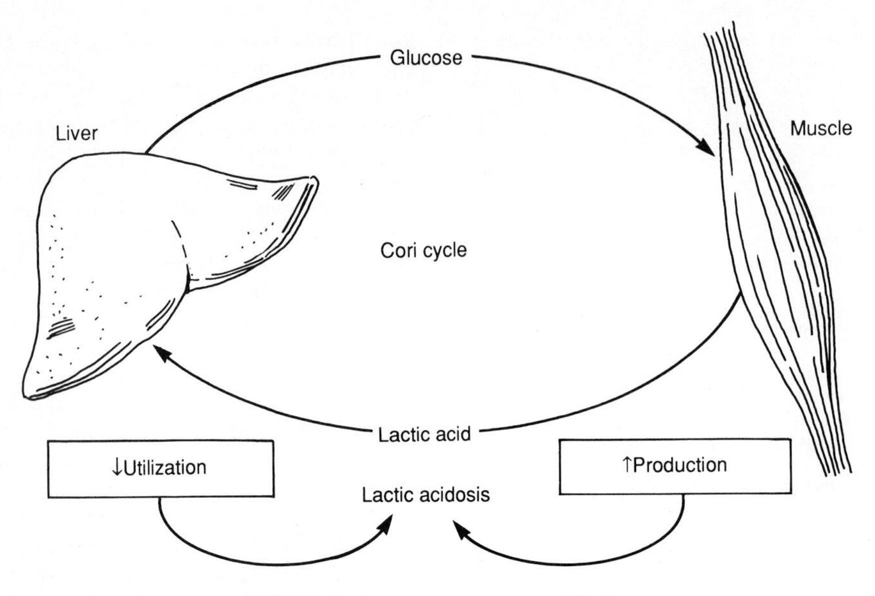

Figure 12-9. Causes of lactic acidosis.

hol consumption can result in ketoacidosis. In contrast to the diabetic form, glucose concentration is only moderately deranged if at all in alcoholic ketoacidosis. β-Hydroxybutyrate is the predominant plasma ketone present, and lactic acidosis frequently coexists. Therapy consists of glucose and sodium chloride administration as D_5-normal saline (but not insulin), with potassium, phosphate, and magnesium repletion as indicated.

Starvation

Mild ketoacidosis occurs within the first 24–48 hours of starvation or caloric deprivation. The ketoacidosis is associated with mild hypoinsulinemia. The severity of the acidosis is worsened by exercise or pregnancy. Because of the ketoacidosis, there is also renal wasting of sodium chloride (Chapter 2), potassium (Chapter 8), calcium (Chapter 16), phosphate (Chapter 19), and magnesium (Chapter 22). Carbohydrate refeeding corrects the acidemia, and other electrolytes should be repleted as needed.

TOXINS

Salicylic Acid

Aspirin overdose usually gives rise to a complex acid-base disturbance, most often respiratory alkalosis (due to primary ventilatory drive) and high anion gap metabolic acidosis. The latter is due only in part to salicylate. For instance, a toxic salicylate level of 100 mg/dL accounts for a rise in anion gap of only 7 meq/L. The rest of the elevation in anion gap is attributable to simultaneous induction of ketoacidosis and lactic acidosis, due to the metabolic consequences of salicylate within cells and the hyperventilation.

Therapy is afforded by alkaline diuresis to allow the relatively impermeant anionic form—salicylate—to be trapped in the tubular fluid and excreted into the urine. Alkaline diuresis may be effected by infusion of half-isotonic saline plus half-isotonic sodium bicarbonate (eg, half-normal saline plus two sodium bicarbonate ampules of 45 meq each per liter) at 250–500 mL/h. The extracellular volume must be expanded and the urine pH caused to rise to more than 7.5. Arterial pH should be monitored and not allowed to rise above 7.55. If the pH is over 7.55, acetazolamide (250–500 mg intravenously every 4–6 hours) should be initiated to sustain the alkaline diuresis.

For severe poisoning and a high salicylate level, hemodialysis is indicated using a highly permeable (high-flux, polysulfone) membrane. Such a newly developed membrane allows better clearance of low- to medium-molecular-weight solutes than the conventional cuprophan or other hemodialysis hollow fiber membrane. Dialysis should be performed with high blood flow (300–400 mL/min) for 4–6 hours and repeated daily until the blood level of the toxin is negligible and acidosis is no longer present.

Ethylene Glycol

Ingestion of ethylene glycol—contained in antifreeze and other solvents—either accidentally or to induce inebriation gives rise to a high anion gap metabolic acidosis. The acidosis is primarily attributable to the metabolism of ethylene glycol to oxalic acid (Table 12–7) as well as secondary lactic acidosis. The plasma oxalate is filtered, poorly reabsorbed, and forms crystals in the urine—an important clue for diagnosis of this form of toxic acidosis. Severe central nervous system, cardiopulmonary, and renal toxicities result from oxalic acid production.

The diagnosis is confirmed by the history, the finding of a high anion gap metabolic acidosis with urinary oxalate crystals, and the presence of an osmolar gap greater than 9 mosm/kg (see Chapter 5). The latter is due to a high plasma concentration of this water-soluble, low-molecular-weight substance (MW 62) which is not included in the usual calculation of the plasma osmolality ($2[Na^+]$ + [glucose]/18 + BUN/2.8). Such a disparity between the measured and calculated plasma osmolalities can be found following ingestion of a large quantity of any low-molecular-weight substance but for practical purposes is seen only with ethanol, ethylene glycol, and methanol.

Treatment of ethylene glycol toxicity includes ethanol administration to shift the intracellular redox state (the $NADH:NAD^+$ ratio) and prevent metabolism of ethylene glycol (which is itself relatively nontoxic) to its toxic metabolites. Ethanol (20% solution) is given intravenously at a rate of 0.6 g/kg over 30–45 minutes followed by a 5% sustaining solution at 110 mg/kg per hour to produce an ethanol serum level of 100–150 mg/dL. Thiamine (100 mg intravenously) and pyridoxine (100 mg intravenously) should also be given to impair ethylene glycol metabolism. To rid the body of the ethylene glycol and oxalate, normal saline is administered briskly to induce extracellular volume expansion, and hemodialysis should be instituted with a high-flux membrane as described above.

Methanol

Ingestion of methanol as wood alcohol following the home brewing of alcoholic beverages or in paint thinners and varnishes can give rise to severe optic nerve and central nervous system toxicities. This organ system damage follows a latent period of about 24 hours in which methanol is metabolized to formaldehyde and formic acid. In addition to formic acid and formate, lactic acidosis and ketoacidosis contribute to the acidosis and high anion gap (Table 12–7).

This form of toxic acidosis should be considered a strong diagnostic possibility in the management of a "blind drunk" patient—an intoxicated individual with visual impairment and a high anion gap metabolic acidosis. There is usually nausea, vomiting, and abdominal pain as well as disordered central nervous system function. The eye often shows evidence of papilledema and retinal edema. There is also a plasma osmolar gap resulting from the high concentration of this small toxin (MW 32).

Therapy is similar to that for ethylene glycol, consisting of ethanol administration, extracellular volume expansion, and hemodialysis.

Other

Paraldehyde, though little used currently, can be metabolized partially to acetic acid from acetaldehyde as well as to other organic acids.

UREMIA

With severe reduction in GFR to values less than 20 mL/min, there is impairment of filtration of organic anions normally produced from metabolism. The subsequent build-up of these anions in the plasma causes a mild metabolic acidosis of the elevated anion gap variety. Plasma bicarbonate concentration is usually over 15 meq/L, and the anion gap is usually less than 10 meq/L.

This form of uremic acidosis should be distinguished pathophysiologically from the type III (impaired buffer delivery) distal RTA associated with GFR 20–30 mL/min. The latter acidosis is hyperchloremic and due to impaired ammonia shunting in the medulla. As chronic renal insufficiency progresses, it is most frequent for the hyperchloremic, normal anion gap form of metabolic acidosis to occur first and to be subsequently joined by the normochloremic, high anion gap uremic acidosis to create a complex or mixed form of metabolic acidosis.

Therapy is indicated to ameliorate the component of renal osteodystrophy due to acidemia. As with type III distal RTA, base therapy need be only modest, 1–3 meq/kg per day—sufficient to compensate for the diminished urinary net acid excretion.

SYMPTOMS

Symptoms of acidemia are often somewhat nonspecific, as shown in Fig 12–10. More prominent symptoms frequently relate to the underlying disorder responsible for the acidosis.

RESPIRATORY

As shown above (Fig 11–1; Table 10–1), acidemia predictably stimulates respiration to lower the P_{CO_2} and ameliorate the drop in arterial pH. Tidal

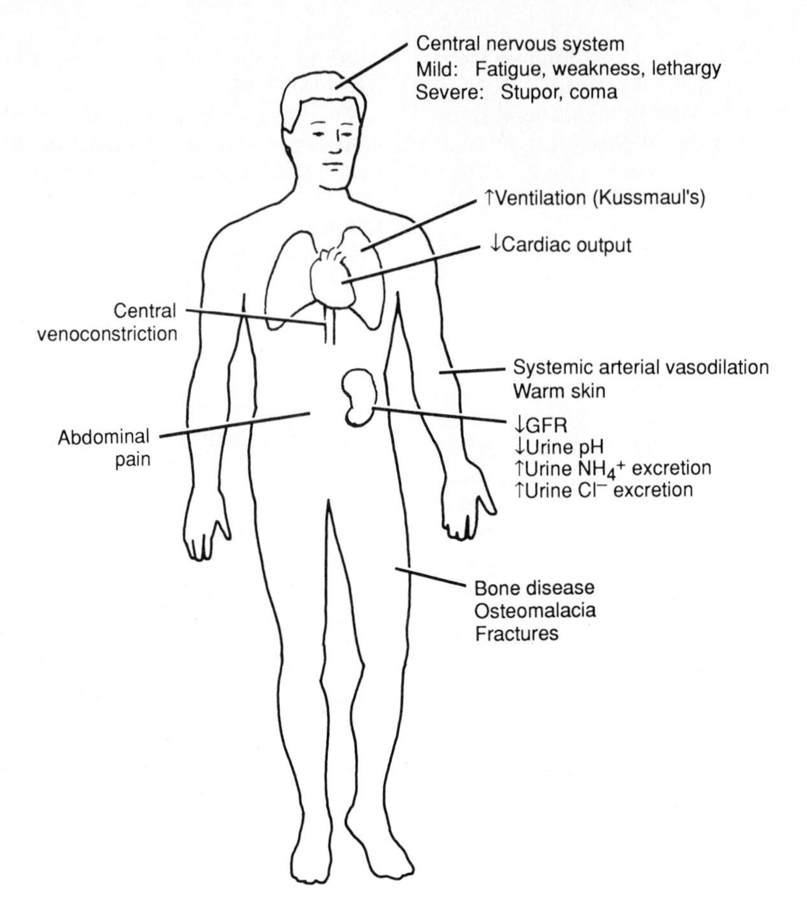

Figure 12-10. Symptoms and signs of acidemia.

volume is increased, manifested as Kussmaul respiration.

There are opposing changes in the oxyhemoglobin dissociation curve. Initially, a shift to the right occurs—the Bohr effect—in acute acidemia. This is followed in chronic acidemia by a shift to the left due to 2,3-diphosphoglycerate depletion.

CARDIOVASCULAR

Vasculature

Acidemia causes peripheral vasodilation with warm skin. This vasodilatory effect of acidemia may even counteract the vasoconstriction usually seen with hypotension, resulting in a form of "warm shock."

Paradoxically, acidemia causes venoconstriction. The decrease in central and pulmonary venous compliance predisposes an acidemic individual to pulmonary edema with only modest volume challenge. This acidemia-induced venoconstriction is the rationale for treating a volume-contracted, acidemic patient with alkali rather than saline first, to correct the diminished venous compliance so that the subsequent saline administration can be tolerated safely.

Cardiac

Cardiac contractility is directly suppressed by acidemia, at least when the arterial pH is less than approximately 7.10. Counteracting this effect is the release of catecholamines induced by acidemia, which serve a positive inotropic function. Enhanced vagal stimulation also occurs. As mentioned previously, there is arterial dilation, venous constriction, and reduction in extracellular volume. The ultimate effects of these complex changes in preload, afterload, and inotropy on cardiac performance are variable. Little change usually occurs in cardiac function until severe acidemia occurs, but decreased inotropy may occur with an arterial pH of less than 7.10.

NERVOUS SYSTEM

Acidemia can depress central nervous system function, ranging from fatigue, weakness, and lethargy to more severe impairment, including stupor and coma. Nonspecific abdominal pain can also occur.

RENAL

A reduction in plasma bicarbonate concentration tends to reduce GFR. Part of the explanation for the glomerular hypofiltration is that there is diminished sodium bicarbonate delivery (due to the low filtered load) and augmented sodium chloride delivery (due to hyperchloremia and suppressed proximal sodium chloride reabsorption) to the loop of Henle. Enhanced sodium chloride reabsorption in the macula densa in response to the high delivery activates tubuloglomerular feedback to reduce GFR (Fig 1–16).

Acidemia causes a rapid adaptive decrease in urinary pH and a slower increase in ammoniagenesis and urinary ammonium excretion (Fig 12–2). Impaired reabsorption of sodium chloride in the proximal tubule during acute metabolic acidosis (Fig 1–23) results in a transient chloruresis. If chloride intake is restricted, negative sodium chloride balance and extracellular volume contraction can ensue. Thus, total body chloride can be diminished despite the fact that chloride concentration may be high in the hyperchloremic forms of metabolic acidosis.

The primary disorder causing metabolic acidosis may also independently cause effective circulating volume to fall (eg, the osmotic diuresis in diabetic ketoacidosis), resulting in further compromise of renal hemodynamics and fall in GFR. In addition, there is usually a mild increase in urinary potassium excretion due to the increase in solute delivery to the cortical collecting tubule and the fact that renin and aldosterone tend to rise. When present, potassium deficiency potentiates the effect of acidemia to stimulate ammoniagenesis. Despite potassium wastage, acidemia if due to a mineral acid causes potassium to shift out of cells, resulting in hyperkalemia (Table 7–1).

BONE & CALCIUM METABOLISM

With chronic acidosis, there is progressive leaching of the body's enormous buffer reserve in bone. Bone carbonate is mobilized along with calcium. Chronic acidemia causes osteomalacia and exacerbation of other bone disease, such as renal osteodystrophy, and fractures can ensue. The hypercalciuria (derived from bone calcium mobilization), phosphaturia, and reduced urinary citrate excretion (due to direct enhancement by acidemia of renal citrate reabsorption) lead to a propensity to nephrolithiasis and nephrocalcinosis in certain forms of distal RTA.

DIAGNOSIS

APPROACH TO ETIOLOGIC DIAGNOSIS

History & Physical Examination

The history and physical examination are essential components of the diagnostic approach to metabolic acidosis. The possibility of diarrhea (primary or laxative-provoked) or an external acid load (parenteral feeding or toxin ingestion) and careful questioning regarding a history of diabetes, alcoholism, renal disease (nephrotic syndrome, other glomerulonephropathies, and tubulointerstitial disease), or hepatic insufficiency are all important.

Physical signs of acidemia itself are somewhat nonspecific, but the physical examination may reveal clues to the underlying disease (diabetes, alcoholism, sepsis, etc). The numerous causes of altered lactate metabolism should be sought, including hypovolemic disorders, intrinsic cardiac disease, primary liver disease, and sepsis.

The urine should be carefully examined for signs of primary renal disease (eg, proteinuria, cellular casts), ketones (ketoacidosis), glucose (diabetic ketoacidosis or type I proximal RTA), or oxalate crystals (ethylene glycol intoxication)

Drug History.

Amphotericin B shunts the pH gradient in the collecting tubule and predictably causes type I classical distal RTA. Drugs that interfere with the renin-angiotensin II-aldosterone system can impair urinary acidification and predispose to metabolic acidosis by creating a type IV generalized distal RTA. Such drugs include beta-blockers and prostaglandin inhibitors (inhibit renin release); converting enzyme inhibitors (impair angiotensin II formation); spironolactone, amiloride, and triamterene (antagonize aldosterone action); and nonsteroidal anti-inflammatory drugs, other analgesics, and penicillin-derivative and sulfonamide antibiotics (tubulointerstitial nephritis involving the collecting tubule with subsequent dysfunction of the aldosterone target cells).

Urinary Electrolytes: The Urine Anion Gap

Sometimes the clinical distinction between diarrheal disorders with normal renal acidification and the RTAs is difficult. Urinary ammonium and net acid excretion easily separates the two classes of disorders (Table 12–1). Unfortunately, urine ammonium is not measured by clinical laboratories.

The urine anion gap $[Na^+] + [K^+] - [Cl^-]$ can be used to estimate urinary ammonium concentration. For this calculation, it is assumed that the urine contains simply sodium chloride, potassium chloride, and ammonium chloride. All other cations (eg, calcium and magnesium) and anions (eg, phosphate, ketones, and organic anions) are ignored. The sum of the cations (sodium, potassium, and ammonium) is equal to the sum of the anions (chloride), or as follows:

$$[Na^+] + [K^+] + [NH_4^+] = [Cl^-]$$

The sodium, potassium, and chloride concentrations are easily measured by the clinical laboratory, but the ammonium is not. By analogy with the plasma anion gap, ammonium thus represents the analytically missing component of the urine electrolytes. The term "urinary anion gap" is really a misnomer, since the major missing electrolyte is a cation. Rearranging the above equation shows that the urine anion gap is proportionate to the negative value of the urinary ammonium concentration:

$$\text{Urine anion gap} = [Na^+] + [K^+] - [Cl^-] = -[NH_4^+]$$

As shown in Fig 12–11, clinical experience has confirmed that the urine anion gap becomes increasingly negative as ammonium concentration increases. In diarrheal disease, the urine anion gap is markedly negative (ie, indicating high urinary ammonium concentration). In contrast, the urine anion gap is zero or positive when the ammonium concentration is low, as occurs in all the RTAs.

Other Serum Values

Ancillary laboratory tests can be diagnostically useful in indicated situations. Tests that can be helpful include BUN and serum creatinine (uremic acidosis and distal RTAs), serum glucose concentration and osmolality (diabetic ketoacidosis), serum ketones (ketoacidosis), serum lactate concentration (lactic acidosis), and serum salicylate level (salicylate intoxication). The simultaneous measurement of serum sodium and glucose concentrations, BUN, and plasma osmolality is useful in an anion gap acidosis to check whether an osmolar gap is present (ethylene glycol or methanol intoxication).

Coincident Acid-Base & Electrolyte Disturbances

A chronic metabolic acidosis invariably associated with hyperkalemia is type IV generalized distal RTA. Acidemia itself tends to cause hyperkalemia (Fig 7–6). The degree of hyperkalemia is greater with hyperchloremic acidosis than with organic, normochloremic acidosis. This tendency to develop hyperkalemia due to acidemia is sometimes counteracted by primary loss of potassium by the underlying disorder, resulting in hypokalemia, as occurs when potassium is lost through the gastrointestinal tract in diarrheal states, or by the kidney in an osmotic diuresis (diabetic ketoacidosis) or in type I proximal RTA or type II classical distal RTA.

Coincident acid-base disturbances with metabolic acidosis—such that a complex acid-base disorder is created—are not uncommon. Primary hyperventilation in excess of that appropriate for the metabolic acidosis—causing respiratory alkalosis—is frequent in febrile illnesses and sepsis, salicylate intoxication, or hepatic disease. Metabolic alkalosis may complicate a metabolic acidosis (so the plasma bicarbonate concentration is not as low as would be predicted from the magnitude of elevation of the anion gap) when there is superimposed vomiting or diuretic administration. Finally, as mentioned above, another type of complex metabolic acidosis (so that the decrement in plasma bicarbonate concentration is more than

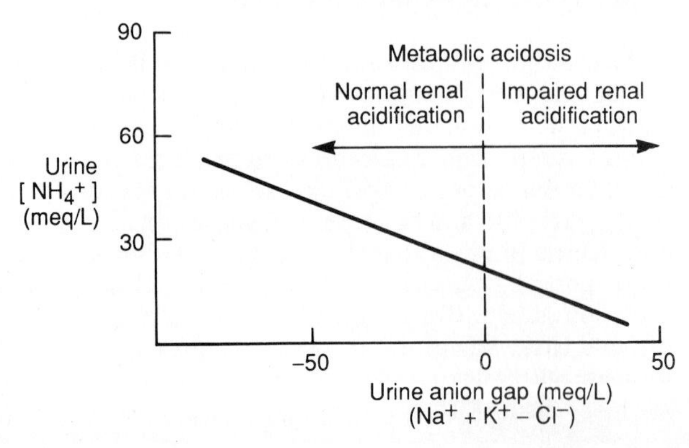

Figure 12–11. Urine anion gap in metabolic acidosis.

predicted from the increment in anion gap) may be seen in advanced chronic renal disease when the non-anion gap metabolic acidosis associated with type III impaired buffer delivery distal RTA is superimposed on the uremic high anion gap form of metabolic acidosis.

THERAPY

DECISION TO TREAT

Of course, since metabolic acidosis is not usually a primary disorder, the underlying cause of the acidosis should be treated whenever possible. The following discussion will address issues of relevance to metabolic acidosis as a general entity. However, comments about treatment for the specific metabolic acidoses have already been provided above. Many factors enter the decision about whether the metabolic acidosis component of the primary disorder should be independently treated.

As a generality, the normochloremic metabolic acidoses are acute, and the decision to treat with alkali rests on the severity of the arterial pH reduction, whether there are symptoms from the acidemia per se, and whether there is potential circulating base (eg, ketones or lactate) that might be metabolically converted back to bicarbonate without supplemental bicarbonate therapy. The RTAs, on the other hand, are relatively chronic disorders in which the acidosis is specifically treated to improve bone metabolism and to accelerate growth in children.

Degree of Acidemia & Symptomatology

In acute acidosis, when the arterial pH is less than 7.10 and there is hemodynamic compromise with hypotension, systemic alkali therapy is indicated. At this level of acidemia, cardiac performance is likely to be impaired. In addition, residual buffer reserves are quite low, so that a further modest increment in acid generation can cause arterial pH to plummet. An arterial pH range from 7.10 to 7.20 is a gray area where therapy depends on the clinical situation. During acute metabolic acidosis, an arterial pH of 7.20–7.40 is usually relatively well tolerated and, in the absence of significant symptoms, need not necessarily be treated.

As mentioned above, there is experimental evidence that sodium bicarbonate therapy of acute lactic acidosis actually promotes lactic acid production and exacerbates acidemia. However, therapeutic options are limited in this situation, and most clinicians would advocate that alkali therapy should still be given when acidemia is extreme. In toxin ingestion, alkali therapy is also useful to limit central nervous system toxicity and to enhance renal excretion of the charged metabolite.

Plasma Ketones & Lactate as Potential Base

There is a greater need for alkali therapy in metabolic acidosis of the hyperchloremic form than for the normochloremic form with a comparable level of acidemia. In the latter, circulating ketones and lactate may be metabolically converted following primary therapy (eg, insulin or hemodynamic resuscitation) to bicarbonate. Thus, the increment in anion gap can be considered equivalent to potential base. On the other hand, there is no potential base in the plasma during a hyperchloremic form of metabolic acidosis. By similar reasoning, rebound alkalosis has a greater propensity to occur in the treatment of normochloremic acidoses, since alkali given as therapy plus metabolic conversion of circulating anions to bicarbonate may together be excessive. During therapy of ketoacidosis, it is often unpredictable to what extent ketones can be converted back to bicarbonate before being lost into the urine due to normalization of GFR.

EMERGENCY ALKALI THERAPY FOR ACUTE, SEVERE ACIDEMIA

Indications

In an emergency, acidemia should be confirmed by blood gas analysis to be due to metabolic acidosis (eg, lactic acidosis versus respiratory acidosis during cardiopulmonary arrest). If acidemia is symptomatic or severe (< 7.10) as a result of reduction in bicarbonate concentration (eg, < 10 meq/L), sodium bicarbonate therapy is indicated. As summarized in Table 12–9, one to three ampules (1–2 meq/kg) of 7.5% sodium bicarbonate (45 meq of bicarbonate per ampule) should be given intravenously followed by rechecking of the blood gases. If more time is available, two or three sodium bicarbonate ampules may be added to 1 L of 5% dextrose in water (to create a near-isotonic solution of sodium bicarbonate) and administered at a rate of 100–250 mL/h.

Quantity of Therapy

The amount of bicarbonate given intravenously in these acute situations is calculated based on the apparent volume of distribution of bicarbonate of half the body weight, which is roughly equivalent to total body water and assumes buffering in both the intracellular and extracellular volumes. A

Table 12-9. Alkali therapy for metabolic acidoses

Alkali	Amount	Indications
Intravenous		
$NaHCO_1$	45 meq/ampule × 1 – 3	Acute, severe symptomatic acidosis
$NaHCO_3$	90–135 meq/L at a rate of 100–200 mL/h	
Oral		
$NaHCO_3$ or Na·citrate	In oral repletion solutions	Diarrhea
$NaHCO_3$	1–3 meq/kg/d	Types II, III, and IV distal renal tubular acidosis
$NaHCO_3$/$KHCO_3$ or citrate	10–15 meq/kg/d	Type I proximal renal tubular acidosis
K·citrate	30–60 meq/d	Nephrolithiasis
Hemodialysis		
Acute: $NaHCO_3$	35–40 meq/L in dialysate	Severe acidosis without renal function
Chronic: $NaHCO_3$ or Na·acetate	35–40 meq/L in dialysate	Chronic renal failure

greater bicarbonate distribution volume (approaching the entire body weight) is actually observed as the acidemia worsens owing to an increased contribution of intracellular buffering. However, because of uncertainties in this calculation, a general guideline is that no more than half of the bicarbonate concentration deficit should be corrected over a 3- to 4-hour period without rechecking the arterial blood gases. Such a calculation also ignores ongoing acid production and bicarbonate titration. Thus, the minimal amount of bicarbonate needed for therapy is 0.5 × body weight × 0.5 × the decrement in plasma bicarbonate concentration. For example, in a 70-kg individual with plasma bicarbonate concentration of 4 meq/L, the bicarbonate needed initially would be $0.5 \times 70 \times 0.5 \times (24 - 4)$, or 350 meq.

Oral alimentation solutions have been successfully used to treat the acidemia, extracellular volume depletion, and hypokalemia associated with diarrhea (eg, cholera). Twenty-five to 40 meq/L of bicarbonate or its metabolic equivalent citrate is present in these oral alimentation solutions to compensate for base loss (Fig 12–3).

In patients without renal function in whom acidemia is severe and progressive, hemodialysis is indicated. The base equivalent in the dialysate should ordinarily be bicarbonate (not acetate) at the highest possible concentration, usually 35–40 meq/L.

Hazards of Alkali Therapy

The potential acute toxic side effects of sodium bicarbonate therapy include hypernatremia, hypercapnia, hypokalemia, and alkalemia. As discussed in Chapter 3, hypernatremia can occur when ampules of sodium bicarbonate are used because of their very high sodium concentration (892 meq/L). Acute hypercapnia can occur because of CO_2 production: the combination of H^+ and HCO_3^- to form H_2CO_3 with subsequent dehydration to CO_2 (Chapter 10). Serum potassium concentration predictably declines with correction of acidemia and, if preexisting potassium deficiency exists, can become frankly subnormal (Chapter 8). Cardiac monitoring for arrhythmias is essential in the correction of metabolic acidosis if preexisting potassium depletion is suspected. Finally, overcorrection of acidosis can cause more symptoms than acidemia itself. The exogenous bicarbonate may be additive to bicarbonate subsequently generated endogenously from circulating ketones or lactate to yield frank alkalemia with the risk of tetany (Chapter 13).

Metabolic Alkalosis 13

INTRODUCTION

DEFINITION

Again—as with all acid-base disorders—arterial blood gases and serum electrolytes are essential in the diagnostic approach to metabolic alkalosis, summarized in Fig 13-1. Metabolic alkalosis is a hyperbicarbonatemic, alkalemic, hypochloremic disorder.

Hyperbicarbonatemia, Alkalemia, & Appropriate Hypercapnia

Arterial pH is essential to ensure that a high blood bicarbonate concentration—greater than 26 meq/L—is associated with alkalemia—arterial pH greater than 7.45—rather than acidemia, suggesting the presence of chronic respiratory acidosis.

As shown in Fig 10-13 and Table 10-1, there is a degree of expected secondary elevation in the arterial P_{CO_2} due to primary suppression of respiratory drive by the alkalemia. A greater or lesser change in P_{CO_2} implicates a superimposed respiratory disorder—respiratory acidosis or alkalosis, respectively. The acid-base nomogram (Fig 11-1) is important in ascertaining that the hypoventilatory response is appropriate.

The acid-base diagnosis is often difficult in a patient with chronic lung disease who has a baseline chronic respiratory acidosis. Concurrent diuretic therapy (eg, for right heart failure) may superimpose a component of metabolic alkalosis, which further increases the plasma bicarbonate concentration, causes P_{CO_2} to rise further with deterioration of P_{O_2}, and may result in a normal or even alkaline arterial pH.

Appropriate Elevation in the Anion Gap

Though metabolic alkalosis is always hypochloremic, the fall in plasma chloride concentration is not exactly the inverse of the rise in bicarbonate concentration; ie, there is an increase in the anion gap. As discussed previously (Table 11-1), the anion gap predictably increases during metabolic alkalosis due to the greater negative charge equivalency on albumin, the major contributor to the anion gap. The anion gap rises by 0.4–0.5 meq/L for each meq/L increment in plasma bicarbonate concentration during metabolic alkalosis. For example, if plasma bicarbonate concentration were 44 meq/L—a rise of 20 meq/L over normal—the anion gap should be 20–22 meq/L (normal 12 meq/L + 0.4–0.5 × 20 meq/L). An anion gap greater than this value would imply superimposition of a normochloremic, high anion gap form of metabolic acidosis.

PATHOPHYSIOLOGY

Generation & Maintenance of Metabolic Alkalosis

There are two important considerations in approaching metabolic alkalosis. First, because the physiologic condition the body usually confronts is that of acid addition, it is unusual for alkali to be added to the body. Thus, the initial issue is to identify the source of the excess plasma bicarbonate concentration in the generation of metabolic alkalosis. As shown in Table 13-1, bicarbonate may come from an exogenous source or from loss of normally acidic fluid, gastric contents or urine. When hydrogen loss from the body is of renal origin, it is due to an increase in one or more of the determinants of acidification in the collecting tubule, including aldosterone, sodium delivery and flow rate, P_{CO_2}, and nonreabsorbable anion (Table 10-3).

Second, the normal renal response to an elevation of plasma bicarbonate concentration from whatever cause is to quantitatively excrete the excess bicarbonate, as was shown in Fig 10-15 and Table 10-4 and summarized in Fig 13-2. If the bicarbonate is instead retained, something has abolished the kidney's normal bicarbonaturic response and thus metabolic alkalosis continues. The most important factors that "reset" the kidney to maintain metabolic alkalosis are extracellular volume and potassium deficiencies.

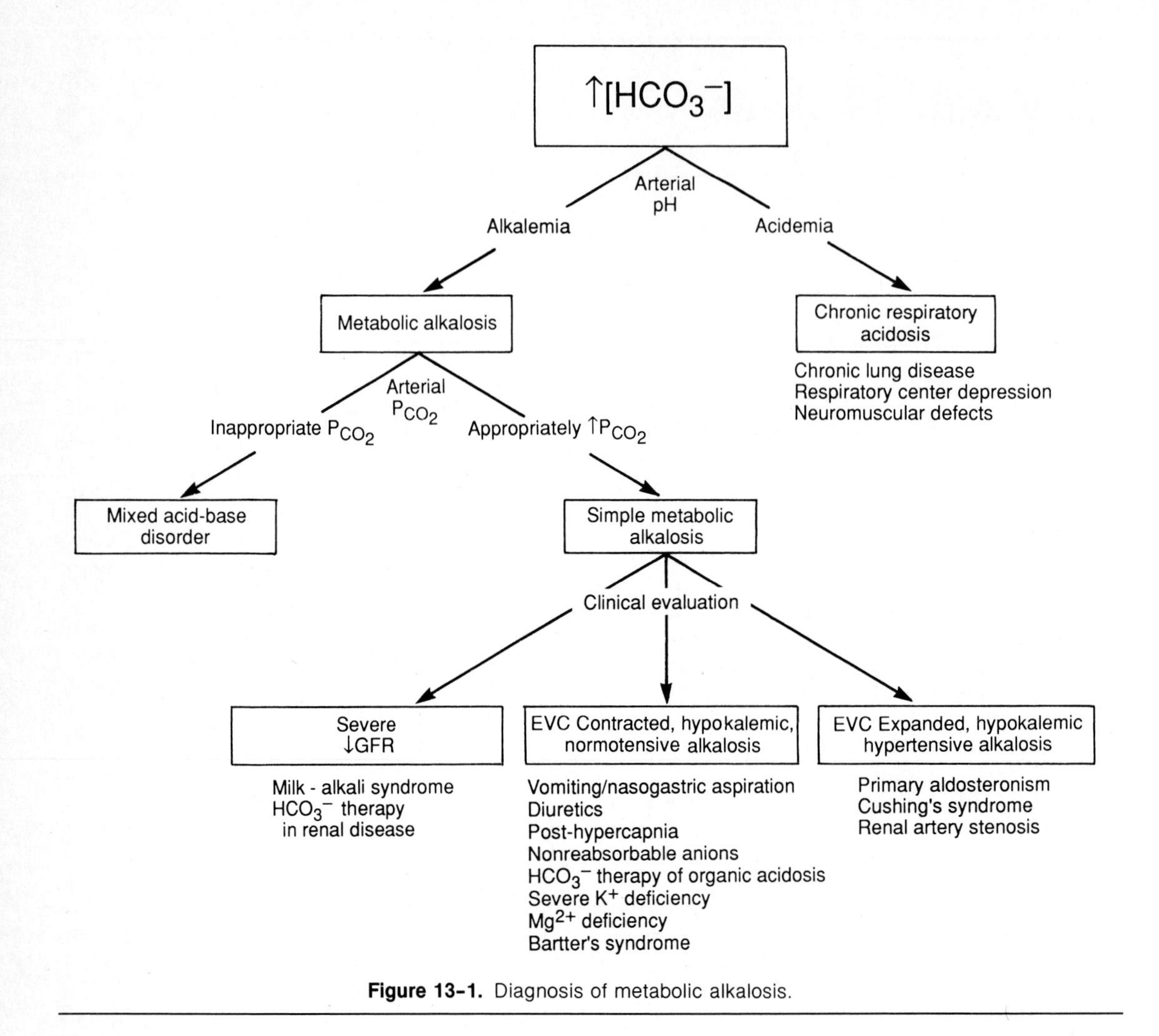

Figure 13-1. Diagnosis of metabolic alkalosis.

Roles of Extracellular Volume & Potassium Deficiencies in Maintaining Metabolic Alkalosis

In most forms of chronic metabolic alkalosis, there is extracellular volume depletion. Extracellular volume depletion is synonymous with chloride deficiency, since chloride—like sodium—is contained predominantly in the extracellular volume and is the main anion in this compartment. For example, during vomiting, the chloride is lost in the form of hydrochloric acid. The hydrogen ion loss generates the alkalosis, while the chloride loss serves to diminish the extracellular volume and, as will be seen subsequently, maintain the alkalosis. With diuretic administration as another example, the chloride is lost in the urine in the form of ammonium chloride, as well as sodium chloride and potassium chloride. Again, the excessive urinary ammonium loss serves to generate the metabolic alkalosis while the chloride loss causes extracellular volume depletion.

Body potassium stores are universally depleted in metabolic alkalosis, usually because of kaliuresis, and hypokalemia is the rule. During vomiting, for instance, transiently increased sodium bicarbonate delivery to the cortical collecting tubule with hyperaldosteronism causes potassium wastage. Similarly with diuretics, the increased sodium delivery and flow with hyperaldosteronism augments potassium excretion.

As was shown in Figs 10-7 and 10-14, alkalemia renders bicarbonate reabsorption static in the proximal tubule, unable to rise above a normal value. In this setting, GFR must be reduced inversely to the rise in plasma bicarbonate concentration (Fig 10-16) to allow the filtered bicarbonate load to remain normal (ie, not elevated). As shown in Table 10-4, in this setting, a normal rate of bicarbonate reabsorption in the proximal tubule suffices to prevent an increase in bicarbonate delivery to the more distal portion of the nephron, the cortical collecting tubule. The collecting tu-

Table 13-1. Generation of metabolic alkalosis.

	Examples
Alkali administration	
HCO_3^-	Therapy of metabolic acidosis
Carbonate	Antacids
Acetate	Hyperalimentation, hemodialysis
Lactate	Parenteral solutions, peritoneal dialysate
Citrate	Blood anticoagulant
Gastrointestinal H^+ loss	
Gastric H^+ loss	Vomiting, nasogastric aspiration
Stool H^+ loss	Congenital chloridorrhea
Renal H^+ loss	
↑Aldosterone (especially with ↓K^+)	Primary hyperaldosteronemic states, primary hyperreninemic states, Mg^{2+} deficiency
↑Na^+ delivery and flow rate	Thiazide or loop diuretics
↑P_{CO_2}	Posthypercapnia
↑Nonreabsorbable anion delivery	Ketones, penicillin-derivative antibiotics

bule is then not overwhelmed with bicarbonate, reabsorbs it normally, and prevents bicarbonaturia. The high plasma bicarbonate concentration is maintained, therefore, by a commensurate reciprocal reduction in GFR.

The key to the maintenance of metabolic alkalosis is attributable to factors that cause GFR to fall as plasma bicarbonate concentration rises, as illustrated in Fig 13–3A. The major factor that causes GFR to fall is chloride deficiency (synonymous with extracellular volume contraction, as discussed above). The consequent hypovolemia sets in motion a series of neurohumoral changes that result in glomerular hypofiltration (Chapter 2). Potassium depletion may also participate by causing an increase in the potent glomerular vasoconstrictors angiotensin II and thromboxane.

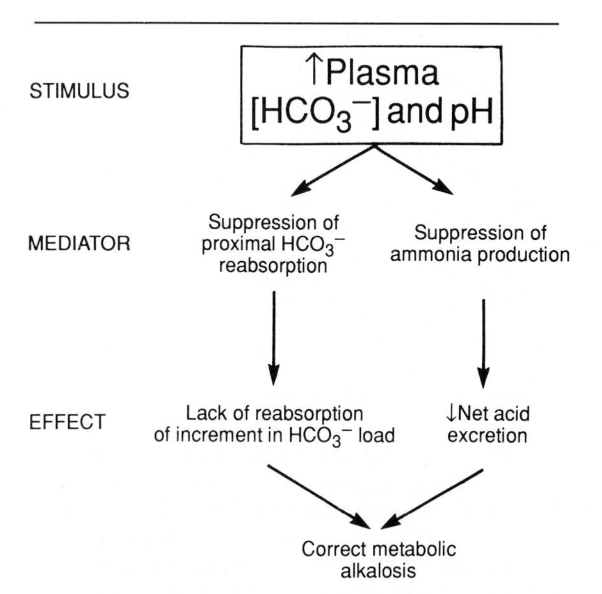

Figure 13-2. Pathophysiology of metabolic alkalosis.

Role of Aldosterone in Maintaining Metabolic Alkalosis

Another pathway for maintaining metabolic alkalosis without reduction in GFR is if there is augmentation of tubular hydrogen ion secretion, as shown in Fig 13–3B. As discussed extensively above (Chapter 10), there is little evidence that the major site of bicarbonate reclamation, the proximal tubule, can be stimulated in the setting of alkalemia. However, hydrogen ion secretory mechanisms in the α-intercalated cell of the collecting tubule can be stimulated both directly and indirectly by aldosterone even during alkalemia. Thus, besides enhanced sodium reabsorption and potassium secretion, hyperreninemia or other primary cause of hyperaldosteronism can at least partially account for increased tubular reabsorption of bicarbonate (Fig 13–3B). Potassium depletion in these conditions might serve to enhance the effect of aldosterone on hydrogen secretion. The capacity for hydrogen ion secretion in the collecting tubule is low, so the degree of elevation in filtered bicarbonate load that can be sustained by this mechanism is relatively small. The overall effect of hyperaldosteronism is thus to induce a mild hypokalemic metabolic alkalosis with sodium chloride retention and hypertension.

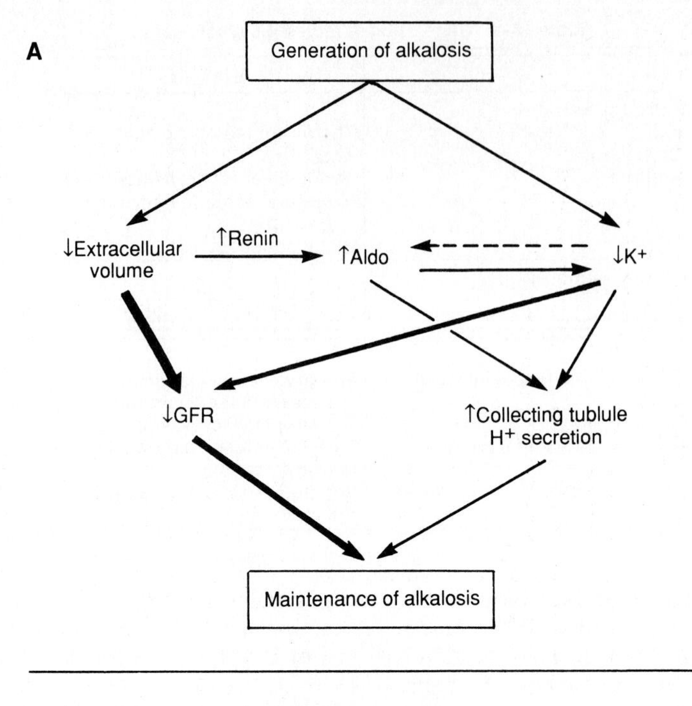

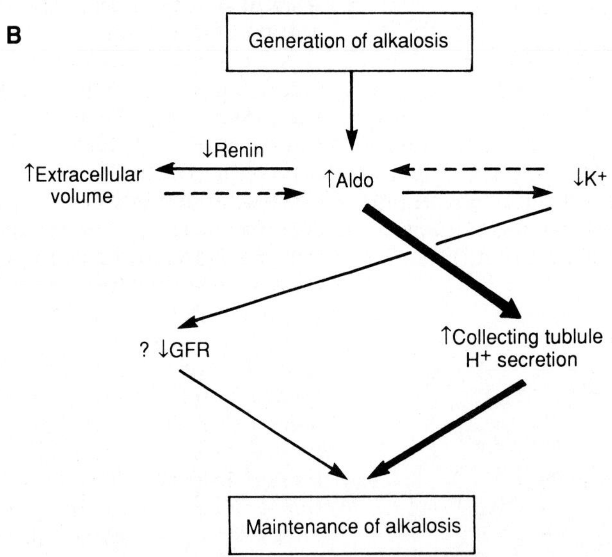

Figure 13-3. Patterns of generation and maintenance of metabolic alkalosis.

EXOGENOUS BICARBONATE LOAD IN THE SETTING OF RENAL DISEASE

The causes of metabolic alkalosis are summarized in Table 13-2.

ALKALI ADMINISTRATION

Alkali Load With Decreased GFR

Bicarbonate administration in a quantity as high as 20 meq/kg per day to an individual with normal renal function and bicarbonaturic response induces little change in the plasma bicarbonate concentration and pH (Figs 10-14 and 13-2). Thus, unless extremely rapid, bicarbonate

Table 13–2. Metabolic alkalosis.

- **Exogenous bicarbonate loads**
  - Acute alkali administration
    - Bicarbonate
    - Carbonate
    - Acetate
    - Lactate
    - Citrate
  - Milk-alkali syndrome
- **ECV contraction, normotension, potassium deficiency, and secondary hyperreninemic hyperaldosteronism**
  - Gastrointestinal origin
    - Vomiting
    - Gastric aspiration
    - Congenital chloridorrhea
    - Villous adenoma
  - Renal origin
    - Diuretics (especially thiazides and loop diuretics)
      - Acute
      - Chronic
    - Posthypercapnic state
    - Nonreabsorbable anion delivery
      - Ketones
      - Penicillin
      - Carbenicillin
    - Following treatment of ketoacidosis or lactic acidosis
      - Carbohydrate refeeding following starvation
    - Magnesium deficiency
    - Potassium depletion
    - Bartter's syndrome
- **ECV expansion, hypertension, potassium deficiency, and hypermineralocorticoidism**
  - Associated with high renin
    - Renin-secreting tumor
    - Renal arterial stenosis
    - Accelerated or malignant hypertension
  - Associated with low renin
    - Primary aldosteronism
      - Adenoma
      - Hyperplasia
      - Carcinoma
      - Glucocorticoid-suppressible
    - Adrenal enzymatic defects
      - 11β-Hydroxylase deficiency
      - 17α-Hydroxylase deficiency
    - Cushing's disease or syndrome
      - Ectopic ACTH
      - Adrenal carcinoma
      - Adrenal adenoma
      - Primary pituitary hypersecretion
    - Other
      - Licorice
      - Chewing tobacco
      - Nasal spray
      - Carbenoxolone
      - Liddle's syndrome

administration does not produce metabolic alkalosis. On the other hand, if GFR is impaired, bicarbonate cannot be filtered and excreted properly so that an alkali load may generate a sustained metabolic alkalosis. As shown in Tables 13–1 and 13–2, an alkali load may be in the form of bicarbonate (therapy of metabolic acidosis, hemodialysate), carbonate (antacids), acetate (hyperalimentation solutions, hemodialysate), lactate (parenteral solutions, peritoneal dialysate), or citrate (blood anticoagulant).

Ingestion of large quantities of milk and antacid can produce a syndrome of renal insufficiency with metabolic alkalosis called the milk-alkali syndrome. Excessive calcium and vitamin D ingestion predisposes to hypercalcemia, nephrocalcinosis and renal insufficiency. The diminished GFR then allows for the retention of the simultaneously administered alkali in the antacid. Discontinuation of the antacid repairs the alkalosis.

HYPOVOLEMIA, POTASSIUM DEFICIENCY, & SECONDARY HYPERRENINEMIC HYPERALDOSTERONISM

Some metabolic alkaloses are generated by gastrointestinal or renal hydrogen ion loss with simultaneous chloride and potassium loss. The consequent extracellular volume contraction and potassium deficiency then diminish GFR or enhance tubular hydrogen ion secretion primarily via hyperaldosteronism to maintain metabolic alkalosis (Fig 13–3A). These factors serve to abrogate the normal renal bicarbonaturic response to a high plasma bicarbonate concentration and pH. The appropriate bicarbonaturic response can be restored, with repair of this form of metabolic alkalosis, by sodium chloride and potassium repletion.

GASTROINTESTINAL HYDROGEN ION LOSS

Vomiting or Nasogastric Aspiration

The loss of hydrogen ion from the body with equal gain of bicarbonate accounts for the generation of metabolic alkalosis induced by vomiting or nasogastric aspiration. Gastric parietal cells form a hydrogen ion and a bicarbonate in a 1:1 stoichiometry from carbonic acid, which in turn is derived from CO_2 and water. The hydrogen ion enters the gastric lumen while bicarbonate enters the blood. Since gastric fluid can achieve a pH as low as 1.0 (hydrogen ion concentration of 0.1 mol/L), simultaneous entry into the blood of a significant quantity of bicarbonate can occur if gastric volume loss is large. Factors that stimulate gastric acid secretion include histamine, cholinergic agents, and gastrin.

Chloride is lost as the accompanying anion with hydrogen ion (ie, HCl is lost) during during vomiting, thereby diminishing extracellular volume. There is little chloride in the urine during this phase of the induction of metabolic alkalosis, since the filtered sodium chloride load is low (hypochloremia) and renal sodium chloride retention is avid (due to hypovolemia).

As shown in Fig 13-4, during the active phase of vomiting, the rise in plasma bicarbonate concentration initially causes a qualitatively appropriate renal response to excrete bicarbonate, albeit insufficient to repair the alkalosis. This bicarbonaturic response secondarily enhances potassium excretion and causes body potassium depletion. Potassium secretion in the cortical collecting tubule is facilitated because aldosterone is rising owing to the chloride loss and hypovolemia and because sodium is delivered to the cortical collecting tubule with bicarbonate, a relatively nonreabsorbable anion, rather than chloride. Chloride excretion falls rapidly to a negligible value owing to the hypochloremia and avid renal reabsorption resulting from hypovolemia. Both the hypovolemia and the hypokalemia cause GFR to fall, and that sustains the alkalosis.

When vomiting ceases, the plasma bicarbonate concentration falls slightly, though it is still elevated. Within 24 hours, bicarbonate disappears from the urine and the urine returns to its usual acidic nature, which is inappropriate for the normal renal response to alkalemia. At this stage, there is the "paradoxical aciduria of metabolic alkalosis." The kidney has then been "reset" by the hypovolemia and potassium depletion to diminish GFR and hence allow complete conservation of filtered bicarbonate despite an elevated plasma bicarbonate concentration and alkalemia. During this phase of metabolic alkalosis, extracellular volume depletion is a strong stimulus to renin secretion, though the expected enhanced aldosterone release is counteracted by the hypokalemia. Lack of bicarbonaturia diminishes the renal loss of potassium in this phase of metabolic alkalosis. There is still little urinary chloride excretion owing to intense renal sodium chloride reabsorption due to the low filtered chloride load and hypovolemia.

Repair of the alkalosis can be effected by reversing those hemodynamic and tubular factors that have prevented the normal bicarbonaturic re-

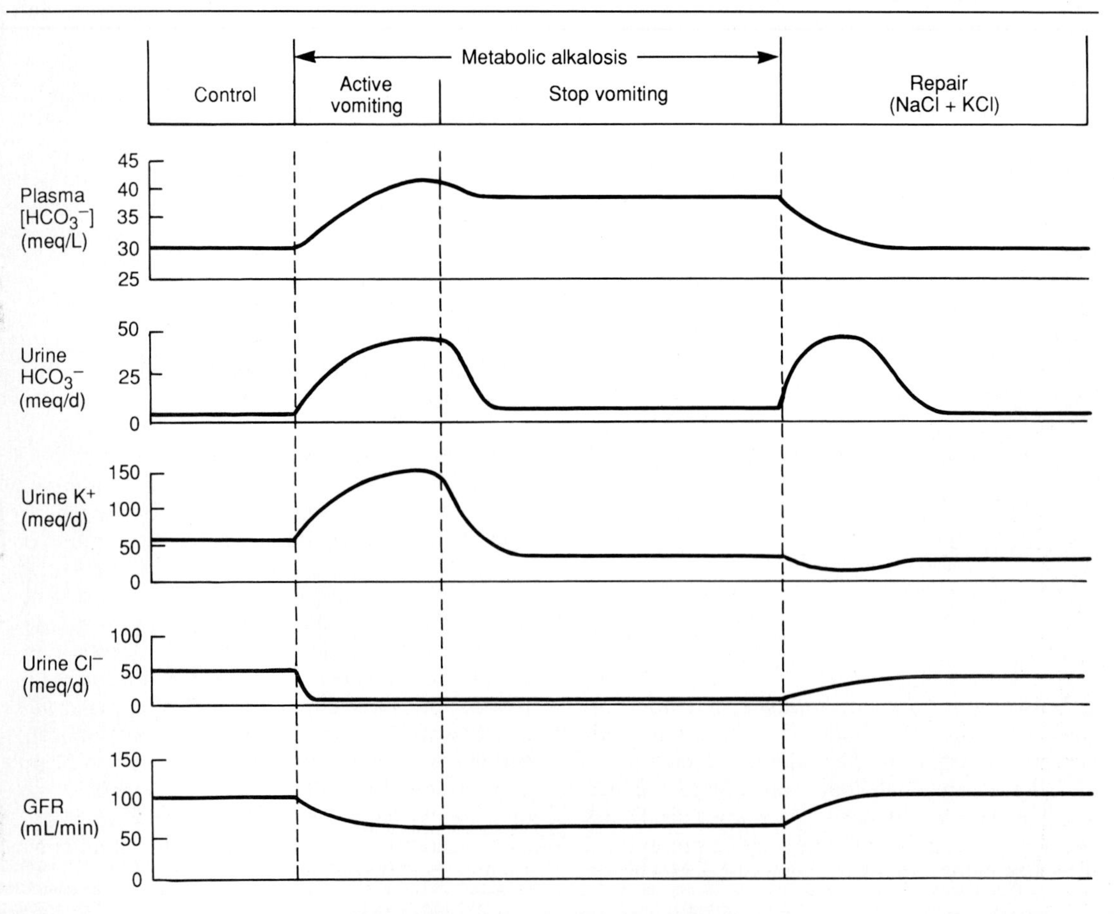

Figure 13-4. Renal response to vomiting.

sponse. When sodium chloride and potassium chloride are given, the kidney appropriately retains both in order to return extracellular volume to a euvolemic condition to replete potassium stores. Restoration of extracellular volume and potassium allows GFR to return to normal and then allow the appropriate corrective bicarbonaturic response to hyperbicarbonatemia and alkalemia. It has been suggested that enhanced chloride availability and delivery to the cortical collecting tubule might also stimulate luminal bicarbonate secretion by the β-intercalated cell resident in this segment (Fig 10–9) to add to the recovery process. The plasma bicarbonate concentration and pH then can fall to normal values, and the metabolic alkalosis is repaired.

Congenital Chloridorrhea

In this rare hereditary disorder, fecal acid loss occurs with enhanced bicarbonate delivery into the blood. The pathogenesis is thought to be loss of the normal ileal (HCO_3^-/Cl^- exchange (Fig 12–3) with continued function of the parallel Na^+/H^+ exchanger. Metabolic alkalosis also occurs in an occasional patient with villous adenoma, but the mechanism is unclear.

RENAL HYDROGEN ION LOSS

Factors that act to increase hydrogen ion secretion in the cortical and medullary collecting tubules and thereby enhance urinary net acid exretion serve to generate new bicarbonate. Stimulation of proximal acidification does not have this effect, since the proximal tubule simply reabsorbs filtered bicarbonate and does not generate new bicarbonate. As shown in Fig 13–5, the hydrogen ion secretory system in the α-intercalated cell of the collecting tubule may be stimulated directly or indirectly by (1) a high aldosterone state (either primary or secondary causes of hyperreninism or primary hyperaldosteronism), (2) increased sodium delivery and lumen-negative potential difference generated in the neighboring principal cell (due to diuretics, especially with a nonreabsorbable anion), (3) increased intracellular CO_2 (hypercapnia), or (4) increased luminal pH and ammonia tension (due to high ammoniagenesis induced by hypokalemia).

Diuretics

Diuretics that act in the loop of Henle (furosemide, ethacrynic acid, and bumetanide) or the distal convoluted tubule (thiazides) can acutely diminish extracellular volume due to natriuresis and chloruresis. There is no accompanying bicarbonaturia, so that plasma bicarbonate content is unaltered and the plasma bicarbonate concentration therefore rises. This acute disorder is called a contraction alkalosis, usually of mild degree due to cellular buffering processes.

Chronic administration of a loop or thiazide diuretic enhances sodium delivery to the cortical collecting tubule, sustaining an increase in the lumen-negative potential difference and in hydrogen ion and potassium secretion. The extracellular volume contraction stimulates renin and hence aldosterone, adding to the increased urinary net acid excretion and kaliuresis. The resulting hypokalemia stimulates ammoniagenesis and hence promotes a rise in ammonium excretion. Thus, new bicarbonate is generated and metabolic alkalosis is created and sustained by these combined effects of enhanced sodium delivery, increased aldosterone, and hypokalemia in the cortical collecting tubule.

The extracellular volume contraction and potassium depletion also maintain the metabolic alkalosis by decreasing GFR (Fig 13–3A). Repair of the alkalosis is achieved by sodium chloride and potassium repletion to normalize extracellular volume and potassium stores.

As summarized in Table 13–3, while metabolic alkalosis is generated and maintained by loop and thiazide diuretics, diuretics that directly interfere with function of those nephron segments involved in hydrogen ion secretion—the proximal tubule and the cortical collecting tubule—have the opposite effect, to induce metabolic acidosis (RTA), as discussed in greater detail in Chapter 12.

Posthypercapnia

Hypercapnia stimulates urinary net acid production (Fig 10–8) and increases plasma bicarbonate concentration, sustained by the adaptive increase in proximal bicarbonate reabsorption (Fig 10–8). Hypercapnia also induces a sodium chloride reabsorptive defect in the proximal tubule (see Chapter 1), which can cause some degree of extracellular volume depletion, exacerbated by the frequent concurrent use of a diuretic.

If the P_{CO_2} is rapidly returned to normal, the kidney normally responds with a brisk bicarbonaturic response to normalize the plasma bicarbonate concentration. If the preexisting chloruresis (direct effect of hypercapnia or diuretic) has caused extracellular volume depletion, however, the renal response may be incomplete. The hypovolemia serves to maintain the high bicarbonate concentration until chloride is administered or diuretics are discontinued to restore euvolemia (Fig 13–3A). Recurrent bouts of hypercapnia followed by normocapnia—as sometimes occurs in mechanically ventilated patients—exacerbate the urinary hydrogen ion and chloride losses and worsen the metabolic alkalosis.

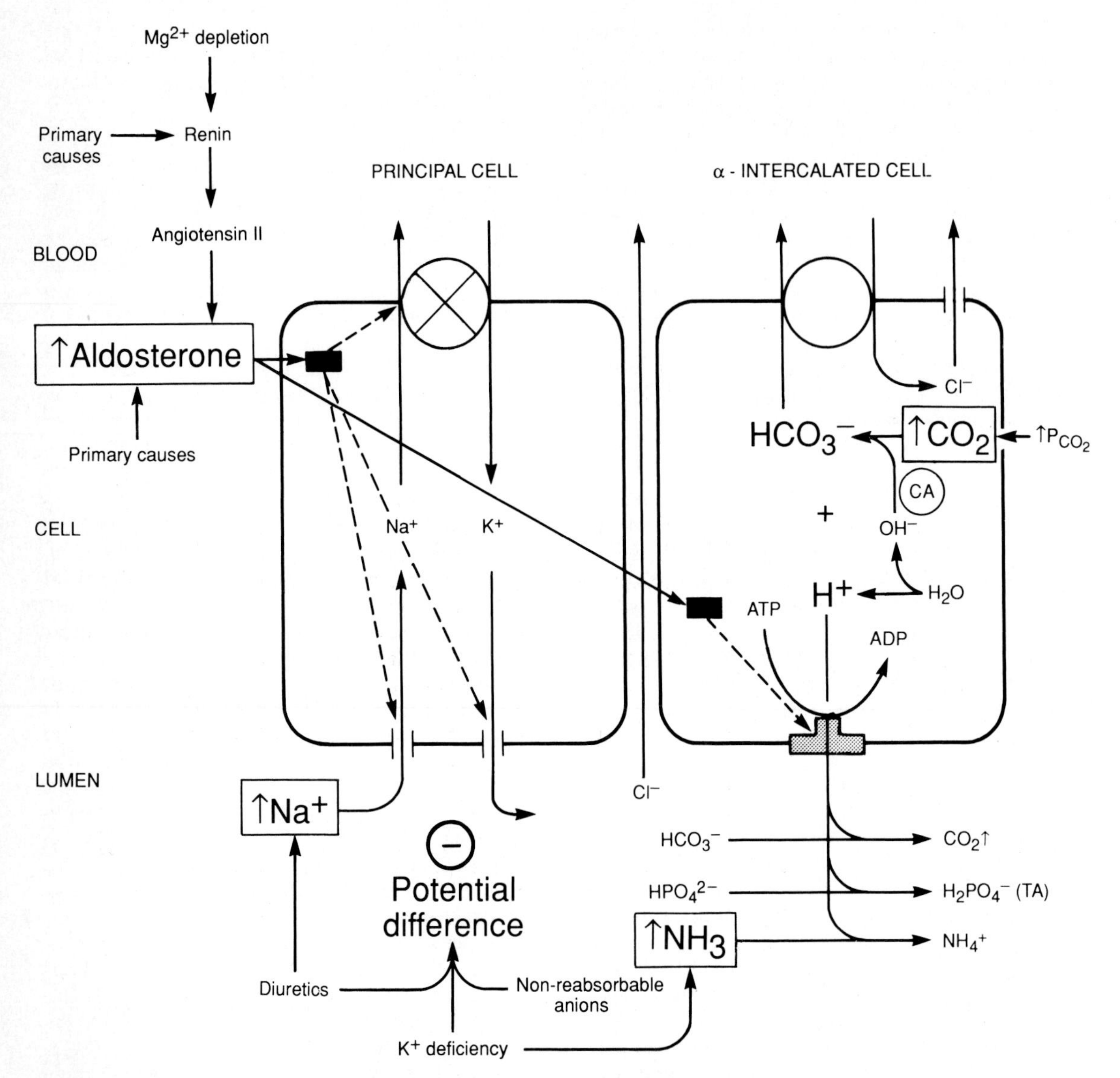

Figure 13–5. Possible cellular mechanisms in the cortical collecting tubule responsible for generation of metabolic alkalosis.

Nonreabsorbable Anion Delivery

If sodium is delivered to the principal cell of the cortical collecting tubule with an anion that is poorly reabsorbed—which for practical purposes is any nonchloride anion—there is enhanced lumen-negative potential difference and hence hydrogen ion and potassium secretion (Fig 7–14) compared to that generated by comparable sodium delivery with chloride.

Examples of nonreabsorbable anions that may reach the collecting tubule in sufficient quantity to generate a metabolic alkalosis include ketones (see below) and some antibiotics, such as penicillin and carbenicillin. The alkalosis is sustained by extracellular volume contraction and potassium depletion.

Following Treatment of Organic Acidosis

When bicarbonate is therapeutically administered during the course of ketoacidosis or lactic acidosis and the circulating ketone or lactate is subsequently metabolized to bicarbonate (eg, by insulin therapy and circulatory resuscitation), a surfeit of circulating bicarbonate can occur. Enhanced urinary net acid excretion and generation of bicarbonate during the acidemic period can exacerbate this process.

Table 13-3. Acid-base disturbances due to diuretics.

Nephron Site	Transport Effect	Example	Acid-Base Consequence
Proximal tubule	↓H^+ secretion	Acetazolamide	Metabolic acidosis (type I proximal RTA)
Loop of Henle	↓NaCl reabsorption	Furosemide, ethacrynic acid, bumetanide	Metabolic alkalosis
Distal convoluted tubule	↓NaCl reabsorption	Thiazides	Metabolic alkalosis
Cortical collecting tubule	↓NaCl reabsorption, ↓K^+ secretion, ↓H^+ secretion	Spironolactone, amiloride, triamterene	Metabolic acidosis (type IV generalized distal RTA)

Another related form of metabolic alkalosis can occur following carbohydrate refeeding after starvation ketoacidosis. During the phase of ketoacidosis, enhanced renal generation of bicarbonate stimulated by the acidemia may prevent the plasma bicarbonate from falling as much as would be predicted by the amount of keto acid generation. When refeeding occurs, circulating ketones can be converted to bicarbonate, adding to that generated during the preexisting ketoacidosis and thereby causing an acute metabolic alkalosis. Maintenance of the alkalosis is effected by the volume contraction and potassium deficiency incurred during the acidemia phase.

Magnesium or Potassium Depletion

Magnesium deficiency increases renin production, though the mechanism is poorly understood. The consequent secondary hyperaldosteronism increases potassium and hydrogen ion secretion in the cortical collecting tubule to generate and sustain a hypokalemic metabolic alkalosis.

A large potassium deficit from any cause—of more than 1000 meq—has sometimes been associated with the generation and maintenance of metabolic alkalosis. There are probably renal and nonrenal mechanisms for the generation of the alkalosis. Maintenance of the alkalosis appears to be unrelated to extracellular volume, since sodium chloride administration is without benefit; it may be due to primary potassium deficiency-induced glomerular vasoconstriction that reduces GFR.

Bartter's Syndrome

Bartter's syndrome is a very rare, usually inherited disorder primarily affecting young males that is manifested by hypokalemic metabolic alkalosis and a very high renin and aldosterone level. Chronic stimulation of renin secretion causes juxtaglomerular apparatus hypertrophy. Frequently there is also magnesium wasting and high renal prostaglandin production. The cause of this syndrome is thought to be a primary intrinsic defect in sodium chloride reabsorption in the thick ascending limb of Henle (see Chapter 2). The clinical expression of this disorder is therefore very similar to that produced by chronic furosemide administration. Treatment is aimed at ameliorating the potassium wasting (potassium supplementation and pharmacologic blockade of the renin-angiotensin-aldosterone axis).

HYPERVOLEMIA, ELEVATED BLOOD PRESSURE, POTASSIUM DEFICIENCY, & PRIMARY STIMULATION OF ALDOSTERONE

HYPERALDOSTERONISM

Primary stimulation of aldosterone, unregulated by extracellular volume status, predictably causes (1) primary sodium retention and expansion of the extracellular volume with hypertension (volume-dependent hypertension, discussed in Chapter 3); (2) potassium wasting with hypokalemia (Chapter 8); and (3) increased hydrogen ion secretory capacity with stimulation of urinary net acid excretion initially to generate metabolic alkalosis (Fig 13–4) and enhanced bicarbonate reabsorptive capacity subsequently to maintain the alkalosis (Fig 13–3B). Potassium deficiency-induced depression in GFR may also serve to maintain the alkalosis. The mineralocorticoid escape phenomenon occurs from all these effects, so that a stable state is rendered of hypertension, hypokalemia, and metabolic alkalosis. The degree of metabolic alkalosis is usually quite mild in these disorders, given the low hydrogen ion secretory capacity of the cellular target of aldosterone in the cortical collecting tubule.

Therapy must be aimed at removal of the source of hyperaldosteronism. Sodium chloride administration with further expansion of the extracellular volume is ineffective in reducing the mineralocorticoid level and only serves to exacerbate the potassium wasting and hypertension.

High Renin, High Aldosterone States

Renin overproduction, unregulated by the extracellular volume, can occur with a renin-secreting tumor of the juxtaglomerular apparatus and intrinsic diseases of the renal vasculature, including renal artery stenosis and accelerated or malignant hypertension. In all these conditions, unregulated stimulation of renin and hence aldosterone occurs. Therapy consists of removal of the tumor or increasing renal arterial flow as appropriate. Palliation may be gained by using a converting enzyme inhibitor to diminish angiotensin II and hence aldosterone levels.

Low Renin, High Mineralocorticoid States

Tumors and hyperplasia of the adrenal gland are associated with primary overproduction of aldosterone and other mineralocorticoid hormones. Excess mineralocorticoid hormone production also occurs with some adrenal enzymatic defects (11β- or 17α-hydroxylase deficiencies) and in Cushing's disease or syndrome. Treatment is effected with spironolactone or surgery.

Ingestion of substances with intrinsic mineralocorticoid activity or with the ability to inhibit 11β-hydroxysteroid dehydrogenase can result in a clinical state of hyperaldosteronism. The 11β-hydroxysteroid dehydrogenase normally metabolizes the abundant endogenous glucocorticoids to prevent them from occupying and activating the mineralocorticoid receptor. This enzyme can be inhibited by substances that contain glycyrrhizic acid, such as licorice, chewing tobacco, or carbenoxolone. A syndrome caused by an unidentified endogenous mineralocorticoid is Liddle's syndrome.

SYMPTOMS

The symptoms of metabolic alkalosis are summarized in Fig 13–6.

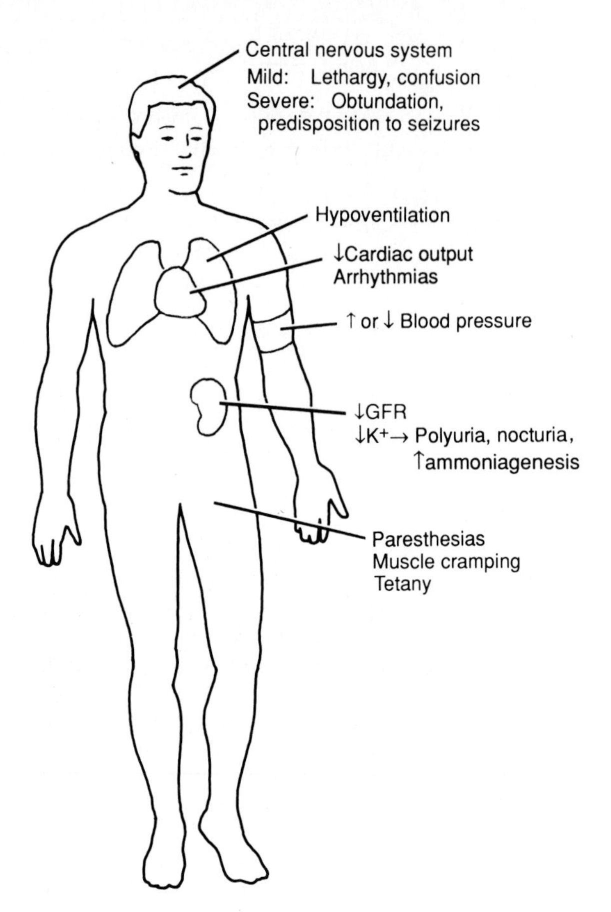

Figure 13–6. Signs and symptoms of metabolic alkalosis.

NERVOUS SYSTEM

To a large extent, the effects of alkalemia on neuromuscular function are similar to those of **hypocalcemia.** Alkalemia increases binding of calcium to protein (especially albumin) and thereby diminishes the free, ionized component of total plasma calcium. Total (bound plus free) plasma calcium remains unchanged.

Alkalemia can potentially cause central nervous system dysfunction, including lethargy and confusion in mild cases and obtundation with predisposition to seizures as the condition worsens. As with the hypocalcemia, alkalemia can provoke paresthesias, muscle cramping, and even tetany (Chapter 16). The Chvostek and Trousseau signs can often be elicited. Respiratory muscle paralysis is one of the most serious consequences of metabolic alkalosis. Potassium deficiency associated with the alkalosis can exacerbate muscle weakness.

CARDIOVASCULAR

In a patient with preexisting heart disease, alkalemia decreases cardiac output. Of great importance is the predisposition to supraventricular and ventricular arrhythmias. The almost universal association of potassium deficiency with metabolic alkalosis increases arrhythmogenic potential. There are no systematic electrocardiographic changes of alkalemia except those attributable to ionized hypocalcemia and hypokalemia (Chapters 8 and 16). The change in blood pressure depends on the cause of the alkalosis and directional change in extracellular volume; either hypotension or hypertension can occur.

RESPIRATORY

The decrease in ventilatory drive induced by metabolic alkalosis and changes in oxyhemoglobin binding results in a predictable increase in P_{CO_2} and decrease in P_{O_2}. If a patient has preexisting pulmonary disease, this hypoventilatory response can precipitate overt respiratory failure.

RENAL

Alkalemia itself has little influence on renal function except to reduce the potential for bicarbonate reabsorption. The secondary potassium depletion causes a decrease in GFR, which depresses concentrating ability and frequently causes increased ammonia production (Chapter 8). There is also a tendency for a decrease in calcium excretion and enhanced citrate excretion.

DIAGNOSIS

APPROACH TO ETIOLOGIC DIAGNOSIS

History & Physical Examination

The history usually provides a reliable indication of how the metabolic alkalosis was generated. Base intake in the form of acetate (hyperalimentation solution) or citrate (in blood transfusions) should be quantitated. There should be a careful search for gastrointestinal disease (for causes of vomiting or magnesium or potassium depletion).

Physical examination is crucial for ascertaining the extracellular volume status and blood pressure, since the metabolic alkaloses are generally divided etiologically on this basis. Signs of glucocorticoid or mineralocorticoid excess or renal arterial disease should be sought in a hypertensive, alkalemic patient.

Drug History

Although diuretic usage is usually known, surreptitious diuretic intake is suspected in some individuals without other cause of hypokalemic alkalosis and can be confirmed by direct analysis of the urine for diuretics and their metabolites. Ingestion or chewing of substances that enhance endogenous mineralocorticoid activity should be sought, such as licorice or chewing tobacco.

Urine Sodium & Chloride Concentrations

As shown in Table 13-4, metabolic alkaloses can be divided into (1) those in which maintenance of the alkalotic state is primarily dependent on extracellular volume contraction and hence will be expected to be repaired with normal saline and (2) those maintained by a high, unregulated aldosterone level. The former condition is saline-responsive, while the latter condition is saline-resistant.

A low urinary sodium or chloride concentration reflects a diminished extracellular volume state and hence a saline-responsive alkalosis. In most circumstances, sodium and chloride concentrations in the urine are similar, and the former is usually used (see Chapters 1 and 2). However, as shown in Fig 13-4, during the active phase of vomiting and for about 24 hours thereafter, there may be an obligate excretion of sodium with bicarbonate. In this case, the contracted extracellular volume is better reflected by the urinary chloride concentration, which remains low. An exception to this rule occurs in the case of active diuretic intake, in which the urine contains both sodium and chloride despite extracellular volume contraction. Discontinuation of the diuretic then causes the urine chloride concentration to fall to an appropriately low value.

In primary hyperaldosteronemic, saline-resistant alkalosis, the patient is in sodium chloride balance (due to mineralocorticoid escape), though at the expense of extracellular volume expansion and hypertension. Urinary sodium and chloride output is thus equal to intake, typically 100-200 meq/d, and the chloride concentration is greater than the 20 meq/L cut-off shown in Table 13-4 for distinguishing the two forms of metabolic alkaloses.

Coincident Acid-Base & Electrolyte Disturbances.

The serum potassium concentration is invariably low in metabolic alkalosis and should be checked for degree of replacement therapy (see Chapter 8). Hyponatremia is a common finding in

Table 13-4. Diagnosis and saline responsiveness of metabolic alkalosis.

	Saline-Responsive	Saline-Resistant
Extracellular fluid volume	↓	↑
Blood pressure	↓ (orthostatic change)	↑
Urine [Cl⁻]	↓ (< 20 meq/L)	↑ (> 20 meq/L)
Examples:	Vomiting, nasogastric aspiration Diuretics[1] Posthypercapnia Nonreabsorbable anions Therapy of organic acidosis	Primary hyperreninism Primary hyperaldosteronism Alkali + ↓GFR

[1]Urine [Cl⁻] ↑ with active diuretic use, ↓ after diuretics discontinued.

those alkaloses associated with hypovolemia and consequent baroreceptor stimulation of ADH release and water retention (Chapter 5). Hypomagnesemia may occur as a result of gastrointestinal or renal losses (see Chapter 22) and contribute to the perpetuation of the alkalosis by causing hyperreninemia.

Vomiting is sometimes superimposed on a preexisting acid-base disorder, such as ketoacidosis or sepsis. In such cases, complex acid-base disturbances are often expected, indicated by the presence of an abnormally elevated anion gap or inappropriate ventilatory response (Chapter 11).

THERAPY

Therapy of metabolic alkalosis is aimed at eliminating the generation of bicarbonate and reversing those factors that have maintained the alkalotic state. The approach is generally dictated by the division of disorders into two groups, shown in Table 13-4: whether the alkalosis is maintained by extracellular volume depletion, and therefore saline-responsive; or is primarily mineralocorticoid-driven and therefore saline-resistant.

A mild degree of alkalemia (arterial pH 7.40–7.50) is usually well tolerated and need not be treated except in patients with preexisting cardiac or pulmonary disease. If alkalosis is severe (pH > 7.60) or there are cardiac, pulmonary, or neuromuscular symptoms of alkalemia, the disorder should be rapidly treated, as summarized in Table 13-5.

TREATMENT OF SALINE-RESPONSIVE METABOLIC ALKALOSIS

In the common forms of metabolic alkalosis, due to vomiting or nasogastric suction or to diuretics, normal saline to repair extracellular volume depletion plus potassium repletion induce a brisk bicarbonaturic response and promptly repair the alkalosis (Fig 13-4). The amount, route, and quantity of saline and potassium depend on the degree of hypovolemia and potassium deficiencies, discussed in detail in Chapters 2 and 8.

Metabolic alkalosis of the saline-responsive variety sometimes develops in a patient in whom saline administration would be hazardous. This may occur in a patient in the intensive care unit with severe cardiac disease or with a pulmonary capillary leak syndrome. In both cases, volume administration to correct the alkalosis might cause pulmonary edema. The optimal therapeutic option in these cases is intravenous acetazolamide, 250–500 mg every 4–6 hours, to disrupt renal bicarbonate reabsorptive capacity. Another approach is the in-

Table 13-5. Therapy of metabolic alkalosis.

Saline-responsive
- Standard
  - NaCl + KCl
- Impaired cardiac or pulmonary function
  - Acetazolamide, 250–500 mg IV every 4–6 hours
  - HCl 0.1 mmol/L in central vein
- Impaired renal function
  - Dialysis with low dialysate [HCO_3^-] or [acetate⁻]

Saline-resistant
- Diagnosis-specific
  - Remove source of ↑mineralocorticoid
  - Converting enzyme inhibitor (primary hyperreninemia) or spironolactone (primary hyperaldosteronism)

fusion of hydrochloric acid, 0.1 mol/L, slowly into a central vein (owing to the sclerosing nature of this solution) to directly titrate plasma bicarbonate. The amount of acid given is calculated to decrease the plasma bicarbonate concentration by one-half over 2–4 hours, assuming a bicarbonate volume of distribution of 0.5 × body weight. Other alternatives include the infusion of ammonium chloride or arginine hydrochloride, but these are dangerous in a patient with renal or hepatic disease. Finally, dialysis can be used in a patient with poor renal function employing a dialysate with reduced buffer content (20 meq/L of either bicarbonate or acetate).

TREATMENT OF SALINE-RESISTANT METABOLIC ALKALOSIS

Removal of Source of Hyperaldosteronism

Since aldosterone excess is independent of extracellular volume, saline does no good in correcting and may actually worsen blood pressure in this group of metabolic alkaloses. Therapy of these saline-resistant alkaloses is aimed at the source of the mineralocorticoid, whether tumor removal (kidney, pituitary, or adrenal), repair of renal arterial perfusion (renal artery stenosis), or discontinuation of hypermineralocorticoid-promoting substance (licorice or chewing tobacco), as indicated.

Blockade of Renin-Angiotensin-Aldosterone Axis

In a situation in which the mineralocorticoid cannot be removed or in preparation for surgery, the effect of increased renin can be diminished by a converting enzyme inhibitor and of aldosterone by spironolactone (Chapter 3). A large dose of these agents is sometimes necessary, titrated to correct the potassium wasting and metabolic alkalosis. Potassium repletion is required to repair the hypokalemia (Chapter 8).

14 Respiratory Acidosis & Alkalosis

BUFFERING & RENAL RESPONSE

The acute nonbicarbonate buffering of the blood and the impact on arterial pH of changes in P_{CO_2} were portrayed in Fig 10–18. The subsequent renal responses and alteration in plasma bicarbonate concentration were shown in Figs 10–18 and 11–1. Hypercapnia stimulates net acid excretion to generate a high blood bicarbonate concentration and elevates bicarbonate reabsorptive capacity in the proximal tubule to sustain the hypercarbonatemia. Hypocapnia does the opposite.

ETIOLOGY & DIAGNOSIS

As shown in Table 14–1, respiratory acidosis can arise from obstruction of the airways, depression of the respiratory center, or neuromuscular or restrictive defects of the chest. As shown in Table 14–2, respiratory alkalosis can derive from primary stimulation of the respiratory center or pulmonary diseases.

The approach to diagnosis of respiratory disorders has been previously summarized in Figs 12–1 and 13–1. It is important to verify that the appropriate renal compensation to the degree of P_{CO_2} change has occurred using the nomogram of Fig 11–1. The anion gap is normal unless there is superimposed metabolic acidosis.

SYMPTOMS

HYPERCAPNIA

Nervous System

Central nervous system manifestations acutely include headache, anxiety, confusion, and psychosis, extending to stupor and coma. A narcotic-like effect is seen chronically, frequently with myoclonus and asterixis. Signs of increased intracerebral pressure may develop (pseudotumor cerebri).

Cardiovascular

Peripheral vasodilation is the prominent vascular effect of hypercapnia. There are inconsistent changes in pulmonary resistance and cardiac conductivity.

Renal

Hypercapnia increases ammonia production and hydrogen ion secretion to augment urinary net acid excretion and to raise proximal tubule bicarbonate reabsorptive capacity (Chapter 10), diminishes proximal sodium chloride reabsorption (Chapter 1), and tends to increase potassium secretion though without causing hypokalemia (Chapter 7).

HYPOCAPNIA

Nervous System

The principal findings relate to the diminution of cerebral blood flow induced by acute respiratory alkalosis and include symptoms of lightheadedness and confusion with a predisposition to seizures. There may also be symptoms of paresthesias, chest tightness, and circumoral numbness.

Cardiovascular

Initially, cardiac output tends to fall with hypocapnia, but later it rises, with a fall in peripheral vascular resistance.

Table 14–1. Causes of respiratory acidosis.[1]

Acute	Chronic
Airway obstruction	**Airway obstruction**
Aspiration of foreign body or vomitus	Chronic obstructive lung disease (bronchitis, emphysema)
Laryngospasm	
Generalized bronchospasm	
Obstructive sleep apnea	
Respiratory center depression	**Respiratory center depression**
General anesthesia	Chronic sedative overdosage
Sedative overdosage	Primary alveolar hypoventilation (Ondine's curse)
Cerebral trauma or infarction	Obesity-hypoventilation syndrome (Pickwickian syndrome)
Central sleep apnea	Brain tumor
	Bulbar poliomyelitis
Circulatory catastrophes	
Cardiac arrest	
Severe pulmonary edema	
Neuromuscular defects	**Neuromuscular defects**
High cervical cordotomy	Polymyositis
Botulism, tetanus	Multiple sclerosis
Guillain-Barré syndrome	Muscular dystrophy
Crisis in myasthenia gravis	Amyotrophic lateral sclerosis
Familial hypokalemic periodic paralysis	Diaphragmatic paralysis
Hypokalemic myopathy polymyositis	Myxedema
Drugs or toxic agents (eg, curare, succinycholine, aminoglycosides, organophosphorus)	Myopathic disease polymyositis
Restrictive defects	**Restrictive defects**
Pneumothorax	Kyphoscoliosis, spinal arthritis
Hemothorax	Fibrothorax
Flail chest	Hydrothorax
Severe pneumonitis	Interstitial fibrosis
Infant respiratory distress syndrome (hyaline membrane disease)	Decreased diaphragmatic movement (eg, ascites)
Adult respiratory distress syndrome	Prolonged pneumonitis
	Obesity
Mechanical hypoventilation	

[1]Adapted from Madias NE, Cohen JJ: Respiratory acidosis. In: *Acid/Base.* Cohen JJ, Kassirer JP (editors). Little, Brown, 1982.

Table 14–2. Causes of respiratory alkalosis.[1]

Hypoxia
- Decreased inspired oxygen tension
- High altitude
- Ventilation-perfusion inequality
- Hypotension
- Severe anemia

CNS-mediated
- Voluntary hyperventilation
  - Anxiety-hyperventilation syndrome
- Neurologic disease
- Cerebrovascular accident (infarction, hemorrhage)
  - Infection (encephalitis, meningitis)
  - Trauma
  - Tumor
- Pharmacologic and hormonal stimulation
  - Salicylates
  - Dinitrophenol
  - Nicotine
  - Xanthines
  - Pressor hormones
  - Pregnancy (progesterone)
- Hepatic failure
- Gram-negative septicemia
- Recovery from metabolic acidosis
- Heat exposure

Pulmonary disease
- Interstitial lung disease
- Pneumonia
- Pulmonary embolism
- Pulmonary edema

Mechanical overventilation

[1]Adapted from Gennari FJ, Kassirer JP: Respiratory alkalosis. In: *Acid/Base.* Cohen JJ, Kassirer JP (editors). Little, Brown, 1982.

Metabolic

Hypocapnia increases organic acid production, such as lactic and citric acids. The anion gap may therefore be slightly increased in acute respiratory alkalosis.

Renal

Hypocapnia reduces bicarbonate reabsorptive capacity and transiently inhibits net acid excretion (Chapter 10), and initially promotes a mild kaliuresis (Chapter 7), though hypokalemia does not occur.

THERAPY

The general approach to respiratory disorders is removal of the underlying cause. In acidosis, therapy is directed at improving ventilation: relieving airway obstruction, improving oxygenation, or diminishing bronchospasm as indicated.

SECTION IV REFERENCES: ACID-BASE HOMEOSTASIS & DISORDERS: ACIDOSIS & ALKALOSIS

Cogan MG: Disorders of proximal nephron function. Am J Med 1982;72:275–288.

Cogan MG: Regulation and control of bicarbonate reabsorption in the proximal tubule. Sem Nephrol 1990;10:115–121.

Cogan MG, Rector FC Jr: Acid-base disorders. In: *The Kidney*, 4th ed. Brenner BM, Rector FC Jr (editors). Saunders, 1990. [In press.]

Hsueh WA: New insights into the medical management of primary aldosteronism. Hypertension 1986;8:76–82.

Schambelan M, Sebastian A: Hyporeninemic hypoaldosteronism. Annu Rev Med 1979;24:385–405.

Sebastian A et al: Disorders of distal nephron function. Am J Med 1982;72:289–307.

Seldin DW, Rector FC Jr: The generation and maintenance of metabolic alkalosis. Kidney Int 1972;1:306–321.

Section V: Calcium Homeostasis & Disorders: Hypocalcemia & Hypercalcemia

Normal Calcium Homeostasis 15

INTRODUCTION

TOTAL BODY CALCIUM

The body of a normal 70-kg adult contains about 1200 g of calcium, of which over 99% is in bone, as shown in Fig 15–1. Of the remainder, most (0.6%) is sequestered within cells, and only about 0.1% (1.3 g) is in the extracellular fluid.

Bone Calcium

Calcium in bone is bound primarily in the form of hydroxyapatite crystal ($3[Ca_3(PO_4)]Ca(OH)_2$). A thin film of water surrounds these crystals, with the calcium and phosphate ions in solution. Bone is in dynamic equilibrium with extracellular calcium by means of osteoclast-mediated resorption (about 10–20 meq/d) and diffusional exchange with the surface film (about 100–200 meq/d).

Intracellular Calcium

Most intracellular calcium is complexed and sequestered by various ligands within membrane structures. Within the cell, the free, ionized calcium concentration—about 20–100 nmol/L—is over four orders of magnitude less than that in the extracellular volume. Strong active transport mechanisms maintain this low intracellular calcium concentration by pumping calcium out of the cell or into intracellular organelles, such as the endoplasmic reticulum and mitochondria. The ionized calcium within the cell serves important second messenger functions and regulates many enzymes.

Extracellular & Plasma Calcium

While only a small fraction of total body calcium exists in the extracellular fluid, plasma ionized calcium concentration plays a critical role in neuromuscular and cardiac excitability.

Under routine conditions, the total calcium content of plasma is measured and is usually reported in milligrams per deciliter. The normal total plasma calcium is **9.0–10.0** mg/dL (or 4.5–5.0 meq/L). However, it is only the free, unbound, ionized calcium that is physiologically active. As shown in Fig 15–1, free, ionized calcium is about 50% of the total plasma calcium, normally averaging 4.5–5.0 mg/dL (or 2.2–2.5 meq/L).

About 10% of total plasma calcium, or 0.9–1.0 mg/dL, is complexed to various anions, including bicarbonate, citrate, and phosphate. The remaining 40% of total calcium, or 3.6–4.0 mg/dL, is bound to plasma proteins, mostly to albumin (75% of the protein-bound calcium, or 30% of the total calcium) and the rest to globulins.

Changes in Total Versus Free Plasma Calcium Concentration

Change in the free, ionized calcium concentration of plasma sometimes is not proportionate to that of total calcium concentration. Since the ionized calcium concentration is the physiologically important fraction, changes in total calcium may

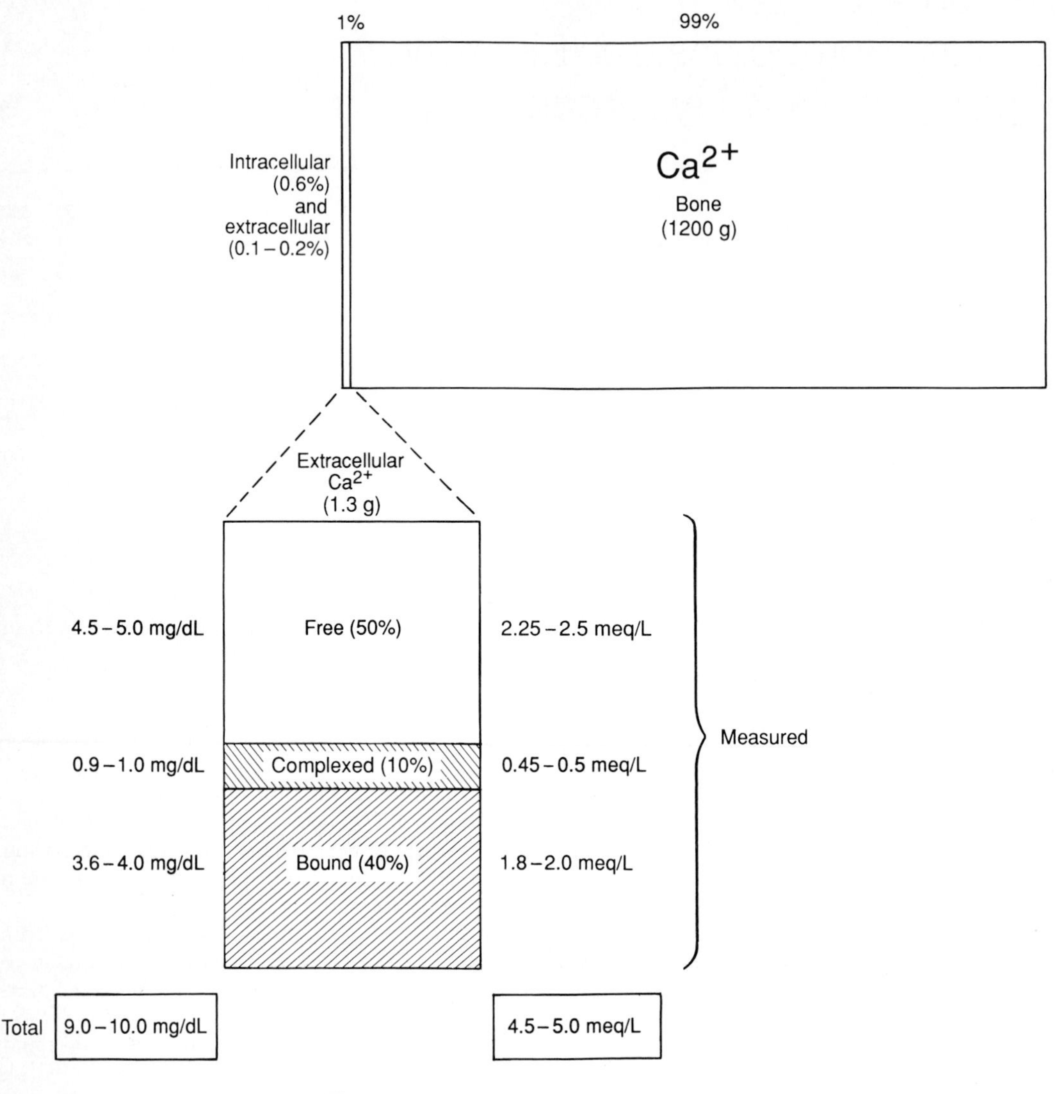

Figure 15-1. Distribution of body calcium.

or may not be reflective of homeostatic disorder, as during alteration in plasma albumin concentration or pH.

Since albumin in plasma has a normal concentration of about 4 g/dL and binds to about 3 g/dL of total calcium, a decrease in 1 g/dL of albumin diminishes bound and hence total calcium concentration by about 0.75 (range 0.7–1.0) mg/dL. For instance, in a hypoalbuminemic state with albumin concentration of 2 g/dL (2 g/dL less than normal), total calcium would be expected to be diminished by about 1.5 mg/dL to 7.5–8.5 mg/dL, but the ionized calcium concentration would remain unchanged. In such a case, a decrease in total plasma calcium would not be indicative of true hypocalcemia (properly referring to the ionized component).

A change in arterial pH shifts the balance of ionized to bound and complexed calcium. An increase in pH increases the fraction of total calcium complexing to albumin and to bicarbonate. The combined effect is that plasma ionized calcium decreases by 0.16 mg/dL for an increment of 0.10 unit rise in arterial pH. For instance, if pH increased to 7.55 (a rise of 0.15 pH units), ionized calcium would be expected to be diminished by 0.24 mg/dL. In this case, total calcium would be unchanged, but the ionized:bound ratio would decrease, physiologically predisposing to symptoms of hypocalcemia.

Plasma sodium concentration within the physiologic concentration range also has a small effect on protein affinity for calcium. Hyponatremia increases albumin binding of calcium, and hypernatremia has the opposite effect.

REGULATORY HORMONES INVOLVED IN CALCIUM HOMEOSTASIS

Control of plasma calcium concentration is brought about by intestinal, bone, and renal mechanisms which by themselves are insufficient to ensure full homeostasis. For this regulatory purpose, two hormones are highly sensitive to and critically important in maintaining normality of plasma calcium concentration: parathyroid hormone and vitamin D.

Parathyroid Hormone (PTH)

PTH is an 84-amino-acid peptide (MW 9500) synthesized in the chief cells of the four parathyroid glands. It is derived from a preprohormone of 115 amino acids. After secretion by the parathyroid gland, PTH is cleaved into two fragments. The N-terminal 1–34 cleavage product is the functional element, with a biologic half-life of about 100 minutes. The other C-terminal 35–84 fragment is biologically inactive and has a longer half-life. PTH assays sensitive to the C-terminal portion of the molecule may be markedly elevated in states in which GFR is reduced, when the C-terminal fragment is preferentially retained, and may not accurately reflect the magnitude of 1–34 fragment elevation.

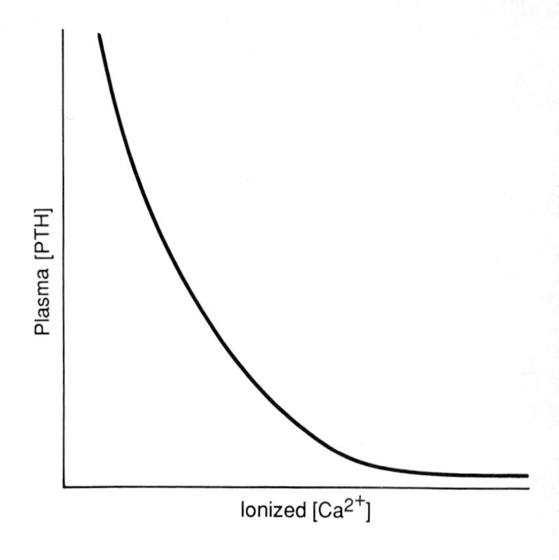

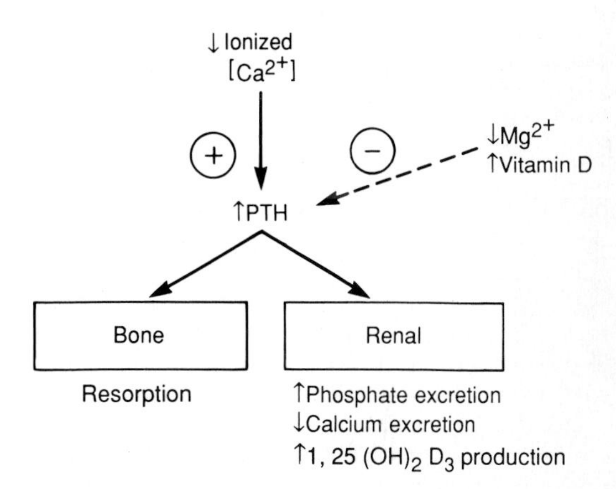

Figure 15-2. Regulation and actions of parathyroid hormone (PTH).

As shown in Fig 15–2 (top), hypocalcemia is a potent secretagogue for PTH, probably by decreasing intracellular calcium concentration in the chief cell. PTH secretory stimulation by low intracellular calcium is opposite to that for all other peptide hormone release (except renin) in which a rise in intracellular calcium concentration causes secretion. PTH release is independently inhibited by the active form of vitamin D (1,25($[OH]_2D_3$) and by hypomagnesemia.

PTH binds to a high-affinity receptor on the plasma membrane of its target cells—predominantly in bone and kidney but also other tissues—to activate a stimulatory G protein (G_s) coupled to adenylate cyclase and thus to increase intracellular cyclic AMP (cAMP) concentration (Fig 10–6). Most but probably not all of the biologic effects of PTH are cAMP-mediated. The urinary excretion of cAMP has been used to reflect the renal activity of PTH. As shown in Fig 15–2 (bottom), a major action of PTH includes resorption of bone that mobilizes calcium and phosphorus. This osseous action of PTH requires normal levels of vitamin D and magnesium. PTH also acts directly on the kidney to increase urinary phosphorus excretion and to decrease calcium excretion, and stimulate production of 1,25$(OH)_2D_3$.

Vitamin D

As shown in Fig 15–3, the active metabolite of vitamin D is 1,25$(OH)_2D_3$. This hormone is derived from conversion in the skin by ultraviolet light of 7-dehydrocholesterol to vitamin D_3, which is also derived from diet. The vitamin D_3 is 25-hydroxylated in the liver to 25$(OH)D_3$, a form of vitamin D with modest biologic activity but which is physiologically unregulated. The further modification by the 1α-hydroxylase in mitochondria of

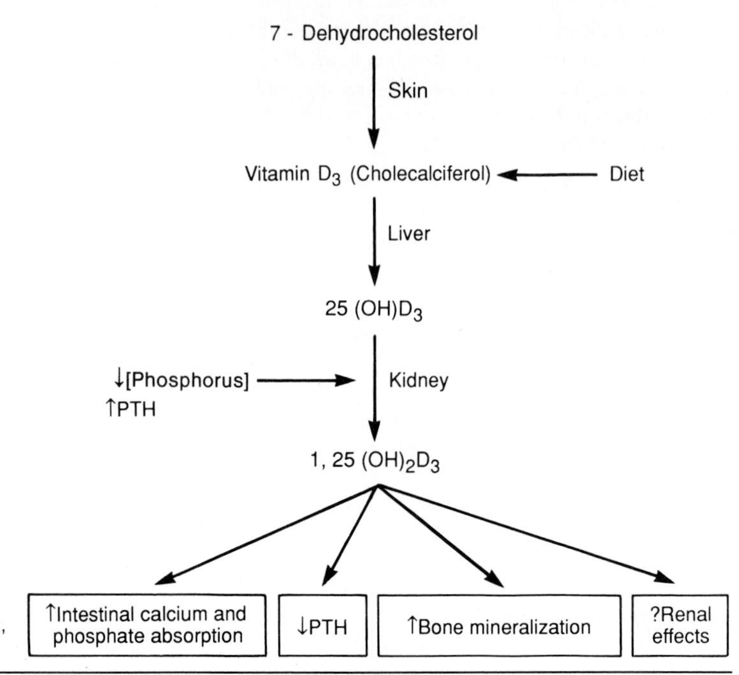

Figure 15-3. Metabolism, regulation, and actions of vitamin D.

proximal tubule cells converts this precursor into the biologically very active form of vitamin D, $1,25(OH)_2D_3$.

Renal 1α-hydroxylase activity and hence $1,25(OH)_2D_3$ production is stimulated by decrease in plasma phosphorus level or increase in parathyroid hormone level.

$1,25(OH)_2D_3$ circulates in plasma tightly bound (> 99%) to protein. Target tissues, such as intestine, bone, parathyroid gland, and perhaps kidney, have cytoplasmic receptors that are involved in transcriptional regulation and perhaps other functions. $1,25(OH)_2D_3$ increases intestinal calcium and phosphorus absorption. As mentioned above, $1,25(OH)_2D_3$ decreases PTH secretion. It may directly increase bone resorption but also plays a permissive role in the PTH mediated enhancement of bone resorption. Finally, $1,25(OH)_2D_3$ may increase renal calcium reabsorption, but this is not proved.

CALCIUM INTAKE, ABSORPTION, & DISTRIBUTION

CALCIUM INTAKE

As shown in Fig 15-4, dietary calcium averages about 900 mg/d in the average adult male, though with a very broad range (95% confidence limits, 200-3000 mg/d). Women ingest less calcium on average—about 550-700 mg/d—again with a very wide range (100-1700 mg/d). Calcium ingestion tends to diminish with age and averages about two-thirds the values listed above for people over 65 years of age. Blacks also tend to ingest less calcium—on average, about two-thirds of what is ingested by whites.

INTESTINAL CALCIUM ABSORPTION

Sites & Mechanism

Under normal conditions, intestinal absorption of calcium is 30-40% of intake. As shown in Fig 15-4, when calcium intake is 900 mg/d, intestinal absorption is about 350 mg/d. Most of the calcium absorption occurs in the proximal small intestine (duodenum and jejunum), though absorption occurs in the ileum and colon as well. Calcium absorption is complete within about 4 hours after intake.

The mechanism of calcium absorption involves passive permeation of calcium from the intestinal lumen into the cell. Once in the cell, the calcium must be actively pumped out across the basolateral membrane by means of an energy-requiring process, probably a calcium-ATPase pump.

There is a simultaneous secretory flux of cal cium into the intestinal lumen of about 150 mg/d,

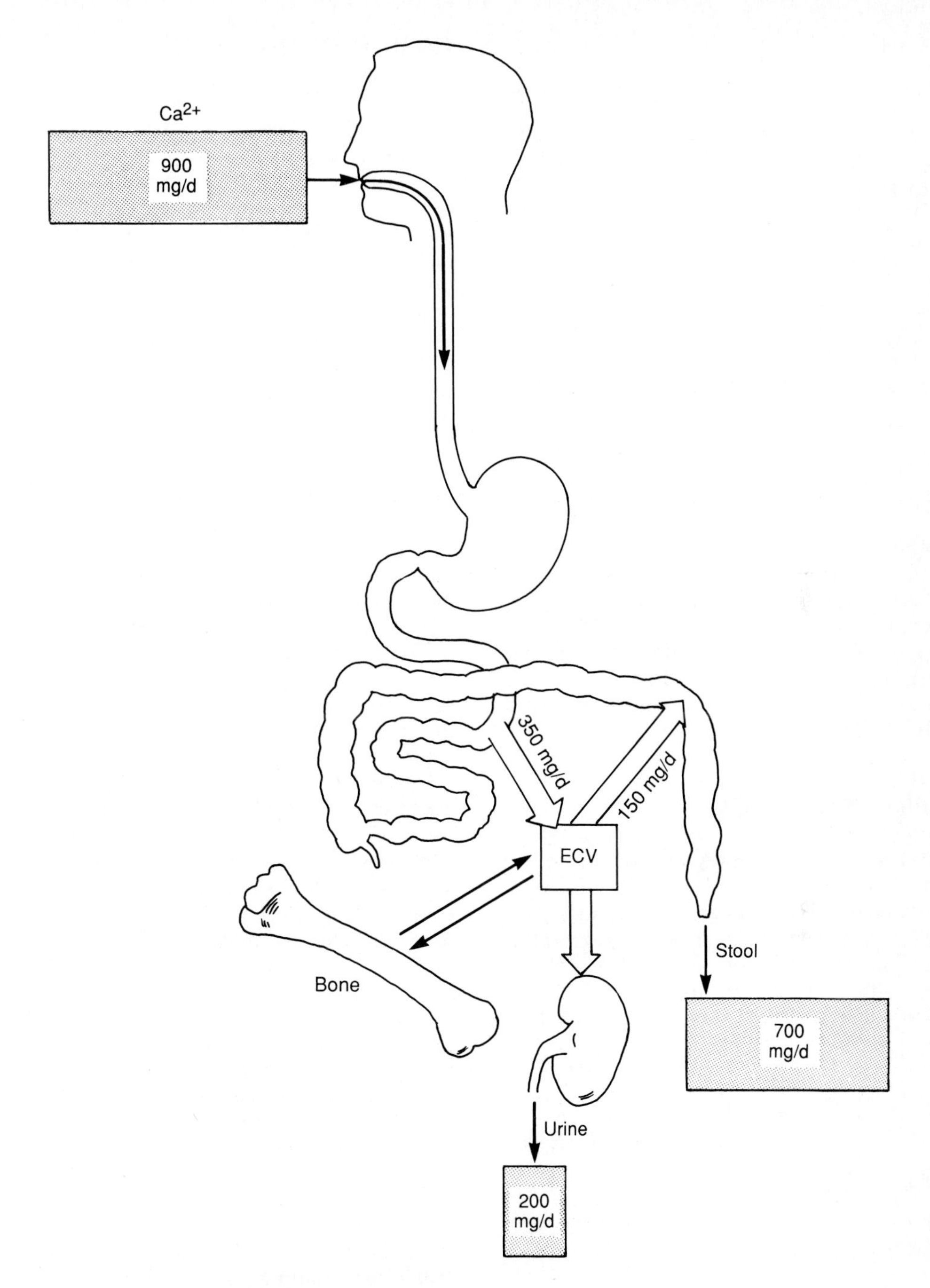

Figure 15-4. Normal calcium intake and excretion.

as shown in Fig 15-4. This process is probably passive and does not appear to be physiologically regulated. Thus, net input of calcium in the extracellular volume is the difference between unidirectional absorption (350 mg/d) and secretion (150 mg/d), or 200 mg/d.

Intestinal calcium absorption rises in proportion to calcium intake, as shown in Fig 15-5. However, calcium balance can become negative with dietary calcium restriction because there is continued obligatory loss of calcium in feces (due to unregulated intestinal calcium secretion) and urine (due to incomplete calcium reabsorption). Therefore, dietary calcium intake substantially less than about 400 mg/d can be associated with negative calcium balance.

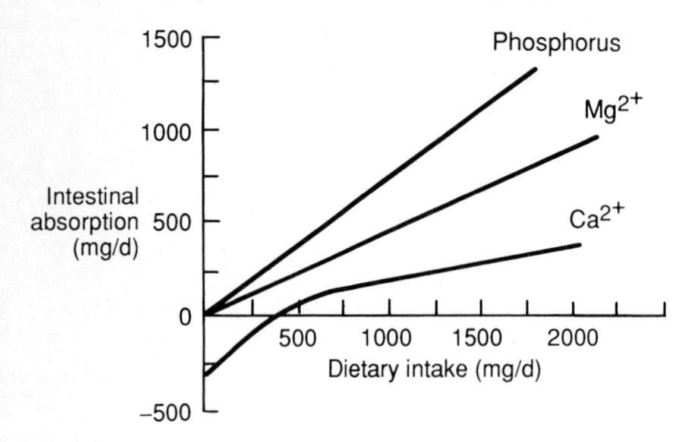

Figure 15-5. Intestinal absorption of divalent ions as a function of intake.

Regulation by Vitamin D

$1,25(OH)_2D_3$ enhances calcium uptake in virtually all segments of the intestine. The mechanism of regulation of transepithelial calcium transport appears to be control of calcium translocation across the luminal membrane of the intestinal cell. Increased basolateral pumping would then be a secondary consequence.

CALCIUM DISTRIBUTION

As shown in Fig 15-1, over 99% of the body's calcium resides in bone, some of which is in dynamic equilibrium with extracellular fluid. This vast reservoir of bone attenuates changes in plasma calcium concentration—acts as a calcium "buffer"—that would otherwise occur as a result of primary alteration in dietary intake or renal excretion.

RENAL CALCIUM EXCRETION

As shown in Fig 15-6, the normal ultrafiltrable plasma calcium (free and complexed) is 6 mg/dL (60 mg/L). With a normal GFR of 150 L/d, there is then 9000 mg of calcium filtered daily. Since net dietary input is approximately 200 mg/d (Fig 15-6), it is clear that most calcium, 97-98% of the filtered load, must be reabsorbed to prevent wastage and negative calcium balance.

TUBULAR REABSORPTION OF CALCIUM

Proximal Tubule

Along the length of the proximal tubule, the calcium concentration remains relatively constant—only about 20% above the value that pertains in the glomerular ultrafiltrate. Since water is progressively reabsorbed along the proximal tubule, stability of calcium concentration implies that calcium is reabsorbed in proportion to sodium and water. Thus, about 50% of the filtered calcium is reabsorbed in the proximal convoluted tubule and another 10% in the proximal straight tubule.

The mechanism of calcium reabsorption in the proximal tubule is predominantly passive, through the paracellular pathway. Both solvent drag and diffusion, driven by the lumen-positive potential difference (the chloride diffusion potential—see Fig 1-23) participate in driving calcium transport.

There is also a minor active transport mechanism that effects transcellular calcium reabsorption. The luminal membrane entry of calcium into the proximal tubule cell is probably passive, down its electrochemical gradient. Whether the active extrusion of calcium across the basolateral membrane is via a $3Na^+/Ca^{2+}$ antiporter mechanism (see below) or a Ca^{2+}-ATPase pump is unknown.

As a generality, factors that diminish sodium and water transport in the proximal tubule (see Chapter 1) proportionately decrease calcium reabsorption, and vice versa. The most important of these factors is extracellular volume. Extracellular volume depletion, by decreasing GFR and enhancing the fraction of proximally reabsorbed sodium and water, also increases the fraction of filtered calcium that is reabsorbed. Extracellular volume expansion has the opposite effect. PTH does not substantially affect transport of calcium in the proximal tubule.

Thick Ascending Limb of Henle

While little calcium transport occurs in the thin limbs of Henle, a substantial amount of the filtered calcium, approaching 15-20%, is reabsorbed in the thick ascending limb of Henle. The major route of calcium transport in this segment appears to be paracellular, driven by the lumen-positive potential difference established by the luminal Na^+-K^+-$2Cl^-$ cotransporter (Fig 1-25). As such, the major determinants of calcium reabsorption are the lumen-to-blood calcium concentration gradient and the magnitude of the trans-

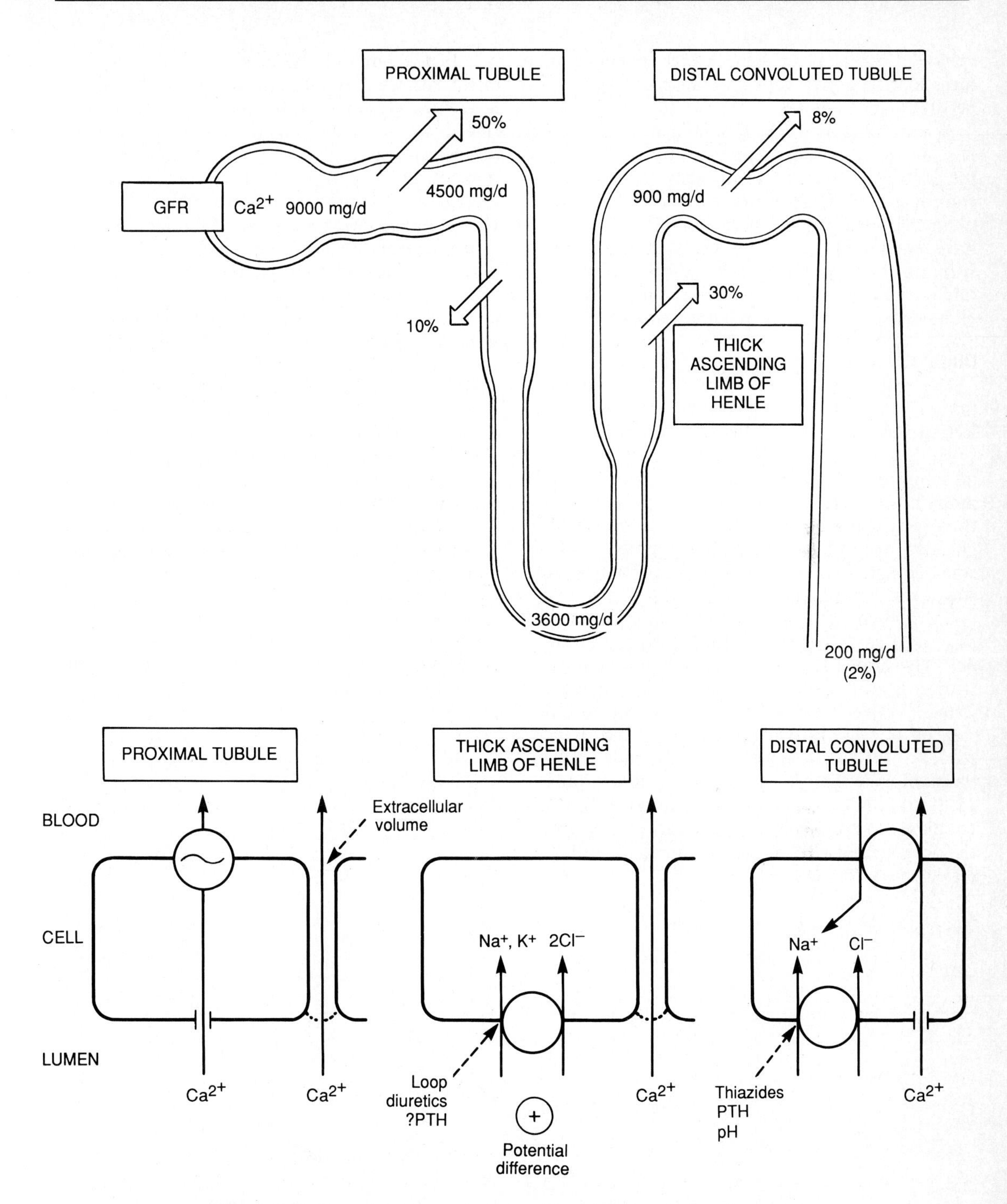

Figure 15-6. Normal amounts of calcium filtered, reabsorbed, and excreted.

epithelial potential difference. A contribution of transcellular, active calcium transport may also occur.

A loop diuretic, which inhibits the luminal Na^+-K^+-$2Cl^-$ cotransporter and thus diminishes the potential difference, secondarily depresses calcium transport. PTH, which tends to augment sodium chloride transport in the loop of Henle, may have the opposite effect, to increase calcium reabsorption. The physiologic roles of hypermagnesemia to diminish and of calcitonin to enhance calcium transport in this segment are unresolved.

Distal Convoluted Tubule

The distal convoluted tubule, with contributions also of the connecting tubule and early collecting tubule, is responsible for reabsorbing between 5% and 10% of the filtered calcium and is an important modulating site for controlling the amount of calcium that enters the urine. In contrast to the proximal tubule and loop of Henle, calcium reabsorption can be demonstrated under some conditions to be inversely proportionate to sodium reabsorption.

Calcium enters the cell of the distal convoluted tubule passively, down its electrochemical gradient. The principal mechanism of basolateral extrusion is an exchanger that moves three sodium ions into the cell for one calcium molecule out of the cell (ie, $3Na^+/Ca^{2+}$ antiporter). The driving force for this electrogenic process depends on sodium, because there is a steep concentration gradient (intracellular sodium concentration is very low compared to extracellular concentration, owing to the operation of Na^+/K^+-ATPase) and the cell interior is electronegative.

A thiazide diuretic affects sodium reabsorption in this segment by interfering with the transport mechanism responsible for moving sodium chloride from the lumen into the cell (Fig 1–27). In the presence of a thiazide, the resulting drop in intracellular sodium concentration provides a more favorable driving force for $3Na^+/Ca^{2+}$ exchange on the basolateral membrane. The accelerated extrusion of calcium serves to lower intracellular calcium and thus to secondarily enhance calcium entry across the luminal membrane. Overall, therefore, a thiazide diuretic decreases sodium reabsorption but increases calcium reabsorption.

PTH stimulates calcium transport in this segment. Calcium transport is also affected by systemic pH, though the mechanism is poorly understood. Calcium reabsorption is increased by metabolic alkalosis and decreased by metabolic acidosis. Other factors that may influence calcium transport in the distal convoluted tubule but are of uncertain physiologic significance include phosphate stores, vitamin D, insulin and glucose, and amiloride.

Collecting Tubule

Available evidence suggests the collecting tubule plays only a minor role in adjusting the amount of calcium which enters the urine.

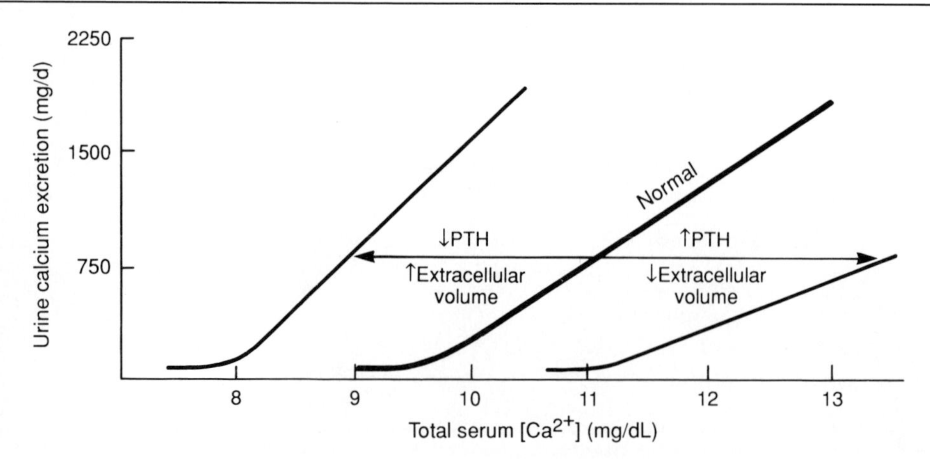

Figure 15-7. Regulation of calcium excretion.

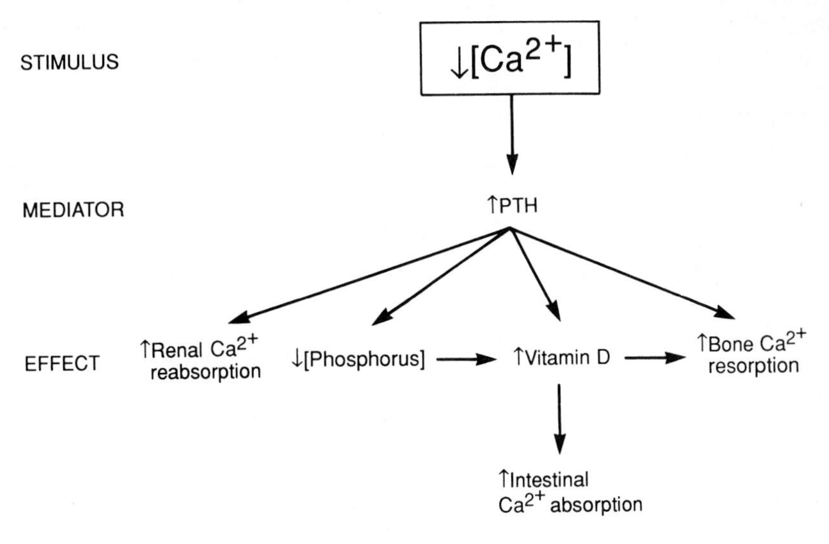

Figure 15-8. Response to hypocalcemia.

REGULATION & LIMITS OF RENAL CALCIUM EXCRETION

Filtered Calcium, Extracellular Volume, PTH

As detailed above, many factors influence renal calcium handling. Of these, urinary calcium excretion is regulated primarily by (1) the amount of calcium filtered (determined by the plasma ionized calcium concentration and GFR); (2) the extracellular volume status (affecting primarily the fraction of calcium reabsorbed in the proximal tubule); and (3) PTH (primarily affecting the calcium reabsorbed in the distal convoluted tubule).

As shown in Fig 15–7, a curvilinear relationship normally exists between urinary calcium excretion and plasma ultrafiltrable calcium concentration. The curve is shifted to the right by PTH excess and by hypovolemia and to the left by PTH deficiency and hypervolemia.

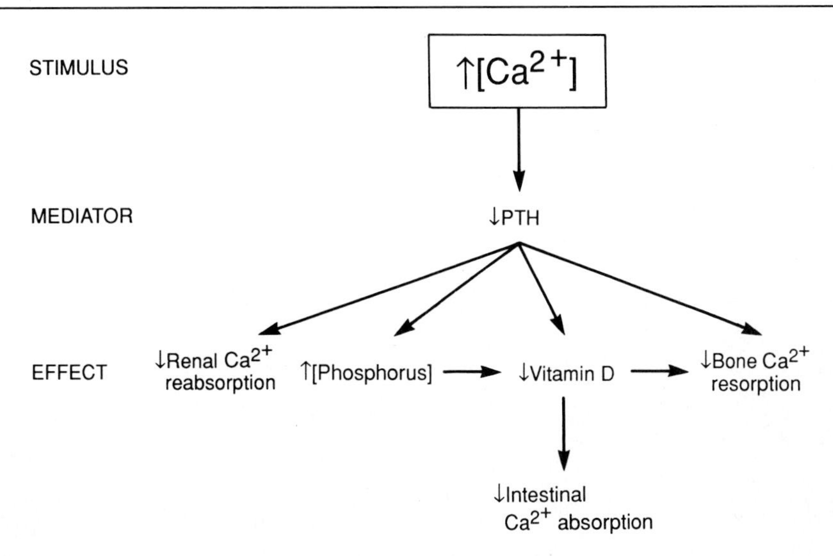

Figure 15-9. Response to hypercalcemia.

Limits of Renal Calcium Conservation & Excretion

As shown in Fig 15-7, under conditions of maximal stimulation by hypocalcemia and hyperparathyroidism, urinary calcium excretion can be as low as 50 mg/d. On the other hand, in response to hypercalcemia with hypoparathyroidism or hypervolemia, it has been difficult to demonstrate saturability of calcium excretion, at least within the relatively narrow sublethal range of plasma calcium concentration. Renal calcium excretory rates as high as 1500 mg/d have been recorded.

RESPONSE TO CHANGE IN PLASMA CALCIUM CONCENTRATION

The stability of plasma calcium concentration is largely dependent on the integrity of PTH action, as summarized in Fig 15-8. When PTH release is stimulated by a fall in plasma ionized calcium (Fig 15-2), it brings into play both directly and indirectly a series of coordinated renal, intestinal, and bone mechanisms to restore plasma calcium concentration. A rise in plasma calcium concentration has the opposite effects, as shown in Fig 15-9.

As shown in Fig 15-6, PTH acts predominantly in the distal convoluted tubule and perhaps the thick ascending limb of Henle to increase calcium transport. Thus, hypocalcemia is physiologically compensated by hypocalciuria and hypercalcemia by hypercalciuria.

Hyperparathyroidism also induces phosphaturia. The consequent reduction in plasma phosphorus concentration—in addition to the direct effects of PTH—serves to increase the renal 1αactivity to increase production of $1,25(OH)_2D_3$, which in turn increases intestinal calcium absorption (Fig 15-3). Thus, hypocalcemia is associated with stimulation of intestinal calcium absorption and hypercalcemia with suppression of intestinal calcium absorption.

Finally, a rise in PTH in response to hypocalcemia increases osteoclastic activity and mobilizes calcium from bone. Vitamin D helps PTH mobilize calcium from bone. Hypercalcemia is associated with the opposite effects.

Hypocalcemia

16

INTRODUCTION

DEFINITION

Reduction in Ionized Calcium

The biologically important component of the total plasma calcium concentration is the free, ionized calcium concentration (Fig 15–1). A change in total calcium concentration due to alteration in the fraction of calcium complexed or protein-bound should not be considered a primary disorder of calcium metabolism.

Although an ion-selective electrode can be used to directly measure ionized calcium concentration, this technique is generally not available in clinical laboratories. Instead, ionized calcium concentration is usually estimated from total plasma calcium concentration, corrected for any aberration in plasma albumin concentration or pH. Total plasma calcium concentration decreases by 0.75 mg/dL for a reduction of 1 g/dL in plasma albumin concentration and by 0.16 mg/dL for each 0.10 unit rise in arterial pH. Given these caveats, hypocalcemia is diagnosed when total plasma calcium concentration is **less than 8.5 mg/dL.**

PATHOPHYSIOLOGY

Calcium Sequestration Versus Functional Hypoparathyroidism

Reminiscent of the situation with potassium, only a minute portion (0.1%) of the total body content of calcium resides in the extracellular fluid (Fig 15–1). Therefore, a relatively small change in the distribution of calcium may easily result in overt hypocalcemia. Alternatively, given the strong dependence on PTH for calcium homeostasis (Fig 15–2), a deficiency of PTH activity (with secondary change in vitamin D metabolism) may affect renal, intestinal, and bone function to upset the normal balance of calcium absorption and excretion.

Thus, as shown in Fig 16–1, hypocalcemic disorders may be generally categorized into (1) those in which ionized calcium has been removed from the blood by various factors or (2) those in which there is deficient PTH action due to diminished secretion or target organ response.

INCREASED SEQUESTRATION OF CALCIUM

Causes of hypocalcemia are listed in Table 16–1.

HYPERPHOSPHATEMIA

These disorders are considered in more detail in Chapter 20.

A phosphorus load may be generated endogenously, as during chemotherapy of a bulky, poorly differentiated lymphoma—part of the tumor lysis syndrome. Exogenous phosphorus is also administered on occasion, as in antacids or hyperalimentation.

Even when the dietary phosphorus load is not excessive, acute or chronic renal failure (GFR < 30 mL/min) is predictably associated with hyperphosphatemia. Diminished $1,25(OH)_2D_3$ production in chronic renal failure aggravates the hypocalcemia. The low rate of $1,25(OH)_2D_3$ production is due both to structural loss of proximal tubule mass (the site of production) and functional suppression of 1α-hydroxylase by hyperphosphatemia and hyperparathyroidism.

Even though plasma calcium concentration becomes depressed, a marked elevation in the plasma phosphorus concentration can increase the calcium × phosphorus concentration product to a dangerously high level. When this concentration product exceeds 60, there is a predisposition to soft tissue deposition of the complex as calcium phosphate. Injury may occur to the heart (conduction system abnormalities), lungs (gas diffu-

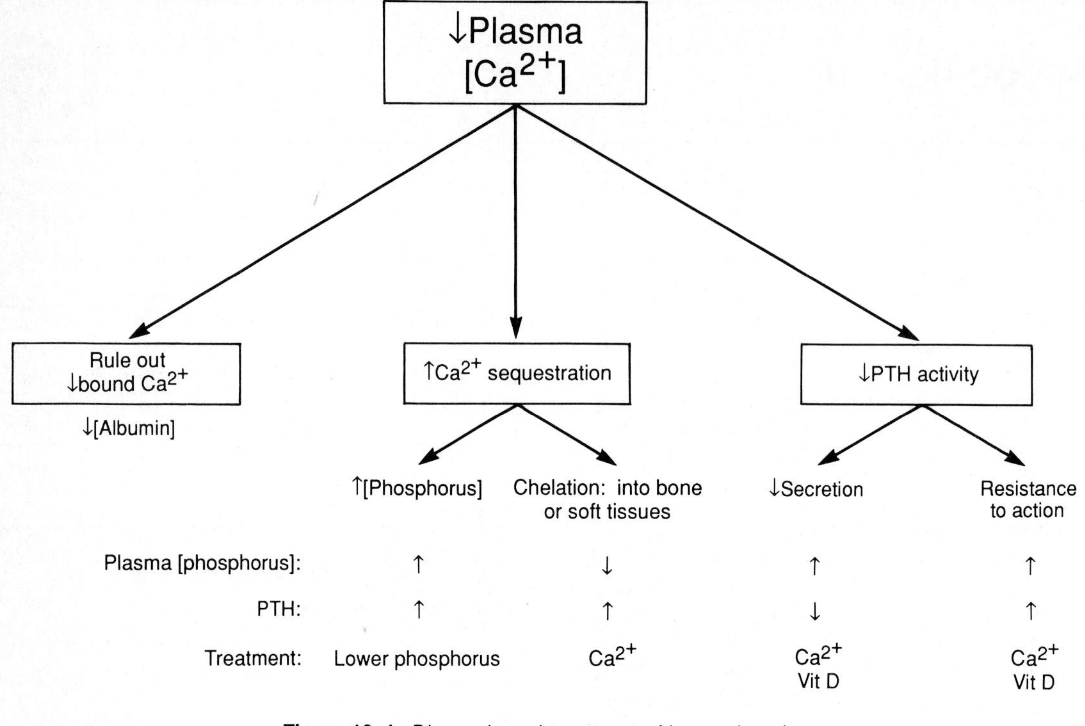

Figure 16-1. Diagnosis and treatment of hypocalcemia.

sion defects), kidneys (nephrocalcinosis), skin (extraosseous calcification), and brain.

Therapy in hyperphosphatemic disorders is discussed in Chapter 20. Strategies include dietary phosphorus restriction or chelation (with antacids), extracellular volume expansion to enhance phosphaturia in the tumor lysis syndrome, and hemodialysis if GFR is intrinsically depressed and cannot be increased by volume expansion. Calcium administration should not be used until the phosphorus concentration is brought under control, especially if the calcium $\times$ phosphorus concentration product is greater than 60.

CHELATION

Citrate or Edetate Calcium Disodium (EDTA)

The anticoagulant in blood is citrate, which is specifically used to chelate calcium, a necessary cofactor for various clotting factors. Each unit of blood contains 6 meq of citrate. Thus, massive transfusions (usually $>$ 10 units) often require concurrent calcium replacement through a different vascular access. Administration of the calcium chelator edetate calcium disodium can also induce hypocalcemia.

SOFT TISSUE DEPOSITION

Acute Pancreatitis or Rhabdomyolysis

Fat necrosis and precipitation of calcium soaps appears to be a major mechanism of hypocalcemia in acute pancreatitis or rhabdomyolysis. There is also induced a poorly understood PTH resistance state, a propensity to vitamin D deficiency, and hypomagnesemia, which all contribute to the hypocalcemia. Symptomatic hypocalcemia is treated with intravenous calcium without vitamin D. Calcium is released from soft tissue sites during the recovery phase of pancreatitis or rhabdomyolysis and can cause a rebound condition of hypercalcemia.

BONE DEPOSITION

Osteoblastic Metastases

Rapid growth of osteoblastic metastases of breast, prostate, or some lung cancers often take up calcium, phosphate, and magnesium faster than can be replenished by gastrointestinal absorption. Hypocalcemia, hypophosphatemia, and hypomagnesemia result.

Table 16-1. Causes of hypocalcemia.

- ↑Calcium sequestration
  - Hyperphosphatemia
    - Exogenous phosphate
    - Chemotherapy of lymphomas, leukemias
    - Acute or chronic renal failure
  - Chelation
    - Citrate (transfusion)
    - Edetate calcium disodium
  - Soft tissue deposition
    - Acute pancreatitis
    - Rhabdomyolysis
  - Bone deposition
    - Osteoblastic metastases
    - Following parathyroidectomy ("hungry bone syndrome")
    - Healing of other bone diseases
- ↓PTH action
  - ↓PTH secretion
    - Congenital, familial, or idiopathic hypoparathyroidism
    - Destruction of parathyroid glands
      - Surgical removal
      - Infiltration (amyloid, tumors)
      - Radiation
    - Inhibition of PTH release
      - Hypomagnesemia
      - ↑Plasma vitamin D
  - Resistance to PTH action
    - Vitamin D deficiency
      - Nutritional deficiency, malabsorption
        - Gastrectomy
        - Intestinal resection
        - Hepatobiliary disease
        - Pancreatitis
        - Sprue
      - Anticonvulsant (phenytoin) therapy
      - Hyperphosphatemia (chronic renal failure)
      - Vitamin D-resistant rickets
    - Hypomagnesemia
    - Pseudohypoparathyroidism
    - Plicamycin (mithramycin) therapy

"Hungry Bone Syndrome" Following Parathyroidectomy or Healing of Other Bone Diseases

Severe osteitis fibrosa cystica can occur as a result of hyperparathyroidism. When the PTH level is acutely normalized by surgery, there is markedly enhanced bone formation and calcium utilization that may cause profound hypocalcemia, along with hypophosphatemia and hypomagnesemia. Therapy consists of intravenous calcium to acutely stabilize plasma calcium concentration followed by oral calcium and vitamin D.

Similar to the postparathyroidectomy state, during the recovery phase of various osteomalacic disorders or following treatment of thyrotoxicosis, bones may be very "hungry" and cause hypocalcemia due to rapid calcium deposition in bone matrix.

DECREASED PTH ACTION

DECREASED PTH SECRETION

Hypocalcemia may be consequent to congenital, familial, or idiopathic hypoparathyroidism or to destruction of the parathyroid glands following surgery, amyloid or tumor infiltration of the glands, or, rarely, following neck irradiation. Functional suppression of PTH release may occur during hypomagnesemia or elevation of plasma vitamin D. Treatment consists of oral calcium supplementation, thiazide administration (to diminish urinary calcium excretion), and, usually, vitamin D supplementation.

RESISTANCE TO PTH ACTION

An important cause of acquired PTH resistance is vitamin D deficiency. Vitamin D deficiency alone would not cause hypocalcemia if PTH action were unimpaired. $25(OH)D_3$ deficiency may be due to defective intestinal absorption, as in malabsorptive, steatorrheic states following gastrectomy, intestinal resection, hepatobiliary disease, pancreatitis, or sprue. Anticonvulsant therapy, notably with phenytoin, can impair the 25-hydroxylation of vitamin D_3. The 1α-hydroxylation can be reduced due to hyperphosphatemia, usually attributable to chronic renal disease, especially if the proximal tubule is specifically disordered (eg, Fanconi's syndrome). A specific defect in 1α-hydroxylation appears to occur in a disorder known as vitamin D-dependent rickets. Calcium and vitamin D supplements are appropriate treatments of these hypovitaminosis D syndromes.

Hypocalcemia predictably occurs during magnesium deficiency, due both to inhibition of PTH release and to impairment of PTH action on bone (Chapter 22). The hypocalcemia is very difficult to treat with calcium supplementation alone and requires magnesium repletion.

Unusual causes of PTH resistance include a rare hereditary syndrome in which the G_s protein that couples the PTH receptor to adenylate cyclase is absent or defective, known as pseudohypoparathyroidism. Plasma PTH is normal or elevated in this disorder but cannot stimulate its intracellular second messenger, cAMP. Finally, a potent drug that inhibits RNA synthesis from DNA and is used to treat hypercalcemia, plicamycin (mithramycin), can in excess cause hypocalcemia.

SYMPTOMS

The symptoms of hypocalcemia are predominantly neuromuscular and cardiovascular, as summarized in Fig 16–2. The severity depends in large measure on the underlying cause of the hypocalcemia, concurrent electrolyte abnormalities, and arterial pH.

NEUROMUSCULAR

Central Nervous System

Symptoms of hypocalcemia range from lethargy and depression to psychosis and dementia. There is also a predisposition to movement disorders and seizures of the grand mal, petit mal, or focal type. Many of these symptoms may be exacerbated by hyperparathyroidism if present.

Peripheral Nervous System

Enhanced neuromuscular excitability underlies many of the symptoms of hypocalcemia. With mild symptoms, there is circumoral and acral paresthesias and hyperreflexia. At this stage, the Chvostek sign may be positive, in which facial

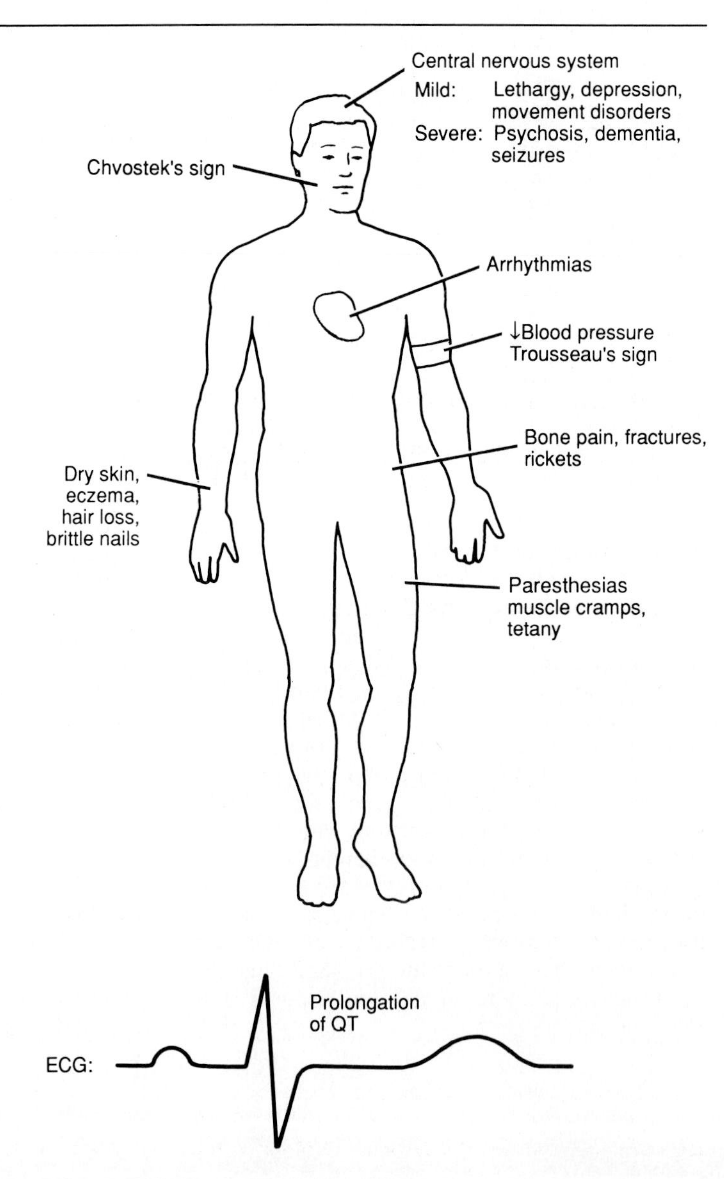

Figure 16–2. Symptoms and signs of hypocalcemia.

twitching occurs when the facial nerve branch is tapped. The Trousseau sign may also be elicited, in which carpal spasm is induced after 3 minutes of ischemia to the arm achieved by inflating a sphygmomanometer cuff above systolic blood pressure.

These symptoms may progress to muscle cramps and finally to laryngeal stridor and overt tetany. These potentially fatal neuromuscular symptoms can be exacerbated by concurrent hypomagnesemia, alkalemia, or hyperkalemia. A primary myopathy can occur with vitamin D deficiency.

CARDIOVASCULAR

A predisposition to arrhythmias occurs, especially heart block and ventricular fibrillation. Hypotension and even heart failure have been reported as a result of hypocalcemia. There may be digitalis insensitivity as well. The principal electrocardiographic manifestation of hypocalcemia is a prolonged QT interval.

BONE

With chronic hypocalcemia, bone pain, fragility, deformities, and fractures may occur, depending on the cause of the disorder. When hyperparathyroidism is also present, osteitis fibrosa cystica can develop whereas with vitamin D deficiency, defective bone mineralization, osteomalacia, and rickets occurs.

CONNECTIVE TISSUE

The skin may become dry, coarse, and scaly with chronic hypocalcemia, with exacerbation of eczema or psoriasis if present. Hair loss and brittle nails are frequent, sometimes due to the underlying disease. Dental abnormalities also occur. Cataracts of the cortical portion of the lens may develop within 1 year after onset of hypocalcemia.

DIAGNOSIS

APPROACH TO ETIOLOGIC DIAGNOSIS

History & Physical Examination

The history and physical examination can be helpful in elucidating the cause of hypocalcemia. Obvious conditions associated with hypocalcemia include recent parathyroid surgery, blood transfusions, phosphate therapy, cancer chemotherapy, acute pancreatitis or rhabdomyolysis, or chronic renal failure. In addition, causes of vitamin D or magnesium deficiency should be investigated (poor diet, gastrointestinal disease associated with malabsorption, history of gastrectomy or intestinal resection or bypass, or hepatobiliary disease). A history of drugs associated with disordered calcium metabolism (phenytoin or plicamycin) should also be sought. Radiographic evaluation is helpful in evaluating the extent of bone disease and excluding cancer.

Plasma Phosphorus & PTH

Both the plasma phosphorus concentration and the PTH level help in characterization of hypocalcemic disorders, as summarized in Fig 16–1. In general, when calcium is sequestered by blood chelation or by deposition into soft tissues or bone, there is a normal PTH response with appropriate elevation of the hormone. There is also an accompanying drop in plasma phosphorus concentration (due to codeposition with calcium in tissues or to PTH-mediated phosphaturia), the exception being hypocalcemia due to primary phosphorus release or infusion. On the other hand, in the hypocalcemic disorders associated with diminished PTH activity, PTH is either reduced if hypoparathyroidism is due to defective secretion or elevated if due to defective end-organ response (pseudohypoparathyroidism). In either case, plasma phosphorus is increased as a result of hypophosphaturia that occurs when PTH activity is low.

Other Tests

Magnesium concentration should be checked to ascertain if hypomagnesemia is the cause of hypocalcemia. If vitamin D deficiency is suspected, $25(OH)D_3$ and $1,25(OH)_2D_3$ levels can be measured. Urinary calcium can be useful in assessing the appropriateness of the hypocalciuric response to hypocalcemia. The physiologic renal response should be reduction in urine calcium excretion to less than 50 mg/d (Fig 15–5).

THERAPY

DECISION TO TREAT

In general, hypocalcemia is an indication for treatment. Even asymptomatic hypocalcemia is treated because of the possibility of developing symptoms, which have the potential of being life-threatening.

Acute symptomatic hypocalcemia, with neuromuscular, central nervous system, or cardiac manifestations, requires urgent treatment. The risk of tetany and death mandates intravenous intervention on an emergent basis. With chronic hypocalcemia with lesser symptoms, oral therapy is often sufficient. The treatment of hyperphosphatemia and hypomagnesemia as causes of hypocalcemia are considered in Chapters 20 and 22.

TREATMENT OF SEVERE, SYMPTOMATIC HYPOCALCEMIA

In the presence of tetany, arrhythmias, seizures, and other serious manifestations of hypocalcemia, systemic administration of calcium should be urgently initiated, as summarized in Table 16–2. Calcium gluconate in 10% solution (93 mg/ 10 mL) should be given as a rapid intravenous infusion (10–20 mL) over 10–15 minutes. Calcium chloride may also be used for the emergent treatment of hypocalcemia, though it is less well tolerated and causes thrombophlebitis.

Since the benefit of acute calcium infusion is often transient, a sustained calcium infusion is usually required. For this purpose, about 10–15 mg calcium/kg body weight, or 6–8 ampules of 10% calcium gluconate (representing 558–744 mg of calcium), can be dissolved in 1 L of D_5W and administered over 4–6 hours.

During the treatment of acute hypocalcemia, plasma calcium concentration should be frequently monitored to guide subsequent therapy. The amount of calcium administered is that which is necessary to normalize the plasma calcium concentration and reverse neuromuscular, cardiac, and electrocardiographic manifestations.

Adjunctively with the intravenous calcium therapy of acute hypocalcemia, concurrent oral calcium and vitamin D supplements should be started if possible. The liquid calcium preparation (eg, Neo-Calglucon, 15 mL 4 times daily) is preferred, as is the rapidly-acting (1–3 days) form of vitamin D ($1,25([OH])_2D_3$; (calcitriol, 0.25-1 μg daily).

TREATMENT OF ASYMPTOMATIC HYPOCALCEMIA

When hypocalcemia is without acute cardiac or neuromuscular symptoms, oral calcium and vitamin D supplements are the cornerstones of treat-

Table 16-2. Treatment of hypocalcemia.

	Amount of Ca^{2+}	Onset	Dose
Intravenous calcium			
Calcium gluconate 10%	93 mg (4.7 meq) per 10 mL	Immediate	93–186 mg over 10–15 minutes; then 10–15 mg/kg over 4–6 hours
Calcium chloride 10%	273 mg (14 meq) per 10 mL	Immediate	
Oral calcium			
Calcium carbonate (BioCal, Cal-Bid, Cal-Plus, Calcet [combination], Os-Cal)	250 mg/625-mg tablet or 500 mg/1250-mg tablet or 600 mg/1500-mg tablet	< 1 hour	0.25–0.5 g calcium 4 times daily
Calcium citrate (Citracal, Calcigard)	200–250 mg/950-mg tablet		
Calcium gluconate	500 mg/tablet		
Calcium glubionate (Neo-Calglucon)	115 mg/5 mL		
Calcium phosphate (Dical-D, Posture)	120 mg/capsule or 300–600 mg/tablet		
Vitamin D preparations			
Vitamin D_2 (Ergocalciferol, Calciferol)		2-4 weeks	50,000–400,000 units/d (50,000 USP units/tablet)
Dihydrotachysterol (DHT)		1-2 weeks	0.2–1 mg/d (0.125, 0.2, or 0.4 mg/tablet)
$25(OH)D_3$ (calcifediol)		< 1 week	50–100 μg/d (20 or 50 μg/capsule)
$1,25(OH)_2D_3$ (calcitriol)		1-3 days	0.25–1 μg/d (0.25 or 0.5 μg/capsule)

Table 16-3. Calcium content of foods.

High (> 250 mg)	Medium (150-250 mg)	Low (< 150 mg)
Milk products	Cottage cheese (8 oz)	Bread (1 slice)
Milk (whole or low-fat) (8 oz)	Tofu (4 oz)	Cereals (8 oz)
Yogurt (8 oz)	Garbanzo beans (2 oz)	Meats (8 oz)
Ice cream (8 oz)	Green vegetables (4 oz)	Chicken
Cheese (2 oz)	Collard greens	Turkey
Chocolate milk (8 oz)	Dandelion greens	Duck
Canned salmon	Kale	Lamb
Canned sardines	Mustard greens	Pork, bacon
Fresh oysters (1 lb with shells)	Turnip greens	Fish
	Breads	Shellfish
	Pancakes (4)	Clams
	Waffle (1)	Crabs
	Corn bread (4 oz)	Egg (1)
	Molasses (1 oz)	Nuts
		Peanuts
		Almonds
		Walnuts
		Rice
		Beans
		Condiments (salt, pepper)
		Most green vegetables
		Fruits
		Potatoes
		Beverages
		Soft drinks
		Coffee, tea
		Beer, wine

ment (Table 16–2). Increased dietary calcium should be ensured. As shown in Table 16–3, milk products are the most important sources of dietary calcium. Dietary calcium should be maintained as high as feasible, at least 1000–1500 mg/d. In general, calcium should also be supplemented as 1–2 g/d of elemental calcium as the gluconate, carbonate, or lactate salts. Hazards of overtreatment with calcium are discussed in Chapter 17. Hyperphosphatemia and hypomagnesemia, if present, should be treated before calcium is supplemented (Chapters 20 and 22).

Many forms of vitamin D can be used when indicated. Older preparations of vitamin D, such as vitamin D itself and dihydrotachysterol, are cheaper but have long durations of action. If hypercalcemia develops, these preparations have the disadvantage of slow subsidence of activity. Newer preparations, such as $25(OH)D_3$ (calcifediol) or $1,25(OH)_2D_3$ (calcitriol) are somewhat safer in having shorter durations of action.

The doses of both calcium and vitamin D should be titrated until the plasma calcium is normalized. Supraphysiologic doses of vitamin D are frequently required as a result of hormonal resistance.

In many patients with hypocalcemia, especially associated with hypoparathyroidism, adjunctive use of a thiazide diuretic (see Chapter 3) may be helpful. A thiazide diuretic increases calcium transport in the distal convoluted tubule and reduce urinary calcium excretion (Fig 15–6).

17 Hypercalcemia

INTRODUCTION

DEFINITION

Increase in Ionized Calcium

As with hypocalcemia, it is important to first exclude conditions that are not associated with abnormality of plasma ionized calcium concentration.

Prolonged drawing of blood with an excessively tight tourniquet may cause hemoconcentration. This artifactual cause of hypercalcemia is called pseudohypercalcemia, as shown in Fig 17–1. Hyperalbuminemia should also be excluded, since it raises the bound component of plasma total calcium concentration but does not affect the ionized calcium concentration. Total calcium concentration rises by 0.75 mg/dL for each g/dL increase in plasma albumin concentration. Calcium binding to protein can also be increased by marked hyperglobulinemia and severe hypernatremia. Acidemia increases ionized calcium concentration without altering total plasma calcium concentration.

When spurious causes are excluded, hypercalcemia is diagnosed when total plasma calcium concentration is **greater than 10.5 mg/dL.**

PATHOPHYSIOLOGY

Hypercalcemia develops when calcium influx into the circulation overwhelms the calcium regulatory hormones (PTH and vitamin D) and renal calciuric mechanisms or when there is a primary abnormality of one or both of these hormones.

Primary Increase in Calcium Absorption Versus PTH- & Non-PTH-Mediated Calcium Mobilization From Bone

Calcium influx may overwhelm renal excretory mechanisms by two general schemes. First, there may be enhanced intestinal calcium absorption due to hypervitaminosis D, with simultaneous abnormality of bone metabolism. Second, given the very large calcium reserve in bone relative to extracellular fluid ($> 99 : < 1$, as shown in Fig 15–1), factors that cause increased bone calcium mobilization can liberate a large quantity of previously sequestered calcium into the extracellular fluid. Such enhanced bone turnover and dissolution may be mediated by an unregulated excess of PTH or by other hormones and substances usually found only in malignant states.

In all of these entities, depression in GFR is either intrinsic to the development or exacerbates the degree of hypercalcemia. A decrease in GFR may be due to the intrinsic disease, to extracellular volume contraction (eg, caused by hypertension and pressure natriuresis), or to hypercalcemia itself.

As with hypocalcemia, the plasma phosphorus concentration and PTH help in separating these entities that predispose to hypercalcemia, as shown in Fig 17–1. When calcium is channeled into the extracellular fluid from gastrointestinal absorption or from bone by a non-PTH mechanism, PTH is physiologically suppressed by the hypercalcemia (Fig 15–2). Hyperphosphatemia tends to occur as a result of the secondary hypoparathyroidism, and there is an appropriate renal hypercalciuric response to hypercalcemia. On the other hand, when hypercalcemia is due to a primary increase in PTH, there is hypophosphatemia and inappropriate hypocalciuria (Fig 15–7).

Intestinal Absorptive, Renal Excretory, & Bone Dissolution Mechanisms Overlap

Although simplified in Fig 17–2, in actuality factors that increase intestinal calcium transport can also increase calcium mobilization from bone and cause renal calcium retention, and vice versa. For instance, while the primary action of vitamin D is on intestinal calcium transport, it also helps PTH augment osteoclastic activity. Similarly, PTH, while primarily causing bone to be demineralized, also stimulates renal calcium reabsorption and production of $1,25(OH)_2D_3$ (both directly and indirectly via hypophosphatemia) to increase intestinal calcium absorption.

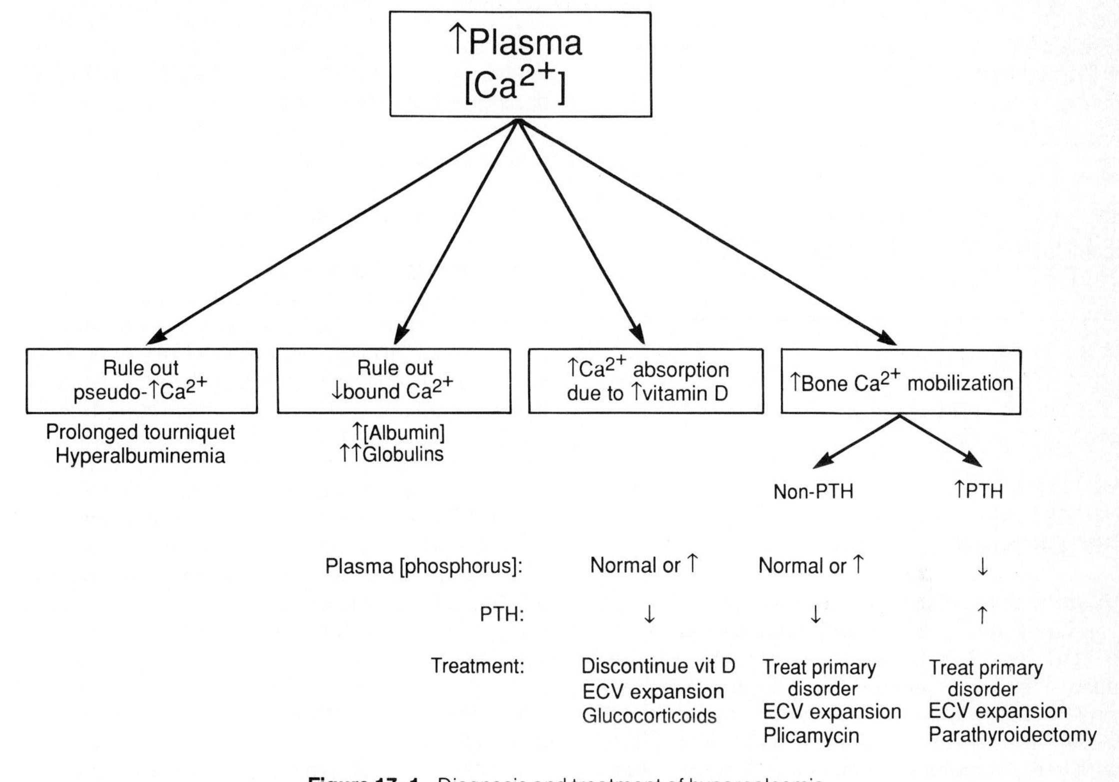

Figure 17-1. Diagnosis and treatment of hypercalcemia.

In fact, if only one mechanism (intestinal, renal, or bone) is involved, hypercalcemia usually does not arise because there are compensations by the other systems. For instance, if there were a primary increase in gastrointestinal calcium uptake due to a change in diet, a normal hormonal response would occur with reduction in PTH and $1,25(OH)_2D_3$ levels that would augment urinary calcium excretion and diminish the fraction of intestinal calcium absorbed, thereby preventing hypercalcemia from developing (Fig 15–9). Similarly, if there were a primary increase in renal calcium reabsorption—due to thiazide administration or extracellular volume contraction—PTH and vitamin D levels would be secondarily diminished, tending to normalize renal calcium excretion and to reduce gastrointestinal calcium absorption. Calcium balance would then resume, and hypercalcemia again would not result.

Thus, it is most common that intestinal, renal, and bone mechanisms are synergistically involved in sustaining hypercalcemia (Fig 17–2). These co-

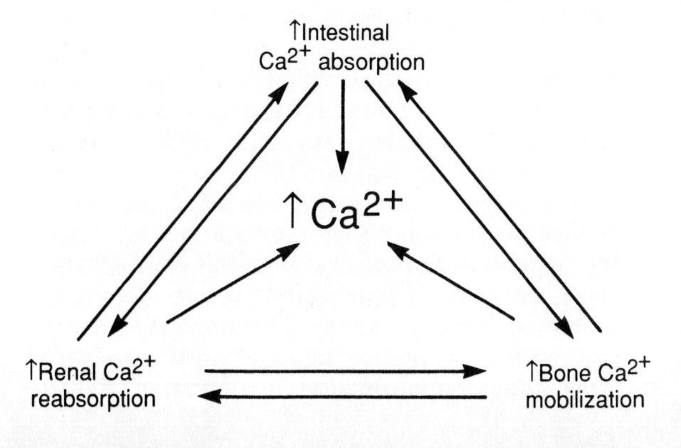

Figure 17-2. Causes of hypercalcemia.

ordinate disturbances are often due to primary disorder of the calcium regulatory hormones, PTH or vitamin D.

INCREASED INTESTINAL CALCIUM ABSORPTION DUE TO HYPERVITAMINOSIS D

A summary of the hypercalcemic disorders is contained in Table 17-1.

INCREASED $1,25(OH)_2D_3$ PRODUCTION

Vitamin D Intoxication

Usually iatrogenic, excessive ingestion of 25 $(OH)D_3$ or $1,25(OH)_2D_3$ may cause hypercalcemia. The hypercalcemia is usually mild and asymptomatic and occurs about 2-3 months after initiation of therapy. A patient with renal failure with preexisting high-normal plasma calcium concentration and severe osteitis fibrosa or osteomalacia due to aluminum toxicity (from antacid therapy to treat hyperphosphatemia) is especially prone to vitamin D-induced hypercalcemia. Owing to the long half-life of some vitamin D preparations (see Table 16-2), hypercalcemia may persist for weeks or even months following discontinuation of the agent. When plasma calcium concentration finally normalizes, the vitamin D preparation can be resumed at a lower dosage.

Table 17-1. Causes of hypercalcemia.

↑Intestinal calcium absorption due to ↑vitamin D
Vitamin D toxicity
Granulomatous diseases
Sarcoidosis
Tuberculosis
Berylliosis
Histoplasmosis
Coccidioidomycosis
Other disorders with renal involvement
Milk-alkali syndrome
Familial hypocalciuric hypercalcemia
↑Bone calcium mobilization
Primary hyperparathyroidism
Adenoma, hyperplasia, carcinoma, multiple endocrine adenomatosis I and II
Lithium-induced
Non-PTH-mediated
Endocrine
Hyperthyroidism
Acromegaly
Pheochromocytoma
Adrenal insufficiency
Other
Malignancy-associated humoral substances
PTH-like protein
Prostaglandin E_2
Transforming growth factor-β
Prolonged immobilization
Vitamin A intoxication
Recovery from pancreatitis or rhabdomyolysis

Granulomatous Diseases

The primary site of 1α-hydroxylase activity and production of $1,25(OH)_2D_3$ is normally the proximal tubule. However, in formation of granulomas, macrophages apparently acquire the ability to produce $1,25(OH)_2D_3$. Hypercalcemia is seen frequently (incidence up to 30% of cases) in a wide variety of granulomatous diseases, including sarcoidosis, tuberculosis, berylliosis, histoplasmosis, and coccidioidomycosis. Therapy consists of avoidance of sunlight (to decrease the available vitamin D_3), a low-calcium diet (Table 16-3), and, if these fail, prednisone (10-30 mg daily), because extrarenal production of $1,25(OH)_2D_3$ is very sensitive to glucocortoid.

Other Disorders With Renal Involvement

In the milk-alkali syndrome, excessive ingestion of milk and antacids is associated with hypercalcemia, metabolic alkalosis, and renal insufficiency (Chapter 13). The hypercalcemia does not develop until the GFR is low—attributable at least in part to nephrocalcinosis. Discontinuation of excess calcium and antacid ingestion is usually sufficient therapy.

INCREASED BONE CALCIUM MOBILIZATION

PRIMARY HYPERPARATHYROIDISM

PTH, if unregulated from calcemic control, can increase bone dissolution while also secondarily augmenting intestinal and renal calcium absorption to cause hypercalcemia (Fig 17-2).

Primary hyperparathyroidism is most often due to a parathyroid gland adenoma (80%) or hyperplasia of all four glands (20%), with fewer than 1% of cases due to parathyroid gland carcinoma. Patients are usually asymptomatic but may present with hypertension, renal calculi (calcium oxalate and calcium phosphate), peptic ulcer, pancreatitis, or bone pain or fracture consequent to osteitis fibrosa cystica. Subtle behavioral and neuropsychiatric abnormalities are common. Associated hormonal hypersecretion due to multiple endocrine adenomatosis (type I or II) is rare. Mild, reversible hyperparathyroidism has also been reported in patients treated with lithium. Primary hyperparathyroidism should be suspected in a patient who develops hypercalcemia in response to thiazide administration.

The diagnosis is confirmed when the PTH level is found to be inappropriately high for the prevailing plasma calcium concentration (Fig 15–2). Plasma phosphorus concentration is usually low, and there is hypercalciuria and a variable increase in plasma alkaline phosphatase activity.

Surgical therapy is usually indicated for primary hyperparathyroidism, with removal of the adenoma or subtotal parathyroidectomy for hyperplasia (sometimes with parathyroid gland autotransplantation in the forearm). Medical treatment alone is usually disappointing.

NON-PTH-MEDIATED

Enhanced bone mobilization can be induced by various substances and hormones besides PTH. In these entities, PTH and vitamin D levels are physiologically suppressed by the hypercalcemia.

Endocrine Causes

Mild hypercalcemia is found in a variety of endocrine conditions, though the mechanisms are generally poorly understood. Hyperthyroidism causes increased bone turnover and hypercalcemia, perhaps by direct stimulation of osteoclastic activity. Amelioration of hypercalcemia usually follows therapy of the hyperthyroid state. Acromegaly, pheochromocytoma, and adrenal insufficiency are also sometimes associated with mild hypercalcemia.

Neoplasms

A variety of humoral factors, some of which are listed in Table 17–1, can be released by various tumors to mobilize calcium. The best-characterized is the PTH-like protein, which bears structural (N-terminal) homology to PTH and binds to the PTH receptor to activate adenylate cyclase. Malignant hypercalcemia due to this or other factors is described in many tumors, especially squamous and oat cell bronchogenic carcinomas, breast cancer, and multiple myeloma.

Miscellaneous

Mild elevation in plasma calcium concentration, of unknown pathogenesis, is occasionally seen following prolonged immobilization following fractures or burns, especially in patients with preexisting high bone turnover syndromes (eg, Paget's disease). Hypervitaminosis A, caused by ingestion of more than 7500 IU/d for prolonged periods, may also increase bone turnover by unknown mechanisms. The liberation of calcium from sequestered sites may occur during the resolution phase of acute pancreatitis or rhabdomyolysis and overwhelm counterregulatory mechanisms to cause rebound hypercalcemia, especially when GFR is still depressed.

SYMPTOMS

NEUROMUSCULAR

As shown in Fig 17–3, hypercalcemia, especially when associated with hyperparathyroidism, may cause significant psychiatric disturbances, ranging from depression and lethargy to confusion or psychosis and even coma. Mild electroencephalographic changes can be documented in hypercalcemic patients even when clinical abnormality is undiscernible. Diminished deep tendon reflexes and muscle weakness typify the hypercalcemic patient, the latter sometimes attributable to PTH-induced myopathy.

CARDIOVASCULAR

Hypercalcemia is moderately inotropic but predisposes to arrhythmias and digitalis toxicity. Shortening of the QT interval is found. There is a high prevalence of hypertension in hypercalcemic individuals, especially those with primary hyperparathyroidism.

GASTROINTESTINAL

Hypercalcemia is associated with a higher incidence of peptic ulcer and pancreatitis. There is a propensity to nausea and vomiting as well to constipation.

RENAL

Hypercalcemia, especially when associated with hyperparathyroidism, decreases GFR, which can be acute and reversible or chronic. Associated hypercalciuria predisposes to nephrolithiasis. Stones usually contain calcium oxalate or calcium phosphate. Calcium is deposited predominantly in renal cortical tissue and causes nephrocalcinosis and tubulointerstitial nephropathy and sometimes type II classical distal RTA (Chapter 12). Hyper-

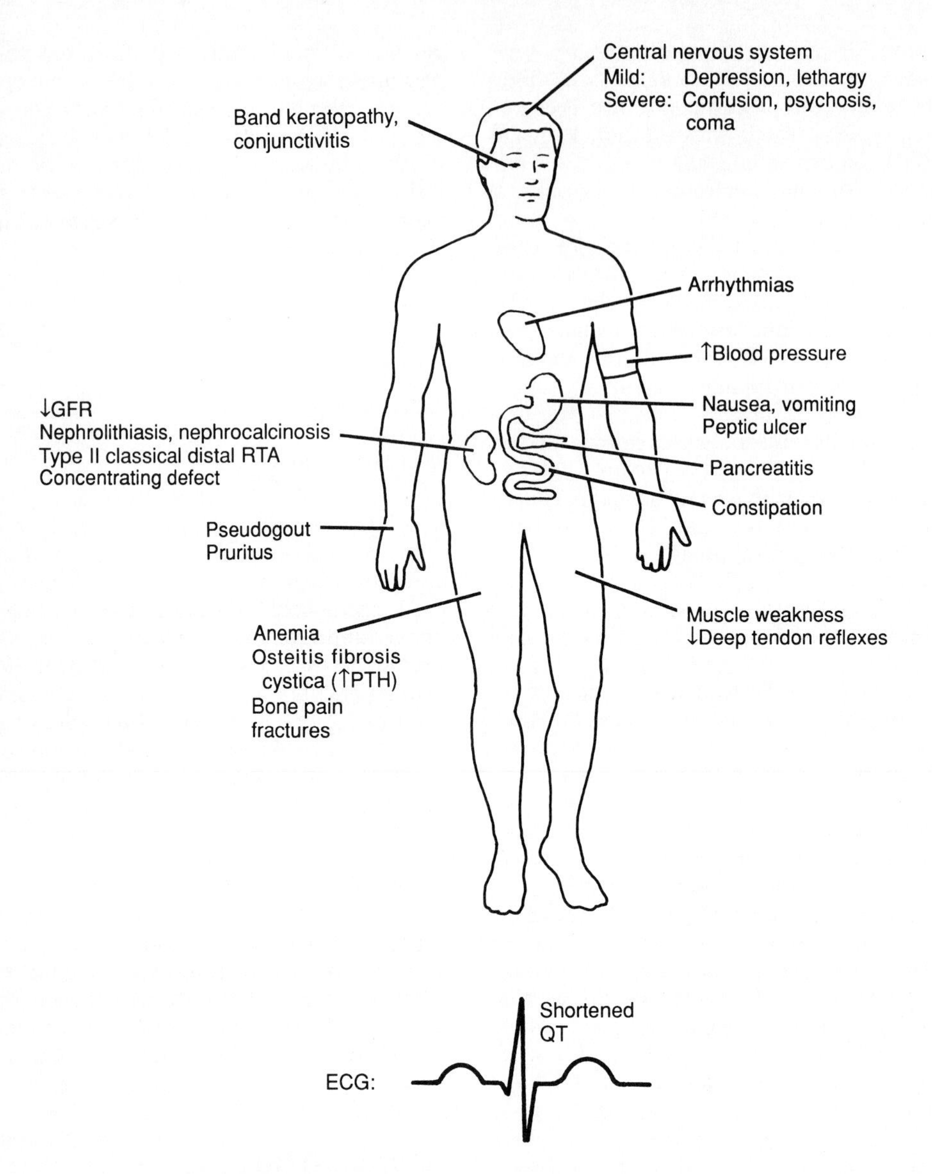

Figure 17-3. Symptoms and signs of hypercalcemia.

calcemia also antagonizes the action of ADH on medullary collecting tubule to decrease urinary concentrating capacity and sometimes even to cause nephrogenic diabetes insipidus (Chapter 6).

BONE & CONNECTIVE TISSUE

Hyperparathyroidism is associated with osteitis fibrosa cystica, with propensity to bone pain, deformity, and fractures. There is also a high incidence of anemia, perhaps due in part to the bone involvement.

Metastatic calcification can be manifested as pruritus, band keratopathy, conjunctivitis, and red eye syndrome. Attacks of pseudogout can also be precipitated.

DIAGNOSIS

APPROACH TO ETIOLOGIC DIAGNOSIS

History & Physical Examination

Features suggestive of primary hyperparathyroidism should be sought, such as nephrolithiasis, peptic ulcer symptoms, neuropsychiatric disturbance, and family history. There should also be a thorough search for signs of cancer, notably lung or breast cancer and multiple myeloma. Pulmonary findings and a high circulating converting enzyme activity suggestive of chronic granulomatous disease should be sought. Dietary intake of calcium and milk should be quantitated (Table 16–3) and ingestion of calcium preparations, vitamins D and A, and lithium documented.

Various radiographic investigations may be useful, including a chest x-ray for cancer, sarcoidosis, or tuberculosis, and bone x-rays (clavicles, hands) to establish the degree of subperiosteal resorption and osteitis fibrosa cystica. Ultrasound and radionuclide scanning examinations may be helpful in revealing parathyroid adenoma or hyperplasia.

Plasma PTH & Phosphorus Concentrations

With hypercalcemia, PTH secretion should be physiologically suppressed. If it is, the hypercalcemia is due to an abnormality of vitamin D metabolism or a malignancy-associated (non-PTH) bone mobilization factor (Fig 17–1).

Conversely, finding that PTH exceeds the expected minimal value (ie, is "normal" or elevated) provides evidence that hypercalcemia is due to primary hyperparathyroidism (Fig 17–1). Plasma phosphorus concentration is concomitantly low in hyperparathyroid states, whereas urinary calcium and cAMP excretion rates are high.

Other Laboratory Tests

When appropriate, measurement of $25(OH)D_3$ or $1,25(OH)_2D_3$ may be helpful to establish the diagnosis of endogenous or exogenous hypervitaminosis D. $1,25(OH)_2D_3$ is also elevated in most hyperparathyroid disorders except when renal failure is present. Urine calcium excretion is less elevated in primary hyperparathyroidism than in hypervitaminosis D but does not usually distinguish between hypercalcemic disorders.

THERAPY

DECISION TO TREAT

Treatment of hypercalcemia is indicated in most cases. Even when asymptomatic, subtle damage may be occurring, such as the neuropsychiatric changes commonly found in mild hyperparathyroidism. In addition, the rate of progression to more severe, life-threatening hypercalcemia cannot be predicted.

As discussed under the individual entities above, withdrawal of milk-alkali, specific treatment of malignant, granulomatous, or endocrine disorders, or parathyroidectomy usually cures hypercalcemia. However, even when the primary disorder is being treated, acute symptomatic hypercalcemia, with neuropsychiatric or cardiovascular manifestations, requires urgent intravenous therapy.

TREATMENT OF ACUTE, SYMPTOMATIC HYPERCALCEMIA

The key to acute treatment of hypercalcemia is to disrupt the mechanisms in the proximal convoluted tubule and thick ascending limb of Henle responsible for most (80%) calcium reabsorption (Fig 15–6). The fraction of calcium reabsorbed in the proximal tubule is linked to sodium, so that it can be diminished by extracellular volume expansion. Calcium transport in the thick ascending limb of Henle is also secondarily linked to sodium transport, so that calcium reabsorption can be blocked by a loop diuretic such as furosemide.

Extracellular Volume Expansion

As summarized in Table 17–2, the primary means of treating acute, symptomatic hypercalcemia is extracellular volume expansion. Intravenous normal saline administration at a brisk rate—5–10 L/d—can increase urinary calcium excretion to as high as 1000–2000 mg/d. This rate of volume expansion is not feasible if significant cardiopulmonary disease preexists. The repair of hypovolemia—commonly present as a result of poor nutrition, vomiting, or urinary solute and water loss—may itself allow the appropriate calciuric response and ameliorate hypercalcemia.

Furosemide

When full overhydration has been accomplished, blockade of sodium reabsorption in the thick ascending limb of Henle adds to the calciuric effect of volume expansion. Frequent dosing (eg, 20–40 mg intravenously every 2–3 hours) is required. It is essential not to let hypovolemia develop, since this will counteract the effects of furosemide and may even exacerbate the hypercalcemia. Therefore, urinary sodium loss as well as potassium and magnesium losses should be repleted quantitatively. Extracellular volume expansion plus furosemide administration usually result in fall in plasma calcium concentration by at least 1–2 mg/dL.

Table 17-2. Treatment of hypercalcemia.

General therapy
Extracellular volume expansion: Normal saline IV, 5-10 L/d
With or without
Furosemide, 20-40 mg IV every 2-3 hours
Natriuresis matched by equivalent normal saline
Replete K^+, Mg^{2+}
Specific therapies
Phosphate, 400-1000 mg IV in normal saline over 8-16 hours, or 1-3 g/d orally in divided doses
Prednisone, 10-30 mg/d orally
Plicamycin (mithramycin), 25 mg/kg in 50 mL D_5W over 3 hours
Calcitonin-salmon, 4-8 IU/kg SC or IM every 12 (6-24) hours
Hemodialysis
Low-calcium diet

Phosphate

In life-threatening hypercalcemia associated with hypophosphatemia, phosphate may be administered slowly intravenously (400–1000 mg in normal saline over 8–16 hours). The risk of soft tissue calcification limits use of this agent. Oral phosphate therapy (1–3 g/d in divided doses) may be used for the hypophosphatemic, hyperparathyroid patient awaiting surgery.

SPECIFIC TREATMENTS FOR CHRONIC HYPERCALCEMIC STATES

Glucocorticoid

Prednisone (10–30 mg/d orally) is useful in diminishing excessive vitamin D production in granulomatous diseases and in reducing hypercalcemia associated with multiple myeloma or lymphoma. It is not helpful in hyperparathyroid states. Correction of hypercalcemia usually takes several days to 2 weeks.

Plicamycin (Mithramycin)

In hypercalcemia of malignancy, plicamycin has been useful in blocking bone resorption by inhibiting RNA synthesis in bone cells. Plicamycin (25 mg/kg in 50 mL D_5W intravenously over 3 hours) may effectively begin to lower plasma calcium concentration within 12 hours, with a peak response in 2–4 days. The drug may be repeated every 3–7 days, usually for 2–3 weeks before renal, hepatic, or hematopoietic (thrombocytopenia) toxicities occur.

Calcitonin

Salmon calcitonin has modest transient efficacy in hypercalcemia associated with carcinoma, multiple myeloma, and primary hyperparathyroidism. Skin testing for allergy should precede dosing. The usual dose is 4 IU (range, 4–8 units) per kilogram of body weight subcutaneously or intramuscularly every 12 hours (range, 6–24 hours).

Other Drugs

Diphosphonates, which are bone crystal formation inhibitors, are currently being evaluated and hold substantial promise for use in hypercalcemic states. Etidronate disodium (EHDP, Didronel) is not currently approved for this purpose but may be useful parenterally.

Hemodialysis

When GFR is very low, hypercalcemia must be treated with hemodialysis. A high-flux (polysulfone) dialyzer membrane, if available, is preferable to a conventional hollow fiber membrane, with enhanced flux of molecules the size of calcium. In any case, dialysate calcium should be reduced as low as possible and preferably removed altogether.

Low-Calcium Diet

For chronic hypercalcemia, a low-calcium diet should be maintained as an adjunct to all the therapies described above, with avoidance of foods high in calcium content listed in Table 16–3, notably milk products. It is also necessary to avoid thiazide diuretics and vitamin D preparations. Encouragement of mobilization also may be a useful adjunctive measure.

SECTION V REFERENCES: CALCIUM HOMEOSTASIS & DISORDERS: HYPOCALCEMIA & HYPERCALCEMIA

Agus AS, Wasserstein A, Goldfarb S: Disorders of calcium and magnesium homeostasis. Am J Med 1982;72:473–488.

Singer FR, Fernandez M: Therapy of hypercalcemia of malignancy. Am J Med 1987;82(2A):34–41.

Section VI: Phosphorus Homeostasis & Disorders: Hypophosphatemia & Hyperphosphatemia

Normal Phosphorus Homeostasis 18

INTRODUCTION

TOTAL BODY PHOSPHORUS

Phosphorus is the fourth most abundant element in the body after carbon, nitrogen, and calcium. Total body phosphorus in the average 70-kg adult male is about 700 g. Approximately 85% of this total is contained in bone and most of the remainder (14%) inside cells, as shown in Fig 18–1. As was the case with potassium and calcium, extracellular phosphorus is a very small fraction (1%) of the total, and only a minute proportion—about 200 mg, or 0.03% of total body phosphorus—is present in plasma.

Intracellular Phosphorus

Within cells, most (90%) of the phosphorus is in the organic rather than the inorganic form. Phosphorus-containing organic compounds are critically important for cell function and include nucleic acids, phosphoproteins, membrane phospholipids, high-energy compounds (eg, ATP), second messengers (eg, cAMP), and coenzymes and regulatory factors (eg, NAD and 2,3-DPG in red cells). In general, the phosphorus content of a cell is proportionate to that of nitrogen, potassium, and magnesium; cell atrophy, as occurs in muscle wasting states, leads to losses of all these elements in a relatively predictable relationship.

Plasma Phosphorus

Extracellular phosphorus exists mainly as **phosphate.** Phosphate is distributed in plasma in both organic, lipid-bound (70%) and inorganic (30%) forms, as shown in Fig 18–2. Only the inorganic phosphate component is routinely measured in plasma.

Most of the plasma inorganic phosphate (85%) is free and physiologically active and exists predominantly as dibasic phosphate (HPO_4^{2-}) and

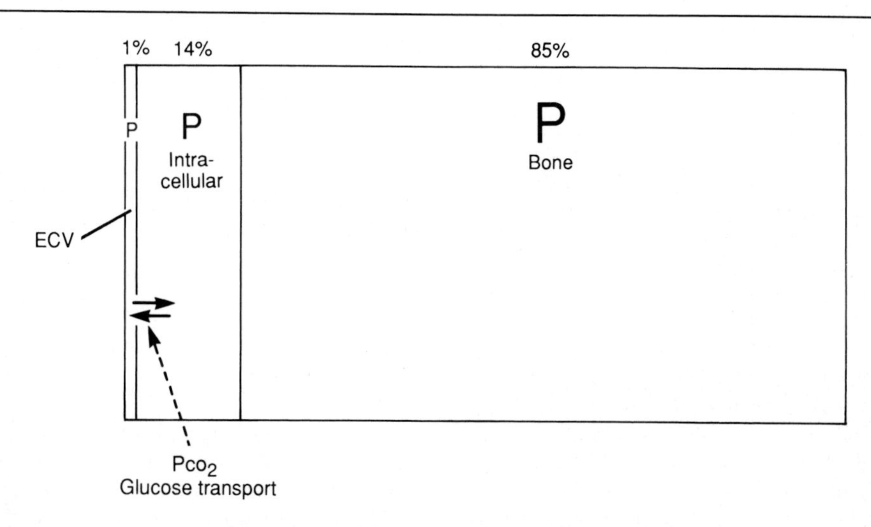

Figure 18–1. Distribution of body phosphorus.

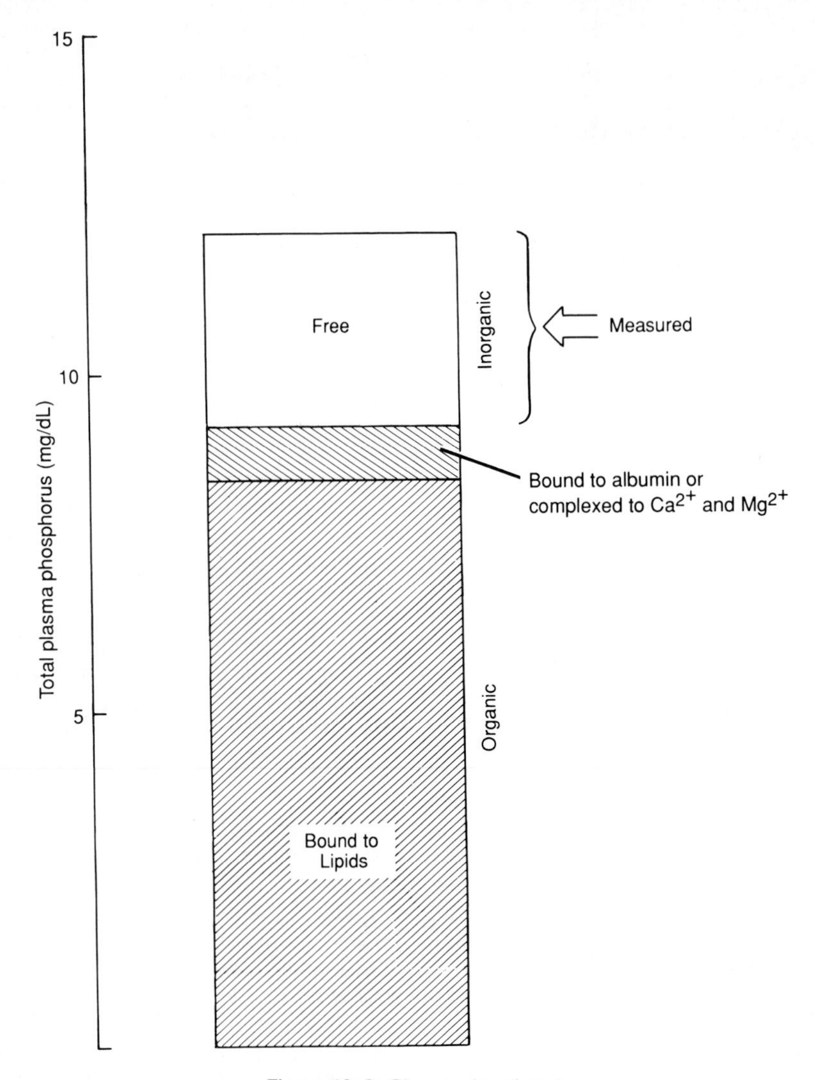

Figure 18-2. Plasma phosphorus.

monobasic phosphate ($H_2PO_4^-$) in a ratio of 4:1. Only a small fraction (15%) of plasma inorganic phosphate is bound to albumin or complexed with calcium or magnesium. Thus, unlike calcium, plasma phosphorus content is little influenced by changes in plasma albumin concentration.

Plasma phosphate concentration is reported in terms of inorganic phosphorus content. The normal plasma inorganic phosphorus concentration is not precisely regulated, averaging 3.5 mg/dL (1 meq/L) but ranging from 2.5 to 4.5 mg/dL. Normal limits of plasma phosphorus concentration are substantially higher in children (3.5–6.0 mg/dL), slightly higher in postmenopausal women, and lower in older men. There is normal diurnal variation of plasma phosphorus concentration, with peaks at 4 PM and 3 AM and a nadir at about 11 AM. The plasma phosphorus level is also influenced by the arterial P_{CO_2} and by dietary carbohydrate intake (see below).

PHOSPHORUS INTAKE, ABSORPTION, & DISTRIBUTION

PHOSPHORUS INTAKE

The average dietary phosphorus intake of adult males in the United States is about 1400 mg/d, with a very broad range (95% confidence limits of 500–3800 mg/d). Phosphorus intake in adult females tends to be less, about 900–1000 mg/d

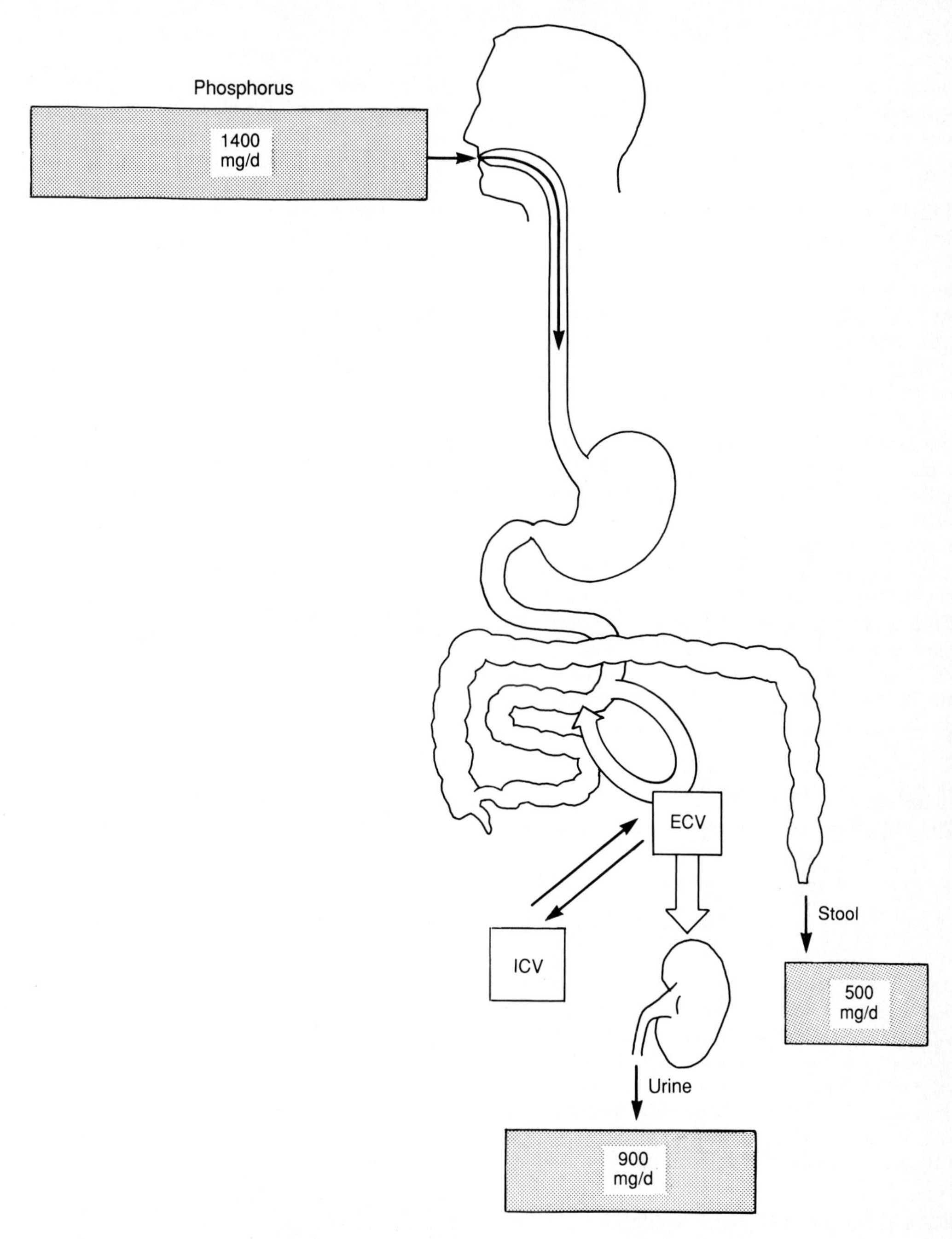

Figure 18-3. Normal phosphorus intake and excretion.

(range 300–2000 mg/d). Phosphorus intake tends to decrease with age, averaging 85% of the above values in both men and women over age 65 years. Blacks also tend to ingest less phosphorus than whites on average.

INTESTINAL PHOSPHORUS ABSORPTION

Sites & Mechanism

As shown in Fig 18–3, of the phosphorus ingested (1400 mg/d), about 80% is absorbed (1100 mg/d), primarily in the jejunum as well as in the duodenum. There is a parallel, small, fixed secretory flux of about 200 mg/d, probably in the jejunum. Because of this bidirectional phosphorus movement, net gastrointestinal phosphorus absorption averages 900 mg/d and remains about two-thirds of intake over a wide range (Fig 15–5). Absorption of phosphorus is diminished by concurrent ingestion of compounds that can bind to it, including calcium, magnesium, and aluminum.

Both active and passive mechanisms contribute to gastrointestinal phosphorus absorption. Phosphorus enters the intestinal cell coupled to sodium, driven by the lumen-to-cell sodium concentration gradient. Phosphorus exits the cell passively.

Regulation by Vitamin D

An important component of phosphorus homeostatic control and a strong stimulant of the rate and maximal capacity of intestinal phosphorus absorption is the active metabolite of vitamin D, $1,25(OH)_2D_3$. As shown in Fig 15–3, hypophosphatemia stimulates renal synthesis of $1,25(OH)_2$-D_3,which in turn enhances intestinal phosphorus uptake. The sites and mechanisms by which $1,25(OH)2_2D_3$ affects phosphorus transport appear to be somewhat different than those on calcium transport.

PTH has little direct action but indirectly affects intestinal phosphorus absorption by its ability to promote $1,25(OH)2_2D_3$ production.

DISTRIBUTION OF PHOSPHORUS BETWEEN PLASMA & CELLS

As a generality, any stimulus that causes intracellular phosphorus deficiency enhances phosphorus uptake into the cell. Distribution of phosphorus between the extracellular and intracellular compartments is governed to a large extent by arterial P_{CO_2} and glucose transport (Fig 18–1).

P_{CO_2}

Shift of phosphorus into cells is potently induced by respiratory alkalosis. The hypophosphatemic effect of acute hypocapnia is quick, within 1 hour. Intracellular alkalosis activates phosphofructokinase, a key enzyme in glycolysis, which increases intracellular phosphorus utilization and lowers intracellular free phosphorus concentration. The intracellular concentration "sink" for phosphorus then increases the driving force for entry of phosphorus into the cell.

As shown in Fig 18–4 (top), the profound effect by which phosphorus is translocated into cells to cause reduction in plasma phosphorus concentration is observed with acute respiratory alkalosis but not metabolic alkalosis. Urinary phosphorus loss does not contribute to this hypophosphatemic effect of hypocapnia, since phosphaturia is greatly diminished (Fig 18–4, bottom).

Glucose

Enhanced phosphorus entry into cells also occurs following cellular glucose uptake. Phosphorus is incorporated into the phosphorylated intermediates in the glycolytic pathway when glucose transport into cells is accentuated by insulin. Similar to the situation with acute hypocapnia, this in-

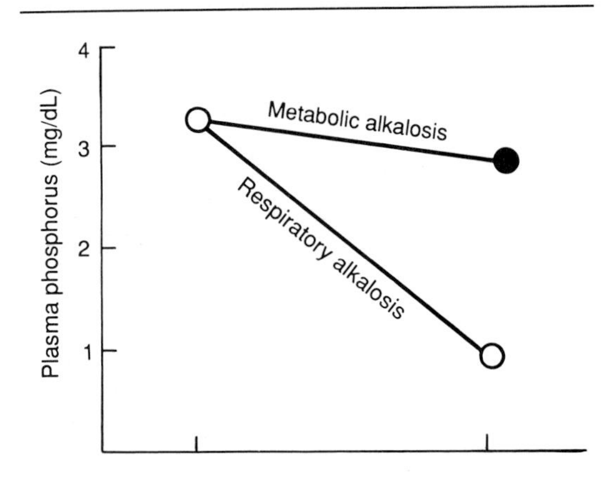

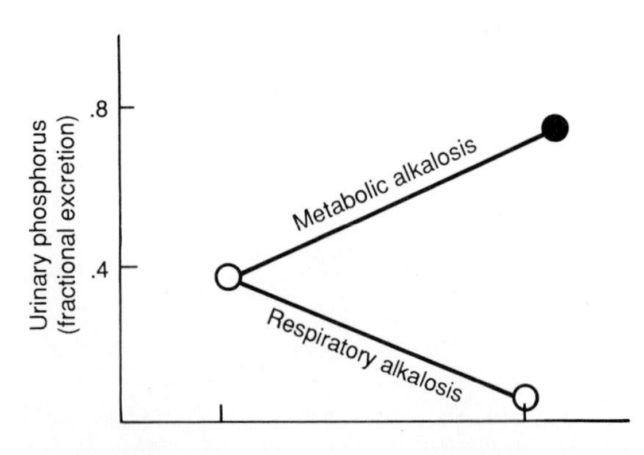

Figure 18–4. Impact of respiratory or metabolic acidosis on phosphorus homeostasis.

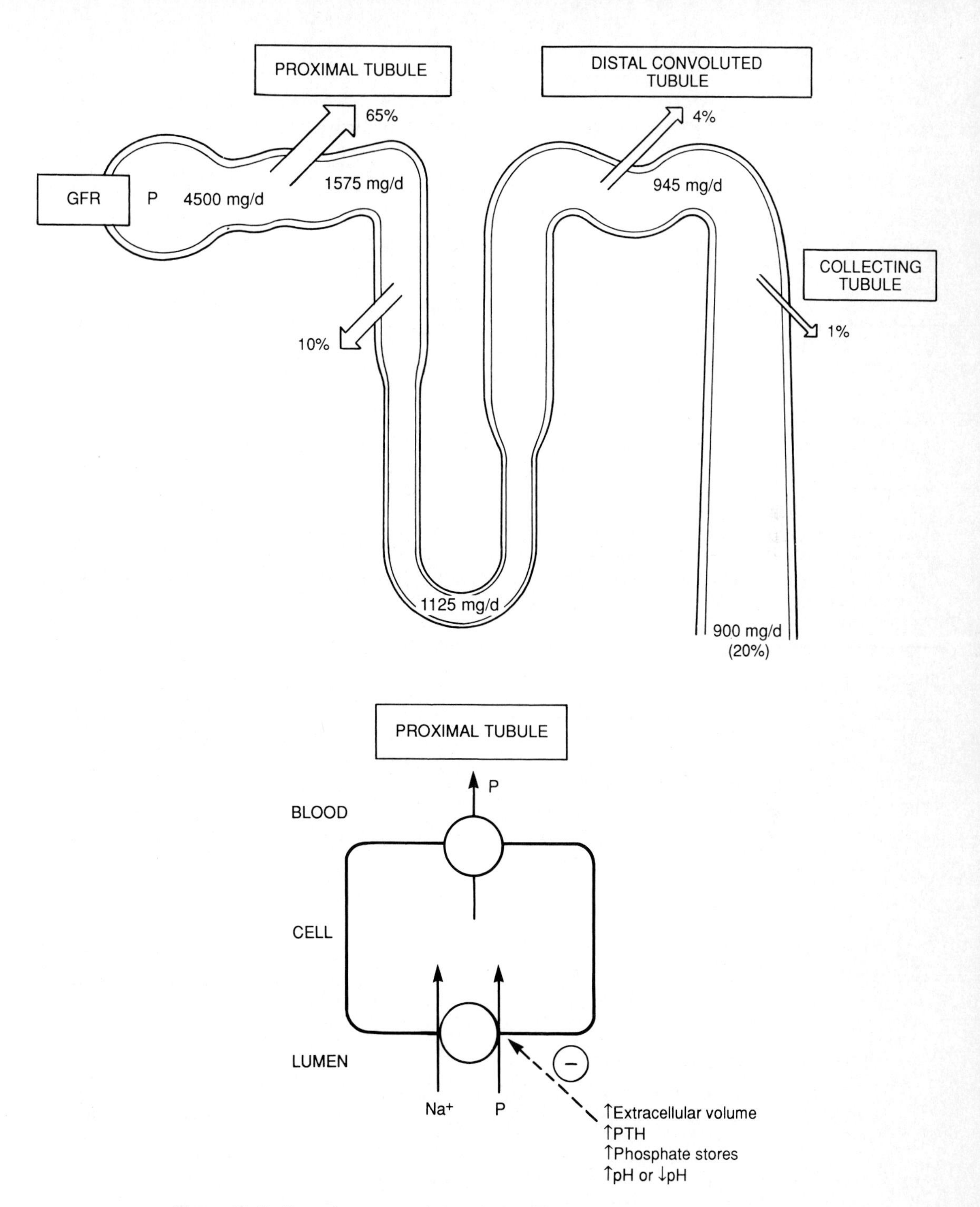

Figure 18-5. Normal amounts of phosphorus filtered, reabsorbed, and excreted.

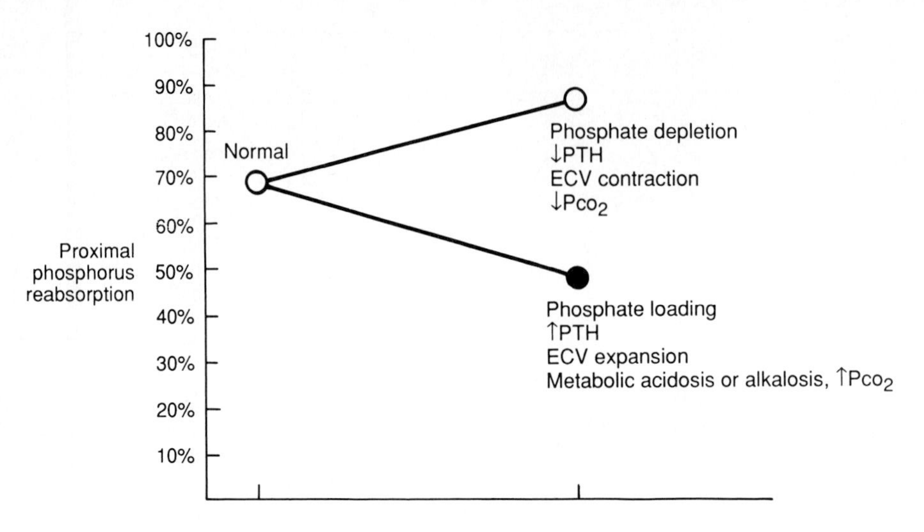

Figure 18–6. Regulation of phosphorus reabsorption in the proximal tubule.

creased utilization of phosphorus causes intracellular free phosphorus concentration to fall, enhancing the driving force for entry of phosphorus into the cell.

RENAL PHOSPHORUS EXCRETION

Although bound or complexed to a small degree (Fig 18–2), most plasma inorganic phosphorus is freely filtered by the glomerulus. Maintaining external phosphorus balance requires that most of this filtered phosphorus be reabsorbed. As shown in Fig 18–5, if plasma ultrafiltrable phosphorus is 3 mg/dL (30 mg/L) and GFR is 100 mL/min (150 L/d), the filtered phosphorus load is 4500 mg/d. Given an average urinary phosphorus excretion of 900 mg/d, the fraction of filtered phosphorus that is normally reabsorbed is about 80% (ie, only 20% is excreted).

Compared to the complexity of sites responsible for transport of electrolytes discussed previously

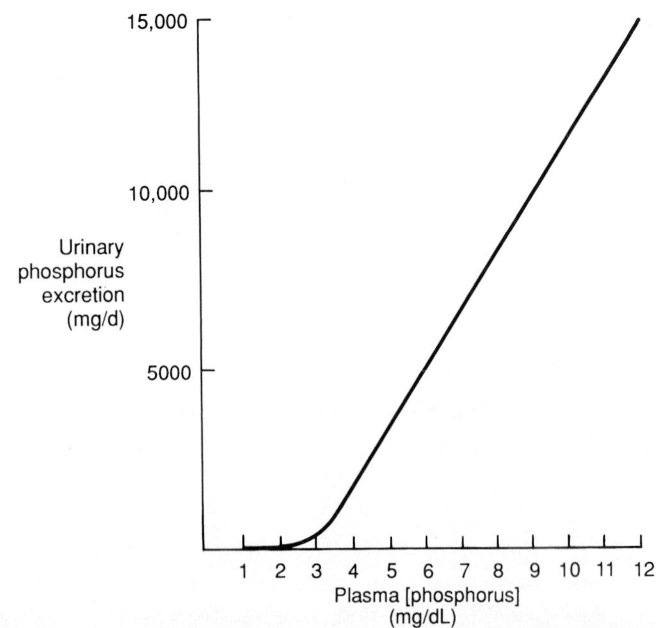

Figure 18–7. Phosphaturic response to hyperphosphatemia.

in this book, the reabsorption of phosphorus is quite simple, taking place almost exclusively in the proximal tubule.

SITES & MECHANISMS OF PHOSPHORUS REABSORPTION

Proximal Tubule

As summarized in Fig 18–5, about 65% of the filtered phosphorus is reabsorbed in the proximal convoluted tubule. Another 10% is reabsorbed by the proximal straight tubule. Phosphorus concentration in the luminal fluid of the early proximal tubule falls rapidly and thereafter remains stable at about 70% of the concentration that was in the glomerular ultrafiltrate. Because of this stability of luminal phosphorus concentration, phosphorus reabsorption generally follows sodium reabsorption in the proximal tubule and therefore is sensitive to extracellular volume status (see Chapter 1).

As shown in Fig 18–5 (bottom), phosphorus crosses the luminal membrane as phosphate coupled with sodium, energized by the lumen-to-cell sodium concentration gradient. Phosphate exit from the cell is by sodium-independent facilitated diffusion, driven down its electrochemical gradient.

As shown in Fig 18–6, transepithelial phosphorus transport is sensitively regulated by the intracellular cAMP level (primarily controlled by PTH) and non-cAMP-mediated factors (eg, due to change in extracellular volume or cellular phosphate content). A rise in PTH level can markedly diminish phosphorus reabsorption in the proximal tubule. Other hormones that can influence proximal phosphorus reabsorption—usually of little physiologic importance—include vitamin D (increase), calcitonin (decrease), glucocorticoid (decrease), insulin (increase), thyroid (increase), and growth hormone (increase). Phosphorus reabsorption tends to be diminished in metabolic acid-base disorders (metabolic acidosis or alkalosis), but the mechanisms are controversial.

Loop of Henle, Distal Convoluted Tubule, & Collecting Tubule

There is no phosphorus transport in the thin limbs of Henle. Under normal circumstances, little phosphorus transport occurs in the thick ascending limb of Henle, though reabsorption can be demonstrated under conditions of dietary phosphate deprivation. Although this segment has PTH-responsive adenylate cyclase activity, the consequences on phosphorus transport are ill-defined.

A small amount—approximately 4% of the filtered load—of phosphorus transport occurs in the distal convoluted tubule, but there is little (0–1%) in the collecting tubule.

REGULATION & LIMITS OF RENAL PHOSPHORUS EXCRETION

When dietary phosphate is restricted, reduction in body phosphorus stores, plasma phosphorus concentration, and PTH level cause phosphorus to virtually disappear from the urine. Within 3–5 days, renal phosphorus reabsorption can approach 100% of that filtered, with urinary phosphorus excretion of **less than 3 mg/d.**

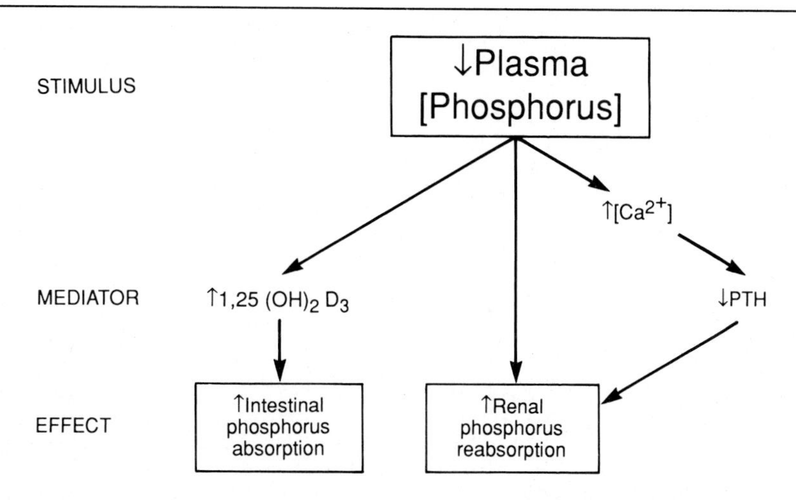

Figure 18–8. Response to hypophosphatemia.

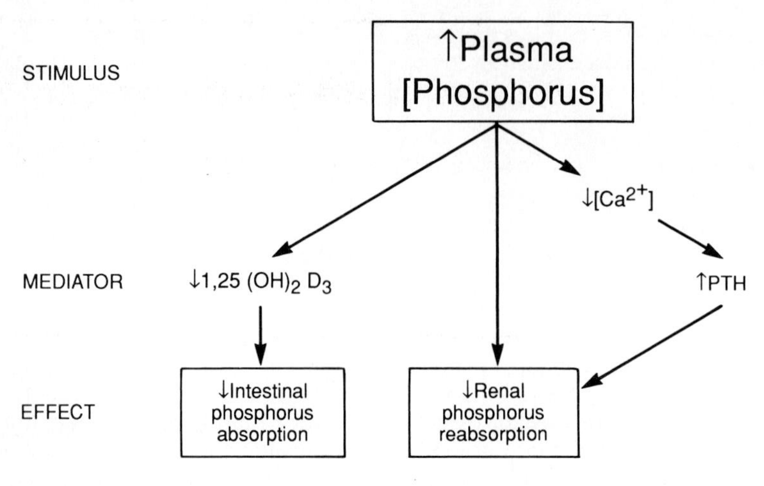

Figure 18-9. Response to hyperphosphatemia.

As shown in Fig 18-7, with intravenous phosphate loading to markedly elevate plasma phosphorus concentration in a normal individual, urinary phosphorus excretion can exceed 10 mg/min, or **15,000 mg/d.** This extraordinary phosphaturic response occurs because renal phosphate reabsorption is intrinsically saturable and is further suppressed by a high cell phosphorus content and PTH level induced by hyperphosphatemia (see below). Thus, because urinary phosphorus excretion can be so brisk, it is normally very difficult to engender hyperphosphatemia in an individual with normal GFR and a normal calcium-phosphorus-PTH metabolic axis.

RESPONSE TO CHANGE IN PLASMA PHOSPHORUS CONCENTRATION

Unless there is abnormal redistribution of phosphorus from extracellular to intracellular stores, hypophosphatemia usually represents, at least qualitatively, a reduction in total body phosphorus content. Reduction in plasma phosphorus concentration causes slight, if undetectable elevation in plasma calcium concentration that diminishes PTH release. At the same time, hypophosphatemia stimulates $1,25(OH)_2D_3$ production (Fig 15-3). As shown in Fig 18-8, hypophosphatemia therefore both directly and indirectly increases renal phosphorus reabsorption and intestinal phosphorus absorption. Thus, fecal and urinary phosphorus losses are minimized. The opposite chain of events occurs in response to an elevation in plasma phosphorus concentration, as shown in Fig 18-9.

Hypophosphatemia 19

INTRODUCTION

DEFINITION

Reduction in Plasma Inorganic Phosphorus

Eight-five percent of inorganic phosphorus is free, not bound or complexed (Fig 18–2). Unlike calcium, plasma phosphorus concentration is little affected by changes in albumin concentration. The only cause of a spuriously low plasma phosphorus concentration is the presence of mannitol, as shown in Fig 19–1. In some laboratory methods of analysis in which molybdate is used, mannitol artifactually lowers the measured phosphorus concentration. With this exception, hypophosphatemia is reliably defined by a phosphorus concentration **less than 2.5 mg/dL.**

PATHOPHYSIOLOGY

Similar to the case of potassium or calcium, only a small fraction of total body phosphorus resides in the extracellular space (Fig 18–1). Small redistribution of phosphorus from the extracellular to intracellular compartments can cause hypophosphatemia without change in total body phosphorus (Fig 19–1). The two most common causes of such redistribution are acute respiratory alkalosis and glucose administration. Alternatively, hypophosphatemia may reflect total body phosphorus depletion, due to insufficient intestinal phosphorus absorption or urinary phosphorus wastage.

In the case of the nonrenal causes of hypophosphatemia—redistribution into cells or poor intestinal absorption—the kidney appropriately avidly retains phosphorus. Urinary phosphorus excretion is close to zero (ie, < 3 mg/d, as shown in Fig 18–7). Alternatively, if renal phosphorus wasting is the cause of the hypophosphatemia, inappropriate phosphaturia is present.

REDISTRIBUTION OF PHOSPHORUS

Causes of hypophosphatemia are listed in Table 19–1.

ACUTE RESPIRATORY ALKALOSIS

Lowering of arterial P_{CO_2} has a profound effect on plasma phosphorus concentration (Fig 18–4). Acute reduction in P_{CO_2} from 38 to 16 mm Hg has been shown to reduce plasma phosphorus concentration from 3.0 to 0.9 mg/dL within an hour. Acute respiratory alkalosis may be due to hypoxia, sepsis, cirrhosis, pharmacologic agents (eg, salicylates), hyperthermia, burns, central nervous system disorders, or other causes (Table 14–2). Neither hyperventilation without hypocapnia nor metabolic alkalosis has this effect (Fig 18–4). Since total body phosphorus is not diminished, symptoms from this redistributional form of hypophosphatemia have not been documented.

CARBOHYDRATE ADMINISTRATION

Glucose

Glucose plus insulin administration to a well-fed, phosphorus-repleted individual has only a transient minor (< 0.5 mg/dL) effect on plasma phosphorus concentration. However, if the individual was previously malnourished and phosphorus-deficient (even without hypophosphatemia), a large reduction in plasma phosphorus concentration can occur. Hypophosphatemia occurs as a result of enhanced cellular phosphorus uptake for phosphorylation of glycolytic intermediates in liver and muscle. Urinary phosphorus excretion is appropriately reduced.

Hypophosphatemia due to carbohydrate feeding is common when nutritional therapy is instituted in an alcoholic. In such a patient who is commonly phosphorus-deficient due to inadequate diet, respiratory alkalosis (due to alcohol

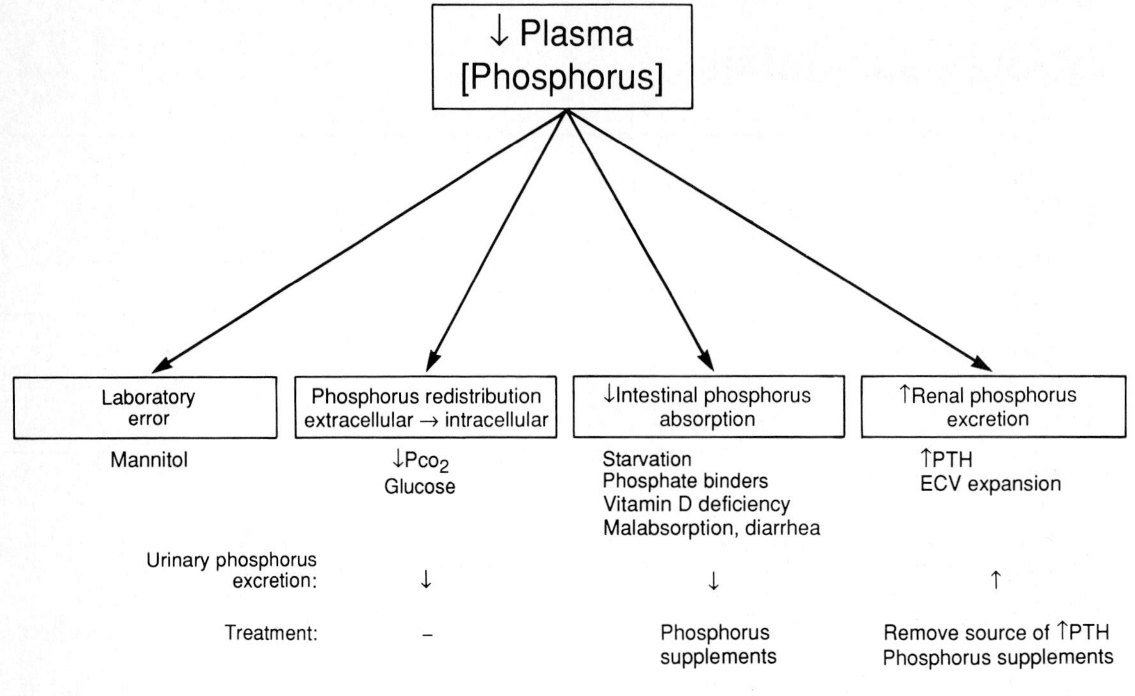

Figure 19-1. Diagnosis and treatment of hypophosphatemia.

withdrawal, cirrhosis, or sepsis) may be an additive factor to glucose administration in development of severe hypophosphatemia.

Insulin Administration in Diabetic Ketoacidosis

Plasma phosphorus concentration during diabetic ketoacidosis may be normal or elevated (due primarily to hypovolemia-induced reduction in GFR). Total body phosphorus may be diminished, however, by poor dietary intake, decreased glucose entry into cells in the hypoinsulinemic state, plus urinary loss of phosphorus caused by osmotic diuresis and metabolic acidosis. The amount of phosphate depletion is usually modest, however, given the relatively short duration of illness in most patients. The presence of hypophosphatemia before treatment therefore indicates severe phosphorus depletion. When insulin is administered, rapid cellular uptake and phosphorylation of glucose may profoundly shift extracellular phosphorus into the phosphorus-deficient cells, causing marked hypophosphatemia.

Table 19-1. Causes of hypophosphatemia.

Redistribution of phosphorus
Respiratory alkalosis (see Table 14-2)
Carbohydrate administration
Glucose
Insulin administration in diabetic ketoacidosis
Hyperalimentation
Fructose administration
↓Intestinal phosphorus absorption
Prolonged starvation
Phosphate binders (see Table 12-2)
Vitamin D deficiency
Malabsorption, steatorrhea, diarrhea
↑Renal phosphorus excretion
Hyperparathyroidism (see Table 17-1)
Hypervolemic states (see Chapter 3)
Acetazolamide administration
Fanconi's syndrome (see Table 12-4)
Metabolic alkalosis or acidosis (see Chapters 12 and 13)

Hyperalimentation

The provision of fructose or another carbohydrate to a previously starved, catabolic and phosphorus-deficient individual can lead to hypophosphatemia by the same mechanisms discussed above for glucose.

DECREASED INTESTINAL PHOSPHORUS ABSORPTION

INADEQUATE PHOSPHORUS AVAILABILITY

Prolonged Starvation

With severe reduction in phosphorus intake, unremitting intestinal phosphorus secretion and phosphorus loss in stool (at least 200 mg/d; Fig 18–3) despite appropriate elimination of phosphorus from urine can cause phosphorus depletion. While phosphorus mobilization from bone mitigates the reduction in plasma phosphorus concentration, negative phosphorus balance and hypophosphatemia eventually occurs. More profound hypophosphatemia may be induced with subsequent carbohydrate refeeding.

Phosphate Binders

Excessive phosphorus entrapment by magnesium-, calcium- or aluminum-containing antacids may diminish dietary phosphorus availability below that necessary to maintain external phosphorus balance. The severity of hypophosphatemia depends on the amount of phosphorus in the diet compared to the dose of antacid.

INADEQUATE INTESTINAL PHOSPHORUS TRANSPORT

Since vitamin D regulates intestinal divalent ion absorption, vitamin D deficiency may cause combined hypocalcemia and hypophosphatemia, leading to osteomalacia. Secondary hyperparathyroidism in response to hypocalcemia can exacerbate hypophosphatemia by preventing renal phosphorus conservation. Disorders of intestinal malabsorption can also impair phosphorus transport directly as well as indirectly by disrupting enterohepatic recycling of vitamin D.

INCREASED RENAL PHOSPHORUS EXCRETION

DECREASED PROXIMAL PHOSPHORUS TRANSPORT

Hyperparathyroidism

Any primary or secondary cause of PTH elevation will predictably enhance phosphorus excretion (Fig 15–2), so that urinary phosphorus excretion is inappropriately elevated (ie, > 3 mg/d) despite hypophosphatemia.

Defective or Inhibited Proximal Transport

Extracellular volume expansion from any cause diminishes the fraction of sodium and phosphorus reabsorbed in the proximal convoluted tubule (Fig 18–6). The phosphaturia is generally mild.

Carbonic anhydrase inhibitors are the only pharmacologic agents commonly used with preferential action in the proximal tubule with associated phosphaturia. Osmotic diuresis also has a small effect. Generalized disruption of all sodium cotransport systems in Fanconi's syndrome predictably results in phosphaturia and hypophosphorusmia (Chapter 12).

Metabolic Alkalosis & Acidosis

Both of these disorders cause a reduction in proximal phosphorus transport (Fig 18–6), phosphaturia, and sometimes mild hypophosphatemia.

SYMPTOMS

To a large degree, symptoms attributable to phosphorus depletion are ascribable to intracellular ATP depletion and impaired oxygen delivery consequent to diminished erythrocyte 2,3-DPG, as shown in Fig 19–2.

NEUROMUSCULAR

Mild encephalopathic symptoms such as irritability and confusion can progress to stupor, coma, and seizures.

Paresthesias, muscular weakness and frank rhabdomyolysis can occur in primary phosphorus deficiency, especially when carbohydrate administration is superimposed. An elevated plasma CPK level and myoglobinuric acute renal failure can ensue. Weakness of the respiratory muscles can cause significant hypoventilation, exacerbating the functional impairment in oxygenation.

HEMATOLOGIC

Red Cell

Erythrocyte phosphorus concentration is proportionate to ATP and 2,3-DPG levels. Hypophosphatemia is associated with reduction in 2,3-

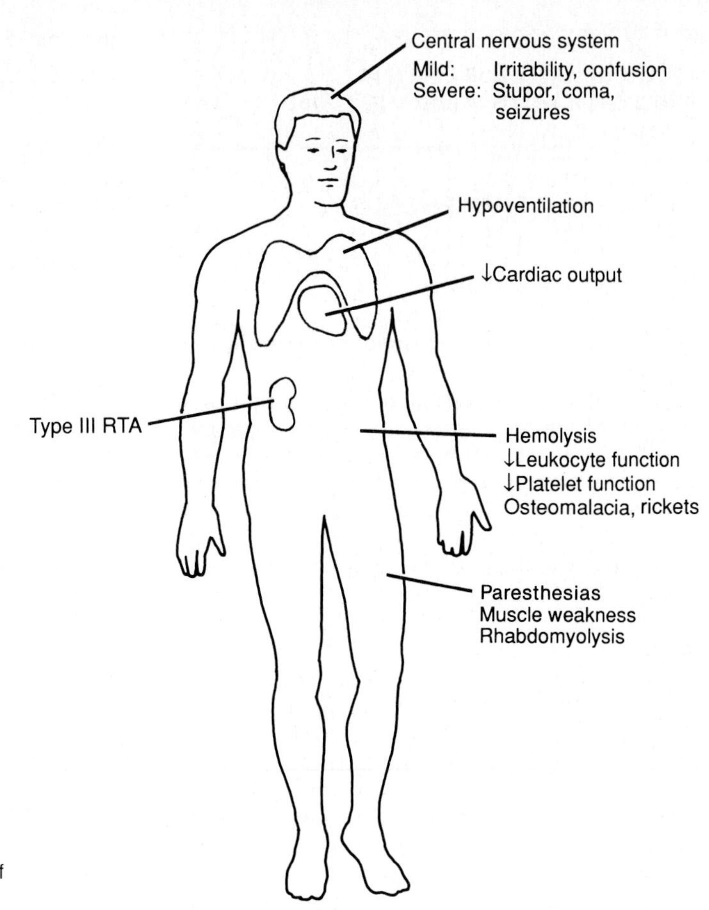

Figure 19-2. Symptoms and signs of hypophosphatemia.

DPG and thus increased affinity for oxygen by hemoglobin. Diminished oxygen delivery to tissues exacerbates local metabolic dysfunction and is particularly problematic in patients with impaired cardiopulmonary disease. Overt hemolysis can occur when plasma phosphorus concentration is less than 1.0 mg/dL.

Leukocyte

Diminished chemotactic, phagocytic, and bactericidal activity of the granulocyte is due to the reduced intracellular ATP level of hypophosphatemia. Enhanced risk of sepsis is a potential complication.

Platelet

Dysfunctional platelet activity and sometimes thrombocytopenia are consequences of hypophosphatemia, with propensity to hemorrhage.

OTHER

Impaired myocardial contractility has been reported in hypophosphatemia. Gastrointestinal manifestations, including anorexia, nausea, and vomiting, may be due to hypophosphatemia. Phosphorus depletion also can contribute to osteomalacia and rickets, especially in the setting of vitamin D deficiency. Phosphorus depletion diminishes titratable acid formation by the kidney to an extent that may cause metabolic acidosis, a form of type III (impaired buffer delivery) distal RTA.

DIAGNOSIS

APPROACH TO ETIOLOGIC DIAGNOSIS

History & Physical Examination

A careful dietary history is important in assessing whether total body phosphorus might be depleted. Recent carbohydrate administration (glucose, fructose, or hyperalimentation solution) to a previously phosphorus-deficient individual should be quantitated if present. Disorders associated with impaired phosphorus homeostasis, such as diabetes mellitus, alcoholism, intestinal disorders, hyperparathyroidism, and hypervolemia, should be carefully sought. Drugs that impair intestinal phosphorus uptake (antacids) or enhance urinary excretion (acetazolamide) should be identified.

Urinary Phosphorus

As shown in Fig 19–1, the level of urinary phosphorus excretion easily distinguishes redistribution syndromes or impaired intestinal phosphorus absorption syndromes in which renal phosphorus retention is appropriate (ie, < 3 mg/d) from renal phosphorus wasting syndromes.

Arterial Blood Gases

When acute respiratory alkalosis is suspected as a primary or contributing cause of hypophosphatemia, arterial pH and P_{CO_2} may help confirm the diagnosis (Fig 11-1).

Plasma PTH & Calcium Concentrations

Measuring PTH when hypophosphatemia is associated with hypercalcemia helps confirm the diagnosis of primary hyperparathyroidism (Chapter 17). Secondary hyperparathyroidism may be diagnosed by PTH measurement when hypophosphatemia is associated with hypocalcemia in intestinal disorders or vitamin D-deficient states (Chapter 16). When appropriate, measurement of $1,25(OH)_2$-D_3 may be helpful to establish the diagnosis of hypovitaminosis D.

THERAPY

DECISION TO TREAT

Assessment of Total Body Phosphorus Deficiency

If phosphorus has merely shifted into cells without evidence of total body phosphorus depletion, therapy is not indicated. The most common cause for this redistribution is acute respiratory alkalosis. Such hypocapnia-induced phosphorus shifts do not produce symptoms of hypophosphatemia such as those that occur when total body phosphorus is diminished because intracellular ATP and 2,3-DPG levels are not altered. Likewise, the mild, transient hypophosphatemia induced by glucose administration in a phosphorus-repleted individual induces no intracellular abnormality or symptoms and requires no treatment.

If total body phosphorus deficiency is suspected because of a history of inadequate diet or documented gastrointestinal or renal losses, phosphorus repletion may be necessary. In all cases, an attempt should be made to remove the cause of abnormal intestinal phosphorus absorption or renal phosphorus wasting, such as discontinuation of antacids or diuretics. Unfortunately, reduction in plasma phosphorus concentration does not correlate well with total body phosphorus deficiency. Plasma phosphorus concentration does not usually become subnormal unless total body phosphorus is reduced by 3 g or more. Symptomatic phosphorus deficiency usually does not occur until total body phosphorus deficit reaches at least 10 g.

Table 19–2. Treatment of hypophosphatemia.

Preparation and Dosage	Phosphorus Content	Na^+ Content	K^+ Content
Intravenous: 2–7.5 mg/kg over 6–8 hours			
Sodium phosphate	93 mg/mL	4 meq/mL	0
Potassium phosphate	93 mg/mL	0	4 meq/mL
Oral: 250–500 mg QID			
Neutral sodium, potassium phosphate (Neutra-Phos, K-Phos-Neutral, Uro-KP-Neutral)	250 mg/capsule or tablet or 250 mg/75 mL	6–13 meq/tablet	1-7 meq/tablet
Potassium phosphate (Neutra-Phos-K, K-Phos)	125 or 250 mg/capsule or tablet or 125 or 250 mg/75 mL	0	4–14 meq/tablet
Sodium phosphate (Fleet's Phospho-Soda)	129 mg/mL	5 meq/mL	0
Rectal			
Sodium phosphate (Fleet's Enema)	43 mg/mL	1.6 meq/mL	0

Table 19-3. Phosphorus content of foods.

High (> 250 mg)	Medium (150-250 mg)	Low (< 150 mg)
Milk products (8 oz)	Beans (8 oz)	Bread (1 slice)
Milk (whole or low-fat)	Shellfish (8 oz)	Most cereals (8 oz)
Yogurt	Tofu (8 oz)	Bacon (1 slice)
Cheeses (2 oz)	Ice cream (8 oz)	Egg (1)
Cottage cheese	Lentils (8 oz)	Most vegetables
Cocoa	Lima beans (8 oz)	Fruits
Meats (4 oz)	Mushrooms (8 oz)	Potatoes
Liver	Peas (8 oz)	Rice
Chicken, turkey, duck	Pudding (8 oz)	Condiments (salt, pepper)
Pork, bacon, ham	Peanut butter (2 tbsp)	Beverages
Lamb		Soft drinks
Beef		Tea, coffee
Fish (4 oz)		Beer, wine
Cereals (8 oz)		
Bran		
Wheat-based		
Nuts (4 oz)		

Assessment of Symptoms Attributable to Phosphorus Deficiency

The urgency and route of phosphorus deficiency depends on whether symptoms are present. If hypophosphatemia is severe (< 1.5 mg/dL) or if symptoms attributable to phosphorus deficiency are present (eg, central nervous system abnormalities or weakness of muscles, especially those involved in respiration), intravenous phosphorus is mandated. On the other hand, mild phosphorus deficiency (1.5–2.5 mg/dL) causing no symptoms can sometimes be treated simply with liberalization of oral phosphorus intake.

TREATMENT OF SEVERE, SYMPTOMATIC HYPOPHOSPHATEMIA

Parenteral supplementation is indicated for severe or symptomatic hypophosphatemia, as summarized in Table 19–2. The minimum starting dose of phosphorus is 2 mg/kg every 8 hours, though higher doses to 7.5 mg/kg every 6 hours may be used for emergencies. Plasma phosphorus concentration should be monitored frequently (every 6–12 hours). Oral therapy be started when symptoms have been relieved and plasma phosphorus concentration is greater than 2.0 mg/dL.

Complications of intravenous phosphorus therapy include hypervolemia, hypernatremia, and hyperkalemia because of the accompanying cation load, metastatic calcification in the presence of hypercalcemia (due to excessive calcium × phosphorus concentration product), hypocalcemia due to precipitation of circulating calcium, and hyperphosphatemia if supplementation is excessive and GFR is reduced. Parenteral phosphorus should not be given if plasma calcium concentration is already low—becausing of the risk of symptomatic hypocalcemia—and should be used very cautiously if renal function is abnormal.

TREATMENT OF ASYMPTOMATIC HYPOPHOSPHATEMIA

Dietary Supplementation

A simple increase in dietary phosphorus ingestion may be sufficient to correct modest hypophosphatemia. A list of phosphorus content in foods is presented in Table 19–3. Milk products and most meats are good dietary sources of elemental phosphorus.

Oral Phosphorus Supplements

If food intake is insufficient, oral phosphorus may be administered in the form of sodium or potassium salts of phosphorus, as shown in Table 19–2. Divided doses may be used to provide a total of 1–2 g/d of phosphorus. It should be remembered that only about two-thirds of ingested phosphorus is absorbed (sometimes less with intestinal disease). These preparations can cause diarrhea, hypervolemia, or hyperkalemia (depending on the cation in the preparation used). Close monitoring of plasma phosphorus concentration is indicated.

Hyperphosphatemia

20

INTRODUCTION

DEFINITION

Increase in plasma phosphorus concentration to **over 4.5 mg/dL** defines hyperphosphatemia.

PATHOPHYSIOLOGY

Redistribution Versus Change in Total Body Phosphorus

In direct contrast with phosphorus deficiency, hyperphosphatemia may arise (1) from rapid redistribution of phosphorus from intracellular to extracellular stores; (2) from an excessive exogenous phosphorus load; or (3) from impaired renal phosphorus excretion, usually due to diminished GFR or hypoparathyroidism, as shown in Fig 20–1. Given the normal extraordinary capacity for urinary phosphorus excretion (Fig 18–7)—over 15,000 mg/d or more—depression in GFR is almost always observed in the first two groups as well as in the third. The reduction in GFR impairs the filtration and hence the absolute amount of urinary excretion of phosphorus even when phosphorus reabsorption in the proximal tubule is suppressed.

The rate of urinary phosphorus excretion distinguishes between these causes of hyperphosphatemia (Fig 20–1). In the case of the nonrenal causes of hyperphosphatemia—endogenous redistribution or exogenous load—renal phosphorus excretion is brisk (ie, should exceed 15,000 mg/d, though it may be impaired somewhat by reduction in GFR). In the case of the primary renal retention of phosphorus, renal phosphorus excretion is inappropriately low.

REDISTRIBUTION OF PHOSPHORUS

Causes of hyperphosphatemia are listed in Table 20–1.

RELEASE FROM CELLS

Hemolysis

Rapid mechanical, thermal, or immunologic destruction of red cells releases their ionic contents, especially phosphorus and potassium. A reduction in GFR due to the hemoglobinuria and hyperphosphatemia exacerbates the condition, preventing the normal phosphaturic response. Treatment consists of expanding extracellular volume in order to maintain a high GFR and promote natriuresis and phosphaturia.

Rhabdomyolysis

Muscle cell injury due to crush trauma, viral or bacterial infections, heat stroke, potassium deficiency, seizures, or preexisting phosphorus deficiency can release muscle ionic contents—again, mostly phosphorus and potassium. Reduction in GFR due to myoglobinuria, extracellular volume depletion, and hyperphosphatemia again exacerbates the condition. The degree of CPK and uric acid elevation in the blood and myoglobin in the urine correlates with the extent of rhabdomyolysis. Treatment is aimed at ensuring extracellular volume expansion in order to maintain as high a GFR and urinary phosphorus output as possible.

Tumor Lysis Syndrome

Chemotherapy of some large, bulky tumors, especially Burkitt's and other poorly differentiated lymphomas, can cause rapid destruction of the neoplastic cells. Lymphoblasts and immature lymphoid cells have a higher phosphorus content compared with mature lymphocytes. Release of large amounts of phosphorus and uric acid may also occur following tumor lysis, both of which can diminish GFR to potentiate the hyperphos-

Table 20-1. Causes of hyperphosphatemia.

Phosphorus redistribution
Hemodialysis
Rhabdomyolysis
Tumor lysis
Excess phosphorus load
Phosphorus-containing antacids
Phosphorus-containing enemas or laxatives
Phosphorus-containing oral agents
↓Renal phosphorus excretion
↓GFR
Primary glomerular or tubulointerstitial disease
Secondary to extracellular volume contraction
↓PTH
Primary hypoparathyroidism
Surgical hypoparathyroidism
Pseudohypoparathyroidism
Tumoral calcinosis

phatemia. Prevention of this syndrome can be achieved by extracellular volume expansion with alkalinization of the urine (to achieve 3–5 L/d of urine output with pH > 7.5) and allopurinol administration before initiation of chemotherapy.

EXCESS PHOSPHORUS LOAD

PHOSPHORUS ADMINISTRATION

Administration of excessive amounts of phosphorus-containing antacids, milk, laxatives, enemas, or other oral agents may cause hyperphosphatemia, especially if GFR is reduced by extracellular volume depletion or primary cause. Laxatives and enemas especially predispose to hypovolemia and reduced GFR by inducing diarrhea.

IMPAIRED RENAL PHOSPHORUS EXCRETION

REDUCED GFR

In virtually any form of acute or chronic parenchymal renal disease (glomerular or tubulointerstitial) or acute renal dysfunction due to a hypovolemic state, reduction of GFR to a value less than about 30–50 mL/min is accompanied by hyperphosphatemia. The degree of plasma phosphorus elevation then is inversely proportionate to the GFR. Insufficient phosphorus excretion occurs as a result of insufficient filtration despite secondary effects of hyperphosphatemia to increase PTH, diminish proximal phosphorus reabsorption, and promote urinary phosphorus excretion and to reduce the $1,25(OH)_2D_3$ level and impair gastrointestinal phosphorus absorption.

HYPOPARATHYROIDISM

Since PTH is the main controller of phosphorus reabsorption in the proximal tubule (Fig 18–6), reduction in PTH—either in absolute terms or in cellular response (pseudohypoparathyroidism)—enhances renal phosphorus reabsorption and prevents adequate urinary excretion. A rare hyperphosphatemic entity of tumoral calcinosis also occurs in which ectopic calcium and phosphate are deposited.

SYMPTOMS

Symptoms attributable to hyperphosphatemia per se are not well described. In fact, acute elevation of plasma phosphorus concentration to 12 mg/dL or more in normal individuals has been performed (Fig 18–7) without adverse consequences.

Symptoms of hyperphosphatemia are generally attributable to secondary changes in plasma calcium concentration and in ectopic deposition of phosphorus with calcium. Plasma calcium concentration falls as plasma phosphorus concentration rises. Manifestations of hypocalcemia are shown in Fig 16–2 and include neuromuscular symptoms (paresthesias, tetany, seizures) and cardiac abnormalities (arrhythmia, hypotension). Despite hypocalcemia, hyperphosphatemia may cause a marked increase in the plasma calcium × phosphorus concentration product. When this product exceeds 60, there is a risk of ectopic calcification in the heart, blood vessels, kidney, gastrointestinal tract, and skin, as discussed in Chapter 16.

DIAGNOSIS

APPROACH TO ETIOLOGIC DIAGNOSIS

History & Physical Examination

Evidence or a history of conditions associated with hemolysis, muscle injury, recent chemotherapy of a large tumor, or primary or secondary (eg, hypovolemia-induced) renal impairment should be sought. A careful dietary history is useful in quantitating phosphorus intake (using Table 19–3). A diligent search for prescribed and over-the-counter phosphorus-containing laxatives, antacids, supplements, or enemas should be carefully performed.

Urinary Phosphorus

The rate of urinary phosphorus excretion—whether appropriately high using Fig 18–7—is useful in differentiating hyperphosphatemia due to redistribution or endogenous or exogenous loading versus renal impairment in phosphorus excretion (Fig 20–1).

Plasma Calcium Concentration

The plasma calcium concentration is useful in assessing the symptomatic consequences of hyperphosphatemia as well as diagnostically, if there is hypocalcemia due to primary or functional hypoparathyroidism.

GFR

Invariably present as a pathogenetic concomitant of hyperphosphatemia, the degree of GFR reduction should be estimated by creatinine clearance or at least plasma creatinine concentration.

Other

Depending on indications, useful tests may include measurement of plasma CPK (rhabdomyolysis), LDH (hemolysis, rhabdomyolysis, tumor lysis), PTH (hypoparathyroidism), or urine nitroprusside reaction (positive for myoglobin in the absence of hematuria during rhabdomyolysis).

THERAPY

DECISION TO TREAT

The urgency of treating hyperphosphatemia depends to a large extent on whether there is symptomatic hypocalcemia. The development of symptomatic calcium-phosphorus deposition, as in the

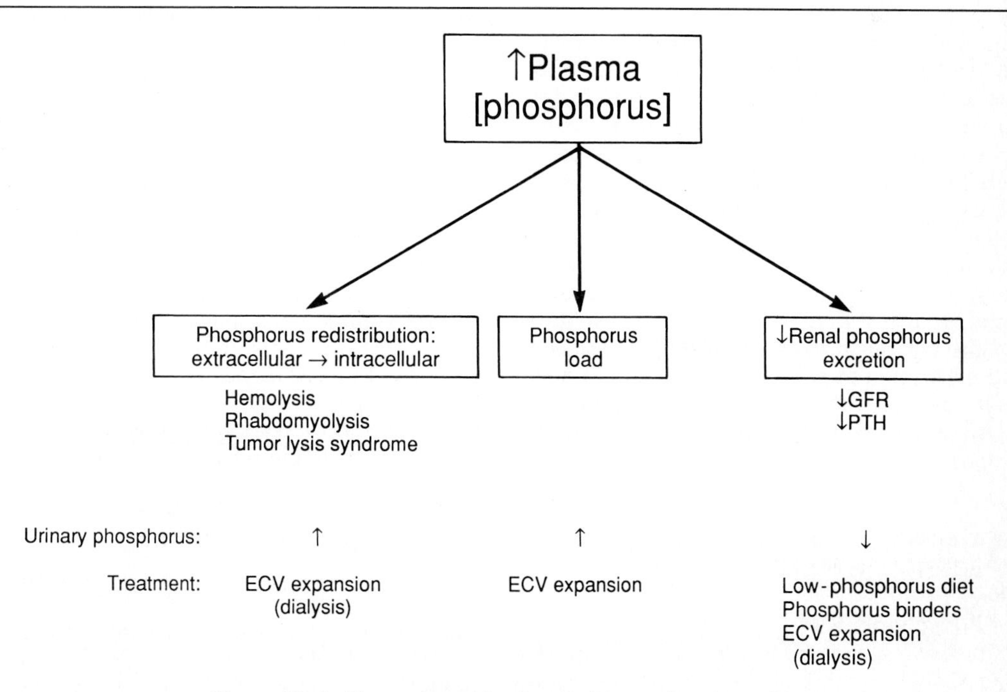

Figure 20–1. Diagnosis and treatment of hyperphosphatemia.

cardiac conduction system with an arrhythmia or in the kidney with acute renal failure, is also a strong indication for rapid and aggressive intervention.

A phosphorus burden causing the hyperphosphatemia should of course be removed if at all possible—by discontinuing oral or parenteral phosphorus supplements and diminishing dietary phosphorus (Tables 19-2 and 19-3).

TREATMENT OF SEVERE, SYMPTOMATIC HYPERPHOSPHATEMIA

Treatment of hyperphosphatemia is summarized in Table 20-2.

Table 20-2. Treatment of hyperphosphatemia.

Acute, symptomatic
Extracellular volume expansion
Hemodialysis (high-flux)
Chronic, asymptomatic
Low-phosphorus diet (see Table 19-3)
Phosphorus binders
Aluminum hydroxide (500-2000 mg with meals)
Capsules, 476-500 mg (Alu-Cap, Dialume)
Suspension, 320-600 mg/15 mL (Alterna Gel, Amphogel, Nephrox)
Tablets, 300-600 mg (Alu-Tab, Amphojel)
Calcium carbonate (500-2000 mg with meals)
Suspension, 1000 mg/5 mL (Titralac)
Tablets, 300-850 mg (Alka-Mints, Amitone, Calcilac, Chooz, Rolaids, Tums)

Extracellular Volume Expansion

The cornerstone of treating hyperphosphatemia is to normalize GFR if depressed and to reduce phosphorus reabsorption in the proximal tubule. These goals can be accomplished by extracellular volume expansion to achieve urine output greater than 3 L/d and preferably 5-10 L/d. Caution must be exercised to limit the magnitude of volume expansion in the setting of cardiopulmonary disorders or intrinsic GFR depression. Alkalinization of the tubular fluid with intravenous sodium bicarbonate administration or by using acetazolamide is of modest additional benefit in promoting phosphorus excretion but should not be employed until extracellular volume is expanded.

Hemodialysis

When GFR is reduced by intrinsic causes or by hyperphosphatemia, a vicious cycle is set in motion in which phosphorus excretion is further impaired. If GFR is very low and cannot be increased by extracellular volume repletion, hemodialysis is indicated for symptomatic hyperphosphatemia or when a plasma calcium × phosphorus concentration product greater than 60 persists for a long time. Newer high-flux hemodialysis membranes are useful in allowing greater phosphorus flux into the phosphorus-free dialysate than conventional membranes. Blood flow should be high, preferably greater than 300 mL/min. Daily hemodialysis is sometimes required to control hyperphosphatemia.

TREATMENT OF ASYMPTOMATIC HYPERPHOSPHATEMIA

Low-Phosphorus Diet

A diet restricted in phosphorus to less than 600 mg/d should be prescribed—and can usually be achieved by limiting milk products (Table 19-3). This may be sufficient if hyperphosphatemia is mild.

Phosphorus Binders

Calcium-, magnesium- or aluminum-based phosphorus binders are useful in chronic hyperphosphatemia especially when associated with chronic renal failure. Binders should be given with meals for the best results in chelating ingested phosphorus.

Magnesium-based binders should be used cautiously in renal insufficiency owing to the propensity for hypermagnesemia (Chapter 23) and avoided completely when GFR is less than about 30 mL/min. Aluminum-based binders are frequently used but impose a risk of promoting aluminum toxicity (central nervous system and bone) with prolonged usage. Hence, calcium-based binders are probably the most useful in patients with chronic renal disease or those with concomitant hypocalcemia such as that caused by hypoparathyroidism. The dose of antacid is adjusted to normalize the plasma phosphorus concentration but without inducing hypercalcemia. When dialysis becomes necessary in the course of chronic renal failure, the extracorporeal removal of phosphorus is usually sufficient to allow marked reduction or elimination of binder therapy.

SECTION VI REFERENCES: PHOSPHORUS HOMEOSTASIS & DISORDERS: HYPOPHOSPHATEMIA & HYPERPHOSPHATEMIA

Slatopolsky E et al: Hyperphosphatemia. Clin Nephrol 1977;7:138-146.

Stoff JS: Phosphate homeostasis and hypophosphatemia. Am J Med 1982;72:489-495.

Section VII: Magnesium Homeostasis & Disorders: Hypomagnesemia & Hypermagnesemia

Normal Magnesium Homeostasis 21

INTRODUCTION

TOTAL BODY MAGNESIUM

The amount of magnesium in the body of an average-sized adult is about 21–25 g (1750–2000 meq). As shown in Fig 21-1, about two-thirds of the magnesium resides in bone and most of the remaining third inside cells. Only about 1–2% of total body magnesium is in the extracellular space.

Intracellular Magnesium

Most magnesium within cells is bound within membranes or organelles to various metalloenzymes and proteins, phosphate compounds such as ATP, nucleic acids, and lipids. Ionized, unbound magnesium concentration is relatively low—0.6–8 mg/dL (0.5–7 meq/L), depending on cell type—and serves important cofactoring and activating roles for many enzymes.

Plasma Magnesium

About 60% of plasma magnesium is free, whereas about 15% is complexed to anions (especially bicarbonate) and 25% is protein-bound (principally albumin), as shown in Fig 21-2. The range of normal magnesium concentration in the plasma is **1.7–2.7 mg/dL** (1.4–2.3 meq/L).

MAGNESIUM INTAKE, ABSORPTION, & DISTRIBUTION

MAGNESIUM INTAKE

The average magnesium intake in the United States is approximately 320 mg/d in adult males and 200 mg/d in adult females. Magnesium intake does not change substantially as a function of age.

INTESTINAL MAGNESIUM ABSORPTION

About 30–40% of ingested magnesium is absorbed, principally in the distal portion of the small intestine—the jejunum and ileum—as shown in Fig 21-3. The fractional absorption is relatively constant as a function of intake (Fig 15-5) but can increase to 70–80% under conditions of severe dietary magnesium restriction. Magnesium is reabsorbed both by a secondary active transcellular transport process and by a passive paracellular process. A very small secretory component of magnesium flux may also occur. It is likely, though controversy still exists, that 1,25$(OH)_2$-vitamin D_3 controls magnesium absorption in the intestine as it does calcium and phosphorus.

MAGNESIUM DISTRIBUTION

Most magnesium is stored in cells (Fig 21-1). However, intracellular-extracellular distribution appears not be under hormonal control.

RENAL MAGNESIUM EXCRETION

Magnesium is freely filtered. At a normal GFR and plasma magnesium concentration, about 3500 mg/d of magnesium is filtered. The average urinary excretion of magnesium is 100 mg/d, which implies that 97% of the filtered magnesium is reabsorbed. The greatest fraction—at least half of this reabsorption—occurs in the thick ascending limb of Henle, as shown in Fig 21-4.

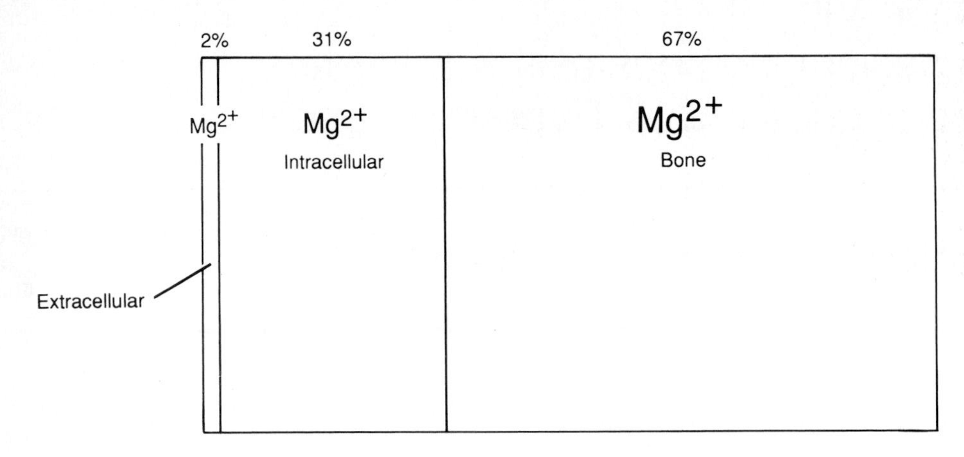

Figure 21-1. Distribution of body magnesium.

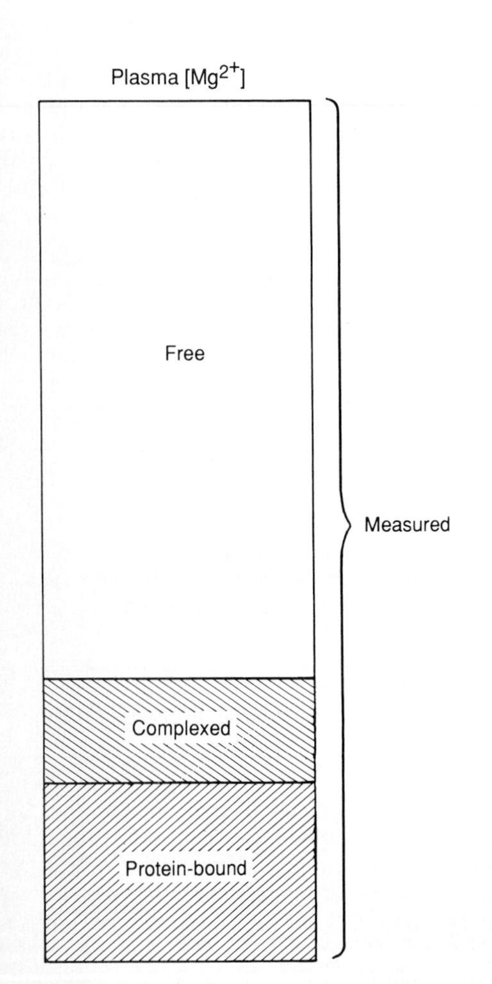

Figure 21-2. Magnesium distribution in plasma.

TUBULAR REABSORPTION OF MAGNESIUM

Proximal Tubule

Magnesium is a unique electrolyte in being relatively poorly reabsorbed in the proximal tubule. By the end of the proximal convoluted tubule, the magnesium concentration in the luminal fluid rises 50–60% above its plasma concentration (whereas most electrolyte concentrations remain stable or fall substantially). Only about 25% of the filtered magnesium is reabsorbed in the proximal convoluted tubule and another 15% in the proximal straight tubule. The mechanisms of reabsorption, whether active or passive, are poorly understood. Proximal tubule magnesium transport is proportionate to sodium transport and thus is influenced by extracellular volume status.

Thick Ascending Limb of Henle

The greatest quantity of magnesium—about 50% of the filtered load—is reabsorbed in the thick ascending limb of Henle. Though there may be a small component of active transport, the principal mechanism for magnesium transport is passive, through the paracellular pathway. The major driving force for this magnesium movement is the lumen-positive potential difference attributable to Na^+-K_+-2Cl cotransport (Fig 1–25). Inhibition of sodium chloride reabsorption and therefore of this potential difference by a loop diuretic secondarily lowers magnesium reabsorption.

Extracellular volume status, by influencing luminal flow rate and fractional sodium reabsorption (Fig 1–26), is an important determinant of magnesium reabsorption in the thick ascending limb of Henle. Another important determinant of

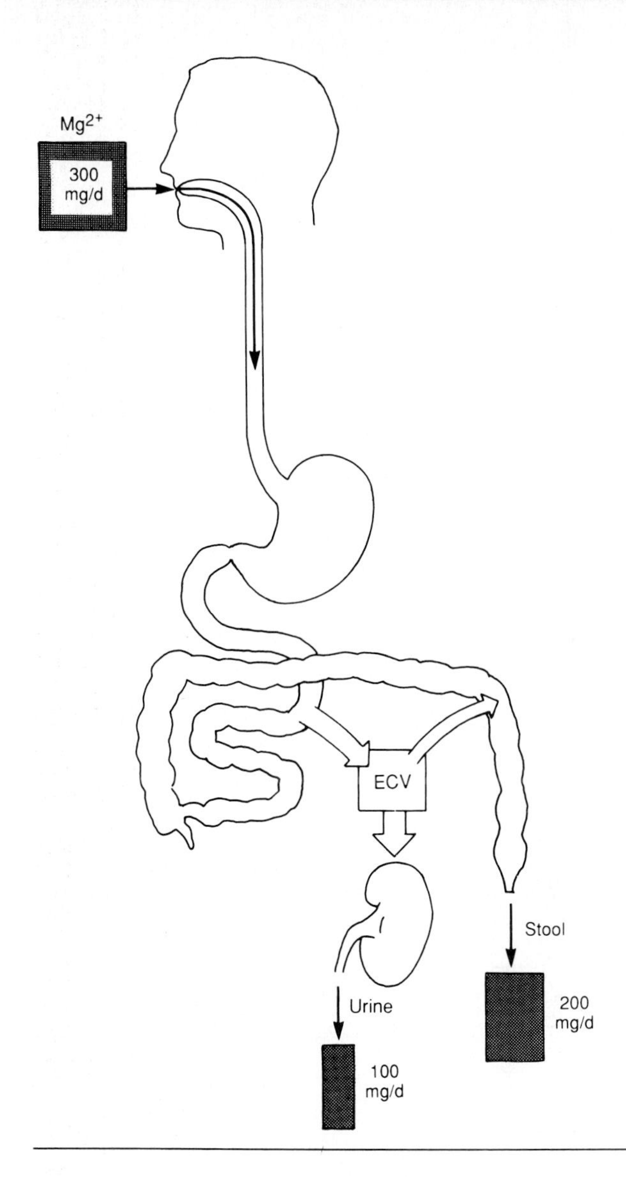

Figure 21-3. Normal magnesium intake and excretion.

magnesium transport is the plasma magnesium concentration. A rise in plasma magnesium concentration depresses magnesium reabsorption. Magnesium reabsorption in this segment is also stimulated by PTH and inhibited by increased plasma calcium level. Transient reduction in magnesium reabsorption can also be induced with acute metabolic acidosis, thyroid hormone excess, or ethanol administration.

Distal Convoluted Tubule & Collecting Tubule

A small percentage (~7%) of the remaining amount of magnesium is reabsorbed by the distal convoluted tubule and collecting tubule by unknown mechanisms. Thiazides, by inhibiting sodium chloride reabsorption in the distal convoluted tubule, cause a small, transient magnesuria.

REGULATION & LIMITS OF RENAL MAGNESIUM EXCRETION

Intrinsic renal transport mechanisms—notably in the thick ascending limb of Henle—appear to be responsible for adjusting urinary magnesium excretion to maintain normal plasma magnesium concentration. In contrast with most of the other electrolytes discussed in this book, hormonal control of renal magnesium transport is not critical for magnesium homeostasis.

When dietary magnesium is eliminated, obliga-

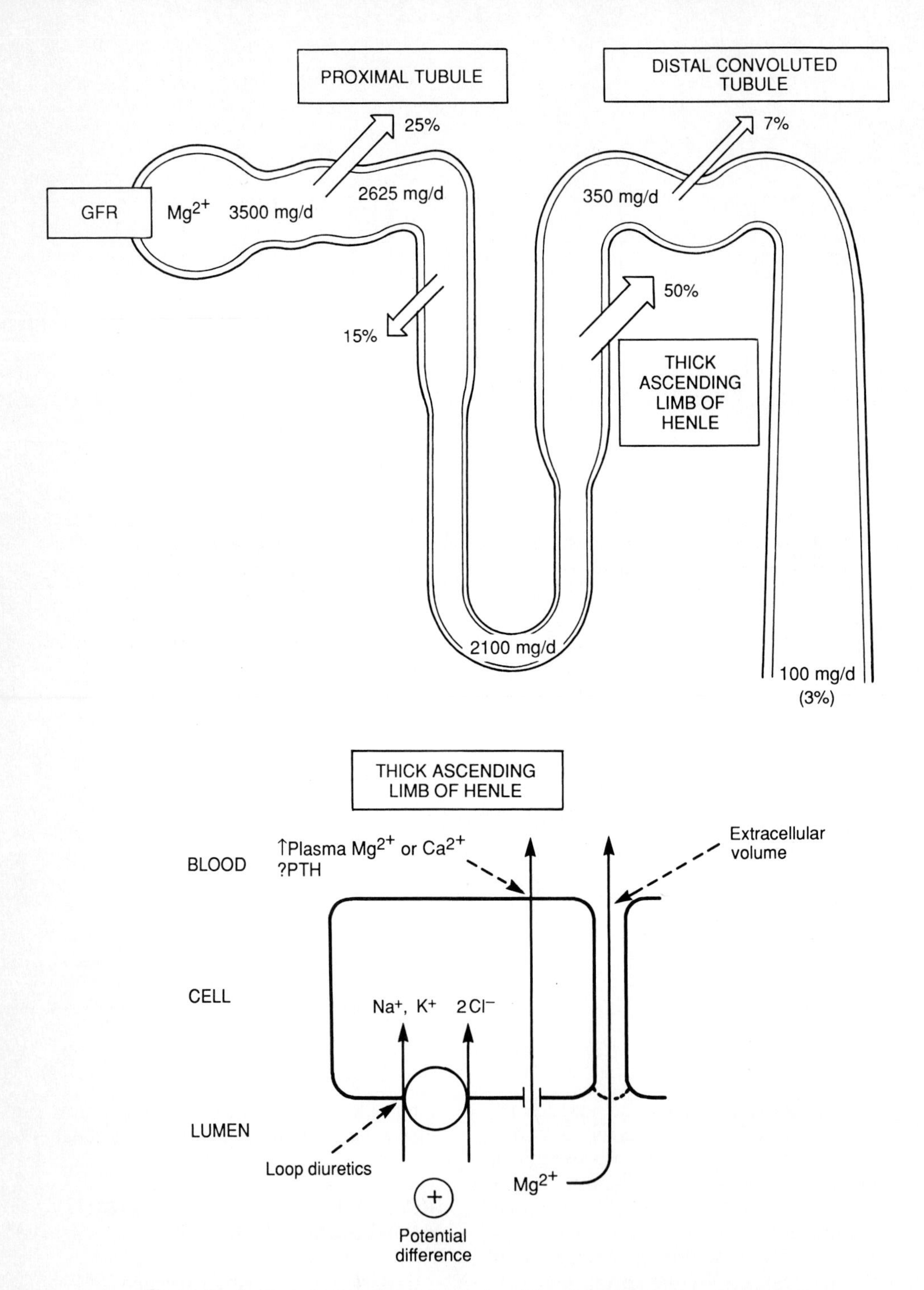

Figure 21-4. Normal amounts of magnesium filtered, reabsorbed, and excreted.

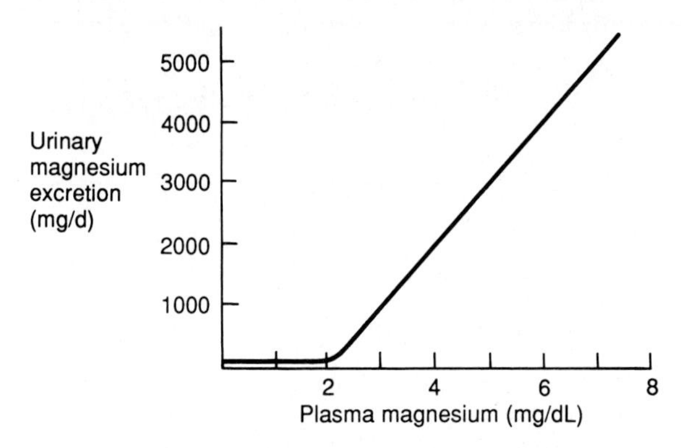

Figure 21-5. Urinary magnesium excretion as a function of plasma magnesium concentration.

tory stool losses fall to about 10 mg/d. The resulting magnesium deficiency induces a fall in urinary magnesium excretion within 3–5 days. The degree of maximal magnesium conservation by the kidney is normally quite good, resulting in urinary magnesium excretion of **10 mg/d or less,** as shown in Fig 21–5.

In response to a magnesium load, the normal capacity for magnesium excretion is very great. With chronic oral magnesium supplementation sufficient to quadruple the usual magnesium intake in an individual with normal GFR, the magnesuric response is brisk and there is no measurable change in the plasma magnesium concentration. Following a large, acute intravenous magnesium infusion, 80–90% of the load is excreted within 8 hours. In fact, when GFR is normal, magnesium loading does not cause sustained elevation in plasma magnesium concentration unless an intravenous magnesium infusion is used that is very rapid and continuous (as sometimes in the therapy of preeclampsia).

As shown in Fig 21–5, in response to rapid intravenous magnesium infusion sufficient to elevate the plasma magnesium concentration, urinary magnesium excretion in a normal individual rises dramatically to a level that can exceed **5000 mg/d.** Within the tolerable limits of plasma magnesium concentration, magnesium excretion does not appear to be saturable. However, this impressive magnesuric response to hypermagnesemia falls off proportionately if GFR is reduced for any reason.

22 Hypomagnesemia

INTRODUCTION

DEFINITION

Magnesium is bound to albumin in plasma, though to a lesser extent (25%) than calcium (50%). Therefore, hypoalbuminemia can reduce the measured total, though not ionized plasma magnesium concentration. When corrected for change in albumin concentration, hypomagnesemia is defined by a plasma magnesium concentration **less than 1.7 mg/dL** (1.4 meq/L).

PATHOPHYSIOLOGY

Change in Absorption & Excretion Balance

Though theoretically possible, shift in the distribution of magnesium from the extracellular to the intracellular compartment is not a common cause of hypomagnesemia. Rather, hypomagnesemia is usually caused by diminished intestinal absorption or enhanced renal excretion. Change in plasma magnesium concentration roughly parallels change in intracellular magnesium content. Urinary magnesium excretion is appropriately reduced—ie, to less than 10 mg/d—when hypomagnesemia is due to impaired intestinal magnesium absorption or inappropriately high—over 10 mg/d —when due to renal wasting.

REDUCED INTESTINAL ABSORPTION OF MAGNESIUM

Causes of hypomagnesemia are summarized in Table 22–1.

STARVATION OR DIETARY INSUFFICIENCY

With most nutritious diets, magnesium input is usually sufficient since it is contained in many green vegetables, meats, fish, and milk products. In specially constructed magnesium-deficient diets or with generalized starvation, body magnesium content may be depleted by as much as 6–10 mg/d (0.5–0.7 meq/d). Dietary magnesium insufficiency is probably the most important cause of hypomagnesemia in alcoholics, a common problem in that group.

STEATORRHEA, MALABSORPTION, OR DIARRHEA

Intrinsic or drug-induced (laxative overusage) disorders of the jejunum or ileum or states of enhanced intestinal volume flow (diarrhea) can diminish magnesium absorption. Magnesium concentration in diarrheal fluid is about 70 mg/dL (60 meq/L).

INCREASED RENAL MAGNESIUM EXCRETION

INHIBITION OF MAGNESIUM TRANSPORT IN THE THICK ASCENDING LIMB OF HENLE

Most states of renal magnesium wasting can be traced to a functional suppression or abnormality of thick ascending limb of Henle function, the major site of magnesium reabsorption (Fig 21–4). Factors that suppress sodium transport in this segment secondarily inhibit magnesium reabsorption.

Table 22-1. Causes of hypomagnesemia.

↓GI absorption
Starvation or dietary insufficiency
Alcoholism
Malabsorption, steatorrhea, diarrhea
↑Renal excretion
Extracellular volume expansion (see Chapter 3)
Osmotic diuresis
Loop diuretics
Furosemide
Ethacrynic acid
Bumetanide
Drug toxicity
Cisplatin
Aminoglycosides
Cyclosporine
Congenital
Bartter's syndrome
Welt's syndrome

Extracellular Volume Expansion or Osmotic Diuresis

Hypervolemia from any cause decreases the fraction of sodium reabsorbed in the loop of Henle (Fig 1-26). Similarly, a marked increase in luminal flow rate due to osmotic diuresis (eg, ketoacidosis) diminishes the fraction of sodium and hence magnesium reabsorbed in the loop.

Loop Diuretics

The loop diuretics furosemide, bumetanide, and ethacrynic acid all predictably induce urinary magnesium wasting. Each impairs Na^+-K^+-2Cl transporter operation and hence the lumen-positive potential difference needed to drive magnesium reabsorption (Fig 21-4).

Toxicity From Drugs

Various drugs, including cisplatin, aminoglycosides, and cyclosporine have been observed to cause renal magnesium wasting and hypomagnesemia. These agents presumably disrupt magnesium transport in the loop of Henle, though the mechanism has not been elucidated.

Congenital

Very rare inherited syndromes that affect all thick ascending limb functions (Bartter's syndrome; see Chapter 13) or specifically magnesium transport (Welt's syndrome) have been described.

SYMPTOMS

SYMPTOMS ATTRIBUTABLE TO HYPOCALCEMIA & HYPOKALEMIA

Hypomagnesemia impairs PTH release (Fig 15-2) as well as the action of PTH on its target organs, bone and kidney. By inducing functional hypoparathyroidism, hypomagnesemia causes hypocalcemia. Symptoms of hypomagnesemia are therefore often difficult to differentiate from those of hypocalcemia (see Chapter 16), as summarized in Fig 22-1. The hypocalcemia cannot be fully corrected without repleting magnesium.

Hypomagnesemia is a strong stimulus for renin release (Fig 8-3). By inducing hyperreninemic hyperaldosteronism, hypomagnesemia causes hypokalemia with its associated symptoms (see Chap-

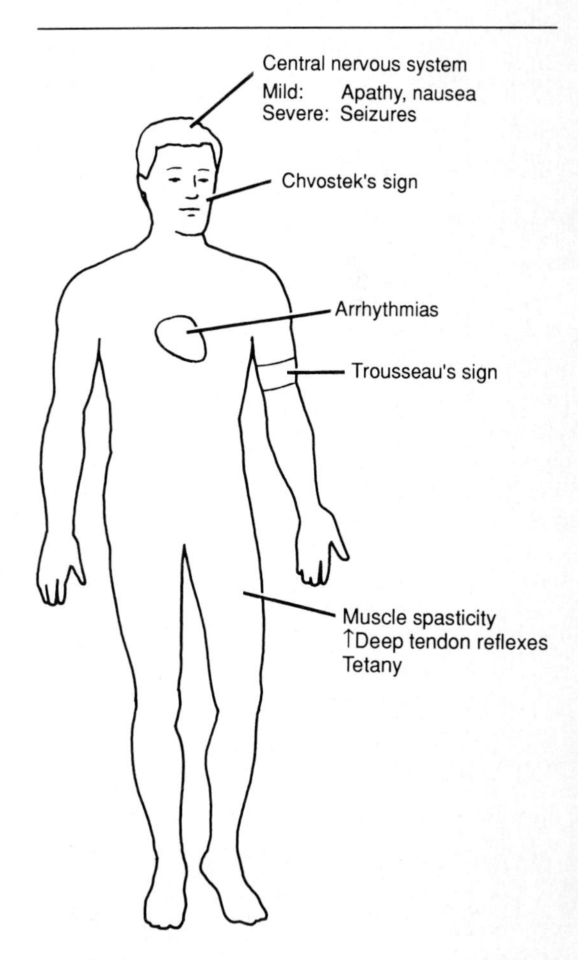

Figure 22-1. Symptoms and signs of hypomagnesemia.

ter 8). The potassium wasting of hypomagnesemia can usually be completely blocked with spironolactone to inhibit the action of aldosterone.

NEUROMUSCULAR

Magnesium deficiency, usually in conjunction with hypocalcemia and hypokalemia, can cause apathy and anorexia progressing to nausea and vomiting and seizures. Magnesium deficiency alone—though usually with hypocalcemia also—causes nerve irritability with a positive Chvostek or Trousseau sign, muscle spasticity, hyperreflexia, and even frank tetany.

CARDIOVASCULAR

Magnesium regulates calcium entry into excitable tissue. Ventricular arrhythmias, especially in the presence of digitalis, are a recognized consequence of magnesium deficiency and have been incriminated as a cause of sudden death. The QT interval is sometimes prolonged on the electrocardiogram. Hypokalemia and hypocalcemia potentiate this arrhythmogenic effect. Arterial resistance may also be enhanced.

DIAGNOSIS

APPROACH TO ETIOLOGIC DIAGNOSIS

History & Physical Examination

A careful dietary history with respect to magnesium intake—especially in alcoholics—and search for possible causes of impaired small intestinal absorption should be performed. The blood pressure and extracellular volume are elevated if hypomagnesemia is attributable to a hypervolemic disorder. Administration of drugs such as loop diuretics, cisplatin, aminoglycosides, and cyclosporine should be identified.

Urinary Magnesium

Estimation of urine magnesium excretion (ie, whether less than or greater than 10 mg/d) helps to differentiate the gastrointestinal versus renal causes of magnesium depletion (Fig 21–5).

Plasma Calcium & Potassium Concentrations

Hypomagnesemia predictably decreases both plasma calcium and potassium concentrations. These concentrations should be measured to assess the magnitude of depletion.

Table 22–2. Therapy of hypomagnesemia.

Drug (and MW)	Mg^{2+} Content	Mg^{2+} Dose
Intravenous		
$MgSO_4$ (247) 50%	50 mg (4 meq)/mL	Emergency: 100–200 mg (8–16 meq) over 10 minutes
25%	25 mg (2 meq)/mL	Less urgent: 12 mg (1 meq)/kg over 24 hours
10%	10 mg (1 meq)/mL	
Intramuscular		
$MgSO_4$ (247) 50%	50 mg (4 meq)/mL	100 mg (8 meq) every 3–8 hours
25%	25 mg (2 meq)/mL	
Oral		
$MgSO_4$ (247)	98 mg (8 meq)/g	250–500 mg (20–40 meq) 4 times daily
$MgCl_2$ (203)	116 mg (10 meq)/g	
Mg citrate (214)	112 mg (9 meq)/g	
MgO (40) (Mag-Ox, Maox, Uro-Mag)	550 mg (46 meq)/g (250 mg or 21 meq/tab)	

THERAPY

DECISION TO TREAT

Plasma Magnesium Concentration as a Reflection of Total Body Magnesium Stores

Since most magnesium resides in cells and bone, a reduced plasma concentration is a priori not necessarily an accurate indication of total body magnesium depletion. In practice, however, the plasma magnesium concentration reasonably reflects total body magnesium stores and can be used as a guide for the initiation of therapy. Symptomatic hypomagnesemia is usually associated with a total body magnesium deficit of 12–24 mg/kg of body weight.

Therapeutic Route Depends on Symptoms

If hypomagnesemia is associated with seizures or other significant neuromuscular symptoms—or an arrhythmia—urgent parenteral therapy is indicated. Otherwise, oral supplementation is sufficient. Magnesium repletion should usually precede calcium and potassium repletion so that the they will be retained.

TREATMENT FOR SEVERE, SYMPTOMATIC HYPOMAGNESEMIA

As listed in Table 22-2, intravenous magnesium should be given for life-threatening symptoms. It may be administered as a 10%, 25%, or 50% magnesium sulfate solution over 10–15 minutes to deliver 100–200 mg (8–16 meq) of elemental magnesium. When less urgent, magnesium sulfate may be given intravenously at a rate of 12 mg/kg per day (1 meq/kg per day) or intramuscularly at a rate if 100 mg (8 meq) every 3-6 hours. Reflexes should be checked frequently and supplementation stopped if hyporeflexia develops. Oral magnesium therapy should be also initiated as soon as feasible.

Table 22-3. Magnesium content of foods.

High (> 50 mg)	Medium (20-50 mg)	Low (< 20 mg)
Nuts (2 oz)	Green vegetables	Most vegetables
Peanuts	Beet greens	Fruits
Cashews	Swiss chard	Egg (1)
	Spinach	Bread (1 slice)
	Meats (4 oz)	Cereals (8 oz)
	Chicken, turkey	
	Pork, bacon	
	Beef	
	Fish (4 oz)	
	Milk products	
	Milk (8 oz)	
	Yogurt (8 oz)	
	Cheese (2 oz)	

TREATMENT FOR ASYMPTOMATIC HYPOMAGNESEMIA

Magnesium may be supplemented in the diet using the guidelines of Table 22-3. If insufficient, oral magnesium supplement as the sulfate, chloride, citrate or oxide salt may be given in a dosage of 250–500 mg (29–40 meq) 4 times daily (Table 22-2). A frequent side effect is diarrhea, but this is less likely with the magnesium oxide preparation.

23 Hypermagnesemia

INTRODUCTION

DEFINITION

An increase in plasma magnesium concentration greater than 2.7 mg/dL (2.3 meq/L) signals hypermagnesemia.

PATHOPHYSIOLOGY

Because of the ability of the normal kidney to excrete enormous quantities of magnesium (Fig 21-5)—greater than 5000 mg/d—very large amounts of magnesium must be administered quickly or GFR must be significantly depressed for sustained hypermagnesemia to develop, as shown in Fig 23-1.

MAGNESIUM LOAD

Causes of hypermagnesemia are summarized in Table 23-1.

Parenteral magnesium is used in the treatment of preeclampsia or eclampsia, occasionally other hypertensive emergencies, and symptomatic hypomagnesemia. If GFR is normal, the elevated plasma level abates within 8 hours when infusion is discontinued.

It is unusual—but has been documented—that a large oral dose of magnesium-containing electrolyte preparations, laxatives, or antacids can precipitate hypermagnesemia in a person with normal renal function. In practice, however, renal function must be reduced for oral magnesium to be retained in sufficient quantity to induce hypermagnesemia.

DECREASED GFR

The most common cause of mild or moderate hypermagnesemia is reduction in GFR and inadequate filtration of magnesium due to any chronic parenchymal renal disease, acute renal failure, or hypovolemia. In the course of chronic renal disease, hypermagnesemia does not usually occur until GFR is less than 30 mL/min, and then it is progressive as GFR declines further

SYMPTOMS

NEUROMUSCULAR

As summarized in Fig 23-2, lethargy and nausea with moderate hypermagnesemia (< 10 mg/dL) can occur, progressing to confusion and frank coma with severe hypermagnesemia (> 10 mg/dL). Hyporeflexia and muscle weakness are also predictable and quantifiable accompaniments of hypermagnesemia.

CARDIOVASCULAR

Hypotension and arrhythmias can be serious consequences of moderate hypermagnesemia (< 10 mg/dL). Very severe hypermagnesemia (> 15 mg/dL) may cause cardiac arrest.

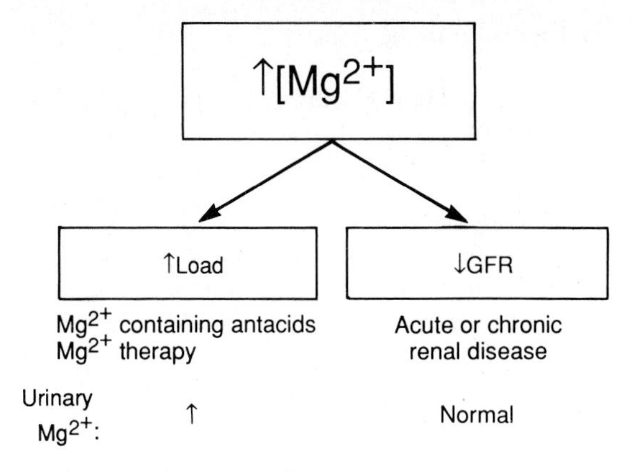

Figure 23-1. Diagnosis of hypermagnesemia.

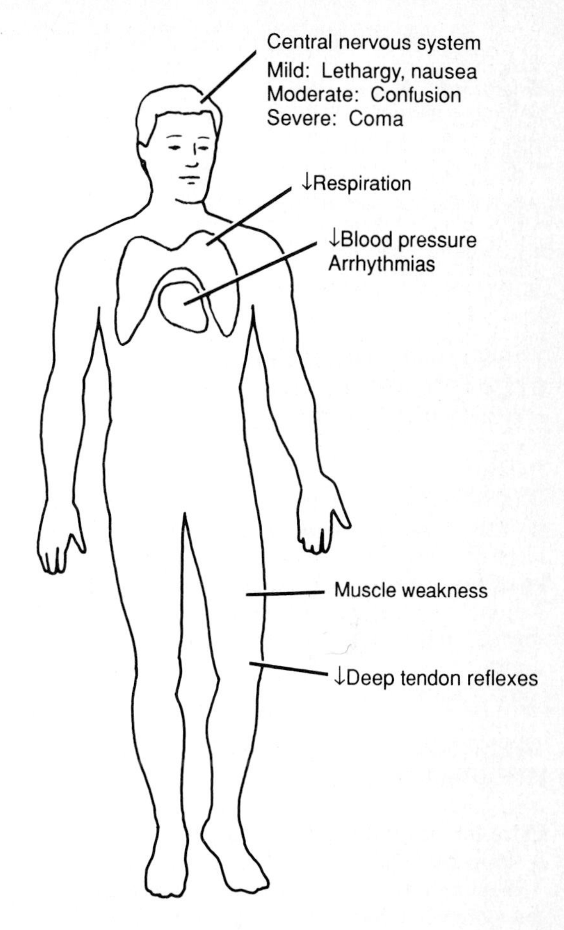

Figure 23-2. Symptoms and signs of hypermagnesemia.

DIAGNOSIS

APPROACH TO ETIOLOGIC DIAGNOSIS

History & Physical Examination

The major factors that should be sought are ingestion or administration of magnesium-containing substances or decrease in GFR. Most magnesium-containing antacids have about 16 mg magnesium per milliliter.

Urinary Magnesium Excretion

Magnesium excretion is appropriately high in cases of excess magnesium administration (using Fig 21–5 as a guide) and less with diminished GFR (Fig 23–1).

Table 23-1. Causes of hypermagnesemia.

Magnesium load
Intravenous magnesium
Therapy of preeclampsia or eclampsia
Oral magnesium
Elemental preparations
Antacids
Laxatives (oral preparations or enema)
↓GFR
Parenchymal renal disease
Glomerulonephropathies
Tubulointerstitial nephropathies
Functional
Acute renal failure (acute tubular necrosis)
Hypovolemia

THERAPY

DECISION TO TREAT

Mild, asymptomatic hypermagnesemia in the course of chronic renal disease may not require therapy. If magnesium elevation is severe (eg, > 10 mg/dL) or symptoms occur, urgent therapy is indicated, as listed in Table 23–2.

In all cases of hypermagnesemia, laxatives, antacids, or other food sources high in magnesium (Table 22–3) should be removed if possible.

Table 23-2. Therapy of hypermagnesemia.

Emergency:
Calcium gluconate, 15 mg Ca^{2+}/kg IV over 4 hours
Less urgent:
Remove magnesium-containing foods, antacids
Extracellular volume expansion
Loop diuretic administration (furosemide, ethacrynic acid, bumetanide)
Hemodialysis

TREATMENT FOR SEVERE, SYMPTOMATIC HYPERMAGNESEMIA

Calcium Gluconate

Intravenous calcium antagonizes and rapidly reverses the neuromuscular and cardiovascular effects of magnesium. Elemental calcium (15 mg/kg of body weight) should be given over 4 hours. Adjunctive measures described below should be started immediately to reduce total body magnesium.

TREATMENT FOR ASYMPTOMATIC HYPERMAGNESEMIA

Extracellular Volume Expansion & Loop Diuretics

If renal function is present, the two factors that can potentially inhibit the main site of magnesium reabsorption in the thick ascending limb of Henle are saline administration sufficient to induce extracellular volume expansion and loop diuretic administration, such as furosemide, bumetanide, or ethacrynic acid (see Chapter 3).

Hemodialysis

When GFR is severely depressed, hemodialysis against a magnesium-free dialysate is an effective and rapid means of reducing plasma magnesium concentration.

SECTION VII REFERENCES: MAGNESIUM HOMEOSTASIS & DISORDERS

Agus AS, Wasserstein A, Goldfarb S: Disorders of calcium and magnesium homeostasis. Am J Med 1982;72:473–488.

Massry SG, Seeling MS: Hypomagnesemia and hypermagnesemia. Clin Nephrol 1977;7:147–153.

Index

Note: Page numbers in bold face type indicate a major discussion. A *t* following a page number indicates tabular material and an *f* following a page number indicates an illustration. Insofar as possible, drugs and chemical agents are listed under their generic or common names.